THE WILDERNESS DEBATE RAGES ON

The Wilderness Debate Rages On

EDITED BY

MICHAEL P. NELSON

AND

J. BAIRD CALLICOTT

Continuing the Great New Wilderness Debate

THE UNIVERSITY OF GEORGIA PRESS ATHENS AND LONDON

Acknowledgments for previously published works appear on pages xii–xv, which constitute an extension of the copyright page.

Athens, Georgia 30602

Set in Granjon text by Graphic Composition, Inc.
Printed and bound by Thomson-Shore

The paper in this book meets the guidelines for permanence and durability of the Committee on Production Guidelines for Book Longevity of the Council on Library Resources.

Printed in the United States of America

12 11 10 09 08 C 5 4 3 2 1
12 11 10 09 08 P 5 4 3 2 1

LIBRARY OF CONGRESS CATALOGING-IN-PUBLICATION DATA
The wilderness debate rages on : continuing the great new wilderness debate / edited by Michael P. Nelson and J. Baird Callicott.
p. cm.
Includes bibliographical references and index.
ISBN-13: 978-0-8203-2740-2 (cloth : alk. paper)
ISBN-10: 0-8203-2740-9 (cloth : alk. paper)
ISBN-13: 978-0-8203-3171-3 (pbk. : alk. paper)
ISBN-10: 0-8203-3171-6 (pbk : alk. paper)
1. Nature conservation. 2. Nature conservation—Philosophy.
I. Nelson, Michael P., 1966– II. Callicott, J. Baird.
QH75.W527 2008
333.78'216—dc22 2008007941

British Library Cataloging-in-Publication Data available

To Nina Leopold Bradley — for a lifetime of good, hard work on behalf of "things natural, wild, and free"

CONTENTS

ACKNOWLEDGMENTS

It is as true as it is banal to say that books do not appear ex nihilo and that authors and editors owe colossal debts of gratitude to those who help along the way. We both wish to thank students, colleagues, and audience members, past and present, who have patiently (most of the time) explored the ins and outs of the "great new wilderness debate" along with us over the last decade.

Additionally, Michael Nelson would like to thank John and Candy Varco for providing the most beautiful setting in which to finish this project (the shores of Little Sand Lake in northern Wisconsin) and Heather Varco for support and patience in circuslike proportions. And Baird Callicott would like to thank his dog, Whisper, whose companionship was constant and usually patient, as well as a few two-legged companions who probably wish to remain unnamed.

Thanks must also go to Regan Huff, Jon Davies, Nancy Greyson, and Andrew Berzanskis of the University of Georgia Press for so promptly and professionally assisting with various details of publication. Finally, a big thanks to Mary M. Hill for wonderfully expert copyediting.

Essays in the public domain include "Animal Life as an Asset of National Parks" by Joseph Grinnell and Tracy I. Storer, originally published in *Science* 44 (1916): 375–80; "The Need for a More Serious Effort to Rescue a Few Fragments of Vanishing Nature" by Francis B. Sumner, originally published in the *Scientific Monthly* 10 (1920): 236–48; "Importance of Natural Conditions in National Parks" by Barrington Moore, originally published in *Hunting and Conservation,* ed. G. B. Grinnell and C. Sheldon (New Haven, Conn.: Yale University Press, 1925), 340–55; selections from *Fauna of the National Parks of the United States: A Preliminary Survey of Faunal Relations in National Parks* by George M. Wright, Joseph S. Dixon, and Ben H. Thompson, which is a National Park Service Publication from 1933; "Big Game of Our National Parks" by George M. Wright, originally published in the *Scientific Monthly* 41, no. 2 (1935): 141–47; and "African-American Wildland Memories" by Cassandra Y. Johnson and J. M. Bowker, originally published in *Environmental Ethics* 26 (2004): 57–75.

"The Importance of Preserving Wilderness Conditions" by Charles C. Adams first appeared in the *New York State Museum Bulletin,* no. 279 (1929): 37–46, and is printed with permission of the NYS Museum, Albany, N.Y. 12230.

"The Preservation of Natural Biotic Communities" by Victor E. Shelford was first published in *Ecology* 14, no. 2 (1933): 240–45. Reprinted with permission of the Ecological Society of America.

"Conservation versus Preservation" by Victor E. Shelford is from *Science* 77, no. 535 (1933). Reprinted with permission from AAAS.

Aldo Leopold's essay "Wilderness as a Land Laboratory" was originally in the *Living Wilderness* 6 (July 1941): 3. It is reprinted here with permission from The Aldo Leopold Foundation.

"The Value of Wilderness to Science" by Stephen H. Spurr was originally published in *Tomorrow's Wilderness,* ed. F. Leydet (San Francisco: Sierra Club Books, 1963), 59–75.

"From Woodcraft to 'Leave No Trace': Wilderness, Consumerism, and Environmentalism in Twentieth-Century America" by James Morton Turner was originally published in *Environmental History* 7, no. 3 (2002): 462–84. "Imagining Nature and Erasing Class and Race: Carleton Watkins, John Muir, and the Construction of Wilderness" by Kevin

DeLuca and Anne Demo was originally published in *Environmental History* 6, no. 4 (2001): 541–60. Each is reprinted here with permission of the Forest History Society.

Lynn Maria Laitala's essay "Jackfish Pete: Pete LaPrairie's Story" was originally published in her novel in stories *Down from Basswood: Voices of the Border Country* (Beaverton, Ont.: Aspasia Books, 2001), 81–91.

"Wilderness Preservation and Biodiversity Conservation: Keeping Divergent Goals Distinct" first appeared in *Bioscience* by Sahotra Sarkar. Copyright 1999 by American Institute of Biological Sciences (AIBS). Reproduced with permission of American Institute of Biological Sciences (AIBS) in the formal Other book via Copyright Clearance Center.

Antonio Carlos Diegues's essay "Recycled Rain Forest Myths" was originally published in *Terra Nova* 3, no. 3 (1998).

"A Willing Benefactor: An Essay on Wilderness in Nilotic and Bantu Culture" by G. W. Burnett, Regine Joulié-Küttner, and Kamuyu wa Kang'ethe was originally published in *Society and Natural Resources* 9 (1996): 201–12. Wayne Ouderkirk's essay "On Wilderness and People: A View from Mt. Marcy" was originally published in *Philosophy and Geography* 6 (2003): 17–32. "Wilderness, Cultivation and Appropriation" by John O'Neill originally appeared in *Philosophy and Geography* 5 (2002): 35–40. All three of these essays are reprinted by permission of the publisher (Taylor & Francis Ltd., http://www.informaworld.com).

"What Is Africa to Me? Wilderness in Black Thought, 1860–1930" by Kimberly K. Smith first appeared in *Environmental Ethics* 27 (2005): 279–97.

Gary Snyder's essay "Is Nature Real?" is from *The Gary Snyder Reader: Prose, Poetry, and Translations* by Gary Snyder. Copyright 2000 by Gary Snyder. Reprinted by permission of the publisher.

"Contemporary Criticisms of the Received Wilderness Idea" by J. Baird Callicott, "The Real Wilderness Idea" by Dave Foreman, and "Changing Human Relationships with Nature: Making and Remaking Wilderness Science" by Jill M. Belsky were all originally published in David N. Cole, Stephen F. McCool, Wayne A. Freimund, and Jennifer O'Loughlin, comps., *Changing Perspectives and Future Directions,* vol. 1, Proceedings RMRS-P-15-VOL-1, "Wilderness Science in a Time of

Change" conference, May 23–27, 1999, Missoula, Mont. (Ogden, Utah: U.S. Department of Agriculture, Forest Service, Rocky Mountain Research Station, 2000). Callicott and Belsky slightly reworked their essays for this volume.

"The Not-So-Great Wilderness Debate . . . Continued" by David W. Orr was originally published in *Wild Earth* (Summer 1999): 74–80.

"Something Wild? Deleuze and Guattari, Wilderness, and Purity" by Jonathan Maskit was originally published in *Philosophy and Geography III: Philosophies of Place,* ed. A. Light and J. M. Smith (New York: Rowman and Littlefield, 1998), 265–83. Maskit reworked the essay for this volume.

"Against the Social Construction of Nature and Wilderness" by Eileen Crist originally appeared in *Environmental Ethics* 26 (2004): 279–97.

Michael McCloskey's essay "Conservation Biologists Challenge Traditional Nature Protection Organizations" was originally published in *Wild Earth* (Winter 1996–97): 67–70.

"Wilderness" by Marilynne Robinson is reprinted here from *The Death of Adam: Essays on Modern Thought* (New York: Houghton Mifflin Co., 1998), 245–54. Copyright © 1998 by Marilynne Robinson. Reprinted by permission of Houghton Mifflin Harcourt Publishing Company. All rights reserved.

"The Implication of the 'Shifting Paradigm' in Ecology for Paradigm Shifts in the Philosophy of Conservation" by J. Baird Callicott originally appeared in *Reconstructing Conservation: Finding Common Ground,* ed. Ben A. Minteer and Robert E. Manning (Washington, D.C.: Island Press, 2003), 239–61. Copyright © 2003 Island Press. Reproduced by permission of Island Press, Washington, D.C.

"Hell, No. Of Course Not. But . . ." by Wendell Berry and "Wilderness as a Sabbath for the Land" by Scott Russell Sanders both appeared originally in *Arctic Refuge: A Circle of Testimony,* ed. H. Lentfer and C. Servid (Minneapolis: Milkweed Editions, 2001), 93–95, 101–4. "Wilderness as a Sabbath for the Land" © 2001 by Scott Russell Sanders; reprinted by permission of the author.

William Cronon's essay "The Riddle of the Apostle Islands: How Do

You Manage a Wilderness Full of Human Stories?" first appeared in *Orion* (May–June 2003): 36–42.

"Letting Nature Run Wild in the National Parks" is a retitle of the final chapter, "Promises to Keep," of Rolf O. Peterson's book *The Wolves of Isle Royale: The Broken Balance* (Ann Arbor: University of Michigan Press, 2007), 165–90. Copyright © 2007 by Rolf O. Peterson. The book was originally published in Minoqua, Wisconsin, by Willow Creek Press, 1995.

"Ecological Theory and Values in the Determination of Conservation Goals: Examples from Temperate Regions of Germany, United States of America, and Chile" by Kurt Jax and Ricardo Rozzi was originally published in *Revista Chilena de Historia Natural* 77 (2004): 349–66.

Kathleen Dean Moore's essay "Wilderness as Witness (Cape Perpetua)" first appeared in *Wild Earth* (Summer 2002): 51–53.

Finally, we offer thanks to Julianne Lutz Warren ("Science, Recreation, and Leopold's Quest for a Durable Scale"), Mark P. Jenkins ("Wilderness Preservation Argument 31: The Psychotherapy at a Distance Argument"), Feng Han ("Cross-Cultural Confusion: Application of World Heritage Concepts in Scenic and Historic Interest Areas in China"), Irene J. Klaver ("Wild: Rhythm of the Appearing and Disappearing"), and John A. Vucetich and Michael P. Nelson ("Distinguishing Experiential and Physical Conceptions of Wilderness") for their original contributions to this volume.

THE WILDERNESS DEBATE RAGES ON

Michael P. Nelson and J. Baird Callicott

Introduction

The Growth of Wilderness Seeds

If you can look into the seeds of time and say which grain will grow and which will not[,] . . . [then] speak.
William Shakespeare

The Great New Wilderness Debate was published in May 1998. A quick glance at the endnotes of the essays in this sequel, *The Wilderness Debate Rages On,* demonstrates that *The Great New Wilderness Debate* has since become the main reference work regarding the currently contested wilderness idea. While we, as that book's editors, are understandably pleased by this, we are at the same time sensitive to the fact that the topic the book tackles remains emotionally highly charged, contested, and controversial. Since that first publication, scores of scholars and wilderness defenders have weighed in on the great new wilderness debate with a considerable number of provocative (sometimes even vituperative) and mostly thoughtful essays. There have been, in fact, far more post–*The Great New Wilderness Debate* contributions to this conversation than could possibly fit into a second volume. *The Wilderness Debate Rages On* collects what we regard as the best of these contributions over the past decade mixed in with a few essays from the mid-1990s that somehow escaped our notice during the preparation of the first volume. This collection also assembles important (yet hitherto unheeded) and timely historical voices of leading ecologists, echoing down to the present, mostly from the 1910s to the 1930s, on the role of wilderness preservation in biodiversity conservation.

THE SIGNIFICANCE OF IDEAS

Ideas matter. Socrates was put to death because of the ideas he critiqued and defended; otherwise sensible people continue to attempt to ban numerous books of fiction because of ideas these books are believed to espouse; and more and more of us realize that our ideas about nature lie at the root of and continue to shape all our land-use decisions. As author Sam Harris sharply states in his recent book *The End of Faith,* "A Belief is a lever that, once pulled, moves almost everything else in a person's life."[1] Requests for a reconsideration of ideas that over time become so sacred that they become fixed and objectified have always subjected those critiquing such ideas to censure and even penalty. And this is certainly the case in the debate over the idea of wilderness. Some of our most renowned and eloquent environmental thinkers and writers have become so spittin' mad about questioning the wilderness idea that they have been reduced to name-calling in defense of what they believe wilderness to be. Other lauded environmental thinkers who offer a critique of the *concept* of wilderness have, for instance, been called "wilderness foes," "faddish philosophers who will soon be forgotten," "anticonservationists," "dubious professors," "antinature intellectuals," "the high end of the wise-use movement," and "high-paid intellectual types . . . trying to knock Nature, knock the people who value Nature, and still come out smelling smart and progressive" by those who see themselves as the embattled defenders of wilderness.[2] Sometimes critique and reconstruction of the wilderness *idea* are lumped in with other forms of nature destruction as just another part of the overall "war against nature."[3] Sometimes good-old-time-wilderness-religion zealots draw suspect analogies premised upon sophomoric logical fallacies.[4] For example, they might say that because the philosopher Socrates makes a comment that evinces contempt for the world outside of the city (the world of nature) and because many of those who critique the wilderness *idea* are philosophers (including the editors of this volume), such critics must believe what Socrates believed about the value of nature.[5] Though such responses are interesting (even, we must confess, somewhat and perversely entertaining at times) and telling, in that they demonstrate the power of ideas and conceptual analysis (i.e., the power of philosophy), they are not arguments; they are, rather, emotive and intellectually empty diatribes, more formally known by philosophers as fallacies of personal attack (most often, ad hominem circumstantial).

Two stories that may be true or may be apocryphal urban legends, we don't know, nicely, but troublingly, illustrate the importance of ideas in this particular debate. The first is a rumor that floated around academic environmental conferences and e-mail correspondence that conservative talk-show bully Rush Limbaugh was regularly citing geographer William Denevan's essay "The Pristine Myth: The Landscape of the Americas in 1492" (reprinted in *The Great New Wilderness Debate*) on his right-wing radio show. According to the rumor, Limbaugh was employing Denevan's work as proof that environmental impact of notable scale has occurred in North America for thousands of years. Supposedly, Limbaugh was suggesting that environmental concern—premised on the idea that nature in the Western Hemisphere was pristine prior to 1492—was ill founded, that the continued currency of this idea among environmentalists demonstrated their naiveté, and that anthropogenic environmental impact was "normal" or "natural" and therefore nothing to worry about. Though this story is unsubstantiated (we declined to listen to hour after hour of Limbaugh tapes to try to confirm it), the fear of what "the other side" might do with the critique of the wilderness idea has been a very common and somewhat understandable response to the critique by some traditional wilderness defenders and environmentalists. Indeed, some of the essays included in this volume (most notably, those of Gary Snyder, David Orr, and Dave Foreman) express that fear.[6]

Second, since the late 1990s it has also been rumored that environmental historian William Cronon, a notorious critic of the wilderness idea, received a telephone death threat from an angry Earth First!er furious over Cronon's (in)famous article "The Trouble with Wilderness" (also reprinted in *The Great New Wilderness Debate*) and the subsequent attention his work on wilderness attracted. A condensed version of "The Trouble with Wilderness" was published in the *New York Times* and was allegedly cited by a member of Alaska's antienvironmental congressional delegation on the Senate floor.

Though we have been unable to verify either of these stories, their mere existence (even if untrue) shows how close to the bone this debate cuts. Name-calling, conference outbursts, accusations of strange bedfellows and political shape shifting, and even rumors of death threats all prove that ideas matter, that philosophical critique is or can be important, that this debate over the concept of wilderness continues to rage, and that a sequel to *The Great New Wilderness Debate* is mandated.

As we insist in our introduction to *The Great New Wilderness Debate,* so we insist here in *The Wilderness Debate Rages On:* we suggest no criticism of the places thought of as wilderness; rather, it is the wilderness *idea* that is problematic. Names matter. They frame what they label and make what they label available for various uses and abuses. By some, for example, well-watered regions of the tropics are framed as "jungles"; by others they are framed as "rain forests." The jungle idea connotes disorder and danger, a place in need of discipline by machete, chainsaw, and bulldozer; the rain forest idea connotes complexity, balance, and harmony, a place in need of nothing but wonder and protection. The reality—and we certainly think that the *places* labeled as jungle or rain forest are real—is the same regardless of the way it is framed conceptually. What does the wilderness idea connote? Part of the reason it is problematic is that the wilderness idea connotes many different and sometimes contradictory things to many different people. To some it connotes a place for a certain kind of physically challenging recreation; to others it connotes a place of solitude and reverential reflection; to still others it connotes a habitat for big, fierce predators. So what are we to do when, say, big, fierce predators make a wilderness area unsafe for wilderness recreationists of either the sporting or the contemplative kind? What are we to do when, say, sporting wilderness recreationists are noisy and irreverent, spoiling the experience of contemplative recreationists, who go to some place designated as a wilderness area to experience solitude and reflect reverentially? This is only a sample of the kind of confusion that the wilderness idea creates when we attempt to operationalize it. These and many other of its problematic connotations are discussed in the essays contained in this volume and its predecessor.

THE PRESENT COLLECTION

Like the pensive traveler in American poet Robert Frost's legendary poem "The Road Not Taken," at a certain point in the course of its development North American conservation took one path instead of another. We contend that, for wilderness preservation, this has indeed "made all the difference." The path that North American conservation chose—for better or for worse—led to the prevailing concept of wilderness (what in *The Great New Wilderness Debate* we refer to as "the received wilderness idea") in the North American mind—and now in the mind of much of the rest of the world. The North American idea of wilderness was fashioned between the 1830s and the 1930s. This received wilderness idea was inspired by the writings of thinkers such as Ralph Waldo Emerson, Henry David Thoreau, John Muir, Theodore Roosevelt, Aldo Leopold, Robert Marshall, and Sigurd Olson—household wilderness names even today.[7] It was this vision of wilderness that was then reflected in and perpetuated by the most important and long-reaching wilderness legislation ever enacted: the U.S. Wilderness Act of 1964. We contend that the course that first U.S. and eventually international wilderness preservation policies actually followed was ultimately set by the received wilderness idea, which itself coalesced from three main sources.

First and foremost was wilderness preservation for human recreational purposes. Theodore Roosevelt, Sigurd Olson, and the young Aldo Leopold were the early architects of this wilderness rationale. Moreover, the types of recreation they had in mind were mainly various sorts of "vigorous" and "manly" recreation that would secure the "virility" of men, namely, wilderness for big-game hunting and primitive travel. Hence, the places that were thought to qualify as a proving ground for manly men needed to be dangerous and rugged, untamed and tough to traverse, and available for manhandling. Environmental historian James Morton Turner's essay in this volume adeptly illustrates how the historical conception of wilderness recreation in the "woodcraft" tradition is ironically in tension with our current take-only-pictures-leave-only-footprints ideas about appropriate wilderness recreation.

Second was the argument—emanating from American Transcendentalists such as Henry David Thoreau and John Muir—that wilderness served not only narrow or immediate human ends like recreation but "higher uses" as well. Thoreau and Muir focused primarily on the spiritual and aesthetic values of what they imagined wilderness to be. Wilderness landscapes were supposed to be awe inspiring, the clear and magnificent handiwork of a beneficent and powerful god, instantiations of beauty as well as the very standard of the beautiful itself, and places providing solitude so as to evoke profound spiritual self-reflection.

Third, and related to the second, was the tradition that focused on American wilderness as a source first of beautiful models for landscape painting and later for nature photography. When wilderness changed from the stronghold of the devil to the handiwork of God, from something viewed with fear and loathing to something lovely and divine, it also moved from the background to the foreground in painting. Through the work of painters such as Thomas Cole, Asher Brown Durand, Frederick Edwin Church, Thomas Moran, George Catlin, and Albert Bierstadt dramatic landscape painting became the visual embodiment of the Transcendental wilderness idea; and the remnants of wilderness in America, which were long gone in Europe, came to represent a new national identity for Americans.[8] In fact, some scholars argue that paintings of wilderness scenery were a stimulus of the movement that established the U.S. national park system.[9] As environmental historian Alfred Runte suggests, the landscape painters "dramatiz[ed] what the nation stood to lose by its indifference, [and] artists contributed immeasurably to the evolution of concern" that underlay the national park idea.[10] As a result, in the American mind wilderness was portrayed (conceptually and now literally—or, rather, visually) as a place of big, dramatic, awe-inspiring, monumental scenery—places that gave Americans a unique national identity. Hence, what America (and perhaps the world) got in a wilderness system was, understandably, land suitable for and consistent with these three sources of the wilderness idea.

Part 1, "The Unreceived Wilderness Idea: The Road Not Taken," of the present collection is an effort to demonstrate that there was an alternate path—"grassy and want[ing] wear"—that the North American conservation movement could have traveled but did not. However, part 1 also represents an effort to demonstrate that, unlike Frost's traveler, we can come, and in fact we may be coming, back to the point of divergence; that we can reblaze, and that we may be reblazing, our intellectual conservation trails and, more specifically, the rationales for a new preservation policy.

This alternate route was first explored by ecologists Joseph Grinnell and Tracy Storer in their 1916 essay "Animal Life as an Asset of National Parks." Tellingly, they introduce their essentially different rationale for wilderness preservation under the umbrella of the recreational value of wilderness perhaps to try to co-opt the prevailing anthropocentrism, but Grinnell and Storer in fact emphasize the role played by wild lands or natural areas ("parks" in this essay) as places that serve as sites for scientific study and as habitat for wildlife. An examination of the subsequent literature on the topic of protected areas at this time reveals that, despite its initial invocation of recreation, this landmark essay greatly influenced those following Grinnell and Storer down the nonrecreational path not popularly taken. The essays by ecologists Sumner; Moore; Adams; Wright, Dixon, and Thompson; Wright; Shelford; and Leopold—all of whose ideas appear to merge in the amazing (for 1963, just one year prior to the signing of the Wilderness Act of 1964) essay by ecologist Stephen Spurr—demonstrate that an alternative path appeared before us at one point in our history. We believe that if the ideas put forward by these early American ecologists regarding those places we now think of as wilderness had informed wilderness preservation policy and the Wilderness Act of 1964, then much of the current brouhaha over the concept of wilderness would never have occurred, and there would be no need for the "rethinking" of the wilderness idea that is going on today. This rethinking suggests that what we currently want in a concept of wilderness is not principally land suitable for manly recreation, higher spiritual or aesthetic uses, and inspiring great landscape art. Rather, current thinking seems to suggest that we now more desperately need the following:

1. Wilderness for science: Early on, this seemed to be a mostly and blatantly self-serving argument by ecologists for study areas. Their felt need for designated wilderness was premised on the prevailing ecological assumptions about climax equilibria that excluded humans. As environmental scholar Julianne Lutz Warren indirectly reveals in her essay, Leopold seems to have picked this theme up from his attempt to ally the recreational preservationists of the Wilderness Society with the ecological preservationists of the Ecological Society of America. However, Leopold transforms the case for preserving places free from human habitation and modification for purposes of ecological study into a case for preserving humanly uninhabited and unmodified areas as "a base-datum of normality," that is, as a control for sustainably managing ecologically similar inhabited and economically exploited lands elsewhere. As philosophers, we would note how the persuasive power of the wilderness-for-science argument increases by the turn Leopold gives it. The value of humanly uninhabited and unmodified places for ecological study seems self-serving, while the necessity for base-data of normality seems to better serve the collective public interest.
2. Wilderness for threatened and endangered species: Some threatened and endangered species require large and unpeopled, unroaded, undeveloped land in which to flourish. Wolves and grizzly bears are leading examples.[11] If they are to survive, such species must have a place to live. Therefore, we must preserve large tracts of land for their habitat. To emphasize the raison d'être for such places, we might cease calling them "wilderness areas," thereby conjuring up images of places to backpack and rock climb or places in which to meditate or view scenery and instead call them "biodiversity reserves," a suggestion developed in this collection by philosopher J. Baird Callicott.

3. Wilderness as preserving representative landscapes and ecosystems: We might urge that all extant kinds of biotic communities, ecosystems, and landscapes be represented on the earth, no matter how unscenic they might be, how uninviting they might be to recreationists and transcendentalists, or, indeed, how marginal they might be as habitat for threatened and endangered species. This preservationist rationale is, like the preceding one, based on the current concern for preserving biodiversity. Biotic communities, landscapes, and ecosystems are levels of biological organization, the variety of which is now included in the concept of biodiversity, as any textbook in conservation biology will attest.[12]

When we (the editors) assert that we (conservationists) now want places serving these three purposes, who are the "we" who want such places? Well, that certainly includes us, the editors. However, our wish list is not just personal or idiosyncratic. We would suggest that this is a growing desideratum on the leading edge of the world conservation movement. From biologist E. O. Wilson to activist Dave Foreman, conservationists appear to be reaching the consensus that biodiversity loss may be one of our greatest environmental problems—if not our single greatest. Reconceiving wilderness preservation in terms of base-data of ecological normality, species preservation, and types of community, ecosystem, and landscape representation seems better to address our most pressing environmental concerns than the erstwhile conception of wilderness preservation in terms of recreation, higher spiritual or aesthetic uses, and viewing scenery, which we believe to be the primary connotations of the received wilderness idea.[13] An examination of the essays in part 1 gives us an alternative historical path to an idea of wilderness preservation better fitting contemporary environmental concerns. Part 1 ends with a fresh essay by philosopher Mark Jenkins. Jenkins's essay articulates and examines an argument (Argument 31) for wilderness preservation missed by Michael P. Nelson in his "amalgamation of wilderness preservation arguments" in *The Great New Wilderness Debate*.

Part 2 of the present collection mirrors part 2 of the previous one. In both *The Great New Wilderness Debate* and this sequel we have attempted to represent the sometimes shocking critique of the received, originally American idea of wilderness as it has been translated for, exported to, and all too often imposed upon the rest of the world. Part 2, "Race, Class, Culture, and Wilderness," adds to the "Third and Fourth World" critiques of the received wilderness idea in part 2 of *The Great New Wilderness Debate.*

In his review of *The Great New Wilderness Debate* and again in the introduction to his coedited (with Marta Ulvaeus) book *The World and the Wild,* philosopher David Rothenberg has defended the importance of the received wilderness idea for humans all over the world.[14] Suggesting that wilderness is a universal, unproblematic, and positively valued good, Rothenberg claims that "wilderness has supporters all over the world, people who come from all levels of education, opportunity, and status. . . . [I]t is clear that wilderness has a place in the environmental philosophies of all cultures. . . . Although many cultures don't have a word for wilderness, when they think about what it means, *they know what to do with it.*"[15]

Indeed, as the essays in this section of the book indicate (and as many of the essays in Rothenberg's own book demonstrate), they do know what to do with the received wilderness idea—or, rather, these essays indicate what many people throughout the world think we Westerners (mainly Americans) can do with it!

Part 2 contains essays by Sahotra Sarkar and by Kevin DeLuca and Anne Demo that offer a more general alternative cultural critique of the received wilderness idea. The section also contains more specific contributions from a wide variety of often underrepresented voices and quarters. Chinese (Feng Han), Bantu (G. W. Burnett, Regine Joulié-Küttner, and Kamuyu wa Kang'ethe), South American (Antonio Carlos Diegues), and African American (Kimberly Smith as well as Cassandra Johnson and J. M. Bowker) perspectives in this volume add to the American Indian, Indian, Central and South American, and Australian Aboriginal critiques that we presented in part 2 of *The Great New Wilderness Debate.* As the reader will note, we have also attempted to broaden the contributions to *The Wilderness Debate Rages On* by presenting essays written in a different, more narrative style than those in

The Great New Wilderness Debate. In this part of the present book we have even included a piece of fiction: a chapter from writer Lynn Maria Laitala's north-woods-gothic book *Down from Basswood,* which critiques the Romantic wilderness ideas of wilderness writer and activist Sigurd Olson and his ilk from an American Indian (Ojibwa)/northern Minnesotan–Finnish perspective.

As it did in *The Great New Wilderness Debate,* part 3 of *The Wilderness Debate Rages On* represents the more mainstream Western and academic debate over the concept of wilderness as it unfolded after the publication of the first volume. To say that this has been a difficult and contentious debate would be like saying that Mike Tyson suffers from "moderate aggression" or that the U.S. presence in Iraq is "somewhat disruptive." The academic (and nonacademic) community has produced many more essays of this kind than we can publish here. In fact, we are herein reprinting only about half of the essays we originally considered for the volume. The core of part 3 (essays by Gary Snyder, J. Baird Callicott, Dave Foreman, and Jill Belsky) is an integrated exchange that took place at the "Wilderness Science in a Time of Change" conference in Missoula, Montana, in the spring of 1999. In addition to reading a selection from his (then) recently released epic poem, *Mountains and Rivers without End,* poet and essayist Gary Snyder also read a version of his contribution to this volume as the conference keynote speech on the opening evening. The following day a session was arranged as a "debate" between Callicott and Foreman, with each presenting the essays included here. The day after that memorable exchange, environmental sociologist Jill Belsky presented an essay that provided a summary and critique of these three essays along with Belsky's own contribution to the discussion (Belsky slightly reworked her essay for this volume, as did Callicott).

The remainder of this part of the volume contains essays from some of our best-known contemporary environmental thinkers and activists. David Orr points to the wilderness critique by writer Marilynne Robinson (reprinted in part 4 of this collection) by way of posting a general warning about the political dangers of critiquing wilderness. (Readers can decide for themselves if Orr gets Robinson's critique right.) Philosopher Wayne Ouderkirk defends the received wilderness idea but occasionally lapses into confusing the wilderness *idea* with the *places* we now associate with that idea.

Until recently, the philosophical side of this debate over the concept of wilderness has been dominated by the Anglo-American philosophical tradition. We ourselves work within that tradition. The essays by philosophers Jonathan Maskit and Irene Klaver, however, nicely represent the contribution of the Continental philosophical tradition—a Continental critique, if you will, of the received wilderness idea. The Anglo-American and Continental traditions of philosophy differ along many axes. One such axis of difference regards science. Generally speaking, Anglo-American philosophy regards science as providing if not certain then at least a body of continually self-correcting and self-refining knowledge, while the Continental tradition has been suspicious if not dismissive of science as a source of human knowledge, aligning more with the ways that creative literature and other cultural productions engage the natural world. Readers will find in Klaver's essay, for example, a more poetic and associative than scientific encounter with the wilderness idea.

Environmental sociologist Eileen Crist and philosopher John O'Neill present excellent but challenging contributions to this debate, though in very different ways. O'Neill's essay provides a needed but oft-neglected philosophical reflection on the justice issues that are central to this debate and that come up repeatedly in parts 2 and 3 of this volume. Instead of simply and emphatically affirming a naive wilderness realism (a view that would assert that wilderness per se exists apart from humans), Crist cleverly subjects the deconstruction of the received wilderness idea by scholars such as Callicott and Cronon to a deconstruction of her own. In the end, Crist provides a brief less for the classic or received wilderness idea than for something that appears to be very similar to Callicott's "biodiversity reserve" idea. Also included is a 1996–97 essay by former Sierra Club Executive Director Michael McCloskey that was somehow skipped over in *The Great New Wilderness Debate.* McCloskey's essay points out the surprising (to some at least) tension between "traditional Nature protection organizations and conservation biologists." In an indirect and ironic way it documents our claim that the biodiversity reserve idea is better suited to contemporary conservation concerns than the classic or received wilderness idea. McCloskey's essay melds nicely with part 1 and with Sarkar's essay in part 2.

Finally, just as we tried to do with part 4 of *The Great New Wilderness Debate,* part 4 of this collection offers ways to rethink, remedy, rehabilitate, or move beyond the received wilderness idea. Though the essays in this section are quite varied both in message and in style, each author or set of authors contributes something significant to this undertaking. These approaches to reconceptualization might be said to fall into two basic categories. On the one side are more radical rejections of the received wilderness idea from writers like Callicott and Robinson. While Robinson proposes a nearly complete abandonment of the concept of wilderness ("I think we must surrender the idea of wilderness"), Callicott proposes the replacement of the idea of wilderness, freighted with all sorts of unhelpful and unjettisonable baggage, with the notion of biodiversity reserves—protected areas named in such a way that the primary purpose they serve is clear and straightforward. Callicott would not prohibit their serving as places to recreate in the form of backpacking or canoeing or as sites for aesthetic and spiritual experience, but the biodiversity reserve idea would make clear what "use" takes priority when other uses conflict with biodiversity conservation in such reserves. Part of Robinson's denunciation of the received wilderness idea—the sense that it can and has served as a crutch and an inappropriate focus for conservation—is echoed strongly in the short essay penned by writer and farmer Wendell Berry in opposition to drilling for oil in Alaska's Arctic National Wildlife Refuge. Though Berry does not necessarily propose the more dramatic abandonment of a focus on wilderness preservation, both he and Robinson warn that we cannot save nature by focusing on protected areas alone, whether we call them wildernesses or biodiversity reserves; in fact, both Robinson and Berry even assert that we may indeed do nature more harm in the long run with such a strategy. An appropriate and comprehensive strategy for conservation would give as much attention to the places we inhabit and exploit as to those we vacate and protect.

The other approach to reconceptualization represented in this part of the book includes authors who, in one way or another, appear to offer some sort of reform of our received wilderness idea. Also writing in opposition to Arctic oil drilling, nature writer Scott Russell Sanders suggests that wilderness areas ought to be thought of in the same way that religious people who honor the Sabbath think of that day of the week—as a space for rest, humility, and reflection and as a gesture of respect toward something larger than humanity. The reader of *The Wilderness Debate Rages On* might, however, note a tension between Sanders's suggestion and the warnings of Robinson and Berry: if we set aside Sunday, or wilderness, as our space for good behavior, then does that imply that the rest of the week, or the rest of nature, is not holy and thus available for profane and inconsiderate uses? Animal ecologist John Vucetich and philosopher Michael P. Nelson jointly suggest a move toward rethinking by pointing to a conceptual muddle that, they argue, clouds our thinking about wilderness. They employ the old philosophical tactic of drawing a distinction when an equivocation or a conceptual confusion is encountered, here tracing the line between a wilderness *experience* and the actual physical *places* that we call wilderness.

Finally, this part of the volume ends with a group of case studies. William Cronon illustrates how the received wilderness idea tragically negated the human stories during the recent movement to establish the Gaylord Nelson Wilderness Area in the Apostle Islands in Lake Superior. Cronon urges a wilderness rethinking that does not do this, that can somehow incorporate and even celebrate these human histories. Wildlife ecologist Rolf Peterson poignantly articulates how a commitment to and policy implementation of the received wilderness idea by the National Park Service could lead to the end of the wolf population in Isle Royale National Park, wolves that made it to the island on their own and wolves that are half of the longest-running continuous predator-prey study in the world. Peterson presents a dilemma that any respectable rethinking of wilderness certainly should address. Finally, conservation biologist Kurt Jax and ecologist-philosopher Ricardo Rozzi employ examples from the United States, Germany, and Chile to argue for the importance of wilderness conceived in accord with current thinking in environmental science, a conception of wilderness that arguably coincides with the kinds of protected areas envisioned in the essays of the early ecologists in part 1.

We end this volume with a beautiful and provocative narrative by philosopher and writer Kathleen Dean Moore. Illustrating the importance of intact and healthy ecosystems (which the reader may or may not think of as wilderness areas), Moore issues a fierce warning about allowing wild lands or healthy ecosystems in general to become degraded over time, thereby affecting our default image of what constitutes a healthy landscape: "This is what we must resist: gradually coming to accept that a stripped down, hacked up, reamed out, dammed up, paved over, poisoned, bulldozed, impoverished landscape is the norm—the way it's supposed to be, the way it's always been, the way it must always be. This is the result we should fear the most."

Both of us editors are environmental philosophers. Both of us have dedicated our lives and our life's work to the attempt to understand human relationships with the more-than-human world and to create systems of ethics that foster appropriate and healthy human relationships with that world and the myriad nonhuman beings with whom we share it. We are not necessarily "for" or "against" the wilderness idea. We are most certainly "for" critical thinking and the clarification of concepts and most certainly "against" muddled or flawed thinking, even if that means that we will at times disagree with people whom we, in nearly every sense, view as our allies and friends, people we respect and admire tremendously. However, we both feel the need to "speak," to critically examine what many of our friends and colleagues consider a holy concept. We both also worry that the seed of the wilderness idea that was planted in North America around the turn of the twentieth century has now sprouted into a plant that has ultimately borne desiccated fruit. We both, however, believe that a neglected conceptual seed can be replanted and nurtured in a different and more productive fashion—even if, in the process, the word "wilderness" is ultimately abandoned as hopelessly tainted and confused. This second collection of essays tracking the ongoing debate about the wilderness idea is offered in the spirit of just such a sowing.

NOTES

1. Sam Harris, *The End of Faith: Religion, Terror, and the Future of Reason* (New York: W. W. Norton, 2004), 12.

2. The first criticism was leveled by Dave Foreman in "All Kinds of Wilderness Foes," *Wild Earth* (Winter 1996–97): 2, 4; the second was a reported yet unverified comment; the remainder of these attacks are from Gary Snyder's "Is Nature Real?" collected in this volume or from earlier versions of that essay.

3. This is the title of Foreman's forthcoming book, which reportedly includes a chapter on wilderness deconstruction as another type of wilderness attack.

4. See Dave Foreman, "The Real Wilderness Idea," in this volume.

5. Socrates' comment about the world outside of the city is found in Plato's *Phaedrus,* 230d–e: "I am devoted to learning; landscapes and trees have nothing to teach me—only the people in the city can do that" (Plato, *Phaedrus,* trans. Alexander Nehamas and Paul Woodruff [New York: Hackett Publishing, 1995]). Of course, all philosophers do not believe what Socrates believed (neither *everything* that Socrates believed nor *this one thing*).

6. See also Foreman, "All Kinds of Wilderness Foes."

7. The essays of these "founding fathers" of the received idea of wilderness are collected in part 1 of *The Great New Wilderness Debate.*

8. Roderick Nash, *Wilderness and the American Mind,* 4th ed. (New Haven, Conn.: Yale University Press, 2001), 78–83.

9. See, for example, Eugene C. Hargrove, *Foundations of Environmental Ethics* (Englewood Cliffs, N.J.: Prentice Hall, 1989).

10. Alfred Runte, *National Parks: The American Experience,* 3rd ed. (Lincoln: University of Nebraska Press, 1997), 25. See also Hargrove, *Foundations,* 77–107.

11. The problem is one of coexistence. As a population biologist colleague who studies wolves once commented, "Wolves basically need two things: something to eat and not to be shot by humans." The same probably goes for grizzlies as well. That is, the main reason big, fierce predators need undeveloped lands to flourish is because of conflicts with people, who accidentally run over predators and intentionally shoot them when the humans feel they or their domestic animals are threatened by predators.

12. See, for example, Martha J. Groom, Gary K. Meffe, C. Ron Carroll, and contributors, *Principles of Conservation Biology,* 3rd ed. (Sunderland, Mass.: Sinauer and Associates, 2006).

13. The linkage between classic wilderness preservation and the preservation of biodiversity is challenged by philosopher of science Sahotra Sarkar in this volume.

14. David Rothenberg, review of *The Great New Wilderness Debate, Environmental Ethics* 22 (Summer 2000): 199–202; David Rothenberg and Marta Ulvaeus, eds., *The World and the Wild: Expanding Wilderness Conservation beyond Its American Roots* (Tucson: University of Arizona Press, 2001).

15. Rothenberg, review of *The Great New Wilderness Debate,* 202, emphasis added.

PART ONE

The Unreceived Wilderness Idea: The Road Not Taken

Joseph Grinnell and Tracy I. Storer

Animal Life as an Asset of National Parks (1916)

Joseph Grinnell (1877–1939) was a professor of zoology and the first director of the Museum of Vertebrate Zoology at the University of California, Berkeley. Grinnell coined the modern ecological usage of the term "niche" to refer to the role that a given species plays in a biotic community. He was also an early critic of "progressive conservation" due to its utilitarian bent and an opponent of predator control in the United States. Tracey I. Storer (1889–1973) was a professor of zoology at the University of California, Davis and served as president of the American Society of Mammalogists. This essay, originally published in *Science,* represents a move from a recreational rationale for the preservation of wilderness (or protected area) to an emphasis on the preservation of wilderness for wildlife and science. Moreover, this is a landmark essay that inspired future ecologists to consider this alternative rationale for wilderness preservation; George M. Wright was, for example, a student of Grinnell at Berkeley.

THE ARGUMENT MOST FREQUENTLY urged in favor of national parks is that they provide on a large scale for the protection of forest areas, and thereby ensure the transmission of a maximum wa-

Contribution from the Museum of Vertebrate Zoology of the University of California.

ter supply from the wooded tracts to the needy lands below. Attention has also been called to their value as refuges for wild life—particularly where the animals to be conserved are useful for game or food. The strict protection they afford enables the birds and mammals within their boundaries to reproduce at a maximum rate, and the surplus thus created, spreading outwards into adjacent unprotected areas, helps to make up for the depletion caused there by excessive hunting. The points mentioned above are fairly obvious. But national parks have other less generally recognized advantages, and among these we consider their potential uses as places for recreation and for the study of natural history, especially worthy of notice. We will here lay particular emphasis on their recreative value because this phase seems to have hitherto been treated only in a cursory way, and with an air of hesitancy, as if it were hardly deserving of practical consideration.

The term recreation is currently applied to any temporary change of occupation that calls vigorously into play latent or seldom used faculties of the mind and body. It is the purpose of this change to restore to the human organs the normal balance which special or artificial conditions of life disturb. As physiologists have long recognized, the interdependence of the various bodily functions is such that the neglect of one is bound to have its effect on the others, and complete health can only be attained when every function is given its adequate share of exercise. In view of this fact and of the general character of urban life at present, it would seem that the type of recreation most urgently needed by the majority of people today is to be found in the open country. The relatively abrupt changes coincident with modern civilization have seriously interfered with the fine adjustments acquired by the human body in the course of long ages; and the modern business man, who may be regarded as the final and typical product of these changes, can now obtain rest in its fullest sense only by resorting for several weeks in the year to the open country or mountains. There he may find entire relief from the nerve-racking drive of city life, and be brought once more into contact with primitive conditions. There he may have an opportunity of reawakening his dormant faculties and of "resetting" his physical "tone," by effecting a readjustment of physiological inter-relations. One of the greatest needs of city dwelling people is to develop objective interests;

"to get out of themselves," as the phrase goes; and a frequently effectual means to this end is a keen interest in outdoor things, encouraging, as it must, a healthy manner of living, an unconfined habit of observation, and a mood unaffected by the nervous tension so peculiar to town life.

If this be true, it follows that the best recreative elements in nature are those which most infallibly tend to revive our atrophied faculties and instincts. Among them the following are important. First: either perfect quiet, or an absence of all save primitive and natural sounds, such as those caused by the wind in the trees, by running or falling water, or by singing birds. Second: landscapes that relieve the eyes from close work by offering distant views, quiet harmonies of color, and a quiescent atmosphere, varied by occasional touches of movement in such objects as running or falling water, scurrying squirrels, or birds in flight. Third: accessible mountains, which encourage climbing and allow the visitor to combine the exhilaration of overcoming obstacles with the physical exercise attending the woodsman's mode of travel. Fourth: natural phenomena that make a purely intellectual or esthetic appeal, as do the conflicts between the great insentient forces of nature, the processes of geological upbuilding and destruction, the intimate inter-relations of plants and animals, and the contentions for mastery that are forever recurring throughout the whole realm of living things. We believe the last, the mental appeal, to be the element of greatest recreative value in nature, but the other three are only of slightly less importance.

The question may now be raised: "Can national parks meet these requirements any more fully than other uncultivated areas?" With the country in its present half developed state the objection has a certain degree of force. In this era one is inclined to think of the unprotected wilds as the silent, virginal and unspoiled regions of the earth, and to regard national parks as comparatively well-peopled areas where plants and animals are subjected to artificial restrictions. To a limited extent, and for the moment, this impression is a true one. But the objection will have less force in the course of a few years, and none whatever if by that time the full recreative possibilities of the parks have been realized. For the commercial exploitation of nature that is now going on so rapidly elsewhere, is daily making the conditions we have described harder to seek, and is confining them more and more closely to the park areas, where the administrators should be taking measures to propagate and conserve them. By this we do not mean that the parks should in any way be conventionalized or transformed. On the contrary, it is their chief function to prevent just that disfigurement of the face of nature by industrial machinery which is being carried on at such a disastrous rate in other localities. We mean rather that the ideal recreative conditions now to be found in them should be preserved, that all factors disturbing to these conditions should be excluded, and that the artificial elements required for the practical work of administration should be disguised or beautified past offense.

Let us, however, take up these points in greater precision and detail. The first necessity in adapting the parks for recreative purposes is to preserve natural conditions. In this respect a national and a city park are wholly different. A city park is of necessity artificial, in the beginning at least when the landscape is planned and laid out; but a national park is at its inception entirely natural, and is generally thereafter kept fairly immune from human interference. Herein lies the feature of supreme value in national parks: they furnish samples of the earth as it was before the advent of the white man. Accordingly, they should be left in their pristine condition as far as is compatible with the convenience of visitors. All necessary roads, trails, hotels and camps should be rendered inconspicuous, or, better still, invisible from the natural points of vantage in the parks. Another reason for retaining primitive conditions is that natural scenery unmarred by man is one of the finest known sources of esthetic pleasure. Any attempt to modify the appearance of a national park by laying out straight roads, constructing artificial lakes, trimming trees, clearing brush, draining marshes, or other such devices, is in the worst of bad taste.

As has already been intimated, the animal life of the parks is among their best recreative assets. The birds and mammals, large and small, the butterflies and the numerous other insects, even the reptiles and amphibians, are of interest to the visitor. As a stimulant to the senses of far sight and far hearing, faculties largely or altogether neglected in the present scheme of civilization, they are of no less consequence than the scenery, the solitude and the trails. To the natural charm of the landscape they add the witchery of movement. As soon as the general surroundings lose their novelty for the observer, any moving object in the landscape will catch his eye and fix his attention. People will walk miles and climb thousands of feet to secure a good view of falling water, and this desire for movement is even more completely satisfied by the sight of animals in motion. The moving deer, passing within range of the stage-coach, rouses exclamations of surprise and delight. Eagles and pigeons in flight overhead readily claim the traveler's notice, and the smaller birds often mingle the fascination of sprightly movement with that of bright color and pleasing song. Considering the predilections of the average visitor, we should perhaps regard these last as the most indispensable creatures in the parks.

The interest of moving objects depends upon a number of elements other than movement among which their color, and especially their size, is important. The chipmunk is more attractive than the ground squirrel, primarily because its movements are more rapid, and secondly because of its more brightly colored markings. But when movement and color are equal the average observer's selection seems to have a quantitative basis, though the rarity of the object, and its romantic or other associations affect the equation. A bear or a deer will elicit more interest than a smaller mammal, even though the latter be of a rarer species. There are exceptional cases where an animal's extreme rarity will make it of exceptional interest in spite of its inferior size, but in general the larger species are the more rare, as they are the first to disappear before human invasion. They have therefore a double claim to consideration, and measures should be taken to prevent their numbers from diminishing. After the visitor's initial curiosity has been aroused and his powers of observation developed, he may be trusted to give a closer study to the smaller species.

To realize the greatest profit, therefore, from the plant and animal life of the parks, their original balance should be maintained. No trees, whether living or dead, should be cut down, beyond those needed for building roads, or for practical elimination of danger from fire. The use of wood for fuel in power stations, or even for cooking and heating in hotels and camps, is made unnecessary by the abundant supply of water power everywhere available, and this may be utilized without marring the scenery in the slightest. Dead trees are in many respects as useful as living, and should be just as rigorously protected. The brilliant-hued woodpeckers that render such effective service in ridding the living trees of destructive insects depend in part on dead trees for a livelihood. In these they find food during the colder months of the year, when the insects elsewhere are in great scarcity. Here, too, they excavate their nesting holes. Some of the squirrels and chipmunks also seek shelter in dead or partially dead trees. Even down timber is an essential factor in upholding the balance of animal life, for fallen and decaying logs provide homes for wild rats and mice of various kinds, and these in their turn support many carnivorous birds and mammals, such as hawks, owls, foxes and martens.

No more undergrowth should be destroyed than is absolutely necessary. To many birds and mammals thickets are protective havens into which their enemies find it difficult to penetrate. Moreover, the majority of the chaparral plants are berry-producers and give sustenance to wild pigeons, mountain quail, robins and thrushes, and to chipmunks and squirrels—this, too, at the most critical time of the year, when other kinds of food are scarce or altogether wanting. The removal of such plant growth will inevitably decrease the native animal life. If any change is to be made at all, it would, indeed, seem preferable to increase the number of indigenous berry-producing plants, especially in the vicinity of camps and buildings. This would compensate for the shrubbery lost in constructing roads and buildings, and would also serve to attract berry-eating species to the point where they might be seen by the largest number of people.

It goes almost without saying that the administration should strictly prohibit the hunting and trapping of any wild animals within the park limits. A justifiable exception may be made when specimens are required for scientific purposes by authorized representatives of public institutions, and it should be remarked in this connection that without a scientific investigation of the animal life in the parks, and an extensive collection of specimens, no thorough understanding of the conditions or of the practical problems they involve is possible. But the visiting public should be warned against injuring, and even against teasing or annoying any of the mammals, against destroying lizards and snakes (except the rattlesnake), and against disturbing the nests of birds, or their young. In the last instance a very slight disturbance will often lead to subsequent destruction. The principle underlying these suggestions is apparent. The native complement of animal life must everywhere be scrupulously guarded, particularly along the most traveled roads and paths, where the animals are likely to be observed by the greatest number of visitors. It is there that each individual animal is of highest intrinsic value from an esthetic viewpoint.

As a rule predaceous animals should be left unmolested and allowed to retain their primitive relation to the rest of the fauna, even though this may entail a considerable annual levy on the animals forming their prey. We, as naturalists, are convinced that the normal rate of reproduction among the wild non-predaceous species, such as mice and squirrels, has adjusted itself to meet a certain annual draft on their population by carnivorous enemies. Another point worth emphasizing is that many of the predatory animals, like the marten, the fisher, the fox and the golden eagle, are themselves exceedingly interesting members of the fauna, and as their number is already kept within proper limits by the available food supply, nothing is to be gained by reducing it still further. Here again may be recognized the special and intimate relations everywhere existing among the various plants and animals.

The rule that predaceous animals be safeguarded admits of occasional exceptions, according to season, place and circumstance. Coyotes and bob-cats, especially the latter, when they are numerous, are likely to kill a great many grouse, quail, rabbits and squirrels. Cooper and sharp-shinned hawks, and, to a lesser extent, blue jays, are proven menaces to small birds, and it might be advisable to reduce them in the neighborhood of camps and much-traveled roads. Caution, however, should be exercised in doing so, and no step taken to diminish the numbers of any of these predators, except on the best of grounds.

We would urge the rigid exclusion of domestic cats and dogs from national parks. Cats are the relentless enemies of small birds; they are forever on the alert, and in ninety-nine cases out of a hundred can not be trusted, however well fed they may have been at home, to let birds alone. The fact that they readily go wild, that is, quickly revert to a feral state, makes it all the more important that they be kept out of unsettled regions. To admit them would mean adding one more predator to the original fauna, and this would tend to disturb the original balance, by making the maintenance of a normal bird population difficult or impossible.

Equal vigilance should be used to exclude all non-native species from the parks, even though they be non-predaceous. In the finely adjusted balance already established between the native animal life and the food supply, there is no room for the interpolation of an additional species. If pheasants be introduced, and allowed to become established in the wild, the native quail and grouse will inevitably suffer from competition with them at the season of minimum food supply, and will be numerically reduced in consequence. The same is true of elk in competition with the native deer, and of many imported small birds in rivalry with the native varieties. In the latter connection we need only mention the well known instance of the English sparrow. Cattle and sheep are also of importance as elements hostile to natural conditions, but their destructiveness has already been emphasized by foresters.

Thus far we have laid chief stress on the importance of the national parks to recreation, and have shown the necessity, in adapting them for this purpose, of retaining the original balance in plant and animal life. But the same necessity attaches to their adaptation for another end, hardly less important than recreation, namely, research in natural history. As the settlement of the country progresses and the original aspect of nature is altered, the national parks will probably be the only areas remaining unspoiled for scientific study, and this is of the more significance when we consider how far the scientific methods of investigating nature then obtaining will be in advance of those now applied to the same study.

As a final requirement, we would urge that provision be made in every large national park for a trained resident naturalist who, as a member of the park staff, would look after the interests of the animal life of the region and aid in making it known to the public. His main duty would be to familiarize himself through intensive study with the natural conditions and interrelations of the park fauna, and to make practical recommendations for their maintenance. Plans to decrease the number of any of the predatory species would be carried out only with his sanction and under his direction. He would be able to establish and supervise local feeding places for birds and mammals during the tourist season, and could do this without in any serious degree altering natural conditions. His acquaintance with the local fauna would enable him to communicate matters of interest to the public in popularly styled illustrated leaflets and newspaper articles, on sign posts, and by lectures and demonstrations at central camps. He would help awaken people to a livelier interest in wild life, and to a healthy and intelligent curiosity about things of nature. Our experience has persuaded us that the average camper in the mountains is hungry for information about the animal life he encounters. A few suggestions are usually sufficient to make him eager to acquire his natural history at first-hand, with the result that the recreative value of his few days or weeks in the open is greatly enhanced.

We have attempted in these columns to emphasize the value of national parks as places for recreation and for scientific research, two of their uses that have been rather commonly overlooked, and to show the importance in both connections of the animal life they contain. If the reasons and instances we have adduced are valid, there is surely ample warrant for saying that the animals in the parks should be given more care and attention than they are now receiving, and that they should be conserved and utilized to a fuller extent.

Francis B. Sumner

The Need for a More Serious Effort to Rescue a Few Fragments of Vanishing Nature (1920)

Francis B. Sumner (1874–1945) was a professor of biology at the Scripps Institute of Oceanography at the University of California. He served as vice president for the American Academy for the Advancement of Science, was a member of the National Academy of Science, and was the joint chair of the Ecological Society of America's Committee on the Preservation of Natural Conditions in the United States (originally the Preservation of Natural Conditions for Ecological Study). Known as a scientist with a "social consciousness," Sumner delivered this paper as a lecture to the California Academy of Sciences in December 1919 (it was later published in the *Scientific Monthly*). In the address (and in subsequent essays) Sumner focuses on two main reasons to set aside certain areas of land: "scientific knowledge" and "highest esthetic enjoyment." The scientific value of set-asides resides primarily in their value as unique sites for botanical and zoological study. In this essay Sumner also admonishes biologists for not doing more to stem the "world-wide assault upon living nature."

IN THESE DAYS OF anxiety and suffering, when large populations are on the verge of starvation, and revolution threatens us within our

Lecture delivered before the California Academy of Sciences, San Francisco, December 3, 1919.

own gates, it might not seem especially opportune to urge the claims of any movement not immediately concerned with the welfare and safety of mankind. This is perhaps particularly true of any proposal to render unavailable for human consumption considerable fragments of what we are wont to term our "natural resources." Indeed, the most plausible objection on the part of those who are engaged in the commercialistic exploitation of these resources, would be this one, that everything in the world must be made of some "use" to humanity. Such talk seems to breathe of the spirit of altruism, and it also harmonizes very well with the dominant utilitarianism of the day.

And so it comes about that most of the "conservation" which is being preached in these days is the conservation of our *resources,* of our coal and our lumber and our water-power and our fisheries and what not. Heaven knows that all these belated reforms are necessary enough. But there is another sort of conservation which has thus far received utterly inadequate attention. I refer to the conservation of nature—nature as a source of scientific knowledge and of the highest esthetic enjoyment of mankind.

Some of you will think at once of our National Parks and our National Forests, of our Audubon Societies and our game-laws, and will conclude that the needs which I speak of have already been sufficiently well provided for. I have little doubt that a large proportion of even that small minority who have any real interest in the preservation of nature and of wild life soothe themselves with this reflection that the existing agencies are quite adequate for the purpose.

We can not give too high praise to those individuals and organizations which have thus far succeeded in checking, however slightly, the destruction of our natural scenery and of the native fauna and flora of our continent. But it is utterly foolish to imagine that these slight beginnings are sufficient to avert a great and irreparable catastrophe.

I think that no harm can come from our endeavoring to indicate, in a general way, the sort of things which we should like to accomplish, had we the power. This, even though our aims, in some respects, may prove to be impossible of realization. There are two equally important lines of effort, prompted by two quite distinct motives. The first of these motives is the scientific one.

Biology is the science of life, of the totality of plants and animals which occupy or have occupied the face of our planet. One might readily conclude, however, that there are zoologists, and some of them occupying high positions, who would not be greatly disturbed if the entire natural fauna and flora of the earth, with a few specified exceptions, should be destroyed overnight. If we left them the various laboratory types which are dissected or sectioned in "Zoology 23," along with the fruit-fly and one or two domesticated mammals and birds, I am not sure but they would rest peacefully in the belief that all future needs of their science had been fully provided for. I can not speak with any such confidence for the botanists, but the fact that neither they nor the zoologists are making themselves heard from at all audibly in this matter, seems evidence of the comparative indifference of both groups of biologists to the worldwide assault upon living nature.

It would hardly seem necessary to justify to scientific readers the importance of saving from destruction the greatest possible number of living species of animals and plants, and of saving them, so far as possible, in their natural habitats and in their natural relations to one another. And this is not primarily in order that the ornithologist, the entomologist, and the rest, shall gratify their traditional passion for naming species. It is because some of the most important problems of biology are concerned with relationships, and can only be solved by comparisons. Comparative anatomy was to Darwin, as it is to us, one of the chief fields to which we must look for our evidences of organic evolution.

Another of these fields is that of geographic distribution. It is not only the broader questions involving the larger subdivisions of the animal kingdom—the orders, families and genera—which are of importance here. One of the most promising lines of attack upon the still unsolved problem of the "origin of species" is to be found in the study of our *incipient species,* i.e., our "subspecies" or geographic races. If most of these are shortly to be exterminated, and many others to be displaced from their natural habitats, we shall have lost a highly important clue to the initial stages in the formation of species in nature.

Then, too, what shall we say of the modern science of ecology, which concerns itself with the totality of animal and plant life in particular regions, viewing these whole assemblages as being themselves organisms, having a definite cycle of development, like an individual animal or plant. It is significant that the ecologists, of all biologists, appear to be most active at present in the campaign for the preservation of natural conditions.

But aside from these broader considerations, can the laboratory biologist feel quite certain that even "types" chosen for his class-work, or the "material" for his own researches will be eternally safe from depletion without any effort on his part. We have for some years been hearing of the difficulty of obtaining such classical objects of research as the sea urchin and some other organisms for the laboratories at Woods Hole. It would seem as if the biologist, like the hunter and the fisherman, had been seriously jeopardizing his own future supply of game.

The second great motive for the conservation of nature is that which, for a better name, we might call the esthetic one. I think, however, that this word, if employed in the present connection, would convey my meaning quite inadequately. For the love of nature includes vastly more than the appreciation of natural scenery. It includes that deep-rooted feeling of revolt—not yet quite dead in most of us—against the noise and distraction, the artificiality and sordidness, the contracted horizon and stifled individuality, which are dominating features of life in a great city. Now it may be that such a feeling of revolt may represent merely a passing phase in human development, indicative of incomplete adjustment to a really higher plane of existence. It may represent the welling up of atavistic, anti-social tendencies, which it is our duty to sternly repress. Perhaps the highly socialized man of the future will gather more inspiration from the mad crush of the home-bound subway crowd in New York City than he will from the solemn majesty of the desert at sundown.

Perhaps all these things are true, but I should be sorry to think so. If this is the real trend of human evolution, we who represent the "unfit" type, may well pray for a speedy extermination. I suspect that for many of us students of nature the appeal of natural scenery is of the same kind, though vastly more compelling, than that of either music or poetry.

I am quite aware that I am exposing myself to the charge of leaving the field of scientific discussion and of resorting to sentiment and rhetoric. This in a sense may be granted. But I doubt whether any harm slow growth, we may expect its practical extinction within large areas in the near future.

Where, then, has our discussion thus far led us? Are we to check the growth of population, arrest the march of progress, and withdraw from a large part of the land which we have occupied? Such a proposal is, of course, ridiculous. What is worth serious consideration is an insistence upon the conservation of our fauna and flora and natural scenery to an extent hitherto not contemplated by our people or our government.

Large tracts of land, representing every type of physiography and of plant association, ought to be set aside as permanent preserves, and properly protected against fire, and against every type of depredation. Here would be included desert and chaparral, swamp land and seashore, mountain and prairie. All this would doubtless cost vast sums of money, but what is money for? The question is really one of relative values.

Instead of game-laws, we should have a nearly absolute prohibition, both within and without these reservations, of the shooting of every wild mammal or bird not definitely known to be harmful to man. Exceptions might be made in certain cases, but only after careful consideration by competent and disinterested persons.

All lumbering operations should be under the supervision of the government. Throughout large tracts, such operation should be absolutely prohibited, and where permitted, the trees should not be removed faster than their natural rate of replacement. Exception is of course to be made of land which is to be permanently cleared for agricultural purposes. But the question as to which areas should become farming lands, and which ones permanent forests ought to be decided by disinterested experts, with sole reference to the higher welfare of the public and of posterity, and not on the basis of purely local circumstances or the accidents of private ownership. Scientific considerations, such as the advantage of maintaining the continuity of a given forest area, should figure here, among other motives.

The fact that *all* of our redwood forests, with two unimportant exceptions, are now in private hands and are fast disappearing, shows how far we are from the realization of such an ideal. Fortunately, in this particular case, a public-spirited group of citizens have taken steps required of a nature reservation in our sense of the words. They are inadequate for our purposes, however, in that they include only areas of exceptional scenic grandeur. The demand for extensive tracts, representing every type of physiography and of plant association, has not thus far been met. Likewise, the national parks are preserved primarily as "public playgrounds," and the public is admitted to every portion of these lands. Such playgrounds are doubtless among the important assets of the nation.[6] But there ought to be still other tracts, in which the fauna and flora are reserved primarily for the studies of the botanist and zoologist—ones in which the native life will be more adequately safeguarded than at present. It does not seem to me utopian to try to have the National Park Service or some other bureau of the government adopt, and publicly and explicitly avow, this additional motive for reserving tracts of land from settlement or depredation—namely, the permanent preservation of the native fauna and flora by reason of their value to science, and to the higher interests of generations to come.

2. The National Forest Service, which administers nearly a quarter of a million square miles, an area about as large as the New England and Atlantic States combined. One might be disposed to think that in setting aside these huge areas, the government had done far more than the most sanguine conservationist would have a right to ask. But we must note several important drawbacks, from the point of view of the scientist. To begin with, the national forests are located chiefly in the more elevated and mountainous regions of the country. Very little of the lowland forests—none, indeed, of the coast redwoods—are thus included. In the second place, lumbering, although on a restricted scale, is permitted in the National Forests, as are also grazing, hunting and camping. All of these last conditions are quite incompatible with the interests of botany and zoology, and they are likewise incompatible with the aims of those who would like to retain great tracts of virgin forest as a heritage for the future. Would it not be possible to reserve certain tracts—wisely chosen by disinterested experts—from which the lumberman, the hunter and the cattleman should be forever excluded?

3. The United States Biological Survey has established a considerable number of bird and game refuges, in which the shooting or molestation of birds is prohibited. There are 74 such refuges thus far es-

tablished, these, in every instance, being selected with reference to their value as nesting grounds. Attention must be called, however, to the lack of any guarantee as to the permanency of such sanctuaries. Two of the most important of them, lying within Oregon and California, were set aside as bird refuges by President Roosevelt in 1908. One of these, Lower Klamath Lake, has been recently seized upon by the Reclamation Service, and is already practically ruined as a breeding ground for water-fowl. This is all the more reprehensible, since the director of the Reclamation Service has recently admitted, after investigation, that the potential value of these marsh lands for agricultural purposes is doubtful.[7] The second of these reservations referred to, that of Malheur Lake, seems about to undergo the same fate.

4. Aside from these branches of the federal government, we have certain departments of various state governments which administer more or less extensive state forests and game refuges.

The problem before those biologists who are interested in the aims above set forth is to develop an organization which will be able to mediate between themselves and the various state and national agencies through which such ends could be accomplished. An entirely new society might be organized for this purpose, but the general sentiment of scientific men at present seems to be against multiplying these societies. It has been suggested by one of my correspondents that a new section of the American Association might be created for this purpose. Another suggestion is that the National Research Council might properly serve as a clearing-house for such efforts. I am not in a position at this time to make a recommendation in the matter. The main thing at present is to induce the various scientific societies, national and local, to take the matter up and discuss it seriously.[8] This might result in the formulation of a wise plan of action and it would at least, serve a good purpose if it succeeded in rousing from their present apathy many of those who could be of considerable service in the movement. Wise leadership is of course necessary, here as everywhere.

Two further comments seem to me worth making before I close this rather long harangue.

One is that we should state our objects and aims with absolute frankness. If we believe, as I hope we all do, that scientific, esthetic and higher humanitarian considerations are strong enough to stand upon their own feet, let us not justify every step that we take by appeals to economic and crassly utilitarian motives. If we favor, as perhaps many of us do, a practically total prohibition of the hunting of harmless species for sport, let us not league ourselves with the sportsmen themselves, and pretend that we are merely trying to save the "game" for future generations of hunters. It is certainly an unfortunate circumstance, to my mind, that most of the present enforcement of game-laws is paid for out of funds derived from the issuance of hunting and fishing licenses.

The second and last of my comments is a repetition of the warning that this question is an urgent one, and that emergency measures are necessary. With an increasing population, modern implements and vastly more efficient means of transportation, more destruction can be wrought in a single year than was formerly possible in a decade. Then too, we have recently had the inevitable raid upon our resources of all sorts necessitated by the late war, and the equally inevitable raid which is bound to result from the post-bellum problems of reconstruction. A bill is now before Congress, setting aside five hundred million dollars from the treasury for the purpose of "reclaiming" a large part of such usable land as has not yet been developed agriculturally. This measure has been introduced ostensibly in the interest of the returned soldiers.

I do not know how many of these deserving men have expressed any desire for farming lands. The sponsor for this bill may have data on the point. But in any case, a certain considerable fraction of our unreclaimed land ought to be reserved from settlement. Such a step would be in the interest both of science and of a truer humanitarianism which sees even greater benefits to our race than those which may be conferred by lumber mills and irrigation projects.

NOTES

1. "Our Vanishing Wild Life," New York, Chas. Scribner's Sons, 1913; "Wild Life Conservation in Theory and Practice," New Haven, Yale University Press, 1914. See also the able and forceful article by W. G. Van Name, entitled "Zoological Aims and Opportunities," *Science,* July 25, 1919.

2. Save the Redwoods League. For information, address Mr. R. G. Sproul, Sec.-Treas., University of California, Berkeley.

3. *Scientific Monthly,* March, 1919.

4. "Wild Life Conservation," p. 184.

5. *Science,* July 25, 1919.

6. Especially to be commended is the plan to have trained field naturalists detailed for duty in these parks during the summer season, for the purpose of giving instruction to such visitors as may be seriously interested in the natural history of the region.

7. Cited by W. L. Finley, state biologist of Oregon, in *Portland Oregonian,* October 26, 1919.

8. I must again urge the fact that the Ecological Society of America has already made a beginning in this direction.

Barrington Moore

Importance of Natural Conditions in National Parks (1925)

Barrington Moore (1883–1966) was the only president of the Ecological Society of America who served twice, and he was the first editor of its journal, *Ecology*. He was also a member of the Ecological Society of America's more activist oriented Committee for the Preservation of Natural Areas in the United States. Moore is also known as a prominent first-generation forest ecologist and an early proponent of the philosophical idea that ecology was as much "a point of view" as a science. Since 1955 the Society of American Foresters has presented the Barrington Moore Memorial Award for outstanding biological research in forestry. Moore originally published this article in George Bird Grinnell and Charles Sheldon's book *Hunting and Conservation*. In this essay Moore begins by focusing on preservation arguments centered on recreation and representation of national character, quickly warns the reader about the shortcomings of these arguments, and then settles on the preservation of set-asides for the purposes of education and scientific study.

THE NATIONAL PARKS MUST be considered in their relation to the broader conservation viewpoint, which requires that every acre of land be put to the highest use of which it is capable without unnecessary waste or injury. This viewpoint, which by some of the friends of the parks has been regarded as too materialistic and opposed to the best interests of the parks, is, whether we wish it or not, the one from which they will be judged in the long run. Fortunately, it is the one on which the strongest defense of the parks can be based.

Fertile and moderately level lands must be tilled; if they happen to bear magnificent forests the trees must be cut, much as we may regret their loss. Lands which are too steep to cultivate (aside from orchard and vineyard lands), or which are too sandy or sterile to produce good crops without undue fertilizing, should be used for growing forests which will be cut at maturity and then replaced by natural or artificial means. There are many million acres of such land, especially in the Lake States and the Coastal Plain of the south. Reckless lumbering, followed by fire, has wrought such havoc that over eighty million acres are now bare of forest.

On certain other lands the natural wonders or scenic beauty are such that their highest use requires preservation for the benefit and enjoyment of all. When the features are of outstanding *national* character, as in the Grand Canyon of the Colorado River, the valley of the Yosemite in California, the Yellowstone region in Wyoming, and others, and at the same time the land happens to belong to the Federal Government, there is no question as to the desirability of withdrawing the land from entry and making it into a national park. Since commercial development is incompatible with and tends to destroy the natural wonders and beauty, it must be rigidly excluded from national parks. The decision as to whether or not any given area should be made into a national park requires careful study and a balancing of the gain in preserving the area, against the sacrifice of resources and possible hardship to local communities dependent on these resources. In some cases the scenery so far outweighs any possible economic use that there is no difficulty in making a decision. In other cases the need of settlers for cattle or sheep range, or for timber to build houses, barns and corrals, or for storage of water required for irrigation or power, may be so great as to balance, possibly to outweigh, the claims of an area for inclusion in a national park. In any case the decision must be based not on a narrow desire for parks because they are pleasant to have, or on the incentive for gain on the part of the few who will benefit by economic development, but on broader grounds of the public interest. Far-sighted vision and fairness are essential.

There is another consideration in the creation of national parks. In a new and as yet sparsely settled region, the need of setting aside and preserving a piece of wild and inaccessible country which nobody wants, or apparently will want, is not at once apparent. Yet this is just the time for creating the national park. Economic development and increase in population are so rapid that before we know it settlements have sprung up, the range is supporting prosperous herds of cattle and sheep, the timber is being cut and the streams dammed. The creation of a national park in the region, however desirable, then involves sacrifices and encounters opposition. Who can doubt the difficulties which would have to be surmounted in creating the Yellowstone National Park now?

At present the safety of the national parks rests altogether too largely on their recreational aspects. They are called "the nation's playgrounds." Although they can claim first rank in scenic beauty, they are by no means the only playgrounds the nation possesses. The 156,000,000 acres of National Forests are just as much open to the people for recreation. In fact, recreation is now considered as one of the important uses of the national forests, and is being developed by the Forest Service to the fullest possible extent. Furthermore, the forests have one great advantage over the parks, because hunting can be enjoyed on them. The wilderness in many parts of the national forest is just as unspoiled as in the parks, and the scenery is very fine, as, for example, the San Francisco peaks in Arizona, and the Sangre de Cristo mountains in New Mexico. The forests on the higher mountains, being needed for watershed protection, are not allowed to be cut; and there is a movement on foot for preserving on the national forests typical bodies of timber in virgin condition for scientific study and education. From the recreational point of view, the national parks, though they hold front rank for tourist travel on account of their scenery and natural features, do not, by any means, enjoy a monopoly.

The maintenance of the national parks in an absolutely natural condition is of the utmost importance for a reason as yet but little understood by the general public, even by those most familiar with parks and most ardent in their defense. This is the opportunity they offer for the study of plants and animals in their natural surroundings. These studies are fascinating in themselves, but besides this they are capable of adding to the sum of scientific knowledge which forms the basis on which rest the practical methods of cultivating crop plants and domestic animals. The national parks are rich fields for the natural sciences included under the term biology, because on them the native fauna and flora may be found more nearly undisturbed than anywhere else. Many interesting plants and animals will survive in the parks long after they have been exterminated over the rest of the country. The opportunity thus afforded for seeing and studying rare forms will not only give pleasure to thousands, but will aid in the solution of difficult scientific problems.

Scientists of recent years have been less and less content with mere collections and identifications (although the importance of correct identification must not be minimized), and have more and more sought to learn how each living thing developed to its present form, and why it lives under one set of surroundings and not under others. It is this type of enquiry which led Darwin to make the observations on which he based his theory of evolution. In recent years it has given rise to the brilliant researches on heredity and on the relation of plants and animals to their environment. In addition to their purely scientific interest, which by itself is worth while as adding to the sum of human knowledge, the results of this work have a distinctly practical bearing. Everyone realizes that the ultimate source of the world's food supply is agriculture, including live stock raising. If we stop to think a moment we will see that in order to improve the strains of our various crop plants we must know something of the way in which heredity works, and in order to plant the right crops in the right soil and climate we must know something of how the different plants respond to different types of soil and of climate. Therefore research in heredity and environment furnishes knowledge which is basic to successful and permanent agriculture, and hence to our future food supply.

Research in heredity and environment—in fact practically all research dealing with problems of life—requires well-equipped laboratories in which the conditions to be studied can be minutely controlled. But while man-built laboratories are essential, they are not in themselves sufficient. The scientist must have an opportunity to study in nature's laboratory, where the processes of adaptation to environment, and of evolution, are constantly going on. He must compare his results with nature, otherwise he will make mistakes. A well-known botanist once collected five specimens of branches with leaves from a single tree which showed a considerable amount of variation in the different parts of the crown. He sent these to a fellow botanist who was thoroughly familiar with the specimens in a large herbarium, but had apparently never considered it essential to see plants growing in nature, and asked him to identify the "five" trees. A few days later he visited his unsuspecting friend, who spread out the five specimens and said quite seriously: "This one is such and such a tree; that one seems to be such and such but I am not sure; while these three others I have never seen before; they must be *new species.*"

This is merely one of the many kinds of errors into which the laboratory student is likely to fall if he does not check his results by observations in the field.

The importance of the national parks to science is bound up largely with the preservation of the balance of nature. The more this balance is kept free from man's interference, the better is the area for scientific study. That is, the less natural conditions within the parks are interfered with the better. This interference might take the form of introducing non-native game animals, or non-native fish in the streams, or of cutting out dead trees.

The processes of nature are so delicately adjusted that when man interferes in one respect he sets up a chain of consequences the end of which no one can foresee. Darwin's chain from cats to clover is one of the best-known examples. Cats eat mice, which eat the eggs and larvæ of the bumblebees, preventing the development of the mature bees which collect the honey and at the same time pollinate the clover flowers. Thus the cat helps the clover crop by destroying the mice that destroy the bumblebees that fertilize the clover blossoms. It is like "the house that Jack built," only the consequences may be almost endless, and sometimes very awkward or dangerous for man. The extinction of native races in contact with the white man is due to upsetting the delicately adjusted balance with nature under which these races have hitherto been able to survive. When we attempt to stamp out an insect pest, like the gypsy moth, by artificial means such as spraying, we also destroy the natural parasites of the insect. Hence, once we have thus upset nature by artificial interference we condemn ourselves to perpetual artificial measures for maintaining control over our natural enemies. Modern civilization depends in a large degree on artificial control of nature. But we are learning that it is both safer and more effective in the long run to work with nature whenever we possibly can.

One of the fruits of modern biological research is a high admiration for the marvelously intricate and delicately adjusted organisms which have evolved from the simplest forms—Mr. Bryan notwithstanding—and a wholesome respect for the inexorable laws of nature. But before we can avoid the fatal consequences of running counter to nature's laws we must know what those laws are; and before we can use nature's help in attaining our ends we must better understand natural processes. All this requires both searching laboratory tests and thorough field studies in areas on which nature has been undisturbed. The realization of this need among those best qualified to understand the situation has reached the point where one of the nation-wide scientific societies, the one dealing primarily with the relation of plants and animals to their environment (the Ecological Society of America), has undertaken a campaign to have set aside and preserved for scientific research and education areas on which the fauna and flora may be found undisturbed by outside agencies. It is aimed to have at least one such area in each of the important types of vegetation and animal life. The National Research Council, realizing the importance of the movement, is giving its cooperation. Fortunately the scientific aims coincide fairly well with the aims of sportsmen in setting aside game refuges, and the two groups of interest are working together.

The largest and by far the most important areas on which natural conditions are to be found, and can most readily be preserved, are the national parks. Scientists realize that in seeking for additional natural areas throughout the country they should not lose those already in hand. Therefore, it is not surprising to find these men among the most ardent defenders of the national parks against commercial invasions.

The protection of the plant life in the national parks involves not merely protecting the forests from fire and lumbering, but leaving them *absolutely untouched.* The average person wants to "tidy up" by taking out the dead and down trees, which are considered as unsightly and as adding to the danger from fire. In some cases this well-meant desire has gone so far as to cause the cutting out of all the undergrowth, including the tree seedlings and saplings on which the forest must depend for its perpetuation. Around the hotels and other much-frequented places a cleaning up of the dead and down trees, and sometimes even of the undergrowth, is permissible. These spots are part of the human habitation, and as such must be more or less artificial. But to go systematically through the rest of the forest in an endeavor to "improve" it, is wholly unjustifiable, and destroys one of the most important aspects of the parks—their natural conditions. In the forest of Fontainebleau in France, there is a small piece of beech woods known as the "virgin forest" or "artists' corner." The trees are all very old. But all the dead and down material, as well as the undergrowth, has been carefully removed so that these trees stand fully exposed in their naked decrepitude, a truly hideous sight. In nature the fallen trees become covered with moss and ferns, and are often the most beautiful part of the forest. These mossy ridges, which once were tree trunks, form admirable seed beds, especially for young spruces, hemlocks and similar kinds of trees. Hence they are often not only beautiful, but play an important part in the perpetuation of the forest. Protection from fire can be secured by well-located fire lines, lookout stations and patrols much more effectively and cheaply than by cleaning out dead trees and undergrowth.

The maintenance of absolutely natural conditions in the national parks will greatly add to the enjoyment to be derived from them; and the privilege of seeing beautiful valleys and streams without dams and power lines will be but a part of this enjoyment. An even greater pleasure to the true lover of wild life will be the opportunity to see wild animals in their native haunts and to observe their ways. The experience of explorers in remote regions to which man has not yet penetrated has shown that the wild animals which have not been hunted or disturbed often pay little attention to man. It is possible to go surprisingly close to them without frightening them away. There is a common belief that all animals are instinctively afraid of man, and will flee on sight. This is strictly true only where they have come to regard man as their enemy because they have been hunted. Otherwise man is no more to them than some other strange animal, and often is an object of their intense curiosity. Often, however, when wild animals detect the scent of man, they show great fear and flee at once. This may be due to the strangeness of this odor rather than to the fact that it is the odor of man.

The results of protection in the older parks, particularly in the Yellowstone, corroborate these experiences as to the tameness of wild creatures which have not been hunted. Everyone who has been in the Yellowstone knows the pleasure of having seen elk, beaver or bears at close range. The animals seem to learn rather quickly that they have nothing to fear. When a game refuge is established it is not long before the partridges, ducks and other creatures, find it out and congregate there in large numbers. Similarly with song birds around orchards and houses.

The sentiment in favor of keeping the national parks absolutely natural was crystallized at the annual meeting of the American Association for the Advancement of Science, in December, 1921, by the adoption of the following resolution without a dissenting voice:

A Resolution Bearing on the Introduction of Non-native Plants and Animals into the National Parks of the United States

Whereas, one of the primary duties of the National Park Service is to pass on to future generations for scientific study and education natural areas on which the native flora and fauna may be found undisturbed by outside agencies;

Whereas, the planting of non-native trees, shrubs or other plants, the stocking of waters with non-native fish or the liberating of game animals not native to the region impairs or destroys the natural conditions and native wilderness of the Parks;

BE IT RESOLVED that the American Association for the Advancement of Science strongly opposes the introduction of non-native plants and animals into the National Parks, and all other unessential interference with natural conditions, and urges

> the National Park Service to prohibit all such introductions and interference.

This resolution unquestionably represents the feeling of a large number of persons outside of the main body of scientists. It expresses a desire which is certain to become practically universal when once the value of the national parks, not only for recreation, but for science and education, is more generally understood. When this time comes the danger of commercial invasions which are at present threatening the parks will be materially diminished, though there will always be the need of constant vigilance.

The broad conservation point of view, instead of undermining the national parks, strengthens them, by showing their true position in the life and welfare of the nation. From this point of view those who have the best interests of the public at heart, the friends of the national parks and national forests and all lovers of the out-of-doors, will join forces, and work together for the protection of our parks, forests and wild life.

To summarize: The broad conservation viewpoint requires using all the land of the country for the highest purpose of which it is capable. Areas containing scenery or other natural features of such outstanding national character as to outweigh the commercial development of their resources serve their highest purpose in national parks. The importance of the parks in science and education requires their complete preservation in an absolutely natural condition. The sciences which carry on research in heredity and in environment, and are constantly increasing the store of knowledge on which depends our cultivation of plants and animals, and hence our food supply, require areas on which the balance of nature may be found undisturbed. The national parks are the largest areas on which natural conditions now exist and can readily be maintained. Hence they are essential in sciences on which the public welfare rests. Plant life should not be disturbed by removing dead and down trees and undergrowth, because to do so interferes with the balance of nature, and removes the natural site of certain beautiful mosses and ferns, as well as the best seed bed of certain forest trees. Preservation of the national parks increases their attractiveness by rendering many of the wild creatures unafraid of man and hence easily approached and observed. The resolution unanimously adopted by the main body of scientists of the country opposing the introduction into the national parks of non-native plants and animals, and all unessential interference with natural conditions, reflects a sentiment which is sure to become practically universal when the value of the parks for science and education is more generally realized. Friends of the parks, forests and wild life will unite to protect the public interest in these matters.

Charles C. Adams

The Importance of Preserving Wilderness Conditions (1929)

Charles C. Adams (1873–1955) was a leading American animal ecologist and a founding member and former president of the Ecological Society of America. Adams was a key element in the ESA's Committee for the Preservation of Natural Conditions in the United States, which actively lobbied on behalf of wilderness reserves for ecological study. Interested in fusing ecology with geography into biogeography, Adams studied with pioneering ecologist Henry Cowles at the University of Chicago. Adams held a variety of academic as well as museum and forest-management posts throughout his career. In this essay, originally published in the *New York State Museum Bulletin,* Adams argues for preserving areas of "wilderness" because they preserve our American pioneering tradition, provide inspiration for various artistic endeavors and unique locations for wilderness recreation, possess newfound economic value, and, finally, hold important scientific and educational value as ecological study sites.

INTRODUCTION

THE SUBJECT OF THE preservation of areas in a natural or wilderness condition has been discussed more during the past ten years than during all of our previous history. The reason is that with our increased population we are becoming definitely conscious of the limitations of our natural resources. As long as these resources were considered boundless and inexhaustible we drifted along with the unconcern of children at play. Now we are beginning to realize that this problem is a phase of the fundamental relation of the people to the national resources of the country, or a phase of the "land problem." With the exception of the very arid regions, most wild land is in forests and therefore special attention is given to the foresters' point of view. Foresters began the study of their problem as an economic one, as a part of the general conservation movement, and are now just beginning to see that grazing and wild life are also a part of their problem, and are becoming conscious of the social aspects of the whole conservation movement. This realization has recently been brought about through an appreciation of the recreational and educational phases of the movement; which are distinctly social rather than economic. The earlier development of forestry so strongly imbued the profession as a group with economic standards and ideals that many are wholly unprepared for the next advance, which is almost certain to be strongly social. Later we may expect to see these two phases properly balanced both in theory and in practice, but many preliminary adjustments are yet to be made before the broader aspects of the problem are likely to be adequately appreciated. Certain conservatives are now preaching the dire results of the recreational wave, and of the menace threatening "economic" forestry from a too extensive development of parks. Future leaders should take a broader and possibly a "higher view" of the whole problem.

This dominance by economic standards of forestry has tended to make the forester desire to cut over, graze or change in some way nearly everything under his control, except where he dared not, as in the case of protection forests. The value of virgin wilderness conditions for the study of forest ecology has not in the past received better recognition partly because forest research has been pushed aside by administrative emphasis. W. W. Ashe was a pioneer among foresters in advocating the preservation of natural conditions in forests for scientific and silvicultural purposes, and later Barrington Moore, through his interest in forest ecology, became a champion of the cause. Then Aldo Leopold enthusiastically urged the preservation of the wilderness from the standpoint of recreation and wild life. Great momentum is developing among other groups of leaders, not foresters, with whom a reckoning must be made if public interests are given a fair hearing. So much for the general background.

WHAT IS A WILDERNESS CONDITION?

In the old days the idea of a condition of nature was not very carefully analyzed or understood. It was supposed to be a "balanced condition," and we were not very clear just what that meant. This often implied that independent of man's influence, nature was always "balanced," although modern ecology has taught us that in natural conditions, as in all others, this is a relative condition and not a fixed or absolute state; one that is a part of a cycle, involving a continuous process of change in response to all sorts of pressures and influences. The relative balancing of these various influences gives us the so-called balanced condition or a "dynamic equilibrium," which includes a summation of the larger and smaller units of dominance, which with every disturbance are followed by other changes.

Thus when ecologists emphasize the need of setting aside reservations for the preservation of natural conditions they do not mean, and certainly do not expect, the conditions to remain indefinitely "balanced," fixed and unchanged or unchanging, because they know that it is utterly impossible, both theoretically and practically, so to isolate a reservation from the rest of the world and keep it free from all outside influences.

Reservations set aside, however, *to allow nature to take her own course, with as little interference by man as is possible,* is quite another matter and is fundamentally all that is desired. To accomplish this is a very difficult undertaking.

THE VARYING DEGREES OF THE WILDERNESS

As the degree of natural conditions is truly measured by its inverse relation to man's interference, we readily see why there are all degrees from natural conditions to the other extreme, such as metropolitan conditions where nature is crowded to the wall. On the one hand we have the different degrees of the destruction of natural conditions, and on the other hand we have a whole series of conditions built up deliberately by man, all stages and degrees toward a *restored condition,* closely simulating natural conditions. We have known for ages that when man abandons the land it will "go back to nature" in a comparatively short time, depending of course on the character of man's interference, but this restoration is not the same thing as virgin or natural conditions. If this were true our problem would be much simpler than it is.

Urban conditions generally result in very great changes of natural conditions and the industrialized, residential and park uses of the land show many degrees of difference in the intensity of interference. The same is true of agricultural lands. The degrees from orchard and vineyard, grain fields, pastures to woodlots and mature forests likewise show many degrees. In the case of large forests, forested parks, game preserves and such uses of the land, we see relatively less disturbance with the natural conditions than in the more intensively cultivated areas. The cutting of a forest and the different systems of management will without question influence the ecological conditions for both plants and animals, but many of these changes are not so drastic, if fires are excluded, as to prevent the preservation of many plants and animals as long as the major habitats are preserved, as in the case where a continuous forest cover is maintained. In relative terms a woodlot is a wild area on a farm, and in the large state and federal forests there are remote areas which are today in a nearly or completely natural condition. Some of our national parks and some national monuments are in a relatively natural condition, but during the past few years the furore of advertising them and the efforts to get millions of persons into them—before an adequate staff and appropriations are available to handle the crowds—as well as certain policies, such as the introduction of exotic plants and animals and grazing, have caused very serious injury to many of these parks. As the public is taught the significance of these facts we hope that conditions will be remedied, because our national parks have given to the world a new and valuable idea and ideal of land use. This conception is also having a beneficial influence upon forest policies, which have been too exclusively economic and very slightly social in their aims.

In the national and state forests we have a major opportunity for the preservation of excellent samples of natural conditions, since there are large areas that must be devoted to protection forests, and there are other large areas that will remain relatively natural for an indefinite period. Because of their remoteness some of these lands will long remain natural reservations in spite of economic policies.

In the national forests lumbering, grazing and particularly the extensive overgrazing have a blighting influence upon the natural conditions of the vegetation, and with poison weed control, reseeding, planting and similar methods the perpetuation of natural conditions becomes very serious indeed. Grazing animals compete directly with the large game animals in many places, not merely for forage and browse but also for water.

THE VALUE OF THE WILDERNESS

The value of the wilderness must be judged ultimately by its contributions to social welfare. We have no better criterion. What therefore are some of the benefits? First of all let us turn to our own history for a few suggestions. Our American public first learned of natural conditions during its pioneer history. Historians have shown us how much our American democratic institutions have been a direct outgrowth of our *pioneering,* and how this has tended to encourage independence, self-reliance and other traits which have contributed so much toward our institutions and our ideals. There is a whole literature built upon this phase of our national life. Without question this background and our public domain, out of which we could with relative ease set aside national parks and national forests, have been dominating influences in acquainting Americans with the charm of the wilderness. Our first and greatest champion of all this was John Muir, who exemplified the benefits derived from the appreciation of the wilderness. He was a naturalist, an artist, and from the wilderness he derived science, art, education, recreation, producing a literature which is a wonderful blend of all these. He thus exemplified the social uses of the wilderness at its best. This great contribution could not come from one dominated by economic ideals. A whole Nation is now becoming educated to the Muir ideal and, as has been said, this is one of America's large and original contributions to the use of the land, as a definite land policy. This is a policy which has since spread to the Old World and seems destined to have a great future there.

We may briefly summarize the value of natural conditions under the following heads: artistic, scientific, educational, recreational and economic, bearing in mind, of course, that these groups grade imperceptibly into one another in various directions.

Artistic. The inspiration and ideals which the painter, the poet, the author and the artistic photographer get from the wilderness is a form of leadership which helps the average person, not so keenly gifted to see and appreciate the beauties and wonders of wild nature. Muir, Thomas Moran and J. C. Van Dyke have opened up a whole world of interest and beauty to thousands of travelers and stay-at-homes. Jens Jensen, the landscape architect, has long preached the need of preserving wilderness conditions for the inspiration of such leaders. A nation can afford to pay a high price for such leaders but they can thrive only in a favorable environment. Sanborn (*The Personality of Thoreau,* 1901, p. 5) has said, "When Emerson said to his young visitor that 'he was always looking out for new poets and orators, and was sure the new generation of young men would contain some,' Thoreau quaintly said that 'he had found one in the Concord woods—only it had feathers, and had never been to Harvard College; still it had a voice and an aerial inclination—and little more was needed.' 'Let us cage it,' said Emerson. 'That is the way the world always spoils its poets,' was Thoreau's characteristic reply."

Scientific. Too often science is looked upon solely as a tool for economic purposes and many overlook the esthetic and intellectual pleasure derived from its study. There is as keen pleasure of understanding a problem as there is in looking at a beautiful scene or picture. Science is also a recreation for many persons, and one that merits much greater attention than is customary these days. The practical advantage of science might seem to need little emphasis to foresters, since forestry depends so much upon biologic, economic and social science, and yet I am convinced that the lack of scientific research is today one of the major limiting factors in the advance of forestry, particularly on the side of forest ecology, including economic and social research. In forest ecology research conducted on wilderness conditions is needed to supplement controlled experimental studies. We need to study not only forests but as well the animals if we are to have a thorough grasp of the forest as a functional biotic community.

Educational. The educational value of wilderness areas, aside from its scientific value, is very comprehensive and far-reaching. This applies to the young and to adults, and covers esthetic, scientific, recreational, economic and social aspects. With the industrialization and urbanization of our people it becomes increasingly urgent that the problems of the rural world be presented to them in concrete form. Otherwise they will certainly lose contact and sympathy with such conditions. I anticipate that this will be one of the major problems of the forester during the next generation. Economists and sociologists are already warning us of the danger of over-industrialization, and of its menace to agriculture and forestry, because the rural and the urban problems should be properly balanced for the welfare of the people as a whole. The general public should be taught that wilderness areas are of such great educational value that they must be preserved, even if public use is rather severely restricted in the immediate utilization of such areas.

Recreational. First of all let us note that recreation is not necessarily synonymous with the Coney Island variety of amusement. Recreation as a psychological and physiological change has come to have a new meaning with the industrialization of our people. Monotonous and mechanical routine makes recreation not a luxury but a necessity to a vast number of people. We have come to look upon play not as a waste of time, but as a normal healthy function, and even in education it is a relatively new idea to grant that the play of children is to be encouraged and not simply to be tolerated.

A complete change from our customary routine is one of the most important elements in recreation. It is for this reason that wild areas have a particular charm for the city dweller, and the more complete the change the better, if one is trained to appreciate the difference and to take advantage of it. A virgin forest of huge trees has an appeal not found in cut-over lands. Of course, there are many who do not know the difference, but there are those who clearly do.

Economic. And lastly the economic value of the wilderness should not be overlooked. Not long ago there were objections to "locking up" the natural resources in the national parks: the forage, the water, the timber, the game etc., and envious eyes looked upon all these as really wasted. We have not yet outgrown all this, yet some of the neighbors of the national parks are now beginning to learn that these so-called locked-up resources bring more money into their region than their own wide open system. The wilderness has thus come to have an economic value, but if we had allowed the economic interests to control (and they have too much influence even today), they would have ruined our national parks before we had learned to appreciate their value. This statement can not be repeated too often. We may safely predict that the value of these areas will increase in direct proportion to our being able to keep them steadily wild and virgin. These must be preserved even at the cost of a very severe struggle. The public should be taught that overcrowding, alienating and tampering with these wilderness areas will ruin them, and that this will ultimately destroy their economic value in most instances. The state parks and state forests must also face this same problem.

METHODS OF PRESERVING NATURAL CONDITIONS

The best methods of preserving natural wilderness conditions are very difficult to accomplish and the importance of this subject is so great that every possible method should be utilized. At present the outstanding and most successful method has been to encourage the ideal which has been slowly developing for our national parks. To live up to that ideal—*to pass on to future generations natural conditions unimpaired*— is a very difficult undertaking. It can not be assured until an eager, intelligent public sentiment is developed to support trained public officials. We can not maintain this standard in overcrowded parks, with inadequately trained staffs, with the importation of exotic plants and animals, fish and game, and with inexpert control of predatory animals, the pollution of streams, grazing and the cutting of timber, or light burning, the excess of roads or even trails. Such measures can be stopped only when an informed public insist upon the maintenance of the ideal which has now been evolving for over 50 years. National monuments need a kind of care similar to that of the national parks.

It is very interesting to note that the Swiss have gone much beyond us in their methods of preservation of their smaller national parks. Thus Dr. Carl Schröter has recently said:

> Shooting, fishing, manuring, grazing and woodcutting are entirely prohibited. No flower nor twig may be gathered, no animal killed and no stone removed; even the fallen tree must remain untouched. Nature alone is dominant! No hotels are allowed to be erected, and, naturally, no routes for motor cars. The whole must conserve an alpine character. Camping and the lighting of fires is prohibited. How different are the American national parks! There, big hotels are erected, automobile routes are constructed, and camping, even fishing is allowed. The American national parks are pleasure resorts, our national park, a 'Nature Sanctuary!' (*Journal of Mammalogy,* 8:350–51, 1927.)

Turning now to the national forests, we find there are vast areas of wilderness in them, probably exceeding in area that of all our national parks, which are approximately as virgin as the parks themselves. But the national forest ideal involves changing these virgin conditions as rapidly as cultural methods can reach their remote or inaccessible areas. The protection forests are not likely to be distributed greatly. In addition to the virgin areas, which should be preserved for the study of forest ecology in the broadest sense, there should be set aside, as advocated by Emerson Hough and Aldo Leopold, certain large areas as true wilderness. Hough advocated the Kaibab Plateau, north of the Grand Canyon in Arizona, and Leopold favored the Gila region in New Mexico and a similar effort has been for the preservation of a tract in the Superior Forest in Minnesota. Certainly we have other suitable areas, of variable size, in Utah, Idaho, Colorado, Oregon and Washington, in addition to Alaska, and possibly in Canada, that merit similar attention. To accomplish these results, however, there will need to be considerable change in the attitude of many officials and foresters. The acreage usually suggested for such reservations has often been very small indeed, due in part no doubt to the fear of not getting a hearing at all if large areas were suggested, and possibly also to the influence of the example of small "sample plots," used in silviculture. These areas should be large enough to preserve fair samples of all the more important types of forest communities under varied conditions, and more attention should be paid in their selection to the welfare of the plants and animals than to administrative *convenience.* Smaller areas also have their value if properly selected.

In the state forests and parks there should be definite provision for the preservation of certain areas in a natural state. In the highly modified conditions of the East this is more difficult, although along the coasts and in swampy, sandy or rocky areas there are yet to be found many valuable sites. In the West, however, the problem is often simpler, except in the very fertile agricultural areas, where progress seems to be relatively the slowest.

Some persons are enthusiastic over the policy of the preserving of such reservations by private individuals, but few of such preserves survive more than two or three generations, and generally the second is the turning point.

While there has been considerable activity for the establishment of game preserves and bird reservations, the primary purpose subordinates natural conditions to certain game or to birds, and control measures destroy the natural conditions in varying degrees. Such organizations are not today a very hopeful source of aid for the preservation of natural conditions, because as a rule the special interest is so strong that they consider their interference negligible or desirable.

One of the most hopeful prospects for the preservation of natural conditions is in connection with various educational institutions and organizations, including museums. Several of our state universities have already acquired lands and water which they use for scientific purposes, and some such areas are being preserved in a wild state. Thus the University of Washington has a marine biological station and has made a preserve area as a part of it. The University of Montana has some lands on Flathead Lake which have this possibility. The University of Illinois has a small reservation. Indiana University has a small tract. Cornell University has a wild flower preserve as well as the McLean Bog for this purpose. The New York State Museum, under the leadership of Dr. John M. Clarke, acquired by gifts several valuable geological reservations, but funds were never available to protect the plants and animals on them. Yale University has recently acquired 200 acres for a preserve. It is not virgin but plans are being made to restore it. The University of North Carolina has a 950-acre wooded tract adjacent to the campus. Some of the western state universities own considerable land, and ought to be able to set aside valuable areas as wild preserves if a serious effort is made to do so, particularly in Minnesota with its state land.

In spite of the preceding remarks, the universities have been particularly backward in acquiring wild lands for both teaching and for research. Without such lands they are likely to lose touch with outdoor natural history, ecology and allied problems of conservation. It is largely in an indoor atmosphere that what Dr. W. M. Wheeler of Harvard University called "academic dry rot" thrives, and biology can not afford to lose contact with the natural conditions of life. The same condition holds for museums. Only the largest metropolitan museums can hope to be world microcosms, and even these grow largely by their active field work and explorations. Smaller museums need to maintain a similar contact with the outer world and direct active interest in carefully selected reservations and field survey in order to retard the growth of "dry rot" which at times infests other educational institutions. The advantage therefore of wild areas is not simply the opportunity for an occasional visitor and student to make some special study—as important as that is—but is the influence upon the workers themselves and their institutions. If our educational institutions do not have this advantage educational leadership in such matters will pass on to other better qualified leaders.

REFERENCES

Adams, Charles C. 1921. Suggestions for the Management of Forest Wild Life in the Allegany State Park, New York. *Roosevelt Wild Life Bul.*, 1:62–74.

———. 1925. The Relation of Wild Life to the Public in National and State Parks. *Roosevelt Wild Life Bul.*, 2:371–401.

———. 1925a. Ecological Conditions in National Forests and in National Parks. *Sci. Monthly,* 20:567–93.

Ashe, W. W. 1922. Reserved Areas of Principal Forest Types as a Guide in Developing an American Silviculture. *Jour. Forestry,* 20:276–83.

Conwentz, H. 1909. *The Care of Natural Monuments with Special Reference to Great Britain and Germany.* 185p. Cambridge, England.

Haddon, A. C. 1903. The Saving of Vanishing Data. *Pop. Sci. Monthly,* 62:222–29.

Hahn, W. L. 1913. The Future of the North American Fauna. *Pop. Sci. Monthly,* 83:169–77.

Leopold, A. 1925. Wilderness as a Form of Land Use. *Jour. Land and Pub. Utility Econ.*, 1:398–404.

Pearson, G. A. 1922. Preservation of Natural Areas in the National Forests. *Ecology,* 3:284–87.

Shelford, V. E., et al. 1926. *Naturalist's Guide to the Americas.* 761p. Baltimore.

Sumner, F. B. 1920. The Need for a More Serious Effort to Rescue a Few Fragments of Vanishing Nature. *Sci. Monthly,* 10:236–48.

———. 1921. The Responsibility of the Biologist in the Matter of Preserving Natural Conditions. *Science,* N.S., 54:39–43.

George M. Wright, Joseph S. Dixon, and Ben H. Thompson

Problems of Geographical Origin (1933)

From *Fauna of the National Parks of the United States: A Preliminary Survey of Faunal Relations in National Parks*

In teaming together for the first field fauna surveys of U.S. national parks in 1933, National Park Service biologists George Wright (1904–36), Joseph Dixon (1884–1952), and Ben Thompson (1904–97) have been called "trailblazers" for the contemporary field of conservation biology. (Later fauna surveys include recognized works such as Adolph Murie's "The Wolves of Mount McKinley" and L. David Mech's "The Wolves of Isle Royale.") Wright was a visionary biologist whose advocacy of scientifically based natural resource management and whose opposition to purely utilitarian and recreational natural resource management within the national park system were far ahead of their time and just beginning to take hold when his life was cut tragically short by an automobile accident. Dixon was a wildlife biologist and naturalist for the NPS. Thompson held many field and administrative positions within the NPS. In this selection from their first of two fauna surveys of U.S. national parks, the authors present one of the first clear statements focused on the preservation of biodiversity as the main reason for preserving parks, perhaps the parks' "greatest natural heritage."

FAILURE OF PARKS AS INDEPENDENT BIOLOGICAL UNITS

The preponderance of unfavorable wild-life conditions confronting superintendents is traceable to the insufficiency of park areas as self-contained biological units. In the present era of park development, this geographical cause ranks as the most important of the three major causes of wild-life problems. If the influx of visitors were to increase in the future at the same rate that it has in the past 15 years, competition between man and animal in the park could easily become more influential in faunal maladjustments than the geographical factor.

At present, not one park is large enough to provide year-round sanctuary for adequate populations of all resident species. Not one is so fortunate—and probably none can ever be unless it is an island—as to have boundaries that are a guarantee against the invasion of external influences. To all this the practical-minded will immediately retort that an area with artificial boundaries can never be a true biological entity, and obviously this is correct. But it is equally true that many parks' faunas could become self-sustaining and independent if areas and boundaries were fixed with careful consideration of their needs. Already many parks are being improved in this regard, and there is a vast amount more that can be done.

When all advisable enlargements and boundary corrections have been made, there will still be external influences and probably some range inadequacies to be reckoned with and some management measures will be necessary for the protection of the affected species. Whereas both general types of problems due to the inadequacies of parks as independent biological units must be met primarily by changing the boundaries, as the only means of dealing with the fundamental cause, success can not be as great in one instance as the other. By his action, man can restore a needed range to a park provided he is willing to do it, but there is absolutely no way he can keep every unfavorable influence out of that park—not so long as boundaries are artificial, and some of them must always be that.

CONDITIONS CAUSED BY FAILURE TO INCLUDE THE COMPLETE ANIMAL HABITAT

Unfortunately, most of our national parks are mountain-top parks. During the summer, game retreats to the higher elevations. With the coming of winter the game drifts down below the park, away from protection. For a wild-life preserve this is obviously an unsatisfactory condition. It is utterly impossible to protect animals in an area so small that they are within it only a portion of the year. It is just as fundamental to protect the whole range of the resident fauna of any park as it is to protect the watershed of any stream for its water supply. One would not think of just protecting a narrow zone across which the water flows, but would extend the protection to the natural boundaries of the watershed. In like manner, it is useless to draw up imaginary and arbitrary boundaries for a park and expect to protect the animal life drifting through. This is exactly what has been done in creating the national parks—a little square has been chalked across the drift of the game, and the game doesn't stay within the square. In order that our parks may be able to adequately protect and preserve their wild-life as part of our national heritage, it is essential that they be formed principally of natural boundaries, and not arbitrary boundaries. Natural boundaries in this case mean natural barriers limiting the range of the wild-life concerned. While the natural boundaries are not definite lines, they are sufficiently tangible in character to be capable of practical establishment. It is now, perhaps, too late to establish natural boundaries completely around all parks, but that is no reason why they should not be established where it is still possible.

If natural boundaries are natural barriers limiting the range of the wild-life of any particular area, more definite designation of what these natural barriers should be is necessary. The natural barriers are different for each park, and must be treated as individual problems in each case. But if there is any value in a generalization, this much might be said: As a natural barrier, a mountain crest is better than a valley or stream, but the lowest zone inhabited by the majority of the park fauna is probably the best of all.

All of the western parks are mountain areas. Some of them are a fringe around a mountain peak; some of them are a patch on one slope of a mountain extending to its crest; and some of them are but portions of one slope. All of them have arbitrary boundaries laid out to protect some scenic feature. But our national heritage is richer than just scenic features; the realization is coming that perhaps our greatest national heritage is nature itself, with all its complexity and its abundance of life, which, when combined with great scenic beauty as it is in the national parks, becomes of unlimited value. This is what we would attain in the national parks. In order to attain it, their boundaries must be drafted to meet the needs of their wild-life. The entire complexity of wild-life can be protected only within natural barriers. In establishing the boundaries, natural barriers should be followed.

Because so many of the national parks are in the high mountains, the seasonal habitat usually lacking is winter range. One exception where the reverse is true is Grand Canyon; but as it is such an uncommon case, it is discussed under that park only. Lack of summer habitat is not one of the major types of wild-life problems.

George M. Wright

Big Game of Our National Parks (1935)

George Wright (1904–36) studied under Joseph Grinnell at the University of California, Berkeley and went on to become a leading American wildlife biologist for the National Park Service. In 1930, at the age of twenty-six, he became the first chief of the wildlife division of the NPS. Although an automobile accident cut his life and career unfortunately short, he published a number of important works on fauna in the national park system and advanced the field of wildlife ecology and wildlands preservation. The George Wright Society, whose mission it is to promote "the scientific and heritage values of protected areas," was created and named in his honor. Originally published in the *Scientific Monthly,* this essay focuses on the importance of national parks and other protected areas for the preservation of "big game" species of wildlife. Wright seems concerned with defending these places less as actual hunting reserves and more as sources of healthy wildlife populations.

THE PRESIDENT OF THE United States designated 1934 as "National Parks Year." But to thousands of big game animals, every year is national parks year, for in the parks they find a perpetual holiday from the warfare that mankind wages against them.

In America, north of Mexico, big game means buffalo—or bison, as they should properly be called—elk, moose, caribou and deer, mountain sheep or bighorn, mountain goats which are not goats at all but really antelope, antelope which are not antelope but a rare form peculiar to the New World, and cougars, wolves and bears. One of these, the Alaskan brown bear, is the largest carnivorous animal in the modern world.

A hundred years ago, wild game still abounded from the Atlantic to the Pacific in numbers that totaled millions. In a century's time, this vast game field vanished like a continent sinking into the ocean, and there remain only small islands of game where once it stretched continuous and flourishing. The best of these island retreats, where big game animals have been spared from the destructive waves of market hunting, sport shooting, lumbering, and agriculture, are the national parks.

The story of the American bison is a tradition of our national life. By 1902 the only wild buffalo remaining in this country were in the hidden fastnesses of the Yellowstone, and they numbered only twenty. Theodore Roosevelt, who has stood forth as the great conservation President until this very day, when another of the same name threatens that distinction, asked Congress to provide for the reestablishment of a bison herd in Yellowstone Park. A few animals were brought in to augment the wild stock. From this start, careful nursing over a period of thirty years has improved the herd in size and quality, until now it is maintained at one thousand head, the very limit of the carrying capacity of the range. This is a remarkable testimonial to the untiring zeal of the buffalo keepers who have had to hold them and feed them at the Buffalo Ranch through the long winters because they could no longer migrate to the plains. Just when the Yellowstone bison seemed to have been brought safely past danger of extinction, outbreaks of the dread disease, hemorrhagic septicemia, resulted in new losses and fresh anxiety. The entire herd was inoculated with an especially prepared vaccine, and now hemorrhagic septicemia is no longer a major menace to the bison of Yellowstone.

Although the high country of Yellowstone, where bands of hundreds of bison laze away the summer, furnishes the real buffalo show of the parks, smaller show herds may be seen at Platt and Wind Cave and Colorado Monument, and there is a project now on foot to reestablish on the east side of Glacier a bison herd which will belong to the Indians of the Blackfeet Reservation, but which will summer under the watchful eyes of Indian riders.

Visitors are safe in buffalo country, for it is not this grass-eating bovine, shaggy and powerful as he appears, but man who has furnished the element of menace in the story of the American bison. It is only the old bulls, kicked out of the herd by their younger rivals and grown cranky in the isolation of their old age, which it is sometimes just as well not to cross.

The story of the bison is far too long to be attempted here, and we may leave it at what we confidently trust may be a happy ending. But many other interesting animal faces crowd around. There are those strange parti-colored antelope of the deserts and the sage brush mountain slopes. Though their horns have a bony core like those of sheep and cattle, they have a side prong, and the outer sheath is shed annually like the antlers of the deer tribe. The swiftness of the pronghorn antelope is a well-merited tradition. Their insatiate curiosity, which is their undoing where they are hunted, brings them no harm in the parks and frequently helps the visitor to a near view of them.

There are nearly seven hundred pronghorn on the northern game range of Yellowstone. Although this area can hardly be considered as being within their optimum range, and in spite of the fact that the hard winters crowd them down to the vicinity of Gardiner where they must compete with thousands of elk for sustenance on that depleted winter range—still they hold their own and even prosper. A recent addition to Yellowstone Park below Gardiner will benefit the pronghorn perhaps more than any other species concerned.

The life of these Yellowstone antelope contrasts sharply with the life of those native to the arid region of Petrified Forest National Monument in Arizona. There the President's recovery program plans to develop water catchments and is already providing a fence to exclude domestic stock and, thereby, to improve the monument's range conditions for the rightful utilizers of its range—the pronghorn which roam the hot dry wastes of Petrified Forest.

In still sharper contrast is the life of Indian Gardens far down inside the walls of the Grand Canyon. This herd, numbering some twenty-five, is being artificially maintained and protected there until such time as suitable range may be procured and made ready for them in Grand Canyon National Monument, where once again they may be safely allowed to forage for themselves in the freedom and beauty of their normal wild state.

True mountain antelopes are our mountain goats. For their welfare the National Park Service need not be concerned. When winter comes, these hardy veterans of the tumbled crags do not forsake the parks for dangerous country outside. They have no need to seek the more abundant forage of lower altitudes, for succulent herbage is always to be found where strong winds have stripped the snow from the steep slopes of the high country which they inhabit. With their short straight black horns, bearded faces and shaggy white uniforms, they stand sentinel on the rock pinnacles like policemen of the Arctic-Alpine zone.

They are plentiful in Mount Rainier, where their number, estimated at four hundred, is believed to be increasing. In Glacier they are even more abundant. No visitor to that park need leave without having seen one of their number, for the new high road across Logan Pass and the mountain chalets built in the midst of the mountain goats' home territory bring the visitor within such close range of his wary hosts that, even from his car, he may often watch them on the peaks that have been their hazardous homes for centuries.

Another dweller in the high places is the mountain sheep. Unfortunately, however, far too little is known of their ways and of the causes which have led, and seem still to be leading, to a general decrease in their numbers. Indeed, one of our chief wildlife problems today is the attempt to discover what may be done to improve their condition.

Mount McKinley is the home of the Dall sheep with the white coats and slender outward-curving horns. From a distance they look like so many white sheets spread out to dry on the fresh green grass. Recently a low ebb in the cyclical variations of abundance was reached by the Dall sheep of Mount McKinley, where their estimated numbers fell to a third of the 1929 level. Though such cycles of abundance are in themselves a natural phenomenon caused in part by bad weather and the depredations of wolves and coyotes, yet man has added so many more hazards that the time may come when recovery is impossible. Increased protection for the sheep outside the park and a limited control of their natural enemies within the park are the only present means of stemming this dangerous ebb, and latest reports of what is believed to be a slight increase may indicate that the tide is already turning.

The tan-colored Rocky Mountain bighorn are found in Glacier, Yellowstone, Rocky Mountain and Grand Teton. In these parks, too, the sheep seem to be slowly declining in number, unable to cope with the many factors hostile to their welfare. Some of these hostile factors which we can already diagnose are their apparently low resistance to disease, range competition with elk (as in Yellowstone) or with domestic sheep (as in Rocky Mountain), coyote depredations and even, it is suspected, poaching. Careful study and zealous care is necessary if we are to save the Rocky Mountain bighorn from all these hazards.

Very little is known concerning the sheep of Grand Canyon. These faded-out desert bighorns, with their very massive, tightly curled horns, live in the canyon below the rim, and, in their practically impenetrable habitat, they are seldom seen. It may be assumed, however, that the recent removal of over a thousand feral burros may react favorably upon their food supply and consequently upon them.

The various members of the deer family are the bright spirits of the park forests. The tall dark stranger with the broad, palmed antlers, who stalks through the willow thickets of Grand Teton and Yellowstone in white-stockinged feet, is that giant deer, the Shiras moose. Since they are hardy and do not migrate in herds to the lowlands in winter, they find ideal protection in the high country of the parks. Nearly every day they may be seen along the main road between Norris and Mammoth Hot Springs, in Yellowstone's Willow Park. One day last summer a moose cow and her calf stood feeding in the Yellowstone River, their demeanor as placid as the dark green water that smoothly swirled around their legs, while from the banks there arose a din of excited exclamations and camera shutters.

Glacier, too, has its American moose, some one hundred and fifty of them—perhaps more, for they have lately been seen there in valleys hitherto unfrequented by them.

A new member of the resident fauna of Mount McKinley is the Alaska moose, the largest and blackest of the moose tribe. It is only since 1926 that they have been more than casual summer visitors to the park, but the recent inclusion of lands of lower altitude within park boundaries, together with the cumulative effects of protection, and possibly persecution outside the park, may have led them to adopt this high refuge where living conditions are otherwise less favorable.

Caribou are the deer of the north, weird animals seen drifting like wraiths across the tundra twilight. The grotesque, uneven antlers are worn by the females as well as by the bulls, and in this they are anomalous in the deer world. A hard battle is being waged to prevent the fine large barren-ground caribou of Mount McKinley Park from crossing with the dwarfish but aggressive reindeer imported from the Old World.

Because it has had so much publicity, the mention of elk usually conjures a picture of famed Jackson's Hole south of Yellowstone Park. But there is also a large herd of more than ten thousand elk that mingles with the southern herd in summer but goes down on the north side of the park to winter. Then there are the smaller herds of Glacier and Rocky Mountain, reintroduced from Yellowstone stock, as were many other herds of elk now found on the national forests and elsewhere outside the parks system.

Summer is lavish in its gifts to the elk. They are well protected in the interior of the parks and the grasses are lush. There they drop their calves, which are soon running gaily beside their mothers. When chilly nights change summer to gaudy autumn, the fat sleek bulls bugle from every hillside and make ardent love. Then one morning the ground is white, and winter with its weakening cold and hunger steals down to deal death among them. Hunters wait below, but this end is a merciful one, for domestic cattle have stripped their winter lands of forage and there is not room enough for all. The government maintains winter feed grounds, but elk, like human beings, are not improved by being pauperized. The only way to save the elk herds is to secure adequate winter range for them.

Deer, including the sturdy Rocky Mountain mule deer and the graceful white-tailed deer with its white signal plume, are well represented in the parks. Glacier is notable for having large numbers of both types, and Yosemite is famed as the Eden-like garden where deer and men live side by side as neighbors and friends. In fact, the ability of deer to adapt themselves to joint residence with man—together with the special protection they have had in the past through control and often even elimination of their natural enemies—has in some instances created a problem of adjustment of numbers to range carrying capacity. The National Park Service and the United States Forest Service favor the policy of managing the Kaibab deer herd to permit its gradual increase as the range is brought back to normal productivity; and a careful and constant study is made of both deer numbers and range conditions there. On the South Rim of the Grand Canyon, water holes for deer are being constructed, and about sixty square miles of range have been fenced to exclude stock. Further game range studies by the use of fenced and unfenced quadrates and checkplots have been proposed for Yosemite, Sequoia, Zion and Mesa Verde, and such quadrates have already been constructed in Rocky Mountain and Yellowstone.

Cougars are seldom seen, but the knowledge that they are present adds interest to the mountain trip. Besides, they are a valuable part of the whole of nature's plan. It is the presence of its natural enemy, the cougar, which has made the deer that swift alert animal which we so much admire. In the parks all native species receive protection, so that nature's balance may to the greatest possible extent be preserved. Even the wolf is given consideration, but unfortunately it has already disappeared from all the national parks of the United States proper. The handsome timber wolves of Mount McKinley, which attain a length of seven feet and more, need special protection, for the wolf, like the cougar, has hardly a friend and is headed for extinction.

Lastly come the bears, the clowns of the animal world, the bears of comic reputation. Black bears encourage intimacies, but the wise visitor will heed the advice of park officials and remember that it is always better to meet the friendly advances of the cheerful black bear a good deal less than half way.

The grizzly bear is an animal of another temperament. Lord of all the lands he roams and the most powerful of the big animals is King Grizzly, although today his park domain is limited to Mount McKinley, Glacier and Yellowstone. No one who has ever seen this magnificent beast visiting the feed platforms at Old Faithful or Canyon can abide the thought that he should ever disappear from the face of the earth. Fortunately, the grizzly seems to be increasing in the parks, even though his numbers are declining elsewhere.

This is the roster of big game animals in the parks. There are, of course, besides these large animals many others of less size but equal interest, including the coyote, beaver, otter, marten, wolverine, badger and a host of others. These, too, find refuge in the national parks, for the money value of their fur, or their interference with man's agricultural pursuits, has made many of them as much in need of protection as the larger targets, the big game.

Conservation-minded citizens have come to depend on the National Park Service to protect the prized remnants of our almost-vanished large mammals. Old-time rangers tell many an interesting story of long hard winters when they fought side by side with the buffalo and elk against the stalking deaths of poaching, starvation and disease. It is apparent that those darkest hours for conservation are past, but there is still the greatest necessity for unrelenting watchfulness and care.

To the vacationist and the seeker after adventure, as well as to the conservationist, the parks have much to offer. They are the last flourishing remnants of a lost game continent, where one who is weary of cities may still be refreshed by a taste of wilderness and thrilled by a glimpse of the rightful masters of wilderness—the big game.

Victor E. Shelford

The Preservation of Natural Biotic Communities (1933)

Victor Shelford (1877–1968) was a preeminent American ecologist best known for his work with Frederick Clements to refocus the field of ecology (what they called "bio-ecology") on the study of the interactions between plant and animal communities as an integrated biotic community, or "biome." A student of Henry Cowles at the University of Chicago, Shelford was also a founding member and the first president of the Ecological Society of America and the driving force behind the ESA's Committee for the Preservation of Natural Conditions in the United States for twenty years. Unhappy with the ESA's decision to move away from active policy work and hence to disband the Committee for Preservation, Shelford and others founded the Ecologists' Union, whose goal was the "preservation of natural biotic communities, and encouragement of scientific research in preserved areas." The Ecologists' Union later morphed into the Nature Conservancy. Shelford penned a number of essays defending the creation and defense of protected areas for scientific reasons. In this essay, first published in the ESA's journal, *Ecology,* Shelford defines a "Nature Sanctuary" as a place where natural forces are allowed "free play," and he details the various types of protected areas that might be set aside.

A. INTRODUCTION

THERE APPEARS TO BE a definite desire on the part of the people of all countries to preserve at least some of the original vegetation and wild animals. The national parks as well as other parks and reservations of Europe, Africa, the United States and Canada are well known examples. However, types of vegetation other than forest have usually not been preserved except incidentally as inclusions within the forested area or in connection with some historical and ethnological remains. The larger wild animals of all civilized countries have been greatly reduced in numbers, some because they are a hindrance to agriculture, others because of their value as food and clothing. Still others have been slaughtered for no good reason. Many people think that the national parks of Canada and the United States, some state and provincial parks, and some national forests have escaped this elimination of large animals. They believe that such areas are examples of primeval nature with the animal life essentially complete. The national parks of both countries doubtless represent the least disturbed series of areas which we have on the continent. Some state and provincial parks and one or two of the national forests in each country may be equally undisturbed at present, but probably less well safeguarded or on a less permanent basis. Yet, many have been surprised to learn of the large amount of modification which has gone on in times past within these park areas, both before and in some cases perhaps after they were set aside.

The whole trend of research and education is toward specialization on particular objects or particular organisms. These are stressed while the assemblage to which they belong is ignored or forgotten, together with the fact that they are to be regarded as integral parts of the system of nature. Outside of modern ecology and geography there has been little or no tendency towards the development of specialists on the entire life of natural areas. Perhaps one reason why nature study has been unsuccessful is because too often it is not the study of nature but of single natural objects or groups of objects which constitute a small part of any natural aggregation. Often this has resulted in the development of emotions relative to the plants or animals singled out for this study, followed by sentimental desires to protect them.

Biologists are beginning to realize that it is dangerous to tamper with nature by introducing plants and animals, or by destroying predatory animals or by pampering herbivores. Much of the so-called "control" of noxious insects, of predatory animals, and of plant diseases is based upon the idea that nature can be "improved." This is, of course, the dominant idea in all agricultural, silvicultural and game cultural practice. The fact remains, however, that wolves and other predators, and herbivores in numbers have lived together for thousands of years without disaster and that in some of our national parks they have lived together with far better maintenance of natural conditions than in areas where control has prevailed.

In general, from a philosophical and practical viewpoint, the unmodified assemblage of organisms is commonly more valuable than the isolated rare species. However, because the significance of the unmodified assemblage is popularly ignored, the whole is commonly sacrificed in the supposed interest of the rare species. Due to the local habitat relations of the rare species, neither need be sacrificed in any large natural area, hence the importance of large sanctuaries.

The principal activity of field ecologists might be stated to be the gathering of data for the interpretation of nature. To the ecologist, it is the entire series of plants and animals which live together in any community which is of primary interest.

The Ecological Society of America has always been interested in the preservation of natural areas with all their native animals. A committee composed of its members published a book entitled *The Naturalist Guide* ('25) listing a very large number of such areas and so is in position to know the extent and present condition of such areas. Only a very few were reported in a pristine condition so far as animals were concerned. The society has recently held a number of conferences with park and forest officials and after careful consideration, prepared the following memorandum dealing with (1) definition, (2) size, and (3) classification of natural areas containing original plant and animal life.

B. NATURE SANCTUARIES OR NATURE RESERVES

1. MEANING AND USE OF THE TERM

Just what original nature in any area was like from a biological viewpoint, is not known and never can be known with any great accuracy. Primitive man, who could not remove the forest or exterminate the animals, is probably properly called a part of nature. At the time of the discovery of America, a scattered population of Indians had locally modified the vegetation, but had not destroyed any of the vegetation types. However, most of the areas which are now available for reservation as nature sanctuaries or nature reserves were probably not much affected by these primitive men. This is the argument for leaving them out of the picture.

"Nature" and "natural" are purely relative terms and can have significance only as averages, because the outstanding phenomenon of biotic communities is fluctuations in numbers of constituent organisms or reproductive stages of organisms over a period of one to thirty or more years. Thus, a Nature Sanctuary is primarily an area in which these fluctuations are allowed free play.

The term *Nature Sanctuary* has been applied to areas covered by natural vegetation, but not containing all the animal species. In Europe, for example, in some of the nature parks no timber is removed and only persons with serious scientific or other scholarly interest are admitted. The Nature Sanctuaries are surrounded by areas in a less natural state, such as nearly natural forest devoted to growing timber, game production, etc. These surrounding lands are called buffer areas of partial protection.

In the United States and Canada areas of nearly natural vegetation are larger than in central Europe and fewer of the animals have been lost. It is possible, therefore, to recognize several classes of Nature Sanctuaries in North America.

II. CLASSES OF NATURE SANCTUARIES

The categories below are arbitrary and merely for the purpose of providing provisional basis ranking of natural areas. The classification of each area should be determined by a committee of competent naturalists.

1. *First Class Nature Sanctuaries.*

   Any area of original vegetation, containing all the animal species historically known to have occurred in the area (except primitive man), and thought to be present in sufficient numbers to maintain themselves, is suitable for a first class Nature Sanctuary.

2. *Second Class Nature Sanctuaries.*

   A. Second growth areas (of timber) approaching maturity, but conforming to the requirement of No. 1 in all other respects.

   B. Areas of original vegetation from which not more than two important species of animal are missing.

3. *Third Class Nature Sanctuaries.*

   Areas modified more than those described under No. 2.

III. OTHER TERMS AND THEIR MEANINGS IN COMMON USAGE

1. *Nature Sanctuary* — This emphasizes not only the stationary (floral) elements but also the motile (faunal) elements. It necessitates buffering and non-interference by man.
2. The only synonym for Nature Sanctuary that has been suggested is *Nature Reserve.*
3. *Research Reserve* (U.S. National Park Service sense) means Nature Sanctuary, as the areas are selected to represent the primitive biological condition and admission is by permit only. The U.S. National Park Service appears to be working toward a three-zone plan: (*a*) a zone of development which is a small portion of the park devoted to hotels, camps, etc.; (*b*) the greater portion of the park open to the public and traversed by trails and roads (in many cases these areas may serve as second or third

class Nature Sanctuaries); (*c*) Research Reserves open to the public only by permit.

4. *Natural Area* (U.S. Forest Service sense)—This emphasizes the stationary elements of nature, hence is primarily floral.
5. *Buffer Area* is a region surrounding a Nature Sanctuary in which the biotic community, especially the vegetation, is only slightly modified by man. It is a region of partial protection of nature and may be zoned to afford suitable range for roaming animals under full protection.
6. *Research and Experimental Area*—This usually implies modification and management of some of the biological elements.
7. *Primitive Area* (U.S. Forest Service sense)—This is defined as an area in which human transportation and conditions of living are kept primitive. Some of the areas are to be cut over periodically.
8. *Wilderness Area*—This is defined essentially as is *primitive area*.

IV. AVAILABILITY OF NATURE SANCTUARIES AND BUFFER AREAS IN NORTH AMERICA

A. Nature Sanctuaries or Nature Reserves.

1. Except for Desert and Tundra areas, conditions suitable for first and second class Nature Sanctuaries are available only in connection with National Parks and National Forests and (in rare instances) State Parks in the United States and in the National Parks and some Crown Lands and Provincial Forests and Parks in Canada.
2. In eastern North America few first class Nature Sanctuaries of national importance can be established because of the absence of the wolf, wapiti and some other species.
3. Status of Natural Areas in the U.S. National Forests. These have not been selected with consideration for animals. All are too small to contain roaming animals, but could suffice as Nature Sanctuaries, if necessary, when surrounded by

high grade buffer areas in which the roaming animals are protected. To be of full value from the standpoint of primeval forest conditions and processes, all roaming animals must be allowed ingress, or abnormal conditions are likely to arise and defeat the purpose of the natural area from the standpoint of forestry.

4. Research Areas or Nature Sanctuaries in the National Parks. The remarks under No. 2 may possibly be applicable to the eastern National Parks.
5. State Parks, National Monuments (U.S.), private holdings, etc., will usually afford third class Nature Sanctuaries, but their maintenance in the condition in which found or with such improvements as may be possible, is all the more important.

B. *Buffer Areas and Modified Sanctuaries in Different Biotic Types.*

1. *Forested Areas.*
   - (*a*) Areas reserved for experimental work for which the Nature Sanctuary serves as a check (*e.g.,* in case of Experimental Forests).
   - (*b*) Areas devoted to recreation or serving as game refuges. (In the larger National, Provincial and State Parks and National, Provincial and State Forests.) There should be a zone of these types surrounding each Nature Sanctuary inside areas described under (*c*).
   - (*c*) Areas in which there has been selective cutting, grazing approaching capacity, etc.
2. *Woodland Areas.*
   - (*a*) Pinyon, Cedar, etc. The same principle holds as above.
3. *Scrub Areas.* Much scrub is said to have been produced by the invasion of grassland by shrubs belonging to arid or xeric areas through over-grazing. The herbs and grass of large areas of semi-desert scrub have been seriously damaged by over-grazing in the western United States. Except in extreme desert these factors make the selection of scrub sanctuaries

very difficult. Ideally, large modified areas should be fenced and allowed to return to the original condition while buffered by an area of restricted grazing.

4. *Grassland.* There are now natural areas of some types of grassland in the National Forests and Range Reserves or Experimental Ranges, but often these cannot be buffered and are usually too small and lack the original large animals. Ideally, large areas should be buffered by grazing and experimental areas.
5. *Semi-aquatic and Aquatic Areas.* The buffer area should consist of developmental stages of terrestrial vegetation as far as a late subclimax stage for the region.

V. SIZE OF SANCTUARIES AND RELATIONS TO NATIONAL PARKS AND FORESTS

1. The reserved areas in the National Parks are possibly too small, but in any event should be zoned about by (buffer) areas of complete or partial protection of the roaming animals. These zones of protection for certain animals would merely restrict occasional control measures to definite territory.
2. The forested natural areas in the National Forests are many of them too small and should be enlarged in some cases, but must in all cases be surrounded by zones of complete or partial protection for the large roaming animals.
3. Areas should not be fenced against any of the larger native animals, as their presence is necessary to make the conditions natural as regards vegetation, etc.
4. The Nature Sanctuary should be protected from fire, exotic organisms and diseases through management and preventive measures *within the buffer area.*
5. Size. The basis for size is purely biological and must be determined by biological conditions. The aims are (1) to preserve all the animals (birds, mammals and lower forms) native in the area and leave them to reproduce within the sanctuary entirely

unmodified, and (2) to prevent tramping and other injury to the vegetation by man. The animals have to receive primary attention, but vegetation types must also be represented. Two types of sanctuary seem possible: 1. First class sanctuaries in which wolves, mountain lions, bob-cats, coyotes, and migratory game are to be protected, and 2. Second class sanctuaries in areas where these animals have been exterminated or never existed (especially in the smaller parks and forests).

(1) FIRST CLASS SANCTUARIES AND BUFFER ZONES FOR ANIMALS

The animals requiring first and most careful consideration are the carnivores, likely to be unpopular with the agricultural (broad sense including game culture) interests outside the park or forest.

The home range of these animals must be considered. That of the wolf is said to be 50 miles, the coyote 20 miles, the bob-cat 10 miles, and the mountain lion 20 miles. These animals are slated for general extermination by some sportsmen and can be held unmolested only in areas within the larger well-buffered parks or remote wilderness areas of the national forests.

A second group demanding careful study is the migratory herbivores. These in combination with the carnivores (wolf, bob-cat and puma) will give most of the difficulty in selecting nature sanctuaries. Each sanctuary will constitute a problem in itself.

The terrain must be selected with great care wherever choice is possible so as to be about equally favorable to all the native species. The area should also meet with approval as a plant ecological reserve. Areas suitable for all the larger animals should be selected in the large parks, notably McKinley, Yellowstone, Yosemite, Grand Canyon, Rainier, etc., in the United States and Mt. Robson, Jasper, Rocky Mt., Strathcona, Algonquin, Quetico, etc., in Canada.

In so far as possible natural topography should be utilized to bound areas. Places remote from tourist travel without approach by roads or trails may be suitable without guards, but frequently some kind of guarding will be necessary. Each area will prove different in the problems encountered.

The sanctuaries may well be surrounded by areas in which there is some visitation by a limited number of persons; each park will again be a special problem.

(2) SECOND AND THIRD CLASS SANCTUARIES

The same principles hold for the smaller sanctuaries as for the large ones, but the problems are much less difficult, because the larger animals cannot be given any special attention. They may be established in parks and forests of various types, but within the small state parks and reserves there can never be true Nature Sanctuaries because of the lack of the large animals.

(*Notes:* 1. It is necessary to recognize that the National Park Service is under some pressure to let everyone go anywhere in a national park.

2. The game of the U.S. National Forests, including that in the Primitive areas and Natural areas, *belongs to the state in which the forest* is located and is subject to trapping in some cases and to state predatory animal control in others. This is the greatest difficulty in making first class Nature Sanctuaries of the National Forest Natural areas. It will probably take some time to clear this up, either through the states or the federal government. Members of the Ecological Society should use every means to educate the public as to the value of sanctuaries so as to reduce and eliminate these difficulties.)

C. METHODS OF SECURING THE ESTABLISHMENT OF NATURE SANCTUARIES

1. The first principle is to point out the need of complete Nature Sanctuaries. They are essential if any of the original nature (broad sense including large animals) in North America is to be saved for future generations for scientific observation of the important phenomenon of fluctuations in abundance of plants and animals, their social life, etc.
2. Due to our lack of knowledge of these fluctuations, each change in abundance is viewed with alarm by custodians. Hence,

constant pressure must be exerted on governmental agencies to prevent the current popular ideas relative to "control," "modification," and "improvement" of natural areas from affecting national parks, provincial and state parks and other reserves containing natural areas suitable as nature sanctuaries and with buffer territory.

(*a*) Facts now available indicate that many of the control measures have been useless because they were applied to animals at their maximum abundance and only hastened the natural decline.

(*b*) The experiment of letting areas essentially alone, so successful in a few of our parks, is worthy of repetition. Many of the areas within which nature sanctuaries and buffer areas may be established have already been modified, but remedial measures directed toward the return to a so-called equilibrium consist chiefly in allowing nature to take its course. Any other measure should be undertaken only after the most careful consideration.

3. This Committee of the Ecological Society urges a sub-division of all but the smaller reserves into (1) sanctuary, (2) buffer area of partial protection, and (3) area of development for human use where this is one of the aims of the reserve. It further urges that these three sub-divisions be so arranged as to give best conditions for roaming animals within the buffer area and the sanctuary.
4. The aim of the Ecological Society of America and cooperating societies is to secure adequate scientific observation bearing on these fluctuations in abundance and other phenomena which will be furthered by skilled watching of the natural course of events in these reserves.
5. They further aim to stimulate cooperation between the controlling agencies in charge of game and vegetation reservations and in nature sanctuaries and buffer areas in order that more logical units be developed and better methods of administration be obtained.

Victor E. Shelford

Conservation versus Preservation (1933)

Victor Shelford (1877–1968) was a preeminent American ecologist best known for his work with Frederick Clements to refocus the field of ecology (what they called "bio-ecology") on the study of the interactions between plant and animal communities as an integrated biotic community, or "biome." A student of Henry Cowles at the University of Chicago, Shelford was also a founding member and the first president of the Ecological Society of America and the driving force behind the ESA's Committee for the Preservation of Natural Conditions in the United States for twenty years. Unhappy with the ESA's decision to move away from active policy work and hence to disband the Committee for Preservation, Shelford and others founded the Ecologists' Union, whose goal was the "preservation of natural biotic communities, and encouragement of scientific research in preserved areas." The Ecologists' Union later morphed into the Nature Conservancy. Shelford penned a number of essays defending the creation and defense of protected areas for scientific reasons. In this brief essay, originally published in the journal *Science,* Shelford draws an important distinction between preservation and conservation and offers a critique of utilitarian conservation and defense of preservation by summarizing and illustrating the work of George Wright, Joseph Dixon, and Ben Thompson, which suggested the national parks as places to protect wildlife species.

THE YELLOWSTONE PARK ACT of 1872 refers to "preservation from injury or spoliation of all timber, mineral deposits, natural curiosities or wonders within the park, and their retention in their natural condition." It was apparently considered illegal to carry on any "control" measures under the old law. The National Park Service Act, 1916, states that the purpose "is to conserve the scenery and the natural and historic objects and the wild life therein and to provide for the enjoyment of the same in such manner and by such means as will leave them unimpaired for the enjoyment of future generations." The Secretary of the Interior was further authorized to remove timber, control attacks of disease and insects and to provide for the destruction of such plant and animal life as may be detrimental to the use of the parks. This was very unfortunate legislation, as the confidence in beneficial results from such "control" has, since that time, materially weakened.

Many people conceive of the National Park Service as a conservation organization. To conserve, as the term is now most frequently used, means to preserve while in use and it often implies ultimate depletion. In actual practise the operations carried on in the name of conservation are not designed to preserve the natural order but to establish and maintain a different order as regards kind and abundance of plants and animals present. The difference between preservation is well illustrated in a recent publication by Wright, Dixon, and Thompson,[1] who advocate the preservation of the birds and mammals in national parks. They point out the importance of dead timber to various birds and mammals, and the need of such timber for numerous invertebrates might well be added. Conservation as usually practised removes dead and mature timber, while preservation lets nature take its course.

In a series of suggestions by the authors mentioned nearly all the ordinary "conservation" views are reversed:

> Every species shall be left to carry on its struggle for existence unaided, as being to its greatest ultimate good, unless there is real cause to believe that it will perish if unassisted.
>
> No native predator shall be destroyed on account of its normal utilization of any other park animal, excepting if that animal is in immedi-

ate danger of extermination, and then only if the predator is not itself a vanishing form.

The authors of the report further advocate the encouragement of visitors to see animals; *e.g.,* bears in their natural surroundings rather than about a garbage pile.

The conservation idea may reasonably be extended to cover the preservation processes described in this recent publication. If so, one may conceive of the maintenance of exotic pheasants in South Dakota as very near the zero point of the conservation of nature with most other so-called conservation measures not far above this level. A nature sanctuary in a national park or national forest in which every effort was made to preserve a sample of original nature without disturbance may well stand at the top of the conservation series.

In nature sanctuaries the natural fluctuations of organisms are allowed free play and serve among other things to show what natural fluctuations in abundance are like. There is or has been so much interference with natural processes in the form of "control" of this and that organism that the student of "wild life" management who would seek a basis for more scientific treatment of the animals in his charge, is left without guiding principles or reliable information and will continue thus until the preservation measures advocated by Wright, Dixon, and Thompson with additional measures of equal importance are put into effect in as many nature reserves as possible.

NOTE

1. "Fauna of the National Parks of the United States," *Contribution of Wild Life Survey Fauna Series 1,* 1933 [reprinted in this volume as "Problems of Geographical Origin"].

Aldo Leopold

Wilderness as a Land Laboratory (1941)

Aldo Leopold (1887–1948) published essays on wilderness during the last thirty years of his forty-year career as a conservationist. Over that span Leopold's justification for wilderness preservation generally moved from more utilitarian arguments focused on the recreational values of wilderness travel and the associated preservation of the unique American character to a focus on the value of wilderness for science. Originally published in the *Living Wilderness,* the journal of the Wilderness Society, this essay represents Leopold's most mature attempt to articulate the value of wilderness for the science of conservation: wilderness areas as a base-datum of normality in the attempt to understand and measure the health of land. Portions of this article are found in his essay "Wilderness" from his posthumously published book *A Sand County Almanac,* representing Leopold's final comment on wilderness.

THE RECREATIONAL VALUE OF wilderness has been often and ably presented, but its scientific value is as yet but dimly understood. This is an attempt to set forth the need of wilderness as a base-datum for problems of land-health.

The most important characteristic of an organism is that capacity for internal self-renewal known as health.

There are two organisms in which the unconscious automatic processes of self-renewal have been supplemented by conscious interference and control. One of these is man himself (medicine and public health). The other is land (agriculture and conservation).

The effort to control the health of land has not been very successful. It is now generally understood that when soil loses fertility, or washes away faster than it forms, and when water systems exhibit abnormal floods and shortages, the land is sick.

Other evidences are generally known as facts, but not as symptoms of land-sickness. The disappearance of plant and animal species without visible causes despite efforts to protect them, and the irruption of others as pests, despite efforts to control them, must, in the absence of simpler explanations, be regarded as symptoms of derangement in the land-organism. Both are occurring too frequently to be dismissed as normal evolutionary changes.

The status of thought on these ailments of the land is reflected in the fact that our treatments for them are still prevailingly local.

Thus when a soil loses fertility we pour on fertilizer, or at best alter its tame flora and fauna, without considering the fact that its wild flora and fauna, which built the soil to begin with, may likewise be important to its maintenance. It was recently discovered, for example, that good tobacco crops depend, for some unknown reason, on the pre-conditioning of the soil by wild ragweed. It does not occur to us that such unexpected chains of dependency may have wide prevalence in nature.

When prairie dogs, ground squirrels, or mice increase to pest levels we poison them, but we do not look beyond the animal to find the cause of the irruption. We assume that animal troubles must have animal causes. The latest scientific evidence points to derangements of the *plant* community as the real seat of rodent irruptions, but few or no explorations of this clue are being made.

Many forest plantations are producing one-log or two-log trees on soil which originally grew three-log and four-log trees. Why? Advanced foresters know that the cause probably lies not in the tree, but in the micro-flora of the soil, and that it may take more years to restore the soil flora than it took to destroy it.

Many conservation treatments are obviously superficial. Flood control dams have no relation to the cause of floods. Check dams and terraces do not touch the cause of erosion. Refuges and propagating plants to maintain animals do not explain why the animal fails to maintain itself.

In general, the trend of the evidence indicates that in land, just as in the human body, the symptom may lie in one organ and the cause in another. The practices we now call conservation are, to a large extent, local alleviations of biotic pain. They are necessary, but they must not be confused with cures. The art of land-doctoring is being practiced with vigor, but the science of land-health is a job for the future.

A science of land health needs, first of all, a base-datum of normality, a picture of how healthy land maintains itself as an organism.

We have two available norms. One is found where land physiology remains largely normal despite centuries of human occupation. I know of only one such place: northeastern Europe. It is not likely that we shall fail to study it.

The other and most perfect norm is wilderness. Paleontology offers abundant evidence that wilderness maintained itself for immensely long periods; that its component species were rarely lost, neither did they get out of hand; that weather and water built soil as fast or faster than it was carried away. Wilderness, then, assumes unexpected importance as a land-laboratory.

One cannot study the physiology of Montana in the Amazon; each biotic province needs its own wilderness for comparative studies of used and unused land. It is of course too late to salvage more than a lop-sided system of wilderness remnants, and most of these remnants are far too small to retain their normality. The latest report from Yellowstone Park, for example, states that cougars and wolves are gone.[1] Grizzlies and mountain sheep are probably going. The irruption of elk following the loss of carnivores has damaged the plant community in a manner comparable to sheep grazing. "Hoofed locusts" are not necessarily tame.

I know of only one wilderness south of the Canadian boundary which retains its full flora and fauna (save only the wild Indian) and which has only one intruded species (the wild horse). It lies on the Summit of the Sierra Madre in Chihuahua. Its preservation and study, as a norm for the sick lands on both sides of the border, would be a good neighborly act well worthy of international consideration.

All wilderness areas, no matter how small or imperfect, have a large value to land-science. The important thing is to realize that recreation is not their only or even their principal utility. In fact, the boundary between recreation and science, like the boundaries between park and forest, animal and plant, tame and wild, exists only in the imperfections of the human mind.

NOTE

1. Adolph Murie, *Ecology of the Coyote in the Yellowstone,* Fauna Series no. 4 of the National Parks of the United States.

Julianne Lutz Warren

Science, Recreation, and Leopold's Quest for a Durable Scale (2008)

At six o'clock in the evening of June 6, 1940, with the world's attention focused on war and Britain's fiasco at Dunkirk, Robert Sterling Yard dispatched a telegram from his Wilderness Society office in Washington, D.C., to Aldo Leopold in Madison, Wisconsin. Yard was president of the Wilderness Society, which he had cofounded with Leopold and others in 1935. In a few days the society would hold its annual meeting. On the agenda: a proposal to forge an alliance among conservation groups to wage a different kind of war, a war to protect the remaining wild places of America. The Wilderness Society was already fighting, as were a few other organizations. But their aims differed, and they were not working together. Aldo Leopold, Yard hoped, could unite their efforts.

Earlier that year Leopold had asked Yard how the Wilderness Society might "tie the wilderness idea in with more disinterested groups." "The Ecological Society," he suggested, "is, of course, one group presumably disinterested which has a narrow but vital interest in the idea."[1] Needed to bring the two groups together, Leopold proposed, were one or a few prominent conservationists—ambassadors of a shared mission—who understood the needs and enjoyed the respect of both organizations. Leopold identified ecologist Robert Cushman Murphy as one possibility. Well known from his articles in the ESA's journal, Murphy might "prepare a special discussion of the wilderness idea aimed toward the fusion" for the Wilderness Society's upcoming council meeting.[2] Leopold continued the discussion in a communication to Yard on June 4. The society, he said again, should "attempt in some way to join forces with the ecologists" who were "conducting an independent but parallel campaign for wilderness areas." "As you know," Leopold reminded Yard, "V. E. Shelford, University of Illinois, is chairman of [the ESA's] committee on Natural Areas."[3] In his June 6 telegram Yard embraced Leopold's idea warmly but had doubts about Shelford's support and in any event needed something in writing to present at the annual meeting. "Your idea is grand," Yard declared, and "we have prominent ecologist members but Shelford does not understand us [stop] If Murphy will mail concise statement from New York on Wednesday, it will reach us in time for meeting [stop] Cannot you support it with something air-mailed say Monday or Tuesday [stop] May start something big [stop]."

Leopold promptly invited Murphy to the society's upcoming meeting in D.C. The society needed "counsel of the kind you can give," Leopold urged. "One of the points I have in mind is this," Leopold continued.

> The [Wilderness] Society as now constituted is interested mainly in wilderness *recreation*. Another group, the Ecological Society, is interested mainly in wilderness *study*. There is little or no cooperation between the two groups, though both need the same changes in public policy.
>
> What needs to be done, I think, is to persuade both groups that wilderness recreation is destined to become more and more "studious," and wilderness studies more and more appreciative of aesthetics, i.e., recreation. Therefore the two groups should get acquainted.
>
> You are one of the few who would understand this.[4]

Leopold encouraged Murphy to send his thoughts to the society in care of Yard. Meanwhile, Leopold prepared his own submission for the upcoming meeting, a proposed resolution for approval by the society's leaders:

> There is little active cooperation between the Wilderness Society and the Ecological Society, despite the fact that both seek the establishment of wilderness areas.
>
> Many members of the Wilderness Society probably think the Ecological Society seeks mainly small areas for scientific study. Many ecologists probably think the Wilderness Society seeks mainly large areas for recreation.
>
> Both should realize that only a nation-wide system of both large and small areas, will serve the needs of the future. Both should realize that ecological observation is one of the highest forms of recreation, while ecological studies without an esthetic appreciation of the biota are dull and lifeless. . . .
>
> *Be it therefore resolved,* that the Wilderness Society hereby invites the Ecological Society to attend its meetings, to use its journal, and suggests to the Ecological Society a possible joint committee on cooperation.[5]

Neither Leopold nor Murphy could attend the annual meeting. Leopold's resolution was discussed, however, and Yard promptly informed Leopold of the favorable outcome. "Your suggestion went big for a combination with the Ecological Society, was universally approved, and your resolution passed."[6]

As Yard understood this project, the aim was to get two organizations to work in concert preserving wild places, big and small, for science as well as recreation. Organizational unity was desired to bring added strength to the work of achieving related goals. Aldo Leopold, though, was at a different place in his thinking and his hopes. As much as anyone Leopold championed coordination among conservation interests; they needed to work together, he said repeatedly, so that efforts by one group did not conflict with the work of others. Yet organizational unity was merely a side benefit that would come from coordination. What Leopold most wanted to merge was not *organizations* so much as *rationales* for preserving wilderness. And his goal was to create a larger, more vital aim that all wilderness groups could help promote. Wilderness preservation should become not an end in itself but part of a more encompassing effort to foster substantial cultural change. It should promote a new ecological perception of the land, new aesthetic preferences that linked beauty with wildness, and a more humble understanding of the human place in the natural order. Ultimately, Leopold wanted wilderness and wilderness preservation to become the base for yet another quest to form a society that could endure; a society founded, as he would ultimately put it, on "a durable scale of values."[7]

Leopold's and Yard's effort to promote joint work on wilderness protection yielded some fruit, although not in the lifetime of either. Leopold's more ambitious vision, however, would languish for decades. Though there was plenty of common ground for conservation groups to stand on, most chose to protect and keep to their own small slice of it; few could understand the rugged cultural path that Leopold wanted to follow; even fewer were willing to pursue it with him.

In the effort to bring the ESA and Wilderness Society together, Murphy's support would be helpful, but it was Victor Shelford whose influence counted, and Shelford would be a harder sell. Among ecologists active in preservation Shelford stood the tallest. An animal researcher at the University of Illinois, Shelford helped found the ESA in 1915 and served as its first president.[8] He also led the ESA's crusade to protect high-quality natural areas for ecologists and others to study. In 1917 Shelford took charge of the ESA's new Committee for the Preservation of Natural Conditions. Immediately, the committee began identifying all types of "preserved and preservable areas in North America in which natural conditions persist." Such a list, believed Shelford, was essential in working toward "the preservation of natural areas with original flora and fauna (or nearly so as may obtain)" and the maintenance of their natural relations and processes.[9] A 1921 progress report identified various reasons for preservation (scientific, recreational, aesthetic, and material), yet it was the scientific rationale that loomed largest.[10] Shelford directed the ESA effort until 1923, when he stepped aside to work intently on the committee's inventory of natural areas, published in 1926 as *A Naturalist's Guide to the Americas*. Shelford chaired the ESA's efforts to reorganize its preservation and study work in 1930, wrote its critical documents on nature reserves in 1931 and 1933, and directed both ESA programs through much of the decade.[11] Shelford was also kept busy addressing conflict within the ESA over the organization's proper role in political and activist land preservation efforts.

After the Wilderness Society meeting in the summer of 1940, little happened in terms of efforts to unite forces with the ESA for the ensuing half year. Early in 1941 Yard returned to the subject, reminding Leopold of his promise to take further action by working with Murphy.[12] Leopold thanked Yard for the reminder but suggested that the matter was too big to discuss with Murphy by mail; it would need to wait until the two could get together. A few weeks later the executive committee of the Wilderness Society took up the issue, according to the minutes of its meeting:

> Last July, Aldo Leopold had undertaken to bring about activity in common with a group of ecologists, and, at the beginning of this year, [council member] Irving Clark had established an understanding with the National Research Council's Committee for Preservation of Natural Conditions. Mr. Yard also had furthered negotiations for activity in common with the Committee for Preservation of Natural Conditions of the Ecologists Society of America.
>
> The majority of the members of both these committees are discovered to be members of The Wilderness Society.
>
> Also we have renewed relationship with Dr. Victor E. Shelford, who first organized preservation in the Ecological Society, a member of The Wilderness Society; and Dr. Robert F. Griggs of George Washington University and the Research Council. Dr. Griggs joined the Wilderness Society.
>
> It is my plan to have the combined ecological activities under Aldo Leopold's chairmanship.[13]

The idea of putting Leopold in charge was obvious, given Leopold's standing as an ecological researcher and his early advocacy of the wilderness cause.[14] As Yard saw things, Leopold had been responsible for "starting the idea and title of the wilderness area, and first spread it broadcast."[15] "It is you who invented the title wilderness areas," Yard wrote to Leopold, "making practical certain ideals which had been in men's minds for many years, and had occasionally crept timidly into print."[16] Leopold had been largely responsible for the establishment of the first U.S. Forest Service wilderness area in the Gila National Forest and penned early influential advocacy pieces in support of the cause.[17] Among the founders of the Wilderness Society, Leopold stood out for his interest in the ecological aspects of wilderness preservation.[18]

Leopold's original advocacy of wilderness had taken place within the Forest Service, where his argument for preservation employed the familiar "highest use" language and focused appropriately on recreation. Wilderness, he urged, was "a continuous stretch of country preserved in its natural state, open to lawful hunting and fishing, big enough to absorb a two weeks' pack trip, and kept devoid of roads, artificial trails, cottages, or other works of man."[19] As Leopold promoted the wilderness cause, reaching out to broader audiences, he augmented his rationales for preservation while defining the term more expansively. Wilderness areas were useful as places to protect wildlife, particularly the species that had trouble living among or near humans.[20] Wilderness areas also had historic and cultural values in that they reminded Americans of their frontier origins and gave them chances to reenact and be reinvigorated by "the more virile and primitive forms of outdoor recreation" that mimicked pioneer life.[21] In his important 1925 writing on the subject, Leopold proposed that wilderness was not an all or nothing condition. It was "a relative condition," he urged, existing in varying degrees and in varying sizes, "from the little accidental wild spot at the head of a ravine in a Corn Belt woodlot to vast expanses of virgin country[,] . . . from the wild, roadless spot of a few acres left in the rougher parts of public forest devoted to timber-growing, to wild, roadless regions approaching in size a whole national forest or a whole national park."[22]

Leopold took his original conception of wilderness as big recreation area to a more refined level in an important presentation. To the 1926 National Conference on Outdoor Recreation Leopold announced: "We come now to a . . . category of outdoor things which are God-made" yet subject to inexorable destruction by normal economic forces, a category that "may be collectively designated as 'wilderness.'"[23] Wilderness was not so much a place, Leopold was now saying, and certainly not just a large place for recreation as it was an element of wildness that dwelt in the landscape. It was simply untrue, Leopold told his audience, that "an area is either wild or not wild, that there is no place for intermediate degrees of wildness." Indeed, it was essential to retain "the small remaining wild spots," particularly in places where large wilderness areas were no longer intact. Also untrue was the assumption that wilderness only included mountains and other lands deemed scenic by the general public; to the contrary, "swamps, lakelands, river-routes, and deserts"—all representative types of "primeval America"—were worth protecting.[24]

Around this time, too, Victor Shelford was heartily advancing the idea of setting aside large "nonscenic" grassland areas as national parks or monuments. There was an immediate need, in the midst of the Great Depression and dust bowl, to protect grassland tracts for ecological research as well as to save for posterity samples of historic conditions of covered wagon days.[25] Grasslands, lacking such popular scenic values, largely had been overlooked by other preservation efforts. In the dominant view, grasslands should be cultivated, not honored and protected.[26]

It was this very attitude—land understood as a resource factory consisting of individual parts and profitable uses, not as an interconnected whole and a place to live—that Leopold increasingly saw as the major obstacle to the developing wilderness idea. Most ecological preservationists, including Shelford, agreed philosophically yet showed little taste for cultural confrontations.[27] They pressed instead for shorter-term answers, cloaking their efforts with publicly attractive labels or simply veering away from preserving lands in locales where criticism would be fierce.[28] They thought to protect land, not to change people.[29]

When the impetus to form the Wilderness Society came in the mid-1930s, the driving organizers were chiefly interested in preserving big wild places for recreational use, just as Leopold had proposed some fifteen years earlier. Now, however, Leopold brought his enlarging ideas to this table. Leopold had not been active in the nature protection work of the ESA, but he certainly knew about it, endorsed it, and viewed the ESA as an attractive ally. When invited by Bob Marshall in 1934 to serve on the founding committee for the proposed new society, Leopold immediately brought up the issue of preservation for science:

> I have no criticism of the invitation, except that I would raise the question of whether we should definitely indicate that the Society includes only those interested in wilderness from the aesthetic and social point of view, or whether it should also include those desiring wilderness for ecological studies. My hunch is that the ecological group should be included,

> but with minor emphasis. Dr. Merriam could tell you who are the most active members of this group. I know it included Shelford of Illinois.[30]

Seeing a need to keep science from getting lost in the new organization, Leopold put the rationale front and center in the now well-known manifesto "Why the Wilderness Society?" which he wrote on behalf of the society and which appeared in the initial volume of its journal:

> The recreational value of wilderness has been set forth so ably by Marshall, Koch, and others that it hardly needs elaboration at this time. I suspect, however, that the scientific values are still scantily appreciated, even by members of the Society. . . . [T]he scientific need is both urgent and dramatic. . . . The long and short of the matter is that all land-use technologies . . . are encountering unexpected and baffling obstacles which show clearly that despite the superficial advances in technique, *we do not yet understand and cannot yet control* the long-time interrelations of animals, plants, and mother earth.[31]

Leopold also pushed his fellow society founders to think broadly in aesthetic terms. As he reminded Bob Marshall in a letter in early 1935, "we are fighting not only mechanization of country, but the idea that wild landscape must be 'pretty' to have value."[32]

It is noteworthy, looking back today, that Leopold moved slowly in trying to bring the Wilderness Society and ESA together and that he put off Yard's inquiry, claiming that he needed to discuss the matter with Murphy in person. The difficulty was not merely one of organizational logistics, which postal communication could have handled. Leopold sensed that deeper work had to come first, and he turned right to it despite his many other activities. A major contribution, intended to solidify the society's interest in the scientific value in preserving wilderness, came in the form of his important essay "Wilderness as a Land Laboratory," which Leopold wrote at this time and published in the society's magazine, the *Living Wilderness.* In the essay later reworked and included in Leopold's *Sand County Almanac* Leopold took an important step in clarifying his newly distilled idea of land health. Long dismayed about fragmentation within the overall conservation cause and on the land itself, Leopold believed that an overall conservation goal was needed, something that could supply the polestar for all conservation work.[33] Leopold was slow in crafting that goal because he first had to gain an overall understanding of how land worked, what he referred to as a "common concept of land."[34] That work had proceeded far enough by 1939 that Leopold was prepared to talk about it in his important presentation to the joint meeting of the ESA and the Society of American Foresters.[35] By the following year Leopold was at work, turning his common, biotic concept of land into the normative conservation goal of land health.

It now was in this context, a time when Leopold's conservation ideas all were coming together, that Leopold sought to unify and redirect the wilderness protection effort. In his 1941 article for the society's journal, Leopold endeavored not just to get society members to see the research values of wilderness areas (reminding them that "all wilderness areas, no matter how small or imperfect, have a large value to land-science") but to get them to think seriously about land health as an overall conservation vision.[36] Wilderness as Leopold broadly defined it needed to dwell within landscapes that included human-altered lands; it needed to be merged with human land uses and interspersed among human settlements and cities.[37] Full conservation had to attend above all to the functioning and composition of these larger mixed landscapes. Just as important, groups such as the Wilderness Society and the ESA needed to recognize that the main impediments to land health were primarily cultural. Reformers had to go beyond influencing specific land-use decisions of federal agencies: they had to initiate and orchestrate a major shift in dominant cultural values. Well-directed wilderness preservation could help accomplish that shift.

While Leopold was following his intellectual and moral journey, Victor Shelford—ten years his senior but due to outlive Leopold by twenty years—was following his own.[38] Having reorganized the ESA's conservation committees in 1930, Shelford turned to work on a "nature sanctuary plan" for adoption by the ESA. Floated initially in 1931, the plan gained formal approval in 1932 and appeared in *Ecology* and *Science* in 1933.[39] A nature sanctuary was defined as "a [plant and animal] community or community fragment covering a certain area within which fluctuations and other natural changes are allowed to go on unmodified and uncontrolled."[40] Wherever possible, sanctuaries would contain large carnivores and other top-of-the-food-pyramid predators. Core research areas would be of value for scientists as well as for serious writers and artists who came to them for inspiration. Importantly (and a source of conflict with other preservation groups), the core areas would be restricted to other users, thus diminishing their recreational values.[41] Surrounding the core areas would be buffer zones where more activities could take place but in which wildlife and ecological processes would be protected. Shelford's sanctuary plan for the ESA was soon endorsed by the National Parks Association.[42] Leopold would take note of it, keeping in his files the 1933 *Ecology* article along with the ESA's 1935 "Confidential Memorandum on Sanctuaries to Include Predatory Animals."[43] So, too, would George Wright and colleagues of the National Park Service, whose efforts would provide important precedents for the emergence decades later of conservation biology.[44]

Wearing his activist hat, Shelford was, of course, every bit as insistent as Leopold that nature preserves include the full array of biotic communities, not just be places ordinary citizens deemed pretty. He was insistent, too, that preservation efforts include landscapes that humans had altered, especially when they were the best remaining examples of particular community types. Shelford's sanctuary plan recommended classifying nature sanctuaries according to their degrees of human modification.[45] Sanctuaries would be of three types—first through third classes—according to how much change had occurred in the plant and animal life.[46] Human uses of first-class sanctuaries would be greatly limited; uses of second- and third-class sanctuaries would be limited to the extent practically possible. All sanctuaries, however, would be biologically dynamic: "'Nature' and 'natural' are purely relative terms and can have significance only as averages, because the outstanding phenomenon of biotic communities is fluctuations in numbers of constituent organisms or reproductive stages of organisms over a period of one to thirty or more years. Thus, a Nature Sanctuary is primarily an area in which these fluctuations are allowed free play."[47]

Like Leopold, Shelford saw a need for ecologists to join forces with other conservation groups to push to protect natural areas. Indeed, a prime aim of the ESA's preservation committee was to reach out to others, government agencies very much included, to help bring about protection for the nation's remaining wild places. During the 1930s and into the 1940s, according to Shelford's biographer, Shelford "considered forming a large union of societies interested in nature preservation."[48] But Shelford throughout entertained doubts about groups that lacked sound grounding in ecology and that were prepared to sacrifice natural conditions to satisfy human wants. Shelford came on strong with his criticism in a brief 1933 publication in *Science* written in his capacity as chair of the ESA's Committee for the Preservation of Natural Conditions.[49] In it Shelford drew a sharp line between preservation, which he strongly favored for nature sanctuaries, and the principles of conservation that appeared to govern federal agencies, including the National Park Service: "To conserve, as the term is now most frequently used, means to preserve while in use and it often implies ultimate depletion. In actual practice the operations carried on in the name of conservation are not designed to preserve the natural order but to establish and maintain a different order as regards kind and abundance of plants and animals present."[50] The problem here for scientific researchers was acute, particularly when they wanted to study how natural areas changed on their own, free of human intervention. In nature sanctuaries of the type supported by the ESA, "the natural fluctuations of organisms are allowed free play and serve among other things to show what natural fluctuations in abundance are like." On federal lands—even national park lands governed under the nonimpairment clause of the National Park Service charter—too much human-initiated change was taking place. "There is or has been so much interference with natural processes in the form of 'control' of this and that organism that the student of 'wild life' management who would seek a basis for more scientific treatment of the animals in his charge, is left without guiding principles or reliable information."[51] For ecologists wanting to study wild nature, conservation was not enough.

Suspicious not just of federal agencies but of conservation groups that seemed to elevate recreation over true preservation, Shelford was slow to form alliances. His preferred option by the early 1940s, according to his biographer, was to have the ESA take the lead with other groups providing assistance. Other societies could "affiliate themselves with and contribute to the ESA, while depending on dedicated professional ecologists with the ESA as the active core." Only the ESA "had the experience, scientific reputation, and sound conservation judgment needed to do the job."[52]

While Leopold turned his own conservation eye to the long term and to cultural change, he knew that many steps were needed to get the conservation movement as a whole to embrace his reformist view. One useful step, he could see, was to merge and elevate the two dominant ideas about wilderness preservation—wilderness for recreation and wilderness for science. The rationales sounded different, but they overlapped considerably. In their overlap Leopold spotted an attractive way for wilderness to help stimulate cultural change.

It was this possibility for synthesizing a shared rationale that Leopold had raised in his 1940 resolution prepared for the Wilderness Society's meeting. Both the ESA and Wilderness Society, Leopold urged, "should realize that ecological observation is one of the highest forms of recreation, while ecological studies without an esthetic appreciation of the biota are dull and lifeless." Leopold extended the idea in an article, "Wilderness Values," which appeared the following year in the National Park Service's yearbook.[53] When it came to nature, Leopold contended, study and sport overlapped, particularly in their higher reaches. Indeed, he proposed, "there is no higher or more exciting sport than that of ecological observation." The "false cleavage" in the public's mind "between studies and sports," Leopold continued,

> explains why the Natural Area Committee of the Ecological Society does not cooperate with the Wilderness Society, though both are asking for the perpetuation of wilderness. "Serious" ecological studies of a professional nature are, of course, important, and they of course have a place in wilderness areas. The fallacy lies in the assumption that all ecology must be professional, and that wilderness sports and wilderness perception are two things rather than one. Good professional research in wilderness ecology is destined to become more and more a matter of perception; good wilderness sports are destined to converge on the same point.[54]

Thus, it all came down to perception, to the need to cultivate a more elevated ecological orientation among the masses of citizens. In drawing this conclusion Leopold reiterated a thesis offered in his 1938 essay, "Conservation Esthetic," published in *Bird-Lore* and later reprinted in his *Almanac.*[55] There he identified, as a key component of outdoor recreation, the fostering of "a perception of the natural processes by which the land and the living things upon it have achieved their characteristic forms (evolution) and by which they maintain their existence (ecology)." This form of interacting with nature, though often called "nature study," was in truth "the first embryonic groping of the mass-mind toward perception." "To promote perception" of this type, Leopold contended, was "the only truly creative part of recreational engineering."[56] This was an "important" truth, Leopold continued, "and its potential power for bettering 'the good life' is only dimly understood."[57]

Leopold's ruminations on the ESA–Wilderness Society interactions thus came together with his worries that careful nature study was rapidly becoming a professional activity engaged in only by a few. Society needed to move in precisely the opposite direction, he urged. More and more people should learn how nature worked and gain an ability to distinguish between sick and healthy lands. Conservation had to take place on all lands, public and private. This could happen only if private landowners came to see and value nature in new ways. If landscapes were to regain health, if conservation, that is, was to succeed in full, then citizens everywhere needed to know what healthy land looked like. They needed new senses of aesthetics that equated the beautiful with the healthy. They needed to sense their membership in the land community and their ethical duty as community members to help sustain the ecological whole. Wilderness could play a critical role in fostering this new cultural orientation, and it could do so, importantly, even when it was too small or too degraded to provide habitat for wilderness game or room for two-week pack trips. Small wilderness areas "are not wild in any strict ecological sense," Leopold observed, "but they may nevertheless add much to the quality of recreation," particularly recreation that helped people see how land worked.[58]

As it turned out, the ESA's nature preservation work was on its last legs. By 1928 ecologists within the society were already raising questions about the propriety of a professional scientific organization getting involved in the dirty work of politics and advocacy.[59] By the late 1930s internal resistance had become substantial, even though Shelford personally was held in high esteem. The issue came to a head in 1945, when the ESA membership, strongly encouraged by its officers, voted to emasculate the Committee for Preservation, disallowing its involvement in direct legislative activity.[60]

Dismayed by what he saw as the ESA's betrayal, Shelford and others founded the Ecologists' Union to carry on their activist work. In 1949 an energetic former student of Shelford, George Fell, gained election as vice president and convinced union leaders that it needed to broaden its membership beyond scientists. In the fall of 1950 the Ecologists' Union reorganized as a not-for-profit corporation and changed its name to the Nature Conservancy, lifted from the similarly named British organization. George Fell, its de facto and uncompensated staff person, was charged by the board to set up the new organization.[61]

So great was the turmoil within the ESA over the nature preservation work that leaders decided to look outside their ranks for a new president, a respected scientist who might keep the factions together. Leopold had just been elected vice president of the Wilderness Society. Now the ESA turned to him as well, asking him in the fall of 1946 whether he would consider running for the presidency. Leopold's qualified consent noted that he had never attended an ESA meeting but had definite views "on how the Society might increase its usefulness."[62] Early in 1947 Leopold received news of his election in absentia. "I am astonished by my election," Leopold promptly responded. "I had supposed, as a matter of course, that any nominee failing to show up at the meeting would automatically be out of the running. I feel deeply the responsibility implied in my being elected despite this failure on my part."[63]

Leopold's election positioned him to continue working to fuse ESA and Wilderness Society interests, even though the ESA had given up its land protection work. In a letter of March 24, 1947, to ESA secretary William Dreyer about the Ecologists' Union, Leopold noted that he knew very little about the turmoil that led to the union's formation. He had just read the EU's first circular and found its plans a bit naive. "Moreover," he wrote, "there is no mention of cooperation with the Wilderness Society, which is working for identical objectives[,] . . . [a]ll of which," he admitted, "leaves me completely puzzled."[64] Even as he worked within the ESA, Leopold was also adding encouragement to the other side. To Howard Zahniser, new secretary of the Wilderness Society, he wrote on August 21, 1947: "I feel that the testimony of the conservation organizations [over an upcoming piece of legislation] will cancel out because it will deal too much with means rather than ends. . . . It has occurred to me that a new organization like the Wilderness Society might have some chance of pulling the older organizations together on a common program."[65] Leopold had raised the issue in broader terms in a letter the previous year to Benton MacKaye of the Wilderness Society. "I am thoroughly convinced of one basic point: that wilderness is merely one manifestation of a change of philosophy of land use and that the Wilderness Society, while focusing on wilderness as such, cannot ignore the other implications and should declare itself on them."[66] By then Leopold was sounding a familiar note. As early as 1925 Leopold understood wilderness as a cultural issue, not just a matter of pragmatic land use. "As a form of land-use," he had suggested, wilderness was "premised on a qualitative conception of progress."[67] Just so, the aim of the Wilderness Society was far-reaching. The Wilderness Society, Leopold announced in its first issue, "is, philosophically, a disclaimer of the biotic arrogance of *homo americanus*. It is one of the focal points of a new attitude—an intelligent humility toward man's place in nature."[68]

The outgoing ESA president traditionally gave a talk at the annual dinner in the year succeeding his one-year term. Leopold was planning his September 1948 talk well in advance. In a letter of January 3, 1948, to game breeder Wallace Grange, Leopold commented on his plan: "I disagree with you that we should make any blank sacrifices of wilderness values simply on the chance that we might need them in another war. If we lose our wilderness, we have nothing left, in my opinion, worth fighting for; or to be more exact, a completely industrialized United States is of no consequence to me. I am expressing these views in more detail in my presidential address to the Ecological Society."[69] Leopold's death in April 1948 kept him from giving his address, but it seems likely that Leopold planned to deliver some version of what became his final writing on wilderness, the essay of that name (probably unfinished) that

Leopold included in his manuscript of *A Sand County Almanac.* The intended scientific audience for the essay might explain why Leopold devoted so much space in it to matters of science and to the linkage between wilderness preservation and the promotion of land health.

Leopold's plan was to end his book manuscript with "Wilderness" rather than with "The Land Ethic."[70] Had his wishes been followed, wilderness might have drawn greater attention over the years from Leopold scholars. More readers might have grasped the importance of wilderness and wilderness preservation in charting a new path for modern culture. In the piece, his final writing on wilderness, Leopold covered familiar ground. He reviewed the rationales for preservation, noted that wilderness reserves "must vary greatly in size and in degree of wildness," and lamented "the under-aged brand of aesthetics which limits the definition of 'scenery' to lakes and pine trees."[71] What it all came down to, though, was culture and the role of wilderness in embodying vital values and fostering them within citizens. The ability to appreciate "the cultural value of wilderness," Leopold announced in what he intended as the last paragraph of his book, came down "in the last analysis, to a question of intellectual humility." More than that, Leopold said, it was "raw wilderness" that gave "definition and meaning to the human enterprise." Modern culture was headed in the wrong direction. It needed to pause, reevaluate, and start again. Wilderness was the place where people—professional scientists and plain citizens alike—could regroup, learn about healthy lands, and thus decide how they might live so that their civilizations could endure. All history, Leopold opined, consisted of "successive excursions from a single-starting point," to which humans returned "again and again to organize yet another search for a durable scale of values." Wilderness provided that starting point, and it would need to do so for future generations, Leopold implied, again and again.

Yard's death in 1945, the termination of the ESA's preservation effort the same year, Leopold's death in 1948—all would delay cooperative work to protect wilderness tracts in North America. But their successors would see the value of cooperation, and many groups would come together to fight for passage of the Wilderness Act of 1964. Leopold's vision had embraced Shelford's and the ESA's scientific ideals and those of Wilderness Society recreationalists, but it had been far broader than either. Leopold's vision had been to elevate humanity, to promote ecological understandings, to instill new ethical and aesthetic standards, and to present a unified front in the quest for land health. His conservation successors did not follow his lead in this work, and so the tasks remain undone.

NOTES

1. Aldo Leopold to Robert Sterling Yard, February 5, 1940, Aldo Leopold Papers (hereafter cited as LP), series no. 10-2, box 9, Steenbock Memorial Library, University of Wisconsin–Madison.
2. Aldo Leopold to Robert Sterling Yard, June 4, 1940, LP 10-2, box 9.
3. Aldo Leopold to Robert Sterling Yard, June 4, 1940, LP 10-2, box 9. Unknown to Leopold, Shelford was no longer chair of the ESA's committee, though he continued to work on the cause, and he was widely viewed as the committee's intellectual leader. See Robert A. Croker, *Pioneer Ecologist: The Life and Work of Victor Ernest Shelford 1877–1968* (Washington, D.C.: Smithsonian Institution Press, 1991), 127–33.
4. Aldo Leopold to Robert Cushman Murphy, June 8, 1940, LP 10-2, box 9, telegram attached.
5. Aldo Leopold, Resolution, LP 10-2, box 9.
6. Robert Sterling Yard to Aldo Leopold, June 18, 1940, LP 10-2, box 9; council members in attendance were Harold Anderson (treasurer), Harvey Broome, Bernard Frank, Benton MacKaye (vice president), and George Marshall. Founder Bob Marshall had recently passed away.
7. Aldo Leopold, *A Sand County Almanac and Sketches Here and There* (New York: Oxford University Press, 1949), 200.
8. The founding of the ESA is recounted in Robert A. Croker, *Stephen Forbes and the Rise of American Ecology* (Washington, D.C.: Smithsonian Institution Press, 2001), 124–25; Croker, *Pioneer Ecologist,* 120–21.
9. Victor Shelford, *A Naturalist's Guide to the Americas* (Baltimore, Md.: Williams and Wilkins Co., 1926), v.
10. Croker, *Pioneer Ecologist,* 123.
11. Ibid., 51, 64, 120–35.
12. Robert Yard to Aldo Leopold, January 28, 1941, and Aldo Leopold to Robert Yard, February 3, 1941, both in LP 10-2, box 9.
13. Minutes of the Wilderness Society, March 19, 1941, LP 10-2, box 9.
14. Leopold's wilderness advocacy is considered in Roderick Frederick Nash, *Wilderness and the American Mind,* 4th ed. (New Haven, Conn.: Yale

University Press, 2001), 182–99; Max Oelschlaeger, *The Idea of Wilderness* (New Haven, Conn.: Yale University Press, 1991), 205–42; Susan L. Flader, "Aldo Leopold and the Wilderness Idea," *Living Wilderness* 43 (December 1979): 4–8; William Cronon, "A Voice in the Wilderness," *Wilderness* (Winter 1998): 8–13; Craig W. Allin, "The Leopold Legacy and American Wilderness," in *Aldo Leopold: The Man and His Legacy,* ed. Thomas Tanner (Ankeny, Iowa: Soil Conservation Society of America, 1987), 25–38. The influence of the "good roads" movement in stimulating the wilderness protection effort by Leopold and others is considered in Paul D. Sutter, *Driven Wild: How the Fight against Automobiles Launched the Modern Wilderness Movement* (Seattle: University of Washington Press, 2004). Leopold's role in promoting an ecological perspective on wilderness within the Wilderness Society is considered in Daniel J. Philippon, *Conserving Words: How American Nature Writers Shaped the Environmental Movement* (Athens: University of Georgia Press, 2004), 159–218. A brief review of Leopold's wilderness writings in the context of his larger conservation thought is offered in Susan L. Flader and J. Baird Callicott, eds., *The River of the Mother of God and Other Essays by Aldo Leopold* (Madison: University of Wisconsin Press, 1991), 24–27.

15. Robert Yard to Aldo Leopold, April 30, 1940, LP 10-2, box 9. A different account of the origins of wilderness preservation is set forth in Donald N. Baldwin, *The Quiet Revolution: The Grass Roots of Today's Wilderness Preservation Movement* (Boulder, Colo.: Pruett Publishing Co., 1972), which dubs Forest Service landscape and recreation planner Arthur Carhart "Father of the Wilderness Concept."

16. Robert Yard to Aldo Leopold, May 9, 1940, LP 10-2, box 9.

17. Curt Meine, *Aldo Leopold: His Life and Work* (Madison: University of Wisconsin Press, 1988), 23–29, 194–228, 243–48.

18. Philippon, *Conserving Words,* 162, 190.

19. Aldo Leopold, "The Wilderness and Its Place in Forest Recreation Policy," in Flader and Callicott, *River of the Mother of God,* 78–81 (originally published in 1921).

20. Leopold would expand upon this rationale in his "Threatened Species: A Proposal to the Wildlife Conference for an Inventory of the Needs of Near-Extinct Birds and Mammals," *American Forests* 42 (March 1936): 116–19, also in Flader and Callicott, *River of the Mother of God,* 230–34. See also Aldo Leopold to Jay "Ding" Darling, November 27, 1939, LP 10-4, box 8: "None of us had any knowledge of the start you had made with Gabrielson and the Biological Survey to inventory the existing and prospective wilderness areas from the viewpoint of particular animal species. It is welcome news that such a project is under way, and it immediately occurs to me that if the USBS already has it in hand, our [the Wilderness Society's] best bet is to stand at their elbow rather than to try to take over the job."

21. Aldo Leopold, "Wilderness as a Form of Land Use," 1925, in Flader and Callicott, *River of the Mother of God,* 134–42.

22. Ibid., 135–36.

23. Aldo Leopold, "Wilderness Conservation," address delivered at the National Conference on Outdoor Recreation, Washington, D.C., January 20, 1926, LP 10-4, box 8, and later published in *Proceedings of the Second National Conference on Outdoor Recreation,* January 20–21, 1926, S. Doc. 117 (Washington, D.C.: GPO, 1926), 61–65.

24. Ibid., 9.

25. Croker, *Pioneer Ecologist,* 137.

26. Victor Shelford, Open Letter from the Retiring Vice-President of the Grassland Research Foundation, January 31, 1948, Victor Shelford Papers (hereafter cited as SP), 15/24/20, box 1, p. 1, University of Illinois at Urbana-Champaign Archives.

27. The ESA's study committee included in 1930 a committee on grasslands. In 1931 several of those involved in the ESA's work and who were active in the National Research Committee (NRC) formed within the NRC the Committee on the Ecology of Grasslands, which in 1933 was appointed to the NRC's Division of Biology and Agriculture. In 1939 Shelford resigned as chair of the Committee on the Ecology of Grasslands and was involved in the formation of a new nonprofit group, the Grassland Research Foundation, which was incorporated that October in Norman, Oklahoma. This foundation was to assist the NRC in obtaining grants. The NRC, however, was discontinued in 1943 upon the resignation of Paul Sears and the inability to find new leadership. The foundation then took on a life of its own, continuing on in grassland preservation work, and then merged in the late 1950s with the Nature Conservancy. See Shelford, Open Letter, and Victor Shelford to Jeff Kendall, September 23, 1957, SP 15/24/20, box 1.

28. Mr. H. G. Crawford of the Canadian Department of Agriculture recommended that the NRC Committee on the Ecology of Grasslands not savor of "experiment stations" so as not to give the impression of competing with "agricultural authorities," who already used that term for their study sites oriented toward research on enhancing crop productivity. If criticism arose regarding grassland "study centers," it was recommended that sites be moved to other locations. See Shelford, Open Letter, 1.

29. Ibid., 1.

30. Aldo Leopold to Robert Marshall, October 29, 1934, LP 10-2, box 9.

31. Aldo Leopold, "Why the Wilderness Society?" *Living Wilderness* 1 (September 1935): 6.

32. Aldo Leopold to Robert Marshall, February 1, 1935, LP 10-2, box 9. Leopold hit the same theme hard in a 1940 talk to the Isaak Walton League during which he lamented the lack of concern about the fragmentation of a

high desert area that did not "conform at all to the classical picture of pine-bordered lakes." The deficiency, Leopold contended, was "a plain lack of education in esthetics." Too many people were "still in the stage of evaluating wild land by its prettiness." In the same talk Leopold also worked in his thoughts on the issue of size: "Surely there is room for some degree of wilderness in any tract of wild land, however small." Leopold's "What Shrinks Wilderness?" March 29, 1940, ms., LP 10-6, box 16. See Leopold's "Origin and Ideals of Wilderness Areas," *Living Wilderness* 5 (July 1940): 7, which lists as a critical job for the Wilderness Society the need "to secure the recognition, as wilderness areas, of the low-altitude desert tracts heretofore regarded as without value for 'recreation' because they offer no pines, lakes, or other conventional scenery."

33. Eric T. Freyfogle, *Why Conservation Is Failing and How It Can Regain Ground* (New Haven, Conn.: Yale University Press, 2006).

34. Julianne Lutz Newton, *Aldo Leopold's Odyssey* (Washington, D.C.: Island Press/Shearwater Books, 2006), chap. 6.

35. Leopold's talk appeared as "A Biotic View of Land," *Journal of Forestry* 37 (September 1939): 727–30.

36. Aldo Leopold, "Wilderness as a Land Laboratory," in Flader and Callicott, *River of the Mother of God,* 289.

37. J. Baird Callicott considers Leopold's vision of blending wildness into human-occupied landscapes in "The Wilderness Idea Revisited: The Sustainable Development Alternative," in *The Great New Wilderness Debate,* ed. J. Baird Callicott and Michael P. Nelson (Athens: University of Georgia Press, 1998), 337–45.

38. Shelford's story has now been ably told by Croker in *Pioneer Ecologist.*

39. Ibid., 128–30.

40. Victor Shelford, "Nature Sanctuaries—a Means of Saving Natural Biotic Communities," *Science,* June 2, 1933, 281–82; see also Victor Shelford, "The Preservation of Natural Biotic Communities," *Ecology* 14 (April 1933): 240–45; "The Nature Sanctuary Program," draft, includes diagrams, ca. 1936, SP 15/24/20, box 1.

41. Croker, *Pioneer Ecologist,* 127–32; Victor Shelford, "Confidential Report on a Proposed Policy for the ESA Regarding Preservation and Study of Natural Biotic Communities," December 29, 1931, SP 15/24/20, box 1; Shelford, "Preservation"; Shelford, "The Nature Sanctuary Program."

42. George M. Wright, Joseph S. Dixon, and Ben H. Thompson, "Fauna of the National Parks of the United States," in *Contribution of Wild Life Survey Fauna,* Series 1, 1933; National Park Service, "Fauna of the National Parks," *National Park Service Bulletin* 2 (1933): 123–30.

43. LP 10-2, box 2.

44. Craig L. Shafer, "Conservation Biology Trailblazers: George Wright, Ben Thompson, and Joseph Dixon," *Conservation Biology* 15, no. 2 (2001):

332–44. The "rise and decline" of ecological understandings within the National Park Service during the 1930s is charted in Richard West Sellars, *Preserving Nature in the National Parks: A History* (New Haven, Conn.: Yale University Press, 1997), 91–148.

45. Victor Shelford and ESA Study Committee, Memorandum on Nature Sanctuaries or Nature Reserves, April 1932, SP 15/24/20, box 1.

46. As its baseline for natural conditions the ESA loosely used landscapes as modified by American Indians. Although Shelford's report surmised that, in terms of overall vegetative types, landscapes "were probably not much affected" by Indians, the chief reason for going back no further in time was because information was lacking: "Just what original nature in any area was like from a biological viewpoint, is not known and never can be known with any great accuracy" ("A Nature Sanctuary Plan," *Ecology* 14 [April 1933]: 240–45, 242).

47. Ibid., 241.

48. Croker, *Pioneer Ecologist,* 138.

49. Victor Shelford, "Conservation versus Preservation," *Science,* June 2, 1933, 535. In Shelford's view, which he presented to the committee, "conservation" could not properly be used to talk about the ESA's kind of preservation, particularly given the ESA's strong support for protecting predators as vital parts of ecological communities. See ESA Committee for the Study of Plant and Animal Communities, To the Advisory Board and Others, n.d., SP 15/24/20, box 1.

50. Shelford, "Conservation versus Preservation," 535.

51. Ibid.

52. Croker, *Pioneer Ecologist,* 138.

53. Aldo Leopold, "Wilderness Values," in Park and Recreation Service, *1941 Yearbook* (Washington, D.C.: National Park Service), 27–29, reprinted in *Living Wilderness* 7 (March 1942): 24–25.

54. Ibid.

55. Aldo Leopold, "Conservation Esthetic," *Bird-Lore* 40 (March–April 1938): 101–9.

56. Ibid., 107.

57. Ibid., 109. Leopold continued this theme in various later writings, including his essay "Wildlife in American Culture," *Journal of Wildlife Management* 7, no. 1 (1943): 1–6, where he presented wildlife research as "a totally new form of sport," the promotion of which was the "most important job confronting [the wildlife management] profession" (5).

58. Leopold, "Wilderness Values," 28.

59. Croker, *Pioneer Ecologist,* 126.

60. Ibid., 138–44.

61. Ibid., 145–46; Arthur Melville Pearson, *A Legacy of Natural Lands:*

George B. Fell and the Natural Land Institute (Rockford, Ill.: Natural Land Institute, 2005), 13–15.

62. Leopold to Alfred C. Redfield, October 4, 1946, LP 10-2, box 2.

63. Aldo Leopold to William Dreyer, January 11, 1947, LP 10-2, box 2.

64. Aldo Leopold to William Dreyer, March 24, 1947, LP 10-2, box 2.

65. Sterling Yard died in 1945 at the age of eighty-three, and Zahniser replaced him. Aldo Leopold to Howard Zahniser, August 21, 1947, LP 10-2, box 9.

66. Aldo Leopold to Benton MacKaye, May 1, 1946, LP 10-2, box 9.

67. Leopold, "Wilderness as a Form of Land Use," 142.

68. Leopold, "Why the Wilderness Society?" 6.

69. Aldo Leopold to Wallace Grange, January 3, 1948, LP 10-1, box 1.

70. Meine, *Aldo Leopold,* 524.

71. Leopold, *A Sand County Almanac,* 189, 191.

Stephen H. Spurr

The Value of Wilderness to Science (1963)

Stephen H. Spurr (1918–90) was a forest ecologist at Harvard University and the University of Minnesota before becoming dean of the School of Natural Resources at the University of Michigan and president of the University of Texas at Austin. An endowed chair at the University of Michigan bears his name. Known as "a pioneer in the development of photogrammetry and forest photo interpretation," Spurr presented this paper to the Eighth Wilderness Conference of the Sierra Club in 1963, and it was subsequently published in the conference proceedings, *Tomorrow's Wilderness.* In this essay Spurr forwards a sophisticated version of the "wilderness for science" argument. He argues that wilderness provides a base-datum for normal ecosystemic functioning and a reservoir of genetic material, and it is "a dynamic and complex community posing the greatest challenge and interest to the modern day ecologist who is concerned with developing modern ecological theory and knowledge." In this essay Spurr also anticipates and offers a critique of the received view of wilderness and wrestles with the implications of disturbance ecology for wilderness preservation. The reader will also note the response to Spurr presented by a member of the Sierra Club printed at the end of the essay.

SOME OF THE VALUES of wilderness to science are so important and so well known that I need only refer to them briefly at the beginning of this paper. Such is the value of the wilderness as a biological standard of comparison for measuring and evaluating those other biotic communities which are much more influenced by man. A tract of wilderness is indeed the yardstick or tape by which we measure and evaluate the mess of landscapes created by man. Perhaps you may think that I meant to say *mass* of landscapes. I didn't!

A second important and well recognized value of wilderness is its capacity as a reservoir and holding ground for biotic germ plasm. The different strains, races, and general attributes of our many animal and plant species can only be preserved by letting them grow, live and compete in a natural setting. A bacterium grown in monoculture on agar in a Petri dish in a bacteriology laboratory will not remain the same organism indefinitely, but will become attenuated and lose the features that characterized it in its wild state. A tree, grown for generations as an ornamental, will eventually lose, at least in part, the genetic characteristics that made it able to thrive and hold its own in the absence of cultivation. So will a wolf; so will a bear; so indeed will a man.

It is not, however, these well known values that I wish to emphasize today, but rather the value of the wilderness as a dynamic and complex community posing the greatest challenge and interest to the modern day ecologist who is concerned with developing modern ecological theory and knowledge. An understanding of the nature and dynamics of the wild community is vitally important to an understanding of life itself. In the last few years, the science of ecology has made enormous strides, throwing off conventional and oversimplified theories which, as knowledge increased, hampered rather than aided man's understanding of nature. Lacking sufficient data concerning the complexities of nature, and lacking indeed the tools and techniques to measure and analyze nature, the ecologist has too often been a philosopher rather than a scientist. After all, ecology deals with life at the highest level of organization and is therefore the most complex and highest of the biological sciences! It is only natural that this science must come to fruition at a later date than those dealing with the more basic building blocks of life. We are now in the midst of a great break-through in bio-chemistry. We can look forward to equally important break-throughs in ecology in the generations to come. For the first time, I think, we are beginning to have the concepts and the tools at our disposal with which to mobilize our forces for this break-through. In this effort, the wilderness areas of the United States, of North America, and indeed of the whole world, will be important battle-grounds. Furthermore, as we apply these modern ecological concepts to the wilderness problem, the problem itself will inevitably change. As the problems change, our answers change, and as these change, our desires change. These changing problems, answers and desires of man with regard to the wilderness constitute my theme for today.

As with so many controversial topics, the wilderness problem derives a large measure of its dialectic value because the term means many different things to many different people. Let us pause for a moment to define the term. The word *wilderness* is made up of three parts. *Wild, deer,* and the suffix *ness* indicating a state or condition. Literally, therefore, its etymological meaning is "*the place of wild deer.*" This meaning actually comes pretty close to the truth. As the Outdoor Recreation Resources Review Commission study report number 3 expresses it, "the word is poetic and imprecise." The wilderness is certainly a pathless and uncultivated tract of land uninhabited by man, but it is difficult to put any precise limits which separate a wilderness from another tract which is not one.

I would like to point out though, that the wilderness has an existence only with relationship to man. A wilderness is pathless only in the sense that it doesn't have paths made for and by man, but it will contain myriads of wild animal paths. It is uncultivated only in the sense that it is uncultivated by man, for the soil is indeed stirred and mixed by millions of animals and roots which work it continuously. Finally it is uninhabited only in the sense that it is uninhabited by any large number of men. In a general biotic sense the forest wilderness at least is densely populated by large numbers of plants and animals of all kinds, sorts and descriptions. In other words, the concept of wilderness is an *anthropocentric* one, if you will pardon me for using a five-syllable word. It is a concept conceived by man, centered in man, and having a meaning only with relationship to man. It refers to that portion of the earth's surface basically unused and uninhabited by man and therefore relatively independent of man's influence.

From an ecological viewpoint, however, we should not look at the wilderness from the viewpoint of man, which is by definition an external view, but from the viewpoint of the wilderness community itself, a view which is internal and self-evaluating. Such is the view of Marston Bates when he suggests that we look at the tropical rain forest from the viewpoint of the mosquito and not of man. So few naturalists have taken this attitude in examining the wilderness problem, however, that I fail to find a suitable term to serve as antonym for *anthropocentric.* I have heard it said that ecology is a disease, the final and extreme symptoms of which are exhibited when the victim begins coining new terminology. Be that as it may I find no alternative in this particular instance except to coin a term to express the viewpoint of the wilderness as seen by the wilderness itself. In fact I'll give you three: the most accurate I suppose is *biocoenocentric.* If that is too much of a mouthful I suggest *biotacentric.* Finally, I suppose that we could usurp an existing term and speak of this viewpoint as being *biocentric* although this particular word originally has had other meanings. Take your choice of the three; only come with me in looking at the wilderness from the viewpoint of the plants and animals that comprise it and excluding the viewpoint of man who is by very definition an interloper in it.

We will all agree that a wilderness consists of an assemblage of plants and animals living in an environment of air, soil, and water; that each of these organisms is interrelated either directly or indirectly with virtually every other organism in the community; that the health and welfare of the organisms are dependent upon the factors of the environment surrounding them; and that the environment surrounding them itself is conditioned to a considerable degree by the biotic community itself. In other words, the plants, the animals and the environment—including the air, the soil and the water—constitute a complex ecological system in which each factor and each individual is conditioned by and in itself conditions the other factors comprising the complex. There is perhaps nothing really new in this concept of an ecological system; but in recent years, when we envisioned it more simply as an *ecosystem,* we have made considerable progress in understanding the complex interrelationships that exist. In contrast, in former years, by trying to simplify particular elements of the ecosystem into simple cause and effect relationships, we too often drew misleading and inaccurate generalizations.

In biocentric terms a wilderness is simply an ecosystem in which man is a relatively unimportant factor. While I realize that wilderness can and does exist in the desert, in the sea, and in the boreal zones, it is with the forest ecosystem that we are largely concerned and, being a forester by training, I will limit my comments, to the forest ecosystem.

The forest ecosystem then is the complex of trees, shrubs, herbs, bacteria, fungi, protozoa, arthropods, other invertebrates of all sizes, sorts and description, vertebrates, oxygen, carbon dioxide, water, minerals, and dead organic matter that in their totality constitute a forest. Such a complex never does and never can reach any balance or permanence. It is constantly changing both in time and in space.

The changes of the forest ecosystem in time take many patterns. First there are diurnal changes. The balance of the forest community at midnight—when the plants are taking up oxygen and giving off carbon dioxide, when some animals are dormant and others are active, when temperature is lowered, and when moisture and humidity are relatively high—is quite different from the forest ecosystem at midday when the reverse of all of these processes is going on.

The forest ecosystem changes seasonally around the year. As the cycle of activity of each of the organisms changes, and as the climate changes, the balance of the ecosystem itself changes. The forest is not the same biotic community in mid-winter as it is in mid-summer or even in one week as it was during the previous week. It is never the same.

Nor should we ignore long term climatic changes. Whether climatic fluctuation is cyclic or not is immaterial; the fact is that it does change from one set of years to another. The relative warmth of the 1950s is in sharp contrast to the relative cold of the early 1800s. The warm xerothermic times of three to five thousand years ago are an even greater contrast to the late Pleistocene of ten to thirteen thousand years ago when the glaciers reached their last maximum. There is no such thing as a constant climate, and there can be no such thing as a constant ecosystem if the former does not exist.

Finally, the plants and animals that constitute the ecosystem never remain the same over a period of time. There is a continual introduction of new species and the elimination of old, rarely when we consider the large and well established trees, mammals, and birds, but commonly and frequently when we concern ourselves with the fungi, the bacteria and the protozoa that far outnumber the larger organisms in the forest and approach them in overall importance. Even if we could prevent new organisms from constantly migrating into the forest and old organisms from constantly being eliminated from it, the organisms that remain there do not stay the same. Evolution is continual. The Douglas fir of today is not the Douglas fir of one hundred years ago and certainly not that of ten thousand years ago. The bark beetle is not the bark beetle of the previous decade. The blister rust of this year is not the blister rust of last year. Through natural selections, through mutations and through new population distributions of the combinations of genes, the organisms constituting the forest ecosystem are continually in a state of flux.

Equally, the ecosystem changes constantly in space. At any given instant in time the forest ecosystems will be different high in the mountains from those in the lowlands. Even on a level plain the ecosystem will vary from north to south and from east to west. The extremes are obvious but the means are equally important. A distance of one hundred yards may not present any noticeable change in the distribution, size and vigor of the more visible components of the forest. The difference is there nonetheless. When one comes to consider the micro-organisms in the forest floor which play such an important part in the total ecosystem production complex, changes in even one foot may be real, measurable and important in affecting the total ecosystem.

In short, the forest ecosystem exists only at a given instant in time and a given instant in space. Regardless of appearances, the ecosystem is never the same on succeeding days, succeeding years or succeeding centuries, up the slope a thousand feet, north ten miles or east or west one hundred miles. Change characterizes the forest ecosystem continuously. Stability is only relative and is only superficial.

If we adopt the biocentric concept of the ecosystem as opposed to the anthropocentric concept of the wilderness, we may draw some ecological generalizations that modify substantially some long standing and accepted ecological principles. Once we accept the fact that the ecosystem exists only at a given instant in time and at a given instant in space, we can take a fresh view of ecological theory.

First, it follows immediately that there is no meaning to the concepts of native and introduced species from a biocentric viewpoint. Characterizing a plant or animal as being exotic or endemic characterizes it only from the standpoint of man's relationship to it. Actually all plants and all animals are introduced or exotic from a biocentric standpoint except at the very point in space where the particular gene combination was first put together. Whether the subsequent migratory pattern of that organism took place independent of man or with the help of man is important to man but not to the wilderness ecosystem itself.

Examples are many. It is of interest to man to know whether he carried the coconut to a given tropical island or not. To the coconut I suspect it is of little importance whether it floated by itself in an ocean current or was lodged in the hull of a native dugout canoe which in turn floated on the ocean current. To a maple growing in a given spot it is immaterial whether its seed flew there on its own wings or whether it was aided and abetted by the wings of an airplane. A wild cherry is unaffected by concern as to whether its seed was deposited by a gull or by a human recreationist. The forest insect is just as much a member of the California ecosystem if it flew in as an endemic insect from Oregon or was brought in as an exotic insect from Aragon. In short, from the viewpoint of the wilderness, there is no distinction between native and introduced species. The wilderness ecosystem consists of all of the plants and animals that are there at a given point in space and at a given instant in time. All were migrants there. Some recently, some from the dim geologic past, some carried in by wind, some by animals, some by water, some by man. Once they are there they are members of the local ecosystem from that time on until they or their descendants are eliminated. The true endemic perhaps is really that plant or animal that prospers in the local environment, that is competitive in the local ecosystem, and that can maintain and reproduce itself where it lives.

Second, it becomes clear in our philosophic approach that there is no such thing as a climax community in the sense of a permanent stable condition. It makes no difference whether man has interfered or not. Change is perpetual and will go on at a rapid or at a gradual rate depending upon the rate of changes in the climate, the soils, the land forms, the fire history and the biotic composition of the ecosystem. No matter how old the trees are, no matter how long fire has been kept out, the relative populations of the insects, the fungi, the plants, the animals all will change, and the ecosystem itself will as a consequence change. The concept of climax may have some residual value in referring to an old and superficially stable community of plants and animals. In actual fact, though, there is every reason to think that ecological change is going on as rapidly if less obviously in an old established forest as on a freshly disturbed site.

Third, since the ecosystem exists only at a given instant in time and in a given instant of space, it follows that natural succession will never recreate an old pattern but will instead constantly create new patterns. After several hundred years the forest of Angkor Wat resembles the surrounding jungle but is quite different in composition and in detail. The vegetation on Krakatoa is developing successionally, but it is already quite apparent that it will never equate itself with that of undisturbed islands nearby. After each retreat of the glaciers in the Lake States in the Pleistocene the forests moved back and reestablished themselves, but always in a new pattern with some new species present, some old absent and the balance in many cases quite different. With widespread farm abandonment in New England the forests have reestablished themselves and have moved on into a new successional pattern with different balances of composition and different stand structures. In short, forest succession will lead to the development of a mature and long lived forest association but one in which change under the surface is still going on and one which will never repeat exactly the pattern of a previous forest developed under a previous forest succession.

Fourth, moving now from the biocentric to the anthropocentric viewpoint, it is therefore impossible to reestablish the wilderness that we in our nostalgia desire. Whether we visualize and wish to create the wilderness of our childhood, or that which our pioneer ancestors saw when they came over the mountains or landed from the ocean, or that under which the Indians lived prior to the advent of white man, it can't be done. There never was a constant wilderness condition and there never will be. The Spaniards who lived in California in the 18th century saw a different wilderness from the gold miners who came in the 19th. The Indians of the 10th century lived in a different wilderness from those in the 17th. We may create a wilderness in the 20th century or in the 21st century, but it will not be the wilderness of the 19th and not that of our own youth. The new wilderness will be a new forest ecosystem relatively undisturbed by man but with a new combination of plants and animals living in a new climate and growing on a new soil. It will be constantly dynamic, it will be undergoing constant change, but it may be equally attractive and indeed it will be equally attractive to the generation which will see it.

The line of argument just presented actually raises some question as to whether there actually is such a thing as a wilderness. To be more explicit, is there such a thing today as a wilderness devoid of man's influence? The answer is clearly no. In one way or another man has likely influenced every acre in temperate North America. At the surface of the earth, in the forested parts of the world, in the temperate zone certainly there is probably no such thing as a true wilderness.

The reasons for this are many. Man has played an enormous role in affecting the migration and distribution of plant and animal species and biotic introductions have influenced virtually every community. In North America man himself was recently an introduction and the very fact that he is present in any place is sufficient to deny the concept of a wilderness at that place. Insects, diseases, herbs, grasses and other weeds brought in by man, parasites, birds and even to a considerable extent trees and mammals introduced by him—some can be found almost anywhere in our American forest.

Second, man has a continual and important effect upon the frequency and occurrence of fire and this effect is omnipresent whether or not the fire was caused by man or by lightning and whether or not man put it out or failed to put it out. Technically, I suppose the lightning-set fire is a natural occurrence and we should not put it out in a wilderness, whereas a man-set fire is an artificial occurrence and we should put it out. But then, if we do put it out we are interfering in the forest ecosystem and the forest ecosystem is no longer devoid of man's influence.

Third, man is influencing the atmosphere to a considerable extent. There is evidence that the production of carbon dioxide by motor vehicles and by heating units has had a material effect upon the carbon dioxide concentration of the atmosphere upon the climate. Certainly, the content of the air in radioactive materials has been irrevocably influenced by man in recent years. Coming down to a local issue, the climatic obscurities of the Los Angeles and San Francisco Bay regions cannot but affect the forest ecosystem in the nearby mountains.

Fourth, the effect of man on the occurrence of insects and microorganisms in the wilderness is omnipresent whether or not the actual organism itself is introduced by man or whether or not it is controlled by spraying and other protection measures. Here too, our view is anthropocentric rather than biocentric. If the population cycle of the native insect carried it to epidemic stages I presume that the wilderness purist would not advocate spraying to control it. On the other hand, if the insect swarm moved across the Nevada state line rather than built up in the California Sierra it would be rightly classified as an invader and subject to an aerial attack by man.

Fifth, the very presence of man himself cannot but affect the ecosystem. The hiker, the trail rider, recreationists of any kind by their mere presence and certainly by their leavings affect the forest ecosystem in ways that are more than evident to those who follow after them. We can eliminate this influence by eliminating man, but since the wilderness is an anthropocentric concept in the first place if we eliminate man we eliminate the need for the concept of wilderness and I suppose by logical extension we therefore eliminate the reality of the wilderness itself.

Thus, it follows that we can only approximate a true wilderness by setting aside a large area, by never putting out a fire, by never stopping an epidemic, by never damming a stream or putting in a reservoir. If we adopt a completely hands-off policy we will create a new "wilderness ecosystem" with a new series of dynamic changes involving a new combination of plants and animals. Furthermore there will be a new and different ecosystem at each new point and each new stage in time, never a return to any previous stage of wilderness. The answer is we can create such wilderness ecosystems in the future. We can even predict from our ecological studies what patterns they will take. The question is, do we want them?

From the biocentric viewpoint, we may have to agree there is no such thing as a true wilderness in the sense of a modern present-day forest ecosystem uninfluenced in any substantial degree by man. At least, such a wilderness does not exist to any real degree in temperate North America.

Yet we all agree we love the wilderness and want to preserve what we can. What do we actually mean? We are talking now from an anthropocentric viewpoint. We are evaluating a forest ecosystem with the eyes, the mind and the heart of man. The answer is necessarily not a biotic one but a sociologic one. It is subjective and we must each plumb our own hearts and our own desires to come up with our own solution. Here, for what it is worth, is mine.

Each of us wants the knowledge that whether we use it or not, whether we visit it or not, there exists available to us a refuge from the sights and sounds and the smells of man. I need not dwell on the spiritual and psychic values of communion with nature. These are what bring us all together today in the most unwilderness-like atmosphere of this conference hall.

We see this refuge as a wilderness and conceive it as a "forest primeval" or a "virgin forest" as we visualize it by harkening back to our youth or to the youth of our local culture. That is, we desire the forest in a condition approaching that which we first saw or which our ancestors first saw. We are really not interested in the forest as it existed further back in geologic times when the climate was hotter or when the land was covered with ice or the plants and animals were completely different. Our interest is in the recent past, not in the past geologic ages.

In conceiving the idea of a forest ecosystem for our refuge, we accept patterns and occurrences of plants and animals that are congruous with our concepts, regardless of whether or not they are by our standards endemic or exotic. These patterns must only be fitting and proper in our own eyes.

Indeed, what is fitting and proper in man's eyes differs from man to man and certainly from generation to generation. You and I have concepts of a wilderness that our fathers did not have and we can be sure that our offspring will have still different concepts. Let me cite a few examples: First take any weed. A weed is but a definition of a plant growing out of place. Few of us I dare say would object to an occasional dandelion brightening up some corner of a sylvan meadow. It is only when the dandelions take over our lawns that they become objectionable. In Michigan, to be a little bit more specific, the pheasant is a recent and deliberate introduction. Yet, even the purest conservationist among us cannot but be warmed by the sight of a male pheasant followed by his harem marching across our back lawn as we sit at breakfast on a wintry morning. Furthermore, I am certain that to my children that pheasant is as much a natural part of the landscape as the raccoon or gray squirrel that raids our bird feeding trays periodically. Again, the Norway spruce is purely an introduction from Europe, but it was so widely and commonly planted in the farmsteads of the American Midwest over the last hundred years that it has become a natural and intrinsic part of the landscape. True, if we see a growth of Norway spruce planted in straight rows in the form of a square or rectangle, our eyes object to the incongruity of the pattern. But we would object to the pattern whether or not the trees were the introduced Norway spruce or native white pine. As an important and integral part of the forest farm homestead ecosystem we have ceased to object to the Norway spruce in the Midwest. Coming a little bit closer to the wilderness problem, there is considerable talk at the present time about preserving the Allagash wilderness in Maine disregarding the fact that this same wilderness has been logged several times in the past two hundred years. Despite this the fact remains that it still provides the feeling of wildness and refuge that we crave, so that all of us are interested in preserving this wildness regardless of the fact that it is not a wildness developed independently of man but rather as a result of the activities of man.

Moving west, few of us would object to the sight of a horse in the High Sierra or Rocky Mountains or in the deserts of the southwest. Yet the horse is strictly an introduced beast and its very presence denies the fact of wilderness. We have come to accept the horse as a congruent part of the landscape and we no longer object to his presence. Finally, as a forester who has worked on two different occasions in New Zealand, I am fascinated to find that there the Monterey pine and Monterey cypress have been introduced so widely and have spread so naturally that they have become a natural and endemic part of the landscape. In fact the ongrowing generation of youth undoubtedly think of them as New Zealand pine and New Zealand cypress. And, I dare say, much the same could be said for eucalyptus in California. A eucalyptus plantation represents an obvious cultural pattern and looks incongruous in any natural landscape, but I dare say the day will come when somewhere in California there will be a blue gum wilderness. In fact, if this blue gum wilderness exemplifies a natural ecosystem developed from the plants and animals having access to the site I believe that it would appear perfectly pleasing to our eyes and that we should enjoy it immensely.

In short, then, the wilderness is a pattern which is acceptable to our taste, our aesthetic desires and our nostalgia in creating the refuge from civilization that we all need to know exists and that we profit from penetrating on occasion. If we must accept the concept that we can never recreate the forest of the past, we should take pleasure in the fact that the oncoming generations will never desire the forest of the past in exactly the same way that we did, and that the new forest ecosystems that we will be able to create will be wildernesses that will be fully acceptable to the generations that will have access to them.

But how then do we go about creating these new forest ecosystems since we can't turn back the clock? I submit that we should identify the values we wish to create and then set ourselves deliberately to create a wilderness with these values. We need to experience solitude and a sense of wildness. Our aesthetic tastes demand a variety of scenery, of plants and animals. We take pleasure in venerable and large trees and a variety of birds, in wild animals of all kinds—in the wild deer which give the wilderness its name. We demand an assemblage, a forest ecosystem which is congruous to us, which disregarding the artificial concept of native and introduced plants and animals consists of plants and animals which seem to us to go together naturally and which form an attractive ecosystem. That is the wilderness we crave.

Certainly, as I intimated earlier it is possible to develop such a forest ecosystem by staying out, by putting up a fence, by locking the gate, by letting lightning-struck fires burn at will, by letting insects go through their natural cycles of abundance and scarcity, and by letting deer and rabbits go through their own equally natural cycles of abundance, starvation and subsequent diminution of numbers.

But few of us will go this far in developing new types of wilderness unaffected as completely as possible by man's activities. Let's face it. We are already in the business of managing wilderness. We do put out lightning fires. We do spray insects of all kinds and descriptions when they reach high population stages to the point that they damage and threaten to kill the forests. We do dam streams and create reservoirs. We do eliminate predators when they become troublesome. We do clear understory regeneration to get a better view of the big trees.

My point is that we should do more. By more, I don't mean timber harvesting, although I have no objection to cutting a tree anywhere if the incongruity of a sawn stump can be hidden and if by our act we increase the values of which I have spoken. I believe in the intelligent use of our technical skills to create desirable wilderness ecosystems of the future. Let me be explicit by citing two examples referring to wildernesses characterized by large trees where we have made a beginning at bringing the resources of science to bear on the wilderness problem.

Some years ago I spent two summers developing a management plan for Itasca State Park which surrounds the headwaters of the Mississippi River in northwestern Minnesota. This is a wilderness park, attaining its chief value from the many acres of old growth red or Norway pine. Incidentally, this particular tree has long grown in the region and derives its common name from having been first described near Norway, Maine. It was not brought in from the old country. Obviously the most important value of the park lay in perpetuating these old stands of red pine, all of which had developed originally following pre-settlement forest fires. Detailed growth studies showed that the oldest trees were in the neighborhood of 250 years old and that the larger and healthier trees had at least another one hundred years of life ahead of them, although their life span was being affected by increasing mortality in the smaller and less competitive red pine in the forest. I came to the conclusion, therefore, that the best way to perpetuate this old growth red pine forest was to move in and remove from time to time occasional suppressed and overtopped red pine before their decline in health and vigor would affect the health of the surrounding trees. Obviously it will be desirable to cut the stumps flush with the ground, to cover them up or to remove them in such a way as to avoid damage to the residual stand. By such delicate harvest methods, however, it would be possible to raise immediately the average size and attractiveness of the forest by the mere fact of eliminating some of the smaller trees. More important, this treatment should substantially lengthen the span of years during which that forest could be enjoyed by the people who visited it.

Much of the rest of the park consisted of stands of jack pine from thirty to fifty years old, again arising as a result of fires but in this case fires dating from the early period of logging. Here the problem was greater because jack pine is a short-lived species and there was ample evidence to show that, if left alone, these stands would disintegrate and disappear in a relatively few years. Inasmuch as most of the roads and most of the recreational facilities were developed among this jack pine type, we were confronted with the probability that the forests would disappear through the natural course of forest succession not in a hundred years but in twenty years or less. The answer here is purely management. Tracts of jack pine remote from the present road systems and recreational facilities should be clear cut and should be replaced with a naturally spaced planting of the longer lived species of the region, chiefly the red pine. Then when the inevitable occurred and the older jack pine stands disintegrated, there would be young natural-appearing stands of pine into which the roads and recreational facilities could be shifted pending the restoration of the former jack pine types elsewhere. By such management practices the desirable and natural pine forests could be maintained. In their absence they would disappear and be followed through natural succession by a forest dominated by hazel, balsam fir and white spruce, none of which would attain much in the way of size or grandeur. Yet, in a very real sense the old growth pine stands are fully as natural a part of the wilderness as the less majestic species which will succeed them if man does not interfere constructively.

As a second example, let me stick my neck out by referring to a study of the Mariposa grove recently completed by Professor Richard Hartesveldt of San Jose State as a doctoral dissertation at my own School of Natural Resources. Dr. Hartesveldt undertook a detailed study of the effect of recreational use upon the health, vigor and general condition of the famous big trees. I refer you to his study for his specific findings. I only add in my own interpretation that a sequoia wilderness must be created by man through constructive measures designed to favor the establishment and success of the big trees themselves. Furthermore, the longevity of the sequoias, as that of any other tree, will require constructive treatment to prevent undue compaction of the ground, undue damage by road cuts, unnecessary and undesirable swamping by inadequate drainage, and other silvicultural practices designed to create rather than to eliminate wilderness conditions.

In conclusion, let me restate some of the concepts put forth on the value of the wilderness to science. First, the concept of the wilderness itself is basically a sociological concept and not a biotic concept. It involves values of high importance to man's spiritual, mental and physical well being and values which we certainly must encourage and develop. These values, however, are anthropocentric and have little relevance to the forest ecosystem itself. From a biocentric viewpoint there is really no such thing as a true forest primeval, a virgin forest or a wilderness. Instead there is an infinite range of ecosystems constantly varying in time and space.

Among these ecosystems we can identify certain types which most nearly meet the sociological requirements which in themselves vary from time to time and from place to place. We can allow such ecosystems to develop by a hands-off policy. This, however, is a slow and generally wasteful and inefficient process. Second, we can continue our present practice of interfering in a limited extent to control fire, spray insects, regulate hunting pressure in wilderness areas, to fish and to regulate water supply. I believe, however, that we should take a more forward looking and constructive view and manage the forest itself skillfully, silently and inconspicuously to modify the forest ecosystem so that it most nearly approximates those types which are desirable from man's viewpoint.

In actual fact, I am not really advocating any change in our present practice. I am simply urging that we all face the fact that we are already managing the wilderness in a very real sense. We should continue to do so. We should openly bring the immense skill of modern science and technology to bear to create more and better wildernesses rather than to lock our minds and fight a losing battle to hold back the inevitable biological change moving the forests continuously away from our nostalgic memories of our childhood. By so doing we can not only maintain our present wilderness; we can create new and more extensive wildernesses that we desperately need to meet the needs of the generations that are on their way.

{Since the principal purpose of the Sierra Club, as publisher, is to help save some living wilderness to augment whatever reasonable facsimile man may contrive, Robert Golden, club staff man most attuned to the ecology of wilderness, was asked to comment upon Dr. Spurr's paper. Mr. Golden submitted similar comment for the discussion period following the paper but the comment was unfortunately passed over. Three months after the conference Mr. Golden summarized his many differences as follows:

"Perhaps Stephen Spurr has introduced the really new element to the Wilderness Conferences, for it is his paper that will probably be most widely read, examined, criticized, quoted, and misquoted. Dr. Spurr has neatly guided the reader and listener down the almost plausible path toward a belief in the skill and wisdom of the resource manager as the saviour and protector of America's wilderness heritage. We think Dr. Spurr did so deliberately, not out of any real malice toward the wilderness idea, or out of opposition to pending wilderness legislation, but simply as an exercise in critical thinking.

"Most of us who listened during the conference were only vaguely disturbed by Spurr's suggestion that new wilderness ecosystems may be created in the future to include plant and animal assemblages chosen for their attractiveness to man rather than as parts of a native flora or fauna.

"If the reader stops to reflect for a moment he will realize that such synthetic ecosystems are exactly what surround the nation's metropolitan centers and what are disaffectionately labeled 'urban sprawl.' A typical synthetic 'wilderness' of this type contains an interesting array of plants, including chickweed, morning glories, and stunted citrus trees. Repeated efforts are required each Saturday morning to subdue the meadowlike areas with power-driven equipment, and in spite of numerous efforts to attract interesting birds through the use of feeding trays and nest boxes, the faunal dominants are all too often English sparrows, garden snails, slugs, sow bugs, earwigs, Argentine ants, occasional brown rats, and stray cats. Towhees may nest infrequently but the environment is usually hostile to their continued occupancy particularly if part of the weekly maintenance includes a liberal dosing with sundry organophosphates or chlorinated hydrocarbons.

"At the Sixth Wilderness Conference Frank Fraser Darling went immediately to the core of the relationship between wilderness and science when he said: 'The value of wilderness to science put baldly—very baldly and not at all sentimentally—is the provision of study areas of pristine conditions.' And later, in discussion, Starker Leopold pointed to the well-documented fact that evolution of natural communities always tends toward natural complexity while human use of such communities inevitably makes them tend toward organic simplification. These are critical ideas which Spurr avoids in his discussion of wilderness-management policy; unless the reader is critical, their absence will go unnoticed. Spurr's penchant for reformulating ecological principles may leave the reader confused until their mirror images are discerned. Perhaps Professor Spurr did some reading of the proceedings of the Sixth and Seventh Conferences before preparing his paper and has rather playfully paraphrased many of their central ideas. In any event, he invites the critical examination of each of his statements. It is not sufficient for conservationists to disagree with Dr. Spurr. It is infinitely more important to know why they disagree."

We hope, with apologies to Professor Spurr for getting into a debate in this way, that debate will continue in future conferences and in conservation and technical journals in the interim. But let us give top priority to the saving, right now, of as much unspoiled and unmanaged wilderness as possible, just in case Professor Spurr happens to be wrong.

There is an outsize difference in magnitude between the kind of influence man exerts as a hiker and as a subdivider. There is certainly room for the kind of management he speaks of, and a great need for more graduates who know how to apply it intelligently—on tree farms, in places like Golden Gate Park and Central Park, and in multiple-use projects in which recreation is one of the uses. But not in wilderness.—D.B. [David Brower]}

James Morton Turner

From Woodcraft to "Leave No Trace" (2002)

Wilderness, Consumerism, and Environmentalism in Twentieth-Century America

In 1983, *Outside* magazine noted a growing number of "cognoscenti known as no-trace or low-impact campers" backpacking into the wilderness. "As their nicknames indicate, [these people] keep the woods cleaner than they keep their [own] homes." The most devoted backpackers fluffed the grass on which they slept, gave up toilet paper rather than burying it, and preferred drinking their dishwater to pouring it on the ground. No measure seemed too extreme in their efforts to protect the wilderness. Carrying packs loaded with modern gear, backpackers prided themselves on traveling through wilderness as mere visitors. In 1991, Leave No Trace became the official ethic for environmentally-conscious outdoor recreation on the nation's public lands.[1]

That announcement brought to a close a long transition in the place of recreation in the American wilderness. In the 1920s, Aldo Leopold first described wilderness areas as a "means for allowing the more virile and primitive forms of outdoor recreation to survive." His vision evoked an early-twentieth-century tradition of "woodcraft." Unlike today's backpacker, the skilled woodsman prided himself on living off of the land: building lean-tos, cooking over an open fire, and hunting for food. Woodcraft was steeped in self-reliance, masculine rhetoric, and discomfort with the modern consumer economy. Leopold envisioned wilderness as a refuge from modernity, where a working-knowledge of nature would reconnect people and the land.[2]

For environmental historians, this transition from woodcraft to Leave No Trace offers a tool to pry apart the modern wilderness ideal. Opening up the backpacks, leafing through the guidebooks, and revisiting the campsites reveals more than just changes in the ways people have returned to nature. Indeed, it reveals the historical pliability of the very ideals to which wilderness travelers have aspired.

Within the environmental historiography, the intellectual paths in and out of wilderness are well traveled. Since the mid 1980s, this traffic has been particularly heavy, as sharp debate over the cultural and scientific roots of wilderness engaged both the academic and scientific communities. The so-called "great new wilderness debate" emerged from conflicting approaches to wilderness. Conservation biologists gave new emphasis to the role of wilderness reserves in protecting biodiversity; Deep Ecologists furthered the scientists' arguments, making wilderness the centerpiece in a biocentric agenda for restructuring modern society; and historians and literary scholars questioned the imperial, racial, and socioeconomic assumptions underlying the wilderness concept.[3]

Several participants in the recent wilderness debate have linked the growth in wilderness recreation with consumer culture's power to repackage nature—refashioning "wilderness" as an accessible and desirable tourist destination. While participants in the debate have examined wilderness from all sides, exposing the many cultural assumptions that layer the wilderness ideal, they have given little attention to the recreation ethics so important to the American wilderness tradition. Thus, this article follows a different set of paths into the wilderness: It follows the hiking trails.[4]

The transition from the heavy-handed practices of woodcraft to the light-handed techniques of Leave No Trace can be read as a logical response to the tremendous growth in wilderness recreation during the twentieth century. This transition, however, represents more than a straightforward reaction to the growing traffic of wilderness visitors. The exigencies of wilderness politics after the passage of the Wilderness Act in 1964, the growth of a consumer-oriented wilderness recreation industry, and the loosening strictures of gender in postwar America all helped shape a modern wilderness recreation ethic that harbored little of the anti-modern sentiments that charged the woodcraft ethic in the early twentieth century.

Following the hiking trails reveals how the modern wilderness ideal was reinvented in the twentieth century. In the years after the passage of the Wilderness Act of 1964, as wilderness grew in popularity, it was most often described in two ways: as a recreational resource for backpackers and as a pristine ecological reserve for posterity. The modern wilderness recreation ethic served to negotiate this tension in the wilderness ideal. Practicing Leave No Trace allowed an ever-growing number of backpackers to visit wilderness, while leaving its ecological integrity intact. It was a pragmatic balance for the wilderness movement to strike. Yet, giving wilderness primacy as a popular recreational resource, emphasizing an aesthetic appreciation of nature, and embracing the consumer culture that enveloped post-1970s wilderness recreation all eroded the social ideals around which the wilderness movement first coalesced. This shift toward a modern, consumer-oriented wilderness ideal calls into question the effectiveness with which some of America's most ardent environmentalists—its wilderness recreationists—have engaged the environmental challenges posed by the consumer economy.

THE WAY OF THE WOODS

Wilderness emerged as an important site of recreation in the early twentieth century. In those years, railroads and automobiles helped expand the nation's leisure time geography, linking the cities with the rural hinterland. The same economy that cast a pall of smoke over growing cities, offered jobs to the surge of immigrants that crowded city streets, and gave new scale to the commerce that dominated urbanites' daily lives also provided more and more of America's city dwellers with the means to quickly remove themselves to the countryside. For many Americans, nature beyond the city limits increasingly promised an antidote to the ills of urban life.[5]

Americans found many ways to get back to nature in the early twentieth century. Woodsmen, walkers, autocampers, and resort-going tourists all followed their own paths out of town and into the woods. But common sentiments animated their visions of the outdoors: They celebrated the simplicity of nature, the rejuvenating power of the mountains, and suspicion of the ill-health pervading the growing metropolises.[6] Walkers, for instance, cultivated an aesthetic appreciation of nature. In *Walking* (1928), George Trevelyan praised foot travel as a spiritual pursuit that encouraged "harmony of body, mind, and soul, when they stride along no longer conscious of their separate, jarring entities, made one together in mystic union with the earth."[7] More popular in the early twentieth century was autocamping. Starting in the 1910s, hundreds of thousands of middle-class families began motoring into the outdoors, camping by roads, streams, and high mountain meadows, and, on big trips, setting their sights on the nation's national forests and national parks.[8] Rolling across the country, these autocampers made the nation's public lands popular tourist destinations. The conflicting demands of autocampers, walkers, and woodsmen would play an important role in shaping the nation's federal land policy.

Of all these ways of getting back to nature, however, one has attracted little historical attention. Between the 1890s and the 1930s, woodcraft formed a coherent recreation ethic, and an important precursor to the modern wilderness movement. For aspiring woodsmen, a selection of manuals promised to reveal the secrets of woodcraft. George Washington Sears penned the first of these, titled *Woodcraft,* in 1891. In these guidebooks, several characteristics distinguish the woodsman from the walkers and autocampers: his practice of woodcraft celebrated a working knowledge of nature; he was preoccupied with an independent masculine ideal rooted in the frontier; and he exhibited strong misgivings for the abundance of consumer goods available to the outdoorsman. This dual concern—for leisure in the woods, and the consumer economy beyond—emerges as a central tension in woodcraft.[9]

For the woodsman, the woods promised a working vacation. Surveying the table of contents of woodcraft guides such as Horace Kephart's *Camping and Woodcraft* (1906), Edward Breck's *The Way of the Woods* (1908), or Elmer Kreps' *Camp and Trail Methods* (1910) reveals common chapters on essential wilderness know-how. In the outdoors, a woodsman judged himself by his skills in hunting, tanning furs, preserving game, building campfires, setting up shelters, and traveling through the forest. Summing it up, Kephart called woodcraft "the art of getting along well in the wilderness by utilizing nature's storehouse." In contrast to the walking or autocamping literature, none of these handbooks sought to teach outdoorsmen how to pass through wilderness as a visitor. The real measure of the woodsman was in how skillfully he could recreate wilderness as his home.[10]

Proficiency in woodcraft required an intimate, hands-on knowledge of the woods. These expectations are best set forth in the early Boy Scout manuals. An ax was the young scout's most important tool. With it, the first edition of *The Official Handbook for Boys* (1912) explained, the scout could furnish his entire wilderness camp. Ten straight tree branches could be assembled into a lean-to; and cut boughs served to thatch the roof. Two lean-tos facing one another made for an especially comfortable camp. In the kitchen, pot lifters, plates, and cups all could be fashioned from sticks and bark.[11] Ernest Seton, in *The Woodcraft Manual for Boys* (1917), challenged the young woodsman to practice "hatchet cookery" and furnish his entire meal with his hatchet alone.[12] Both the Boy Scout manuals and other woodcraft guides assumed that the skilled woodsman could identify trees (particularly those that burned most cleanly), knew the habits of animals (so he could kill them), and could identify a buffet of edible plants. The woodsman demonstrated his working knowledge of nature by using nature to his own ends.

Nothing troubled the woodsman more than being labeled a tenderfoot. Indeed, two types of tenderfeet stood out—those who took too much equipment into wilderness, and those who took too little equipment into wilderness. Much of the woodcraft literature emphasized striking a balance between these two extremes. This required not only the skills of a woodsman, but of a consumer. As Emerson Hough warned, "there is no purchaser on earth whose needs and notions are better studied or better supplied than are those of the American sportsman." Sears noted in *Woodcraft* that "The temptation to buy this or that bit of indispensable camp-it has been too strong, and we have gone to the blessed woods handicapped with a load fit for a pack-mule. This is not how to do it." In the woodcraft literature, the woodsman knew not only what tools and trinkets he could discard, he also could find the resolve to discard them. For those given to temptation, Stewart White suggested divvying up the gear into three separate piles: the essential, semi-essential, and unessential. Then, he implored the woodsman: "no matter how your heart may yearn over the Patent Dingbat in [pile] No. 3, shut your eyes and resolutely discard the latter two piles."[13]

Similar statements regarding careful packing can be found in 1970s backpacking guides too. But in the 1920s, this preoccupation with consumer goods emerged as a central strand of woodcraft's reaction to modernity. Relying on too many consumer goods not only weighed down the woodsman's pack, more important, it threatened to erode traditional skills, distance the woodsman from nature, and implicate him in a consumer economy preoccupied with profit. In *Woodcraft for Boys,* Seton held up the scouts' grandfathers as the "true Woodcrafters" and lamented that so many of the skills they mastered had become superfluous in the age of the factory. He urged the aspiring woodsman to "know the pleasure of workmanship, the joy that comes from things made well by your own hands." Otherwise, he warned, the camp, and the home, would become little more than an accumulation of artificial, manufactured goods.[14]

New Englander Joseph Knowles gathered up many of these sentiments in 1913, giving both the woodcraft ethic and its critique of modernity its fullest expression. Citing all of the concerns that spurred the back-to-nature movement, Knowles captured national attention when he walked buck naked into the Maine woods. He explained, "there was too much artificial life at the present day in the cities." For the next two months, he made an experiment of himself, trying to live off of the land. When Knowles reemerged, he strutted out of the woods strong and healthy, clad in self-made clothes, and claimed to be no less than the model American man. Much pageantry surrounded Knowles' stunt, and although doubts as to its veracity were immediate, Knowles struck a chord with the American public. In subsequent articles, speeches, and a book, Knowles came the closest to making the woodcraft tradition a vessel of reform. Foreshadowing Benton MacKaye's proposal for the Appalachian Trail and the modern wilderness movement itself, Knowles proposed setting aside "wild lands" and establishing outdoor communities where Americans could retreat from the "commercialism and the mad desire to make money [that] have blotted out everything else [leaving us] not living, but merely existing."[15]

Central to Knowles' adventure, and the woodcraft tradition, was a preoccupation with masculinity common in turn-of-the-century America. Woodcraft promised to return the enervated city-dweller to the mythical frontier, allowing him to play out in leisure Theodore Roosevelt's "wilderness hunter," reaffirming both his masculinity and his Americanness. Of all the woodcraft guides, none were more preoccupied with masculinity than the Boy Scout manuals: "Make it yourself. A real red blooded, HE boy would make his crotched supports, tammel bar, tongs, pot hooks, forks and spoons when and where he needs them."[16]

Knowles expanded upon this concern for masculinity, suggesting its deeper implications for American democracy: "From wilderness life to the simple country life, and then up through the life of a great city liberty gradually decreases." Woodcraft offered a virile form of recreation that distanced the urbanite from a leisure class that hired guides for their wilderness trips or spent their vacations in effeminate mountain resorts. Although those who patronized the woodcraft literature likely hailed from America's upper class, the literature itself idealized woodcraft as a vacation for men of all means "who sorely need and well deserve a season of rest and relaxation at least once a year." In an increasingly urban society, rife with concerns over American civilization and waning masculinity, this opportunity to return to nature and demonstrate one's virility was a powerful salve.[17]

Among early-twentieth-century outdoor recreationists, the rhetoric of woodcraft most closely figures the place of recreation in the early wilderness movement. Other important traditions, such as John Muir's ascetic trips to the High Sierras and Bob Marshall's leg-stretching romps through the Adirondacks, evoke the tradition of walking which also informs the modern wilderness ideal. But the spiritual reverence for nature important to walking gave voice to few of the social concerns important to the woodcraft literature or to the nascent wilderness movement. Indeed, it is the woodcraft literature's preoccupation with the frontier, masculinity, and modernity that all suggest a key place for woodcraft in the heritage of American wilderness thought.

To varying degrees, Aldo Leopold, Bob Marshall, and Benton MacKaye—three founding figures in the modern wilderness movement—all marshaled familiar woodcraft language and themes in setting forth a broader wilderness idea. In the 1920s, Leopold emphasized wilderness not as a place, but as a means to allow Americans to test themselves "living in the open" and "killing game." When Benton MacKaye outlined his 1921 vision of an Appalachian Trail, he described more than a simple trail linking Georgia to Maine. Rather, his Appalachian Trail was a reform project that promised to weave new connections between American society and the land through a working knowledge of nature. Marshall prized wilderness for its "fundamental influence in molding American character" and because only in wilderness could one be completely self-sufficient. For all three men, recreation was an important strand of the wilderness idea. For Leopold, in particular, wilderness recreation, in the tradition of woodcraft, promised to foster a self-sufficient, intimate knowledge of nature.[18]

Automobiles and new roads posed a real threat to the nation's public domain during the interwar years. In 1928, a government report warned that the nation's remaining wilderness was "disappearing rapidly, not so much by reason of economic need as by the extension of motor roads and the attendant development of tourist attractions." Paul Sutter's rich study of the formation of the Wilderness Society locates the origins of the wilderness movement not so much in a concern for protecting land from extractive industries, such as logging or mining, but in a deep aversion to automobile-based recreation. Leopold was leery, in Sutter's words, of the "dysfunctional leisure-based relationship with nature" facilitated by the automobile. Most troubling was that both the Forest Service and the National Park Service increasingly catered to autocampers, building roads and establishing camp grounds, all the while eroding the nation's reserve of roadless, uncommercialized land. The Wilderness Society (founded in 1935) aimed to protect at least a portion of the nation's public domain from the automobile. In Sutter's analysis, the early wilderness advocates' enthusiasm for this practical aim was fired by more significant, and much deeper, misgivings over the emerging consumer economy. In this way, the anti-modern currents running through woodcraft served as a precursor to the broad critique of modernity that inspired the interwar years wilderness movement.[19]

Although the woodcraft ethic helped give voice to the wilderness movement, by the 1930s wilderness advocates had already begun to broaden their arguments, offering the recreational, scientific, and moral justifications that historians now identify as the foundation of the modern wilderness movement. Drawing on this philosophy, wilderness advocates mounted a small-scale campaign before World War II to secure administrative protection for wilderness in the national forests, under the L-20 and later U regulations, and in a few national parks, such as Kings Canyon in California. These advances in Forest Service and National Park Service policy promised to protect at least a portion of the nation's remaining public domain from the hordes of autotourists motoring their way out of the cities.

After World War II, however, the threats to wilderness appeared on additional fronts. Increased pressure for resource development and tourist facilities on the nation's public domain tested the agencies' commitment to protecting forests and parks from development. At the same time, a growing number of wilderness recreationists formed a new constituency for wilderness preservation. By the 1950s, the growing threats to wilderness and this growing constituency of recreationists combined to transform the small-scale interwar years wilderness movement into a national political campaign for wilderness preservation. Yet, with the new popularity of wilderness came an unexpected threat: how to protect wilderness from the backpackers themselves.

THE PARADOXES OF POPULAR WILDERNESS

After World War II, a new breed of outdoor recreationist—the backpacker—heralded a shift in both wilderness recreation ethics and the popular politics of wilderness. Between the late 1950s and the early 1970s, the ideas central to the wilderness movement began to migrate away from the social concerns that informed the interwar years wilderness movement. Two competing notions of wilderness—first, as a recreational resource, and second, as a scientific reserve—served as the dominant themes from which the postwar wilderness ideal was molded. In fact, sharp debate between the Forest Service and two factions of the wilderness advocacy community emerged over whether wilderness recreationists should have unfettered access to wilderness areas, or whether strict limitations should be established to protect wilderness as a biological reserve. This debate became heated after the passage of the Wilderness Act, and it tugged on the very stitching that held the wilderness idea together. These debates laid the groundwork for the rise of a minimal-impact camping ethic in the 1970s that would displace woodcraft as the dominant wilderness recreation ethic.

Historians have long noted the many forces that combined in the postwar era to make it ever easier for Americans to reach and enjoy the nation's backcountry. A new system of federal interstate highways linked urban America and its rural hinterland, a dense network of federally-subsidized logging roads opened up more and more of the national forests, and new technology and abundant information made it easier to plan a trip to the wilderness. As more people began visiting the nation's backcountry, the government began conducting surveys of wilderness visitors. A 1960 survey conducted at seven wilderness areas revealed that on average visitors were overwhelmingly male, married, and college-educated. Most lived in urban areas and fell between thirty and fifty years of age. Notably, less than half reported belonging to a conservation organization and income levels ranged evenly from the modest to the wealthy. Although the survey never asked the question, it is likely that a growing number of these wilderness visitors would have identified themselves as backpackers. In the 1950s, backpacking was becoming an increasingly popular way to enjoy the self-sufficient exploration of wilderness. In the Sierra Club's first wilderness manual, *Going Light* (1951), a young club staffer named David Brower promised to open up the "challenge of wilderness" to all those eager to escape the crowds and head into the backcountry. By the mid-1970s, as outdoor recreation grew in many directions, it was the backpackers who increasingly claimed wilderness as their privileged province.[20]

In the years after World War II, economic development began to encroach on the national parks and national forests. Timber companies looked to the national forests for wood and engineers sized up the nation's rivers for hydroelectric development. Building on the momentum of the national debate over a dam at Echo Park, the wilderness advocacy organizations began to push for a federal law authorizing a congressionally-sanctioned system of wilderness areas on the nation's public lands. Central to their strategy was promoting wilderness as a recreational resource. In the 1960s, the Sierra Club and Appalachian Mountain Club encouraged hikers out into the mountains with tactics such as new trail guides, club-managed huts, and sponsored outings. The Wilderness Society soon followed their lead. These organizations hoped that as more backpackers visited wilderness, many of them would file out of wilderness ready to spearhead the push for a congressionally-sanctioned wilderness preservation system. The subsequent legislative campaign culminated in the passage of the Wilderness Act in 1964. Yet, even as the Wilderness Act passed, the problem posed by wilderness recreation already formed an important sub-theme to the debate.[21]

A 1947 *Sierra Club Bulletin* article, "Saturation of Wilderness," and the club's newly begun Biennial Wilderness Conference in 1952, both asked if backcountry visitors might already be over-using wilderness. In the late 1950s, a growing number of Sierra Club members began writing letters to the club, worried that its tradition of one-hundred-person trips into the high country represented an "over-assault on the wilderness area." To study this possibility, the Sierra Club marshaled its own leaders as data collectors, seeking to document their impact on the wilderness. As the fight for the Wilderness Act heated up in 1960, the *Saturday Evening Post* reported on new threats to the proposed National Wilderness Preservation System. Instead of portraying a wilderness threatened by loggers or miners, the article described a New Mexico wilderness area beset by hikers, motorized recreationists, and Boy Scouts trying to live off the land. With such articles in hand, Brower worried that "chainsaw-toting" wilderness opponents could argue that the "greatest threat to wilderness is from the wilderness lovers themselves." Publicly, wilderness proponents dismissed such concerns. They had to: Wilderness recreationists formed a key segment of the growing wilderness movement.[22]

The Wilderness Act of 1964 established the National Wilderness Preservation System and created new management imperatives for the nation's wild lands. Despite the growing concerns regarding the overuse of wilderness, the act was most significant for what it left unsaid. A paradox underlay the newly-established wilderness system: How could these areas be made available for public use with minimal restrictions, while also preserved as a resource for posterity?[23] Richard Costley, who oversaw the Forest Service's wilderness program, called this the "basic riddle inherent" in wilderness.[24] Solving this riddle saddled the Forest Service with difficult decisions. Restrictions on wilderness recreation threatened the freedom long associated with wilderness travel; unlimited recreation itself could permanently alter the ecology of wilderness areas; and simply putting wilderness behind legislative boundaries threatened to tame it, diminishing the wildness that the act originally aimed to preserve. Ultimately, Costley explained, wilderness could not be all things to all people. "We've got to make the point . . . forcefully that the 'recreational values' are not the only values in wilderness. After all, wilderness areas were set aside to protect them—even from recreationists."[25] Despite all the ambiguities of the Wilderness Act, the Forest Service faced one constant. As Costley put it, "the multitude is at the gates."[26] Between 1960 and 1965 wilderness visitation increased fivefold to 3.5 million visitor days per year.[27]

Faced with rapidly increasing numbers of wilderness visitors and the demand for additional wilderness areas, the Forest Service's wilderness policies began to coalesce in the late 1960s, and the agency began actively promoting a wilderness system that was small, pristine, and managed to limit wilderness recreation. To Costley, William Worf, and others in the Forest Service's recreational division, such a "purity" policy heightened the value of wilderness, making it the exceptional system that Congress had intended. As Worf explained, "We are concerned with laying a solid foundation for a high quality Wilderness System which will withstand the pressures [of time]." Among the greatest threats to that vision was the booming number of backpackers. Following the purity policies, the Forest Service took steps to limit the recreational use of wilderness in the late 1960s. Interpreting the Wilderness Act as strictly as possible, the Forest Service resisted installing bridges, maintaining trails, or providing primitive sanitary facilities in designated wilderness areas. In 1972, the Forest Service further tightened its regulations, arguing that each wilderness area had a fixed "carrying capacity," and could accommodate only so many visitors. Soon, backpackers visiting popular wilderness areas from New Hampshire to California started their trips at the ranger station, seeking a mandatory permit for their wilderness trip.[28]

The Forest Service's purity policies came packaged with another agency proposal: "backcountry."[29] In lieu of additional congressionally-designated wilderness, the Forest Service promoted backcountry as an alternative form of land protection. According to Worf, backcountry areas would accommodate much higher levels of use, and meet the needs of a public "who did not need or desire the kind of wilderness atmosphere or natural conditions described in the Wilderness Act."[30] Unlike wilderness areas, backcountry areas would permit proactive management of non-motorized recreation, including designated campsites, picnic tables, outhouses, better developed trails, and limited logging. The proposed backcountry areas, however, enjoyed none of the congressionally-sanctioned permanence of the wilderness system. The Forest Service couched its argument for backcountry in terms of its purity policies, arguing that new backcountry areas would help protect the wilderness system. Shifting hiking into backcountry areas, explained W. E. Ragland, would reduce recreational pressure on wilderness, thus preserving the "wilderness areas for limited use and maximum protection of the natural ecology."[31]

The Forest Service's argument, however, sounded hollow to many wilderness advocates. During the 1950s and 1960s the timber industry increasingly looked to the national forests as an important source of timber. In these years, industry lobbying led to tighter connections between the industry and the agency, greater reliance on clear-cutting for timber harvests, and increased estimates of the allowable cut permissible on the national forests. In the late 1960s it became apparent that more than 50 million acres of Forest Service land (known as roadless areas or "*de facto*" wilderness) remained up for potential wilderness review. With so much at stake, even Worf acknowledged that many people thought the Forest Service was "promoting the 'backcountry' concept to kill the wilderness movement." Some Forest Service employees, such as Worf, did believe wholeheartedly in the importance of a high-quality wilderness system. Nevertheless, the combination of strict limits on wilderness access, an unbending position on wilderness facilities, and the alternative backcountry designation (which enjoyed none of the statutory protection of the wilderness system) suggested that other Forest Service officials envisioned not only a pristine wilderness system, but a very limited wilderness system that left most of those 50 million acres available for future logging.[32]

Within the wilderness advocacy community, the Forest Service's purity policies and the threat of growing recreational use of wilderness became contentious issues in the late 1960s. At first, the staffs of the Sierra Club and the Wilderness Society openly acknowledged the importance of limiting wilderness access to keep the nation's wilderness system from becoming pockmarked by over-used campsites, riddled with trails, and congested with visitors. But as the Forest Service began to implement its strict wilderness management policies, limiting recreational access to wilderness, a whole range of reactions began to emerge from the wilderness community. The biologist Garrett Hardin suggested that wilderness access should be limited to "physically vigorous people." Paul Petzoldt, the founder of the National Outdoor Leadership School, proposed a meritocracy, giving priority to those best educated in wilderness skills. David Brower offered a reminder "that wilderness will be preserved only in proportion to the number of people who know its values first hand." And a young math professor from California, Theodore Kaczynski, viewed all the visitors and the new regulations as the seeds of demise for wilderness. "In short the so-called 'wilderness' preserves will turn into artificially maintained museum-pieces," wrote the future Unabomber, "with 'do not touch' signs all over them. Real wilderness *living* will be impossible." From these many currents of concern, two dominant approaches began to emerge in this debate over the management—and the ultimate purpose—of the wilderness system.[33]

The Sierra Club's Wilderness Classification Study Committee, a volunteer committee that helped develop the club's wilderness policy, emerged at the center of the first of these views. Francis Walcott, the committee's iconoclastic chairman, argued fervently for the primacy of the biological value of wilderness. "The primary human benefit of wilderness is ecological," he explained; "wilderness will help to preserve endangered species and all other endemic wildlife, to protect fish habitat, and provide clear water, stable stream flows for downstream communities, islands of clean air, and, in general terms, maintain a livable environment for man." Although Walcott's arguments for a biological wilderness were not unique in the late 1960s, he was among the few supporters of the Forest Service's backcountry proposals and sharp limits on recreational access to wilderness. In terms of wilderness management policy, this placed the committee's view in an uncomfortable alliance with the Forest Service—both supported a pristine wilderness system little used by the public.[34]

But the Wilderness Committee moved well beyond the Forest Service's approach to wilderness by elevating the wilderness system as the first tangible realization of a new land ethic which harbored the revolutionary promise of redefining the relationship between society and the land. "The establishment, maintenance, and management of wilderness is a means towards achieving such a land ethic which provides a real respect for the land," Walcott explained. He believed through wilderness the environmental community was "pushing for . . . the integrity of the earth." In its scope and its forward-looking vision, the committee evoked the interwar years wilderness movement, yet its biocentricity suggested a distinct wilderness philosophy. It demanded that the Sierra Club, and the rest of the wilderness advocacy community, demonstrate a measure of restraint, in the form of limited access, as the moral foundation for a truly visionary wilderness system.[35]

A second view of wilderness management emerged from the ranks of professional wilderness advocates who oiled the legislative gears so important to the wilderness movement's success. Brock Evans, the Sierra Club's Pacific Northwest Representative, warned: "from a political standpoint, once we accept the idea of a backcountry system, there will be few, if any, new areas *ever* added to the Wilderness System."[36]

Evans reminded both the club's Wilderness Committee and the Forest Service that "recreation is one of the *named* purposes of the Wilderness Act. In fact, it is the first named purpose—out of alphabetical order."[37] Although the Sierra Club and the Wilderness Society endorsed the possibility of limits on wilderness access in the 1960s, this new debate called into question how tenable that earlier position was. Dick Sill, of the Sierra Club, explained the conundrum: "[I]f restrictive permits are adopted . . . many of the present supporters of wilderness—particularly the recreationists—will turn against wilderness entirely."[38] In 1970, the wilderness system measured only 10.5 million acres (of which 10.2 million acres were under Forest Service jurisdiction), less than one-tenth of its present-day size.[39] Instead of ratcheting down access to wilderness, promoting it as an overly pure system, and jeopardizing its political support, the pragmatic alternative was to encourage the recreational use of wilderness, while fighting to make the system as large as possible. As Sill noted, at the end of the day "the impact of thousands of Sierra Club hikers cannot begin to approach that of one work crew with chain saws, a bulldozer or two, and some appended logging trucks."[40]

In the first decade after the passage of the Wilderness Act, the wilderness advocacy community entertained these two sharply different visions of the future of the wilderness system. This debate, however, makes little sense in the context of the existing wilderness historiography. First, it reveals a wilderness advocacy community unsettled by a philosophical disagreement over the purpose of wilderness at a time when most historians have assumed that the wilderness community was unified in opposition to the federal land agencies and resource industries. Second, much of the historiographical debate has swirled around the assertion that environmentalists focused their energies on an idealized notion of wilderness freighted with romanticism and preoccupied with pristine nature. That argument, however, engages only one strand of a much more diverse and pragmatic wilderness discourse. All told, this late 1960s debate, which swept through the professional and volunteer wilderness advocacy community, marks an important and overlooked junction in the trails of wilderness history. In one direction lay a wilderness system protected by strict visitation limits, dedicated largely as a biological reserve, and demanding a great deal of self-restraint on the part of the wilderness community. In the other direction lay a wilderness system that compromised the biological integrity of wilderness, prioritized human recreation, and promised to command political popularity. By the mid 1970s, it became clear that the wilderness advocacy community, along with a growing number of hikers, had chosen the latter path.[41]

MINIMAL-IMPACT CAMPING OR "THE ART OF USING GADGETS"

Because many wilderness advocates believed that maintaining popular support for wilderness meant supporting liberal access for wilderness recreationists, the movement sought a pragmatic balance between use, political support, and preservation in the early 1970s. Central to that strategy was a new wilderness recreation ethic. As one author noted, the wilderness system could no longer tolerate an "old-style pioneer encampment" like the one his wilderness survey trip discovered in 1972 with "felled trees, a couple of shelters built of boughs cut green and, lying in the middle of it all, a Boy Scout Fieldbook." Rather, the wilderness advocacy community began to promote a new wilderness recreation ethic—minimal-impact camping—that promised to prop the doors to wilderness wide open for a better-educated wilderness visitor.[42]

This new wilderness recreation ethic made it easier for the Wilderness Society and Sierra Club to relax their support for limits on wilderness access. Stewart Brandborg, the Wilderness Society's executive director, noted in an internal paper: "Controls should be as flexible as possible to permit maximum freedom; dispersal of users should be encouraged; rationing: adopt politically acceptable forms." And, with emphasis, he wrote that "carrying capacity should be *increased*" by "training users in wilderness use to reduce impact." Wilderness advocates hoped many more recreationists, if properly educated, could be crowded into the wilderness "before we reach the ecological carrying capacity of intelligent use." New books such as Harvey Manning's *Backpacking, One Step at a Time* (1972), Paul Petzoldt's *The Wilderness Handbook* (1974), and John Hart's *Walking Softly in the Wilderness* (1977) represented a new genre of wilderness manuals that aimed to reeducate wilderness visitors, weaning them off woodcraft, and teaching them the new skills of minimal-impact camping.[43]

To the degree that this new minimal-impact ethic made sense, however, it also reflected the erosion of the skills and anti-modern concerns embedded in woodcraft. No longer did a working knowledge of nature anchor the wilderness recreation experience—the new literature aimed to replace woodcraft, which it dismissed as "old-style" camping. Hart, updating the Sierra Club's wilderness guide, captured the new backpacker's ethic. "[G]oing light" had new meaning, he wrote. "[T]oday it refers less to the load in your pack than to the weight—the impact— of your passage on the land." As David Brower explained in 1971, backpackers aimed to visit wilderness without "leaving perceptible traces." Echoing the language of early-twentieth century walkers, he continued, that was backpacking "in harmony with the spirit of wilderness." In general, the new hiking guides dropped sections on building lean-tos, trapping, and hunting, and replaced them with instructions for selecting minimal-impact campsites, slowing trail erosion, and traveling as discreetly as possible. Hart explained the backpacker's new challenge and reward: It is "quite something . . . to know that you might have harmed a place and that you did not."[44]

The 1970s wilderness recreation ethic required that backpackers enter wilderness with more than a new set of camping skills. They also had to have the right gear. Because backpackers built no fires, hunted no wildlife, and constructed no make-shift shelters, they became increasingly dependent upon what they carried into wilderness with them. To reduce firewood consumption and fire rings, backpackers started carrying small portable stoves. Bedding down no longer meant gathering moss, grass, or tree boughs for a soft night's sleep. Lightweight foam pads offered an environmentally friendly alternative. Canvas tents and hastily assembled lean-tos disappeared in favor of nylon tents complete with metal poles and nylon stakes. To keep warm and dry, backpackers shelved their wool and fur and instead wore new garments made from engineered fabrics such as polypropylene and GoreTex. Marking the gulf separating the anti-modernist woodsman and the modern backpacker, the well-equipped backpack had become a showcase for advanced consumer technology.[45]

For those practicing the minimal-impact ethic, springing a temporary camp from a backpack of modern gear changed the dynamics of wilderness travel. The skills and risks at play in the 1970s backcountry differed significantly from those of the prewar years. Replacing campfires with stoves, twine with plastic fasteners, or a lean-to with a tent all diminished the wilderness travelers' immediate knowledge of the land around them. Aldo Leopold foresaw the emerging trend in *A Sand County Almanac:* "A gadget industry pads the bumps against nature-in-the-raw; woodcraft becomes the art of using gadgets." Lighting a stove or pitching a storm-worthy tent required new skills, but these skills did not promote the same hands-on knowledge of nature celebrated in the woodcraft handbooks or the early Boy Scout manuals. Instead, the modern wilderness ethic cultivated an aesthetic appreciation of wilderness. Long hikes, temporary camps, and an effort to leave no trace increasingly made backpackers transients in the wilderness landscape, observing, appreciating, visiting, but above all else, leaving wilderness unchanged.[46]

The new hiking guides not only abandoned the skills of woodcraft, they also abandoned its masculine rhetoric, supplanting it with language and metaphors that appealed to women and men alike. In part, this reflected a demographic shift, as more women ventured into wilderness in the 1970s, and carved a broader role for themselves in American society as a whole. But equally important, the ebb of frontier rhetoric reflected the decline of the masculine ideal that had been so potent in the early twentieth century. In the wake of the social and cultural upheavals of the 1960s and early 1970s, the image of the woodsman exercising dominion over the natural world with ax in hand struck a discordant note in a society preoccupied with the specter of the atomic bomb, Vietnam, and the unanticipated environmental consequences of the modern economy. For many Americans, particularly those involved in the 1960s social movements, the domineering language of the woodsman no longer offered solace. The changed language of the wilderness experience recast the promise of wilderness—for challenge, self-realization, and escape—in ways that emphasized minimizing one's impact on wilderness, and by analogy, the environment as a whole.[47]

The minimal-impact recreation ethic arrived along with a wave of literature, consumer goods, and marketing campaigns aimed directly at the growing backpacking market in the 1970s. The old woodcrafter's criticism of the consumer economy seemed to disappear in the face of an industry that commanded a $400 million market by the mid-1970s. Sophisticated marketing campaigns for outdoor gear first swept across the pages of the backpacking magazines and journals at this time. Full-page color ads, using a strategy familiar today, framed outdoor gear and models against snow-covered mountain peaks, serene lakes, and streamside camps—the iconography of the National Wilderness Preservation System—advertising a seemingly endless array of outdoor gear. *Backpacker* magazine, founded in 1973, prided itself on a commitment to "spread the new ethic of clean, environmentally aware camping and hiking," but gear reviews quickly became its best-known features, and no doubt manufacturers' advertisements provided a major source of revenue. *Backpacker,* like the rest of the industry, worked to balance a genuine concern for protecting wilderness with the financial imperatives of the market place. Yet, as the advertisements, and the backpacker's ethic as a whole suggest, many modern backpackers were becoming increasingly fluent and comfortable with a powerful language of consumerism: Nature, in the form of wilderness, would stand out, and the equally real nature woven into the sleeping bag, tent, or GoreTex parka would be overlooked, effectively categorized as not-nature.[48]

These changes in wilderness recreation ethics suggest the difficulty the 1970s environmental movement had in fostering a sustained critique of the consumer economy. Even as the environmental movement branched out, taking on energy conservation, air and water pollution, and other problems of the modern industrial society—at least some of the most ardent environmentalists, the backpackers, seemed to give little consideration to the consumer economy. Unlike the woodcraft literature, which engaged both the practice of consumerism and the practice of wilderness recreation, the backpacking literature's zeal for minimal-impact camping eclipsed any dialogue regarding the new ethic's dependence on consumer goods, the waning knowledge of woodcraft, or the shortcomings of the backpacker's wilderness ideal. A few exceptions exist, among which Andrew Abbott's article in *Appalachia* stands out. In 1977, Abbott worried that environmentalists increasingly saw nature only in wilderness. To make his point, Abbott subverted the language of the trail guide to offer a comical, yet sobering, ten-mile hike through urban Los Angeles. At the hike's conclusion, Abbott offered backpackers a warning that surfaced again in the 1990s wilderness debates: "in the 'real' wilderness, we deny categorically that men, and their works that we leave behind, are a part of nature. The more we run back to nature, the more we run from it."[49]

Considered as a whole, however, the backpacking literature gave little attention to the plethora of consumer goods marketed to backpackers or the importance of the new gear to the backpacker's ethic. Most guidebooks offered nothing more than a complacent paragraph-long acknowledgement of the irony that underlay using ever-more modern technology to get back to nature. Hart pointed out that "the stronger our wish to preserve the wild places, the less we can meet them on their own terms; the more sophisticated, civilized, and complex become the gadgets we must bring into them." Similar statements, with the same tone of resignation, can be found sporadically in internal documents at the Sierra Club or the Wilderness Society, in magazine articles in *Appalachia* or *Backpacker,* and in narratives of wilderness adventure. In the 1970s, the new wilderness recreation industry appeared as an inevitable and seemingly unimportant symptom of backpacking's growth. Amidst all the material wealth of the modern consumer economy, the backpacking literature gave voice to little of the social concern that had been so important to the woodcraft literature.[50]

Together, the minimal-impact wilderness ethic, the rise of the wilderness recreation industry, and the wilderness advocates' relaxed policies on wilderness visitation collectively helped forge a powerful coalition in support of the National Wilderness Preservation System in the 1970s. For the minimal-impact wilderness ethic to make sense, however, the meaning of wilderness itself had to change. The modern wilderness movement embraced an aesthetic appreciation of wilderness which denied a working knowledge of nature, implicated the wilderness movement in the consumer economy, and held out wilderness as an ideal to be visited, but above all else, not altered. Many of the promises wilderness recreation offered in the interwar years remained—challenge, restorative experience, and retreat from modernity. Yet the discomfort with consumerism important to woodcraft, and the broader economic critique central to the wilderness movement, had both been diminished—dulling the social critique that once animated the wilderness ideal.

The rise of the minimal-impact wilderness recreation ethic both complements and challenges the recent debate over the place of wilderness in American environmentalism. Hoisting a backpack and returning to wilderness on the hiking trails reveals a sharp shift in the way recreationists have approached wilderness. At the center of this transition is a changing concept of work in nature: the woodsman's ax-swinging wilderness ethic evoked a nostalgia for a hands-on knowledge of nature that gave way to the modern backpacker's skills in insulating themselves from the land. This shift from a prosaic knowledge of the land to a more sublime appreciation of wilderness bolsters Richard White's charge that environmentalists all too frequently appear as a "privileged leisure class"—a charge the woodsmen themselves once sought to evade. The aesthetics of Leave No Trace also fit neatly into William Cronon's charge that environmentalists have elevated wilderness as an ideal, to the detriment of more challenging environmental problems closer to home.[51]

These recent critiques of the modern wilderness ideal, however, make little sense in the context of the interwar years wilderness movement. As the woodcraft literature suggests, and Sutter's study of the interwar years wilderness movement explains, many of the concerns central to the "great new wilderness debate" were precisely the concerns that once empowered the wilderness movement itself. Wilderness coalesced first as a social ideal, not the environmental ideal that distinguishes it today. Despite the power of the recent critiques of wilderness that have roiled the academic and environmental community, these arguments have done little to explain how that interwar years vision of wilderness evolved into its contemporary form. Examining the evolution of wilderness recreation ethics only begins to map out these trails through wilderness history. But the same forces that transformed wilderness recreation played an equally important role in transforming the politics, science, and scope of the wilderness legislation that has expanded the National Wilderness Preservation System to its present-day size of 105.8 million acres. Historians have only begun to blaze these trails through wilderness.[52]

LEAVE NO TRACE

The genesis of Leave No Trace emphasizes the power of the new wilderness ideal. The development of the minimal-impact wilderness ethic in the 1970s laid the groundwork for the codification of a federally-sponsored Leave No Trace program in the 1980s. In the intervening decade, slowing visitation, new scientific research, and the influence of the wilderness recreation industry all helped reshape federal wilderness recreation policies. After consistent growth since World War II, backpacking's popularity appeared to level off in the early 1980s. Despite this temporary lull in visitation, the demographics of wilderness visitors continued to change. A survey revealed the trajectory of the changes taking place. Wilderness visitors were growing older: at one wilderness area the average age jumped from twenty-five to thirty-seven years. The number of women visiting wilderness continued to rise. More wilderness visitors claimed post-graduate education (at one wilderness area the figure jumped from 15 percent in 1969 to 41 percent in 1991). And more wilderness visitors claimed experience from previous wilderness trips, placed more emphasis on the importance of "pristine" wilderness, and belonged to environmental organizations. While surveys do not cover racial or economic profiles, the vast majority of wilderness visitors likely remained white, urban Americans.[53]

For wilderness managers, however, most important was new research that indicated visitation to some western wilderness areas in the early 1980s averaged only one- to two-thirds that of a decade before.[54] Citing the success of minimal-impact camping and predicting a plateau in wilderness visitation, land managers eased up on restrictive backcountry rationing programs.[55] Additional scientific research indicated that the 1970s "carrying capacity" theories had been misleading, which called into question the effectiveness of quotas and gave new emphasis to properly educating wilderness visitors.[56] If all backpackers could be encouraged to embrace minimal-impact ethics and to concentrate their travel and impact in already established sites (reversing an earlier emphasis on dispersing users), the new research indicated that most wilderness would be left relatively unharmed. Wilderness managers adopted a new strategy, called the "limits of acceptable change," which emphasized monitoring ecological conditions in wilderness ecosystems and limiting access only when impact exceeded set standards.[57] A 1985 national conference on wilderness management, drawing on this research, recommended that the four federal agencies overseeing wilderness cooperate in creating a standard, federally-sanctioned education program for wilderness users. The better educated the users, they hoped, the more traffic wilderness could withstand.[58]

In the early 1990s, the public land agencies joined with the National Outdoor Leadership School to start Leave No Trace, a non-profit organization promoting environmentally-sound travel throughout the National Wilderness Preservation System. The wilderness recreation industry soon helped fund the organization: companies like North Face, Gregory Mountain Products, and Mountain Safety Research, which market clothing, backpacks, and stoves respectively, all signed on as Leave No Trace sponsors. Leave No Trace drew heavily on the 1970s wilderness recreation ethic, setting forth a concise set of guidelines in pamphlets, books, and teaching curricula. It urged hikers to comply with six voluntary restrictions on their wilderness behavior. And it explained why: to save wilderness from overuse. For many backpackers the program became a mantra for environmentally-conscious outdoor recreation. Revised wilderness guides helped spread the message. Leave No Trace instructions appeared sewn into the fabric of backpacks. New maps included them alongside the legend. Rick Curtis, author of *The Backpacker's Field Manual* (1998), urged backpackers to "look for the Leave No Trace logo on outdoor equipment and reading material." Even the Boy Scouts joined the program: In 1998, the Scout handbook urged scouts to follow the principles of Leave No Trace and "do your part in protecting our Earth."[59]

Even as the mainstream wilderness community gave official sanction to the Leave No Trace ethic, the ongoing emphasis on wilderness recreation it represented incited the animus of a small contingent of wilderness advocates. In fact, Earth First! revived much of the vision that had distinguished the Sierra Club's National Wilderness Classification Study Committee in the late 1960s, mixing their radical vision of a biocentric wilderness with a new commitment to activism and publicity in the 1980s. Although some Earth First! articles opposed recreation entirely, George Wuerthner, a contributor to the *Earth First! Journal,* expressed a more commonplace frustration with the rules and regulations that confined the modern wilderness experience. Instead of rules, Wuerthner recommended, "limit the technology which has mitigated the natural elements of the land that made it rugged, inhospitable and inaccessible in the first place." In 1989, William Worf, the Forest Service's foremost advocate of the purity policies in the 1960s, helped found Wilderness Watch—an advocacy organization committed to maintaining the ecological integrity of the nation's wilderness system, which has frequently made it one of the few organizations willing to challenge the federal land agencies' commitment to wilderness recreation. And in 1996, Jack Turner wrote a philosophical meditation on wilderness titled *The Abstract Wild,* in which he directly challenged the "'fun hog' philosophy that powered the wilderness-recreation boom for [the past] three decades." Collectively, however, these voices of dissent were overwhelmed by the sound of footsteps, as backpackers continued to visit wilderness in large numbers.[60]

By the mid 1990s, Leave No Trace signs and literature greeted hikers at trailheads and outdoor shops nationwide. In many ways, Leave No Trace represented a logical response by the nation's federal land agencies and the wilderness community to the long-standing problem posed by the popularity of wilderness recreation. It reduced the 1970s minimal-impact camping ethic to an easily digestible code that could be advertised to all wilderness visitors. Yet, to the extent that backpackers actually embraced the notion that they "Leave No Trace," they risked divorcing themselves from their actions as consumers outside wilderness—the actions which had been at the heart of early wilderness recreation ethics. The new ethic focused the attention of backpackers largely on protecting wilderness as a recreational landscape, in turn dismissing larger questions of the modern economy, consumerism, and the environment. In its simplicity, Leave No Trace elided the calculus behind those minimal-impact ethics: the politically pragmatic trade-off between sharp limits on wilderness recreation and the wilderness system's popularity; the exchange of external resources (such as petroleum-fired stoves) for wilderness resources (such as wood-depleting camp fires); or the rise of an aesthetic appreciation of nature in place of a hands-on knowledge of wilderness. All this disappeared into the beguiling notion that anyone can "Leave No Trace." Rather than engaging these questions, the new code helped ally the modern backpacker with the wilderness recreation industry—encouraging backpackers to practice Leave No Trace in the wilderness and keep an eye out for the Leave No Trace logo in the shopping mall. Only in the convoluted logic of modern consumer culture did it make sense that those actions in the shopping mall were the best way to save wilderness beyond.

NOTES

1. Laurel Sorenson, "Camping to Extremes: Here, Eat Your Tinfoil," *Outside,* July/August 1983, 59.

2. Aldo Leopold, "Wilderness as a Form of Land Use," *Journal of Land and Public Utility Economics* 1 (1925): 401.

3. The most important arguments to the debate are compiled in J. Baird Callicott and Michael P. Nelson, eds., *The Great New Wilderness Debate* (Athens: University of Georgia Press, 1998). This compilation offers a cogent reinterpretation of the nationalistic approach to wilderness celebrated in Roderick Nash's thrice-revised *Wilderness and the American Mind,* 3rd ed. (New Haven, Conn.: Yale University Press, 1982).

4. Several historians and environmentalists have given attention to the "packaging" of wilderness. Catton argues this began in the 1930s with Robert Marshall in Theodore Catton, *Inhabited Wilderness: Indians, Eskimos, and National Parks in Alaska* (Albuquerque: University of New Mexico Press, 1997), 135. Other historians observe it later: Ramachandra Guha, "Radical American Environmentalism and Wilderness Preservation: A Third World Critique," in *The Great New Wilderness Debate,* 239; Carl Talbot, "The Wilderness Narrative and the Cultural Logic of Capitalism," in *The Great New Wilderness Debate,* 328; Jack Turner, "In Wildness Is the Preservation of the World," in *The Great New Wilderness Debate,* 622.

5. For an overview of the movement, see Peter J. Schmitt, *Back to Nature: The Arcadian Myth in Urban America* (New York: Oxford University Press, 1969). For an introduction to the movement's preoccupation with wilderness see Nash, *Wilderness and the American Mind,* chapter 9.

6. A growing body of scholarship on twentieth-century tourism lays out the

variety of ways people returned to nature. See Cindy S. Aron, *Working at Play: A History of Vacations in the United States* (New York: Oxford University Press, 1999), chapter 6; John A. Jakle, *The Tourist: Travel in Twentieth-Century North America* (Lincoln: University of Nebraska Press, 1985), chapter 3; Scott Norris, ed., *Discovered Country: Tourism and Survival in the American West* (Albuquerque, N.M.: Stone Ladder Press, 1994); Chris Rojek, *Ways of Escape: Modern Transformations in Leisure and Travel* (Houndmills, Basingstoke: Macmillan, 1993), chapter 1; Hal Rothman, *Devil's Bargains: Tourism in the Twentieth Century American West* (Lawrence: University of Kansas Press, 1998); Marguerite S. Shaffer, *See America First: Tourism and National Identity 1880–1940* (Washington: Smithsonian Institution Press, 2001).

7. George Macaulay Trevelyan, *Walking* (Hartford, Conn.: E. V. Mitchell, 1928), 26–27.

8. The historians Warren Belasco and Allan Wallis offer scholarly accounts of autocamping's origins, tracing autocamping's influence on the rise of the modern motel industry and mobile homes respectively. Warren Belasco, *Americans on the Road,* 2nd ed. (Baltimore: Johns Hopkins University Press, 1997); Allan D. Wallis, *Wheel Estate: The Rise and Decline of Mobile Homes* (New York: Oxford University Press, 1991).

9. George Washington Sears, *Woodcraft,* 14th ed. (New York: Forest and Stream Publishing Co., 1891). Breck specifically cites Sears as the beginning of the genre in Edward Breck, *The Way of the Woods; a Manual for Sportsmen in Northeastern United States and Canada . . . With 80 Illustrations* (New York: G. P. Putnam, 1908), 10.

10. Breck, *The Way of the Woods;* Horace Kephart, *The Book of Camping and Woodcraft* (New York: Outing, 1906), xi; Elmer Harry Kreps, *Camp and Trail Methods* (Columbus, Ohio: A. R. Harding, 1910). Note that walking manuals gave little attention to wilderness camping skills. For a selection of walking manuals, see Bayard H. Christy, *Going Afoot, a Book on Walking* (New York: League of Walkers, 1920); Stephen Graham, *The Gentle Art of Tramping* (New York: D. Appleton & Co., 1926); Elon Huntington Jessup, *Roughing It Smoothly: How to Avoid Vacation Pitfalls* (New York: C. P. Putnam, 1923); Trevelyan, *Walking.*

11. Boy Scouts of America, *The Official Handbook for Boys,* 1st ed. (Garden City, N.Y.: Doubleday, 1912).

12. Ernest Thompson Seton, *The Woodcraft Manual for Boys,* 1st ed. (New York: Doubleday, Page & Co., 1917), 184.

13. Emerson Hough, *Out of Doors* (New York: D. Appleton, 1915), 31; Sears, *Woodcraft,* 4; Stewart Edward White, *Camp and Trail* (New York: Outing Publishing Company, 1907), 26.

14. Seton, *The Woodcraft Manual for Boys,* 216–217.

15. Knowles, *Alone in the Wilderness,* 4, 193. Nash offers an engaging ac-

count of Knowles' popularity and the allegations of fraud in Nash, *Wilderness and the American Mind,* 141–143.

16. Regarding the frontier and masculinity, see Gail Bederman, *Manliness & Civilization: A Cultural History of Gender and Race in the United States, 1880–1917* (Chicago: University of Chicago, 1995), chapters 1 and 5; Richard Slotkin, *Gunfighter Nation: The Myth of the Frontier in Twentieth-Century America* (New York: Harper Perennial, 1992), chapter 1; Theodore Roosevelt, *The Wilderness Hunter, an Account of the Big Game of the United States and Its Chase with Horse, Hound, and Rifle* (New York: The Current Literature Pub. Co., 1893), 24–25. Similar statements are found in many of the early Boy Scouts manuals. This one came from Boy Scouts of America, *The How Book of Scouting* (New York: Boy Scouts of America, 1938), 277.

17. Knowles, *Alone in the Wilderness,* 98; Sears, *Woodcraft,* 3.

18. Leopold, "Wilderness as a Form of Land Use," 404; Benton MacKaye, "An Appalachian Trail: A Project in Regional Planning," *The Journal of the American Institute of Architects* (October 1921). On MacKaye's vision and its connection to the interwar years wilderness movement, see Paul Sutter, "'A Retreat from Profit': Colonization, the Appalachian Trail, and the Social Roots of Benton MacKaye's Wilderness Advocacy," *Environmental History* (1999), 553–577. Robert Marshall, "The Problem of the Wilderness," *Scientific Monthly,* February 1930, 147.

19. This report outlines the state of federal lands, threats posed by tourism, and potential management responses. It is, in effect, the government's reckoning with the surge in interwar years tourism. Joint Committee on Recreational Survey of Federal Lands, *Recreation Resources of Federal Lands* (Washington, D.C.: GPO, 1928), 90. Bob Marshall voiced similar concerns in Marshall, "The Problem of the Wilderness," 46–147. Paul Sutter, *Driven Wild: How the Fight against Automobiles Launched the Modern Wilderness Movement* (Seattle: University of Washington Press, 2002).

20. This extensive survey also reveals a great deal of variety in the responses from the seven different wilderness areas. I have focused on the common characteristics of wilderness visitors. Outdoor Recreation Resources Review Commission, "Wilderness and Recreation—a Report on Resources, Values, and Problems" (Washington, D.C.: GPO, 1962), appendix B. David R. Brower, ed., *Going Light—with Backpack or Burro* (San Francisco: Sierra Club, 1951), xi.

21. For an excellent account of the Echo Park controversy, see Mark W. T. Harvey, *A Symbol of Wilderness: Echo Park and the American Conservation Movement* (Albuquerque: University of New Mexico Press, 1994). The environmental historiography includes numerous accounts of the subsequent fight for wilderness. For a start, see Stephen Fox, *John Muir and His Legacy: The American Conservation Movement* (Madison: University of Wisconsin, 1981), 281–289; Nash, *Wilderness and the American Mind,* chapter 12; and Hal K. Rothman, *The*

Greening of a Nation? Environmentalism in the United States since 1945 (New York: Harcourt Brace, 1998), chapter 2.

22. Many letters regarding overuse can be found in the papers of the Sierra Club Outing Committee. For instance, see Philip Hyde to Stewart Kimball, Chairman, Outing Committee, SC, 1 November 1956, folder "Outing Committee, Impact Study, 1956–1975," Box 43:3, Sierra Club records, BANC MSS 71/103 c, Tile Bancroft Library, University of California, Berkeley [hereafter, SCR]; John Bird, "The Great Wilderness Fight," *Saturday Evening Post,* 8 July 1961, 74–76; David Brower, "Wilderness Is for People, Too!" *Sierra Club Bulletin,* February 1962, 3. For a brief history of wilderness management, see Roderick Nash, "Historical Roots of Wilderness Management," in *Wilderness Management,* ed. John C. Hendee (Golden, Colo.: North American Press, 1990), 38–40.

23. Commentary on the paradox inherent in wilderness can be found in John C. Hendee, *Wilderness Management,* 15; and Mark Woods, "Federal Wilderness Preservation in the United States: The Preservation of Wilderness?" in *The Great New Wilderness Debate,* 134–146.

24. Director Richard J. Costley, Division of Recreation, "Proposed Regulation of the Secretary of Agriculture Governing the Administration of National Forest Wilderness," 12 July 1965, folder "December 1965 Meeting," Box 1, Forest Service Records, Record Group 95, Entry 174/A1, National Archives II, College Park, Md., 3 [hereafter, FSR].

25. Costley, "Comments on the Review Draft of 'Outdoor Recreation in the National Forests,'" 30 March 1964, folder "Outdoor Recreation in the NFs," Box 2, FSR, 8.

26. Costley, "Proposed Regulation of the Secretary of Agriculture Governing the Administration of National Forest Wilderness," 12 July 1965, FSR, 6.

27. These visitation statistics are for the national forests only and can be found in Hendee, *Wilderness Management,* 380.

28. Chief William A. Worf, Recreation and Lands, Region 1, USFS to Brock Evans, Northwest Representative, SC, March 1972, folder "Wilderness Correspondence, 1972–1973," Box 6, MSS 2678-1, Sierra Club Pacific Northwest records, University of Washington Library, Seattle [hereafter, SCPNWR]. United States Forest Service, "Recreation Management, Region One Supplement No. 42," February 1972, folder "RARE and FS Manual," Box 4, MSS 2678-6, SCPNWR, 1.

29. United States Forest Service, "Title 2300 Management—Chapter 2350—General Forest Environment Areas," 1972, Document No. 89, Arthur Carhart National Wilderness Training Center Records, Missoula, Mont.

30. Worf, et al., "Wilderness Policy Review Committee Report," 19 May 1972, folder "Wilderness Primitive Areas," Region 1 Forest Service Archives, Missoula, Mont.

31. Recreation Staff Officer W. E. Ragland, Snoqualmie National Forest, "The Place for Recreation Backcountry on the Continuum of Forest Uses," 30 December 1971, folder "Wilderness," Box 5, MSS 2678-1, SCPNWR.

32. See Paul W. Hirt, *A Conspiracy of Optimism: Management of the National Forests since World War Two* (Omaha: University of Nebraska Press, 1994), especially chapters 6 and 10. Worf to Evans, 9 March 1972, SCPNWR, 2.

33. Statements supporting limits can be found sporadically in the Wilderness Society and Sierra Club files. For instance, Stewart Brandborg, The Wilderness Society, "Some Thoughts Regarding the Carrying Capacity of Wilderness Areas," 26 August 1963, folder "Stewart M. Brandborg," Box 3:100, Wilderness Society Records, Western History and Genealogy Department, Denver Public Library, Denver, Colo. [hereafter, TWS]; Garrett Hardin, "We Must Earn Again for Ourselves What We Have Inherited: A Lesson in Wilderness Economics," in *Wilderness: The Edge of Knowledge,* ed. Maxine E. McCloskey (San Francisco: Sierra Club, 1969), 263–265; Paul Petzoldt, *The Wilderness Handbook* (New York: Norton, 1974), chapter 13; David Brower, ed., *The Sierra Club Wilderness Handbook,* 2nd ed. (New York: Sierra Club/Ballantine, 1971), 147; and Theodore J. Kaczynksi, "Letter to the Wilderness Society," 26 February 1969, folder "California: Correspondence, 1969," Box 7:107, TWS.

34. Chairman Francis J. Walcott, Wilderness Classification Study Committee, "Sierra Club Policy for Wilderness Management and Preservation," 25 November 1969, folder "Wilderness Management," Box 6, MSS 2678-9, SCPNWR, 2.

35. Walcott to John Tuteur, Jr., folder "Wilderness Classification Study Committee, 1970," 12 October 1970, Box 222:26, SCR; Walcott to Shelley McIntyre, folder "Wilderness Correspondence, 1972–1973," 18 December 1972, Box 6, MSS 2678-1, SCPNWR.

36. Evans to Holly Jones, Northwest Chapter Chairman, Wilderness Committee Chairman, 10 November 1972, folder "Wilderness Correspondence, 1972–1973," Box 6, MSS 2678-1, SCPNWR.

37. Evans to Harry Crandell, Director of Wilderness Reviews, TWS, 15 December 1972, folder "Wilderness management: 1972," Box 2:5, Papers of Harry Crandell, Western History and Genealogy Department, Denver Public Library, Denver, Colo.

38. Dick Sill, Impact Dialogue Committee, SC, "Memo Re: Wilderness Entry Permits," folder "Wilderness Committee," 12 June 1972, Box 53:38, SCR.

39. Statistics on the growth of the wilderness system are accessible through the database available at http://nwps.wilderness.net/advsearch.cfm (accessed April, 2002).

40. Sill, "Memo Re: Wilderness Entry Permits," 12 June 1972, SCR.

41. For instance, see William Cronon, "The Trouble with Wilderness; or,

Getting Back to the Wrong Nature," in *Uncommon Ground: Rethinking the Human Place in Nature,* ed. William Cronon (New York: Norton, 1995), 69–90.

42. John Hart, "Wilderness Adventure," *Sierra Club Bulletin,* March 1972, 7.

43. Stewart Brandborg's handwritten notes can be found on Harry Crandell, "Memo to All Reviewers Re: Initial Draft—Wilderness Management Policies," 27 April 1972, folder "Wilderness Management Policy: 1972," Box 2:1, TWS, 32. John B. Nutter, "Towards a Future Wilderness: Notes on Education in the Mountains," *Appalachia,* December 1973, 89. The following are three of many books published in the early 1970s on backpacking. John Hart, *Walking Softly in the Wilderness* (San Francisco: Sierra Club Books, 1977); Harvey Manning, *Backpacking, One Step at a Time* (New York, N.Y.: Vintage Books, 1972); Petzoldt, *The Wilderness Handbook.*

44. Hart, *Walking Softly in the Wilderness,* 35, 230; Brower, ed., *The Sierra Club Wilderness Handbook,* 55. Backpacking articles and books frequently dismissed "old style" woodcraft specifically, even referring to the Boy Scouts on occasion: Hart, *Walking Softly in the Wilderness,* 36; Hart, "Wilderness Adventure," 7; Wilfred Kerner Merrill, *The Hiker's and Backpacker's Handbook* (New York: Winchester Press, 1971), 47; Petzoldt, *The Wilderness Handbook,* 57; Ruth Rudner, *Off and Walking* (New York: Holt, Rinehart and Winston, 1977), 30; and Guy Waterman and Laura Waterman, "The Clean Camping Crusade," *Appalachia,* June 1979, 90.

45. Gear descriptions drawn from Collin Fletcher, *The New Complete Walker,* 2nd ed. (New York: Knopf, 1974). On foam pads, see Petzoldt, *The Wilderness Handbook,* 57. Regarding GoreTex, see "Rainwear," *Backpacker* 7 (1979), 69+.

46. Leopold, *A Sand County Almanac and Sketches Here and There* (New York: Oxford University Press, 1949; repr., 1968), 166. On this point, a comparison between the 1933 Boy Scout Manual's wonderfully detailed instructions on making a pack from scratch (which included which types of wood to select for the thwarts versus the ribs) gave way to the 1979 Boy Scout Manual's imperative "buy the best . . . you can afford." Boy Scouts of America, *The How Book of Scouting,* 320; Boy Scouts of America, *Official Boy Scout Handbook,* 9th ed. (New York: Boy Scouts of America, 1979), 64–68. Historians Susan Strasser and Shosana Zuboff have noted sharp shifts in the repertoire of skills used at home and work during these same years, as changing technologies insulated homemakers and laborers from a hands-on knowledge of nature. Susan Strasser, *Waste and Want: A Social History of Trash* (New York: Metropolitan Books, 1999); Shoshana Zuboff, *In the Age of the Smart Machine: The Future of Work and Power* (New York: BasicBooks, 1988), chapter 2.

47. While men continued to outnumber women in the wilderness, female

visitation did increase between the 1970s and 1990s. Alan E. Watson, David N. Cole, and Joseph W. Roggenbuck, "Trends in Wilderness Recreation Use Characteristics," in *Proceedings of Outdoor Recreation and Tourism Trends Symposium,* ed. Jerrilyn Lavarre Thompson (St. Paul, Minn.: University of Minnesota, 1995). On the 1960s, see David Farber, *The Sixties: From Memory to History* (Chapel Hill: University of North Carolina Press, 1994); Nash, *Wilderness and the American Mind,* 251–255.

48. On the gear market, see Hart, *Walking Softly in the Wilderness,* 15. Regarding *Backpackers* mission, see Gary Braasch, "You & Your Head," *Backpacker* 2 (1974), 98; William Kemsley Jr., "Editorial: A New Voice for Backpackers," *Backpacker* 4 (1976), 5. Jennifer Price explores the divisions between nature and not-nature in *Flight Maps: Adventures with Nature in Modern America* (New York: Basic Books, 1999). In the same ways the consumer economy obscures environmental costs, it also obscures labor costs. On labor, see Talbot, "The Wilderness Narrative and the Cultural Logic of Capitalism"; and Richard White, "'Are You an Environmentalist or Do You Work for a Living?': Work and Nature," in *Uncommon Ground,* 171–185.

49. Andrew Abbott, "Fleeing Wilderness," *Appalachia,* June 1977. See also the introduction to Manning, *Backpacking, One Step at a Time.*

50. Hart, *Walking Softly in the Wilderness,* 36.

51. White, "Are You an Environmentalist or Do You Work for a Living?"; Cronon, "The Trouble with Wilderness."

52. Sutter, "'A Blank Spot on the Map': Aldo Leopold, Wilderness, and U.S. Forest Service Recreational Policy, 1909–1924," *Western Historical Quarterly* 29 (1988), 187–214.

53. David Cole, "Wilderness Recreation Use Trends, 1964–1990" (Washington, D.C.: Intermountain Research Station, Forest Service, 1996). Wilderness survey data is available but remains inexact. Important variables such as race, economic standing, and occupation are unavailable for long-term surveys. This data compared 1960s surveys to late 1980s and early 1990s surveys. Watson, Cole, and Roggenbuck, "Trends in Wilderness Recreation Use Characteristics." For more information on the changes in wilderness demography (and the accuracy of surveys) see Alan E. Watson, "Wilderness Use in the Year 2000: Societal Changes That Influence Relationship with Wilderness," in *Wilderness Science in a Time of Change,* ed. David N. Cole (Ogden, Utah: Rocky Mountain Research Station, Forest Service, 2000), 55.

54. Hendee, *Wilderness Management,* 282–383. More recent research indicates that this lull in wilderness visitation was only brief. For more detailed analysis see Cole, "Wilderness Recreation Use Trends, 1964–1990."

55. Nash, "Historical Roots of Wilderness Management," 40.

56. Society of American Foresters, "Report of the Society of American

Foresters' Wilderness Management Task Force" (Bethesda, Md.: Society of American Foresters, 1989), 19.

57. J. Douglas Wellman, *Wildland Recreation Policy: An Introduction* (New York: John Wiley, 1987), 249.

58. Wilderness Research Center, "Wilderness Management: A Five-Year Action Program" (Moscow, Idaho: University of Idaho, 1985).

59. Jeffrey L. Marion and Scoff E. Reid, "Development of the U.S. Leave No Trace Program: An Historical Perspective" (Boulder, Colo.: Leave No Trace, 2001). Rick Curtis, *The Backpacker's Field Manual: A Comprehensive Guide to Mastering Backcountry Skills* (New York: Three Rivers Press, 1998), 100. Boy Scouts of America, *Boy Scout Handbook,* 11th ed. (Irving, Texas: Boy Scouts of America, 1998), 219.

60. George Wuerthner, "Managing the Wild Back into Wilderness," *Earth First!,* 21 December 1986. See http://www.wildernesswatch.org (accessed April 2002). Jack Turner, *The Abstract Wild* (Tucson: University of Arizona Press, 1996), 87.

Mark P. Jenkins

Wilderness Preservation Argument 31 (2008)

The Psychotherapy at a Distance Argument

Michael P. Nelson's "An Amalgamation of Wilderness Preservation Arguments" "summarize[s] in one place the many traditional and contemporary arguments proffered on behalf of 'wilderness.'"[1] An original contribution to *The Great New Wilderness Debate,* "an expansive collection of writings defining wilderness," the essay attempts to "integrate and reconcile" a host of "disparate compilations," to "rename and recategorize" earlier arguments, and to add "hitherto unexplored wilderness preservation rationales" (154–55).[2] It is, then, a decidedly critical amalgamation—and a decidedly successful one, rendering a significant and distinctive service to the broader environmental philosophy community.

Nelson arranges a total of thirty arguments for preserving wilderness along a rough continuum according to the nature of value each appeals to, "from narrowly instrumental, egocentric, and anthropocentric values" at one end "to broader social, biocentric, and even intrinsic values" at the other (155). The essay makes no attempt to formulate a strict definition of wilderness, and there is no guarantee that each of the arguments aims to preserve quite the same thing; indeed, as Nelson points out, the very concept of wilderness "is correctly subject to intense debate" (155). A number of the arguments are exceedingly well known, requiring little elaboration beyond their titles (e.g., "The Hunting Argument," or, switching sides, "The Animal Welfare Argument"), while others (say, "The Salvation of Freedom Argument") appear relatively obscure. Nelson admirably captures arguments for wilderness preservation, familiar or not, and, by extension, the causes and constituencies behind them. He does so admirably but not exhaustively, for at least one important argument appears to have escaped Nelson's attention, "The Psychotherapy at a Distance Argument" (hereafter Argument 31), which it is the purpose of this essay to present.[3] In the following sections I attempt to establish the force and legitimacy of this new argument largely contrastively (or through a process of elimination) by comparing it with other arguments, dutifully cataloged by Nelson, with which it might be confused or that might be thought to somehow contain it.

By "psychotherapy" I mean nothing especially technical or ambitious, nothing beyond the treatment of psychic discomfort, particularly that occasioned by the exigencies of modern life, whether in the form of anxiety or stress or mild depression, by psychological means, in this case, focused attention on wilderness. By "at a distance" I mean simply to mark the fact that such psychic benefits in no way depend on any direct contact with wilderness, past or present. The argument is straightforwardly and unapologetically instrumental: wilderness is a crucial psychotherapeutic instrument, useful in countering certain pervasive threats to mental health posed by the demands of modern life, and so ought to be preserved.

By "wilderness" I mean any physical location on earth not screwed up by humans. I take this definition to be consistent with the spirit and to merely update the letter of the Wilderness Act's invocation of "an area where the earth and its community of life are untrammeled by man," thereby shifting the interpretive challenge from "untrammeled" to "screwed up."[4] Certainly, the difficulty of this challenge should not be underestimated; intelligent people of good faith argue interminably, for example, about whether this or that (or, for that matter, any) federally protected wilderness area in the United States is truly wilderness. For this reason let me simply stipulate that, at least as far as developing Argument 31 is concerned, the Arctic National Wildlife Refuge (ANWR) counts as wilderness, being pretty uncontroversially a physical location on earth not screwed up by humans.[5]

In no way am I claiming originality for Argument 31, which I take to be basically of a piece with the idea that wilderness has value even for those who may never experience it. In a memorable passage from *Desert Solitaire,* Edward Abbey likens this value, really the value of this value, to hope:

> A man could be a lover and defender of the wilderness without ever in his lifetime leaving the boundaries of asphalt, powerlines and right-angled surfaces. We need wilderness whether or not we ever set foot in it. We need a refuge even though we may never need to go there. I may never in my life get to Alaska, for example, but I am grateful it's there. We need the possibility of escape as surely as we need hope; without it the life of the cities would drive all men into crime or drugs or psychoanalysis.[6]

Notwithstanding the irony that, in my view, wilderness might actually be a tool of psychoanalysis, I think Abbey's message is largely the message of Argument 31: the very existence of wilderness confers benefits even at a distance, and those benefits, far from incidental, may be vital for coping with modern life. As it happens, Abbey made it to Alaska, but I have not, and there's no guarantee I will, yet the Arctic National Wildlife Refuge, as a paradigm of wilderness, possesses value and ought to be preserved not only for the refuge it provides caribou but for the refuge it provides my own troubled mind.

SIMILAR INSTRUMENTAL ARGUMENTS

As mentioned above, Nelson's thirty arguments come loosely arranged according to the flavor of value they appeal to, from instrumental to intrinsic. Since I claim that Argument 31 is straightforwardly instrumental, one might expect that if any of the arguments in Nelson's amalgamation come close to capturing its content, they would tend to occur earlier in his essay rather than later, and such is indeed the case. In fact, three of the first eight arguments Nelson presents appear to have at least some affinity with Argument 31: number 1, "The Natural Resources

Argument"; number 4, "The Service Argument"; and number 8, "The Mental Therapy Argument." I discuss each briefly in turn.

As Nelson is quick to point out, his first argument, the natural resources argument, "is the most narrowly anthropocentric, instrumental, and simpleminded preservation argument that one could advance" (156). It goes like this: "Certain designated wilderness areas are great repositories of a wide variety of natural resources and we humans can render our future more secure by preserving these resource reserves" (156).[7] So what about the therapeutic value of wilderness, the improvement in mood and outlook that results when, stuck in the second or third cycle of the average arterial, six-way set of stoplights during suburban rush-hour traffic, one's focus turns to ANWR, to the fact that it exists, is up there somewhere, and is decidedly not screwed up like this intersection? Is this a resource to be preserved in the same way that oil and bauxite and fresh water are resources to be preserved? I think not, for at least two reasons.

First, the therapeutic value of wilderness attaches to wilderness as a whole rather than to any of the specific resources that may repose there. It is the existence of this place, ANWR, a wilderness, a world that, whatever it contains, is not yet screwed up by humans, and not the existence of any particular ANWR resource, whether wolves, wolverines, or oil, that I rely on to calm my frayed nerves when I am assaulted by successive between-innings pharmaceutical ads questioning my happiness, continence, and virility. Second and perhaps more tellingly, however, even if ANWR's capacity to calm nerves four thousand miles away could be plausibly construed as a natural resource, it cannot be construed as the exhaustible resource the natural resources argument seems to require. Much of Nelson's discussion, in fact, concentrates on a certain conceptual instability he detects in the natural resources argument's juxtaposition of wilderness and resources, inasmuch as "it would seem that if we use an area as a goods resource then we are no longer entitled to call it wilderness" (156). But the sort of instrumental value provided by wilderness in Argument 31 differs from the sort of value provided by the goods of Argument 1. However much you use wilderness to calm your ragged nerves, the resource you tap into, being decidedly renewable or, put differently, being inherently nonextractive, is quite unlike the sorts of resources implicit in the natural resources argument.

The second argument that might be thought to bear more than a passing resemblance to Argument 31 is Nelson's Argument 4, "The Service Argument," a not-so-distant cousin of Argument 1: "In addition to the natural resource goods provided by certain putative wilderness areas, innumerable and invaluable services are said to be provided by many of these areas as well" (160). Nelson identifies the ability of unbroken forests to cleanse the air and of oceans to moderate temperatures as "ecological services" that are "vital to our continued existence," in virtue of which, we may argue, the wilderness areas that provide them should be preserved (160). Now it seems to me that the psychotherapy afforded by wilderness is also a service vital to our continued existence; might not Argument 31, then, turn out to be merely a recapitulation of Argument 4? The answer, I think, is no.

Just as the sense of resources in the natural resources argument differs in kind from the sense in Argument 31, so does the sense of service in Argument 31 differ from that in the service argument. Most basically, while the processes of converting carbon dioxide into oxygen and cooling the atmosphere via evaporation are certainly vital services provided to humans by wilderness and, moreover, services provided at a distance, they are clearly processes that would continue even if there were no humans on earth; they are, then, only coincidentally services in no way dependent on and essentially indifferent to the needs of humans. By contrast, the process of mentally focusing on wilderness in order to access the therapy it provides fundamentally depends upon humans and their need for psychic comfort. In short, while both Argument 31 and Argument 4 depict services provided to humans by wilderness, the former involves a process in which humans must actively participate, while the latter involves a process in relation to which humans are effectively passive.

There is a further reason for separating Argument 31 from the service argument. As Nelson points out, "wilderness is indeed a sufficient condition for the performance of these services, but it does not seem to be a necessary one. That is, these services are not unique to uninhabited or uncultivated places; they are performed by non-wilderness ecosystems as well" (160). In support of this point he cites the superior oxygen-generating power of industrial tree farms over old-growth forests.[8] Clearly, certain vital services provided by wilderness may also be provided, even better provided, by nonwilderness sources. The question, then, becomes, Can the vital service that is the therapeutic effect of wilderness on those paying the day-to-day psychic toll exacted by urban blight, suburban congestion, and exurban sprawl be provided by some nonwilderness source?

I don't think it can, basically on what one might call etiological grounds. As I see it, wilderness is a necessary condition for the performance of this psychotherapeutic service because attention to wilderness is precisely the conceptual counterweight to the root causes underlying the psychopathology it treats, namely, the stress, anxiety, and mild depression occasioned by airport security lines, mobile paper shredders, motor vehicle departments, smog alerts, terror alerts, Amber alerts, leaf blowers, eleven varieties of Colgate toothpaste, and "press 0 for an operator at any time." Focusing the mind on some location unscrewed up by humans helps relieve the tension caused by locations screwed up by humans; focusing the mind on some undeveloped location helps relieve the tension caused by overdeveloped locations; focusing the mind on some unpopulated location helps relieve the tension caused by overpopulated locations; focusing the mind on some natural location helps relieve the tension caused by artificial locations; focusing the mind on some wild location helps relieve the stress caused by tamed locations; in sum, focusing the mind on some wilderness location helps relieve the tension caused by civilization. As Abbey says, "we need a refuge." Prozac is not that refuge; "Life Is Good!" is not that refuge; a week at the shore is not that refuge; Hollywood's latest hit is not that refuge; wilderness is that refuge, for the very idea of wilderness acts as a balm on the psychic scrapes and bruises that inevitably come with navigating this modern world.

To this point I have argued that despite certain similarities neither the natural resources argument nor the service argument can replace my argument for the preservation of wilderness owing to its value in (and as) mental therapy. Surely it must be admitted, however, that the chances of Nelson's eighth argument successfully subsuming Argument 31 look pretty good; after all, he actually calls it "The Mental Therapy Argument." But despite its promising title, dismissing this argument as a rival to Argument 31 makes for comparatively quick work.

"Wilderness experiences can be psychologically therapeutic and can even significantly help treat psychologically disturbed persons" (164). So goes the mental therapy argument. Nelson cites a number of people who "have argued that civilization represses, frustrates, and often breeds unhappiness and discontent in humans that can best be alleviated by periodic escapes . . . to wilderness" (164). While I am pleased to join those prescribing wilderness experience as an antidote to civilization's psychic toxins, the key word serving to definitively differentiate my psychotherapeutic value at a distance argument from Nelson's mental therapy argument is "experience": as with jobs at Taco Bell, so with Argument 31, no experience required. Where Argument 8 suggests that "over-stressed executives can and do often benefit from an occasional dose of wilderness experience," Argument 31 suggests that those same executives can and do often benefit from an occasional dose of wilderness, period, albeit a psychologically self-administered dose requiring no physical contact. "At a distance" says it all.

Nelson goes on to raise two potential difficulties for the mental therapy argument, neither of which, it seems to me, applies to Argument 31. He writes: "None of the proponents of the mental health argument for wilderness preservation explain exactly how the existence of wilderness areas contribute to sanity. But even if we did grant that the existence of designated wilderness areas is a sufficient condition for instilling mental health, is it a necessary condition?" (165). I have already made the case, in connection with the preceding service argument, that wilderness is indeed a necessary condition so far as Argument 31 goes, inasmuch as a lack of wilderness in the form of various psychic pressures exerted by places screwed up by humans constitutes the itch for which, quite naturally, the thought of wilderness provides the scratch. I think I have also addressed the first issue, the question of exactly how wilderness areas contribute to sanity, at least if we assume that "contributing to sanity" in this context refers more to something like "restoring subjective well-being" than "eliminating schizophrenic delusions." ANWR effectively rebuts the argument posed by cage fighting and interest-only mortgages and the status of Pluto and acid reflux and five-bladed razors and UN sanctions; that is to say, insofar as just thinking of these sorts of things can damage one's psyche, just thinking of ANWR can help repair the damage.

INTRINSIC VALUE?

I now want very briefly to turn from arguments clearly steeped in the instrumental value of wilderness to consider the very last entry in Nelson's amalgamation, "The Intrinsic Value Argument," not because I think Argument 31 can plausibly be interpreted as in any way appealing to the intrinsic value of wilderness but rather because the language Nelson employs in introducing Argument 30 may seem, rather peculiarly, to echo my own in Argument 31. Here is that language: "Many, many wilderness boosters claim that simply knowing that there exist designated wilderness areas, regardless of whether or not they ever get to experience such areas, is reason enough for them to want to preserve them. For these people 'wilderness' is valuable just because it exists, just because it is. . . . 'Wilderness,' then, is said to possess intrinsic value" (191). Now, just agreeing on the concept of intrinsic value, let alone locating that concept in nature, has proven to be as difficult a task as it has been crucial to a generation or more of environmental philosophers. As J. Baird Callicott writes, "How to discover intrinsic value in nature is the defining problem for environmental ethics."[9] Unfortunately, I have nothing to contribute to the problem's solution, although I do say a little more about intrinsic value in the concluding section. Here I want simply, by way of clarification, to locate the link Nelson forges between intrinsic value and people "simply knowing that there exist designated wilderness areas, regardless of whether or not they ever get to experience such areas," with respect to the instrumental value Argument 31 apparently ascribes to that very same knowledge.

Both Argument 31 and Argument 30 claim that knowledge of wilderness is sufficient to warrant its preservation. Argument 31, however, claims that the value in virtue of which wilderness should be preserved resides in that knowledge itself. The knowledge that there exist designated wilderness areas has value as a psychotherapeutic tool. Were we to lack wilderness we would lack a useful sort of knowledge, the sort we can use to improve our mental well-being in certain contexts. By contrast, Argument 30 claims that the value in virtue of which wilderness should be protected resides not in the knowledge of wilderness but in wilderness itself. Although both arguments claim that one who simply comes to know about wilderness will thereby recognize the value of wilderness and want to protect it, regardless of whether or not that knowledge comes from direct experience with wilderness, only Argument 31 cites the usefulness of that knowledge as its ultimate rationale for protection, while the intrinsic value argument sees that knowledge as merely reflecting such values as wilderness areas possess "in and of themselves."

FURTHER CONSIDERATIONS

Having differentiated the psychotherapy at a distance argument from other instrumental arguments for wilderness preservation, and having disambiguated certain claims made about intrinsic value and its relation to "simply knowing" about wilderness, and thus having hopefully presented a robust case for Argument 31's amalgamation-worthiness, I want to briefly raise and respond to four considerations that might be thought to undercut its plausibility or to blunt its force. No question but that much more could be said about all four; I hope to say enough in each case to keep Argument 31 persuasive.

First, who says we need actual wilderness to provide the psychotherapeutic benefits I have described? Won't imaginary wilderness do as well? This rejoinder may seem initially promising. After all, we regularly seek psychic solace in all kinds of fantasy worlds, so why not a fantasy world of wilderness? Go ahead and destroy ANWR, take the refuge and the wildlife right out of it, I can always conjure up some similar place in my imagination. And won't my imaginary wilderness do just as good a job of soothing my nerves the next time I'm stuck following a loose-gated gravel truck through eleven miles of single-lane interstate highway construction as ANWR would have done in similar straits? I think not. To begin with, there is Abbey's point, a point unstressed until now, that one of the main reasons "we need wilderness whether or not we ever set foot in it" is because it provides "the possibility of escape." In Abbey's view, significant therapeutic work is done by the fact that whether or not I ever do decide to take off for ANWR, leaving behind the crack dealers and scam artists and panhandlers and worse who make the two-block walk through my "transitioning" neighborhood to buy a newspaper a regular occasion for stress management, I could so decide. I probably won't, but I could. And it is this possibility—again, for Abbey, akin to hope—that underlies the attitude improvement afforded by wilderness. Take away the possibility of escaping to wilderness by taking away the wilderness, and one takes away the therapeutic value of wilderness as well. But there is also another, even more basic point to be made here: reducing actual wilderness will almost certainly tend to ratchet up the very anxiety, stress, and depression that the existence of wilderness palliates.

Second, I claim (as does Abbey) that one need never have visited wilderness to enjoy the psychotherapeutic relief that thinking of wilderness can provide, but this may seem a stretch. After all, what will the content of someone's thoughts be who has never experienced wilderness? Won't such thoughts necessarily be imaginary and so psychotherapeutically impotent in the way I have just suggested? If all my experience is of locations screwed up by humans, how can I formulate a mental picture of a location not screwed up by humans? While perhaps not quite frivolous, such questions seem pretty thin. How do virgin adolescents conceptualize coitus for purposes of relieving stress? They do so with the assistance of testimony, the Internet, literary examples, the latest PG-13 movie, and so forth. I am tempted to say that, in (only somewhat) similar fashion, I learned everything I know about wilderness, certainly everything necessary to form consolatory wilderness images, from Walt Disney's films *Beaver Valley* and *The Living Desert,* one or the other of which I was shown during every rain-canceled outdoor recess from first through sixth grades.[10] A single Sierra Club coffee table book, Shackleton's or Stanley's personal narrative, Discovery or OLN programming,

March of the Penguins or *Grizzly Man,* an "anwr photos" Web search, natural history museums, botanical gardens, zoos, aquariums all provide resources, no matter how tainted by civilization, for bootstrapping oneself into a position to be able to form a therapeutic idea of wilderness; once again, no experience required.

Third, Argument 31 is elitist in at least the following two senses: one, the therapeutic value I claim wilderness possesses is a value that can only be appreciated and so can only be accessed by some fortunate elite possessing a certain intellectual capacity or aesthetic sensibility or ecological sensitivity; and two, Argument 31 has no resonance in the Third World, maybe not even among the lower economic strata of the First. I want to flatly dispute the first sense of Argument 31's alleged elitism. We can all be, in Roderick Nash's wonderful phrase, "intellectual importers of Alaska's wilderness."[11] Perhaps those pushing this line of criticism have also in mind Nelson's Argument 10, "The Inspiration Argument," whereby "putative wilderness areas are important to maintain because they provide inspiration for the artistically and intellectually inclined" (167). Although I do not for a moment doubt the power of wilderness to catalyze genius, neither of Argument 31's preconditions—a mental malaise tied to the exigencies of contemporary civilization and an idea of some location not screwed up by humans—would appear particularly difficult to satisfy.

As to the alleged inapplicability of Argument 31 to the Third World, I am rather more circumspect. Perhaps, in truth, Argument 31 contains three, not two, preconditions: civilization-induced psychopathology, an idea of wilderness, and, going back to Abbey, some real possibility of escape, of swapping civilization for wilderness. Who but a privileged American would commit the obvious blunder of ascribing that possibility to all? As geographically close to genuine wilderness as people of Rangoon or Kinshasa or Jacksonville, Florida, may live, their possibilities of escape are only as real, so this line of thought goes, as their credit card spending limits. Indeed, some might insist that if these people had the means to warrant issuing them credit cards, they wouldn't be so anxious and stressed and depressed in the first place, and so they wouldn't need wilderness therapy. One tempting response, that psychic comfort is inversely proportional to the number of credit cards in one's wallet, as it is to the number of keys on one's key ring, will itself invite charges of elitism.

Fourth, and with apologies to Tolstoy, how much wilderness does a man need?[12] Put another way, how much wilderness is actually required to fulfill its therapeutic promise? On this question I am sure there is much room for disagreement, even among those firmly persuaded by Argument 31. Just for starters, it is the type of question that invites worries concerning the Sorites paradox, for we may be reasonably certain of our ability to identify both sufficient and insufficient amounts of wilderness for purposes of Argument 31 yet be completely incapable of specifying the point at which sufficient wilderness becomes insufficient.[13] The designated wilderness portion of ANWR now stands at roughly eight million acres. If we start removing acreage, at what point will it cease to be viable wilderness? It turns out to be almost impossible to say (or at least to agree upon). What can be said with some assurance is that more is better.

It might be argued that we already lack the wilderness necessary to provide psychotherapy. Although I would not go that far, a point raised in regard to the first consideration above bears repeating: the eradication of wilderness exacerbates just the ills that wilderness is meant to address. "Pave paradise, put up a parking lot," and you can bet that parking lot will not be responsible for lost or stolen items, will charge you the maximum for a lost ticket, will not replace burned-out security lights, and will have an automatic payment machine that eats your ticket five minutes after the attendant leaves for the day; that is, you can bet it will add to rather than subtract from your everyday cares, these being, of course, precisely the sort of cares that focusing on a wilderness paradise can ease. It seems reasonable, then, to think that at some point the capacity of whatever wilderness remains to assuage the anxiety caused by previous wilderness destruction will be exhausted. But besides this, there is Abbey's requisite possibility of escape to consider. ANWR's total of nineteen million acres can accommodate a lot of escapist fantasies from the lower forty-eight, but what if the only remaining wilderness is the fifty thousand acres of Oregon's Cascade-Siskiyou National Monument? In this case, it seems to me, the therapeutic value of wilderness—*per hypothesi* all the wilderness there is—might be close to zero, basically because the possibility of it seriously providing refuge should any significant number of people seek it is also close to zero.

In an article that, like Nelson's, should be required reading in all environmental ethics classes, Andrew Light claims that "environmental ethics is, for the most part, not succeeding as an area of applied philosophy." His rationale for this provocative claim is worth quoting in full:

> While the dominant goal of most work in the field, to find a philosophically sound basis for the direct moral consideration of nature, is commendable, it has tended to engender two unfortunate results: (1) debates about the value of nature as such have largely excluded discussion of the beneficial ways in which the arguments for environmental protection can be based on human interests, and relatedly (2) the focus on somewhat abstract concepts of value theory has pushed environmental ethics away from discussion of which arguments morally motivate people to embrace more supportive environmental views.[14]

Roughly speaking, "a philosophically sound basis for the direct moral consideration of nature" is a roundabout way of referring to intrinsic value, the grail of environmental ethics. In a nutshell, then, Light claims that by devoting so much of their philosophical energy to squaring the circle that is Callicott's "defining problem" of intrinsic value, environmental ethicists have failed to pursue the more practical as well as more important task of promoting environmental protection by impressing upon people the extent to which their own personal interests are bound up with, say, wilderness preservation or ecosystem health or an endangered species.

Styling himself as a "methodological environmental pragmatist," Light concludes: "If environmental philosophers could help to articulate moral reasons for environmental policies in a way that is translatable to the general anthropocentric intuitions of the public, they will have made a contribution to the resolution of environmental problems commensurate with their talents."[15] As I say, I think this essay deserves a wide readership, not because Light's claims are obviously true (they are not) but because it does a remarkable job of constructively problematizing the aims, tasks, methods, and audience of environmental philosophy. In any case, this is certainly not the place to take on the subject of the optimum relationship between environmental philosophy and public policy. I have broached it here merely to highlight, by way of conclusion, the extent to which Argument 31 eludes Light's two principal indictments of contemporary environmental ethics.

First, Argument 31 is nothing if not "based on human interests." It involves a direct appeal to, in Light's terms, the indirect moral consideration of nature. Nature, specifically wilderness, should be preserved not because it is directly valuable in itself but for the value it provides indirectly in the form of psychotherapy to those who focus on it in times of mental distress, especially distress of the sort created by so many of the ultimately inconvenient conveniences of modern life. Second, Argument 31, while not, I will admit, completely without recourse to abstract thought, is certainly an argument capable of "morally motivat[ing] people to embrace more supportive environmental views."[16] As I have tried to show throughout this essay, the psychotherapy at a distance argument relies on a metaphorical connection between lack of wilderness as disease and the idea of wilderness not as cure, unfortunately, but as palliative. To the extent people can (1) be brought to form potent images of wilderness, (2) be brought to appreciate the efficacy of such images in countering psychic distress, and, especially, (3) be brought to understand the relation between decreased amounts of wilderness and increased psychic stress, they will be more likely to "embrace more supportive environmental views." And whether or not increased environmentalism should be the principal goal of argument in environmental philosophy (or a goal at all, for that matter), it is one I am more than happy to assist.

NOTES

1. Michael P. Nelson, "An Amalgamation of Wilderness Preservation Arguments," in *The Great New Wilderness Debate,* ed. J. Baird Callicott and Michael P. Nelson (Athens: University of Georgia Press, 1998), 154–98. Further references to this essay are given parenthetically in the text by page number.

2. The first quote is from the front cover of *The Great New Wilderness Debate.*

3. Not that Nelson takes his amalgamation to be exhaustive or that this essay should be read as in any way denigrating Nelson's accomplishment.

4. "The Wilderness Act of 1964," reprinted in Callicott and Nelson, *The Great New Wilderness Debate,* 120–30, 121.

5. It may be a bit risky to use ANWR as my key example of wilderness. First, ANWR's wilderness status is quite fragile, very much under threat from a host of political and business interests. Second, not all of ANWR is technically wilderness, not by a long shot: only slightly more than 40 percent of ANWR is officially protected as wilderness. Third, even if everyone could agree that some specific subsection of land within ANWR was true wilderness, a few would no doubt still object that the very political and scientific and legal and technological jockeying involved in creating that subsection constitutes sufficient screwing up by humans to taint the resulting product. In any case, readers inclined to reject ANWR as wilderness should feel free to substitute their own preferred un-screwed-up locations. Perhaps I should also admit to feeling no little guilt about freely employing the acronym ANWR instead of repeatedly spelling out Arctic National Wildlife Refuge. I agree with those environmentalists who suggest that using the acronym conveys something of strategic value to those intent on development, inasmuch as writing or speaking "ANWR" excuses one from writing or saying "wildlife refuge." Obviously, it may be easier to argue, for example, for the relatively benign impact of oil and gas exploration on animals and their habitat in something called ANWR than in something called a wildlife refuge.

6. Edward Abbey, *Desert Solitaire* (New York: Touchstone, 1968), 129–30.

7. Unsurprisingly, the argument for drilling on the Alaskan coastal shelf is virtually the inverse of Argument 1: ANWR is a great resource repository, and we humans can render our future more secure by exploiting these resource reserves.

8. Nelson poses this necessary/sufficient distinction as a challenge to the force of several different arguments, yet I worry about this move in the context of the essay as a whole, since, in purporting to be nothing more (or less) than a compendium of arguments for the preservation of wilderness, one might think, to take the present case, that simply the capacity of wilderness to provide these services rather than its exclusive capacity is sufficient to allay criticism, again, given the context. Another way to express this worry might be to ask, with apologies to W. D. Ross, whether Nelson's essay aims to be an amalgamation of prima facie reasons for wilderness preservation or (whatever this might mean) categorical ones.

9. J. Baird Callicott, *Beyond the Land Ethic* (Albany: SUNY Press, 1999), 241. The best article I know of that sorts out the many issues that surround this "defining problem" is John O'Neill's "The Varieties of Intrinsic Value," reprinted in *Environmental Ethics,* ed. Andrew Light and Holmes Rolston III (Malden, Mass.: Blackwell, 2003), 131–42.

10. I should admit that, where I grew up in Southern California, rain-canceled recesses were few and far between.

11. Roderick Nash, *Wilderness and the American Mind,* 3rd ed. (New Haven, Conn.: Yale University Press, 1982), 272.

12. In "How Much Land Does a Man Need?" Tolstoy gives one of the best literary accounts ever for why we need ethics.

13. See http://plato.stanford.edu/entries/sorites-paradox/.

14. Andrew Light, "Contemporary Environmental Ethics: From Metaethics to Public Policy," *Metaphilosophy* 33 (2002): 426–49, 427.

15. Ibid., 427.

16. Ibid., 444.

PART TWO

Race, Class, Culture, and Wilderness

Kevin DeLuca and Anne Demo

Imagining Nature and Erasing Class and Race (2001)

Carleton Watkins, John Muir, and the Construction of Wilderness

In 1990, the southwest Organizing Project (SWOP) sent a series of letters to a coalition of mainstream environmental groups known as The Group of Ten. In what was termed "the letter that shook the movement," SWOP charged the coalition with racism and classism in their perspectives, issue selection, and hiring practices: "Your organizations continue to support and promote policies that emphasize the cleanup and preservation of the environment on the backs of working people in general and people of color in particular."[1] The SWOP letters were not the first to articulate the role of race and class in environmental politics. The modern environmental justice movement starts for many in 1982 when more than five hundred predominantly African-American protestors took to the streets, literally lying in roads, to prevent the relocation of hazardous waste into their community in Warren County, North Carolina. Others point to 1978, and the efforts of working-class families to protect themselves from the toxic dump contaminating their Love Canal neighborhood and school. In 1987, a Commission for Racial Justice report, "Toxic Waste and Race in the

United States," further documented the role of racism in perpetuating environmentally destructive practices. Although not the first to link class and race to environmental injustice, the SWOP letters represented a turning point in the environmental justice movement and environmental politics. The letters served as a turning point because they placed racism and classism at the heart of the environmental movement itself. In addition to addressing issues of staff composition and hiring practices, the letters pointed to fundamental problems within mainstream environmental organizations. "The Achilles heel of the environmental movement," activist Pat Bryant noted, "is its whiteness." Bryant's comment, while critical of the movement's racial homogeneity, refers less to the biological incident of skin color than the deep-seated cultural beliefs that blinker major environmental groups in ways "that make it very difficult to build a mass-based movement that has the power to change the conditions of our poisoning." In other words, whiteness as a social condition—Eurocentrism—enables environmental groups to see environmental issues through a frame of pristine wilderness while blinding them to issues in environments where people live. This perspective hinders the ability of environmental groups to forge coalitions across race and class lines, coalitions that are necessary to challenge the practices of industrialism. As activist Carl Anthony summarizes, "With its focus on wilderness, the traditional environmental movement on the one hand pretends there were no indigenous people in the North American plains and forests. On the other, it distances itself from the cities, denying that they are part of the environment."[2]

The charges have hit home, and some of the mainstream environmental groups spent the 1990s fostering productive partnerships with grassroots environmental justice groups with varying degrees of success. At a Sierra Club centennial celebration, then Executive Director Michael Fischer called for "a friendly takeover of the Sierra Club by people of color . . . [or else] remain a middle-class group of backpackers, overwhelmingly white in membership, program, and agenda—and thus condemn[ed] to losing influence in an increasingly multicultural country. . . . The struggle for environmental justice in this country and around the globe must be the primary goal of the Sierra Club during its second century." Although race has received more publicity and the by-word of environmental justice is "environmental racism," as the SWOP letters suggest, class is equally central to the issue of environmental justice. For example, the infamous Cerrell Associates report to the California Waste Management Board on how to overcome political obstacles to siting mass-burn garbage incinerators is framed in terms of class: "All socio-economic groupings tend to resent the siting of major facilities, but the middle and upper socio-economic strata possess better resources to effectuate their opposition." African-American environmental justice activist Alberta Tinsley-Williams puts it plainly; "The issue is not Black and white. The issue is the haves vs. the have-nots, because poor people, I don't care what color you are, suffer in this country."[3]

During its first one hundred years, the environmental movement has been concerned, almost exclusively, with preserving pristine places. This narrow, class- and race-based perspective of what counts as nature leads the environmental movement to neglect people and the places they inhabit, thus isolating the movement from labor and civil rights concerns and rendering it vulnerable to charges of elitism and misanthropism.[4] Grappling with such a legacy remains a difficult process, for its roots in environmentalism run deep. Class and race are both etched and elided in the early "texts" of the environmental movement. Examining Carleton Watkins's 1860s Yosemite photographs and John Muir's 1890 *Century Magazine* essays on Yosemite, show that a classed and raced notion of sublime wilderness comes to stand for "Nature" and defines what is worth preserving. Yosemite occupies a privileged place in the American imagination and the environmental movement. At the height of the Civil War, it was designated the world's first wilderness park. John Muir discovered his calling and voice in Yosemite and founded the Sierra Club and the environmental preservation movement. As hallowed ground, Yosemite is a rich site for considering both the constructedness of nature and consequences of a pristine wilderness ideal.[5]

Nature has long been understood to be unnatural. Karl Marx argued in 1846 that nature was "an historical product, the result of the activity of a whole succession of generations." Raymond Williams characterized it as "perhaps the most complex word in the language."[6] Under the rubric of thought that can be tentatively titled postmodernism and that would include versions of poststructuralism, cultural studies, and feminism in various disciplines, inquiries into the constructedness of nature and its consequent deconstruction have proceeded apace.[7] Until recently, the related term "wilderness" has remained relatively unscathed; the "trouble with wilderness," like the unnaturalness of nature, now constitutes a flash point for environmental historians and activists alike.

"Wilderness," Roderick Nash noted in the opening lines of his classic study, "has a deceptive concreteness at first glance." One might add that this deception remains at last glance, too, since its concreteness has beguiled even those who suggest otherwise. In his closing discussion of national parks, Nash writes, "Essentially, a man-managed wilderness is a contradiction because wilderness necessitates an absence of civilization's ordering influence." Both Nash's *Wilderness and the American Mind* and Max Oelschlaeger's *The Idea of Wilderness* make evident the constructedness of wilderness yet read as descriptive rather than critical histories. Both authors implicitly grant wilderness a core essence. William Cronon's more recent polemic grants no quarter. In contrast to prevailing notions that depict the wilderness as a transcendent space distinct from humanity, a "pristine sanctuary," Cronon's wilderness "is quite profoundly a human creation," indeed, "a product of that civilization." By mistaking wilderness as an anecdote for the ills of civilization, we fail, Cronon warns, to recognize that the "wilderness is itself no small part" of "our culture's problematic relationships with the nonhuman world."[8]

The deconstruction of wilderness has provoked a firestorm of protest, including special issues of *Environmental History* and *Antipode* as well as the edited volumes *The Great New Wilderness Debate* and *Reinventing Nature?: Responses to Postmodern Deconstruction.* Critics accuse Cronon of dealing in academic abstractions and insist on their ability to point to a thing out there and call it wilderness. Samuel Hays writes, "Cronon's wilderness is a world of abstracted ideas, real enough to those who participate in it, but divorced from the values and ideas inherent in wilderness action." Critics also decry the political consequences of challenging the sanctity of wilderness. Michael Cohen asks, "How is it possible to offer a constructive critique of environmentalism, of the past and present, especially of its 'save the wilderness' version, without damaging viable parts of the movement, and without offering an argument largely usable by the opponents of environmentalism who are motivated only by narrow economic gain?"[9]

Cronon is not alone in attempting to answer Cohen's question. A growing body of work suggests the constructedness of wilderness as well as nature.[10] In deconstructing wilderness, these scholars often point out the implicit race, class, and gender connotations of wilderness and note the unintended political consequences. In particular, through close readings of founding texts of the preservation movement, we attend to the mechanics of creating wilderness. In studying what can be termed the "rhetoric of wilderness," the debate moves from the abstract realm of ideas to a focus on the prosaic practices that construct wilderness on the ground, that turn a stretch of rocks and trees into a place of sublime wilderness. By linking wilderness to whiteness, the universality of the idea of wilderness as a natural fact comes under scrutiny and its cultural specificity and social and political history can be traced.

WHITENING THE WILDERNESS

The ills of the environmental movement diagnosed by activists such as Pat Bryant and groups like SWOP are not only an effect of contemporary shortsightedness within the movement but also reflect a legacy of Eurocentricism. The most obvious example, well documented by environmental historians, concerns the forced removal of native peoples from Yosemite, Yellowstone, and Glacier National Parks between the 1870s and 1930s.[11] The "justifying myth" or ideological narrative that rationalized government policy toward native people from the antebellum era forward characterized Indians as "primitives" obstructing the progress of the nation's destiny.[12] As historian Mark Spence notes, Yosemite advocate Samuel Bowles typified this position and its effects on wilderness discourse. During his 1865 visit to Yosemite, Bowles remarked that the decision to "preserve such areas" was both "a blessing to . . . all visitors" and "an honor to the nation." The preserved wilderness area championed by Bowles privileged a conception of wilderness predicated on Indian displacement. In his 1868 best seller, *The Parks and Mountains of Colorado: A Summer Vacation in the Switzerland of America,* Bowles wrote: "We know they are not our equals . . . we know that our right to the soil, as a race capable of its superior improvement, is above theirs; [therefore,] let us act directly and openly our faith . . . Let us say to [the Indian] . . . you are our ward, our child, the victim of our destiny, ours to displace, ours to protect."[13]

The Bowles passage functions as an appropriate entry into an analysis of whitening the wilderness. First, the passage introduces the notion of race as a category for classification. Indeed, Bowles suggests a stability with regard to the notion of race that our analysis will demonstrate to be an effect of rhetoric rather than science. Second, by classifying a "race," in this case Indians, by their "barbaric" relationship to the land, Bowles articulates the "self-reciprocating maxim" that defined both colonial- and antebellum-era conceptions of the Indian wilderness: "forests were wild because Indians and beasts lived there, and Indians were wild because they lived in forests."[14] The binary logic (civilized/savage) implicated in such maxims is not limited to native peoples, however. Rather, the "we" to which Bowles refers—and that Watkins anticipates and Muir inscribes—naturalized a particular definition of whiteness that also excluded ethnic groups not fully assimilated to the Victorian norms of the period. Finally, Bowles's title, *The Parks and Mountains of Colorado: A Summer Vacation in the Switzerland of America,* suggests the ways that wilderness discourse served nationalist exigencies. Yosemite's mountain cathedrals and majestic redwoods offered cultural legitimacy to a nation seeking a heritage that could compete with the cathedrals and castles of Europe. The wilderness vision dramatized by Bowles, and immortalized in Watkins and Muir's Yosemite views, hinges on Eurocentric dichotomies that privilege particular relationships with nature over others.[15] An analysis of the pristine wilderness ideal depicted by Watkins and Muir is as much a study of whiteness as an examination of preservationism. The growing literature on the history of race as a social category demonstrates that whiteness, much like nature, is neither a self-evident or fixed term. Whiteness, indeed the white race, is defined as "a historically constructed social formation . . . corresponding to no classification recognized by natural science." Whiteness is reconceptualized as a learned "knowledge system" rather than a neutral physiological referent to skin color. Such a shift foregrounds the ways in which discourse, such as Watkins's photographs and Muir's writing, contributes to making whiteness the environmental movement's "Achilles heel." As a complement to the historiography on Anglo dispossession of Indian land, this rhetorical critique seeks to reveal how wilderness imagery engendered a commensurate gain: the colonization of Victorian imaginations.[16] Writing about white America has never been a simple task. Historian Noel Ignatiev notes that in 1790, when the first Congress of the United States voted that "only 'white' persons could be naturalized as citizens," "it was by no means obvious who was 'white.'"[17] In an analysis of changing Anglo-American perceptions of indigenous peoples, historian Alden Vaughan found that "not until the middle of the eighteenth century did most Anglo-Americans view Indians as significantly different in color from themselves."[18] During the period between Watkins's first Yosemite series in 1861 and Muir's Yosemite essays, the meaning of whiteness conformed to the changing political climate as nativist hostilities gave way to a backlash over Reconstruction. Even in its most provisional formulations, whiteness has been defined by its opposition to blackness. Throughout "the century of immigration" (1820–1924), associations with blackness served to mark a group as unassimilable.[19] An analysis of whiteness during the nineteenth century must account for how immigrant populations such as Irish, Italians, and even Swedes attempted to assimilate to whiteness by adopting values, affiliations, and discourses that fostered "a solidarity based on color." Despite working conditions that encouraged "white workers to at least entertain comparison of themselves and slaves . . . the continuing desire not to be considered anything like an African-American," David Roediger notes, discouraged a solidarity based on class.[20]

This is not to suggest that nineteenth-century whiteness was not constituted by a class component. It emphasizes that the construction of whiteness is already inscribed with WASP values. If whiteness is not a biological condition, not all whites are white. In addition to race, class and ethnicity are also crucial markers of whiteness. For most of the history of the United States, "white" has meant "WASP."[21] When turn-of-the-twentieth-century urban environmentalist Alice Hamilton expressed skepticism about an apothecary's claim of no lead poisoning cases in a neighborhood near smelters, the apothecary qualified his claim: "Oh, maybe you are thinking of the Wops and Hunkies. I guess there's plenty of them. I thought you meant white men."[22] The connections drawn between blacks and Irish did not always favor the Irish. Between the 1840s and 1920s, the Civil War, the end of slavery, waves of mass immigration, and industrialism produced enormous social and cultural dislocations. The deployment of electoral and cultural politics to shore up whiteness against these forces of dislocation became a national project and preservation politics, inspired by the images of American landscape photographers like Watkins, played an important role in this project.

CAPTURING THE VIEW: WATKINS'S ICONIC WILDERNESS LANDSCAPES

By the time Ralph Waldo Emerson deemed the mid-nineteenth century an "ocular" age, photography had evolved from a scientific novelty to a commonplace in Victorian parlors on both sides of the Atlantic. William Brewster's 1849 invention of the mass-produced stereoscope, which gave paired images a three-dimensional effect, garnered such international popularity that the London Stereoscopic Company advertised its 1854 opening with the following slogan: "No home without a stereoscope."[23] Related innovations fueled the burgeoning profession of photography, producing not only "daguerreotype factories" that glutted urban markets during the early 1850s but also a generation of landscape photographers who defined the American West with their iconic views. Outdoor view photography, or what art historian Peter Hale characterizes as the "view tradition" within photography, visually enacted the colonizing imperative of manifest destiny by domesticating the American landscape for Victorian sensibilities.[24] Commissioned by the government, industry, and independent patrons, photographers such as Carleton Watkins, Timothy O'Sullivan, William Henry Jackson, and A. J. Russell made sites like Yosemite Valley, Mariposa Big Tree Grove, Yellowstone, the Colorado Rockies, and the Grand Canyon accessible to the nation. Reverend H. J. Morton was both a frequent contributor to the *Philadelphia Photographer* and one of the first to draw public attention to Carleton Watkins's Yosemite photographs. He remarked that "Without crossing the continent, we are able to step as it were, from our study into the wonders of the wondrous Valley, and gaze at our leisure on it amazing features."[25] The Edenic wilderness depicted by view photographers such as Watkins not only displayed humanity's command over nature but also calmed national anxieties by countervailing the stigma of America's Indian wilderness with a visual tradition of sublimity.

Watkins was the apotheosis of the landscape photographers associated with the "view tradition." His mammoth-plate Yosemite images served as models of outdoor view photography for critics in his own time. After viewing Watkins's Yosemite views in the wake of the Yosemite legislation, *Philadelphia Photographer* editor Edward Wilson declared the camera "mightier than the pen."[26] Hale labels Watkins the photographer most representative of the interrelated "production, distribution, consumption, and interpretation of views" that occurred during the 1860s. During an 1858 deposition regarding a boundary dispute, Watkins defined his occupation as a "photographicist" with the ability to find "the spot which would give the best view."[27] His preoccupation with the "best view" is reflected throughout his career—most directly in the descriptors he used to entitle a photograph. Using stereography, a panorama format, or mammoth-plate camera, Watkins depicted wilderness landscapes with an exactness and transparency instrumental to the geographical and geological surveys established in the late 1860s. California state geologist Josiah Whitney's 1868 *Yosemite Book* included twenty-four of Watkins's photographs. Watkins's photographs not only served land sciences such as geology and botany but nascent environmentalism as well. Watkins's 1861 photographs, according to the oft-repeated account of 1864 legislation that consigned Yosemite Valley and Mariposa Big Tree Grove to the state of California "for public use, resort and recreation," served as the only evidence needed to convince California Senator John Conness and his congressional colleagues as well as President Lincoln to "prevent occupation and especially to preserve the trees in the valley from destruction."[28] The immediate political impact of the Yosemite photographs notwithstanding, the vision of nature institutionalized through Watkins's lens may constitute the photographer's most lasting legacy to American environmentalism. Watkins's career as a photographer began as most did, working with daguerreotypes. By 1858, four years after starting in a daguerrian gallery, Watkins was commissioned to photograph outdoors. Although Watkins's pre-Yosemite images featured commercial rather than wilderness landscapes, such as the Guadaloupe Quicksilver Mine and Mariposa quarries, they reveal a developing technical commitment to transparency, visual density, and the sublime spectacle. These early but "deliberated" outdoor views, according to art historian Mary Warner Marien, are defined by a "breathtaking vantage point" that "energizes the image, suffusing it with a sense of the sublime, which would not be lost on the investors for whom these pictures were formulated."[29] The compositional depth and suspended perspective constitutive of the "breathtaking vantage point" that Marien highlights is perfected in Watkins's magisterial Yosemite photographs.

If Watkins's style can best be described as magisterial, then it is through the composition, perspective, and texture of his photographs that ideologies of whiteness and Manifest Destiny were enacted. Reactions to his 1861 and 1865–66 Yosemite series dramatized the effect of Watkins's colonizing vision. Described as a "commanding view" and "transcendent eye," Watkins's way of seeing positioned viewers to encounter the Yosemite landscape as a sublime site. Treatments of the sublime by both contemporary and Victorian authors often begin with Edmund Burke's 1757 treatise, *A Philosophical Enquiry into the Origins of Our Ideas of the Sublime and Beautiful.* Burke's distinction between the sublime and beautiful illuminates the dramatic foreground/background contrasts in many of Watkins's Yosemite photographs. Victorian references to the term often reduced Burke's association of the sublime with terror and pain to a worldly embodiment of the divine. Watkins's technical virtuosity, particularly his ability to depict the landscape with "pictorial transparency," or what Rebecca Solnit describes as "superhuman eyes," dramatizes the scale, amplitude, and sense of perpetuity in Yosemite's sublime landscapes. Reactions to Watkins's photographs echo the awe and wonder that rhetorical critic Robert Harriman argues is constitutive of sublime encounters, which reflect the "paradoxical simultaneity of seeing beauty and experiencing power: we see an aesthetic object, separate from us because so beautiful, and we feel an enormous transfer of energy that sweeps us into a transformed world."[30] Indeed, the "paradoxical simultaneity" that Harriman describes is a recurrent theme in both scholarly interpretations of Watkins and in nineteenth-century reactions to his Yosemite series. Edward Wilson's 1866 essay in the *Philadelphia Photographer,* for example, noted that "Each pebble on the shore of the little lake . . . may be as easily counted as on the shore in nature itself . . . We get a nearer view of the mountains, only to make their perpendicular sides look more fearful and impossible of ascent."[31] Photographs such as Watkins's "Yosemite Valley #1" offer a transcendent view of the valley that not only situates the landscape on an Edenic space-time continuum devoid of human markings but also positions viewers as if to be teetering on a precipice. In beholding the "great power and force exerted" by nature from this seemingly precarious vantage point, Watkins's photographs simulate, and thereby commodify, the "paradoxical simultaneity" of terror and astonishment that defines the sublime.

The sublimity of photographs such as "Yosemite Valley #1" naturalizes a construction of wilderness devoid of humanity. Within this perspective, nature functions as spectacular object rather than as inhabitable space. In addition to commodifying the sublime for East Coast urban patrons, Watkins privileges a particular human/nature relationship. Weston Naef, curator of photography at the J. Paul Getty Museum, notes that "the balance of the one hundred stereographs of Yosemite are totally without figuration," that is, without markers of "human habitation" such as cabins or even wildlife. The cultural legacy that defines wilderness landscapes in opposition to "human habitation" is among the most important aesthetic effects of such compositional decisions. Watkins's pristine wilderness views enact a mode of envisioning nature suited to prevailing national, racial, and class ideologies.[32]

The appeal of Watkins's Yosemite views partly resulted from national anxieties regarding the status of the American landscape in relation to Europe. His Yosemite photographs served as the defining pictorial evidence for American claims to a mental landscape that would effectively counter notions of European cultural superiority, quieting fears that the nation lacked both the history and divine favor associated with the castles and cathedrals of the Old World. Prior to Watkins's photographs, the debate over cultural superiority between the New and Old Worlds was structured by a binary logic that associated the

Old World with sublimity and the New World with ignobility. Photographs such as Watkins's "River View," "Cathedral Rocks," "Yosemite" and "Yosemite Falls" embody the Gothic landscapes heralded by no less than Keats in his 1818 letter to John Reynolds: "I am going among Scenery whence I intend to tip you the Damosel Radcliffe—I'll cavern you, and grotto you, and waterfall you, and wood you, and water you, and immense-rock you, and tremendous sound you and solitude you." Each photograph is composed on two planes: a beautiful foreground and sublime background. The dominant positioning and light shading of the rocky sentinels overshadow the serene foreground in each image. The immense monoliths centered at the focal point of the photographs signify power and dominance. Intimidating in their sheer verticality, the rocky sentinels of Yosemite Falls inspired nationalist comparisons from visitors who remarked upon seeing the falls, "we behold an object which has no parallel anywhere in the Alps," and "I question if the world furnishes a parallel . . . certainly there is none known."[33] Watkins's use of a deep-focus extreme long shot taken from a low angle dramatizes the verticality of the landscapes that inspire such awe.

The awe produced by Watkins's photographs not only evidenced the nation's cultural legitimacy but also naturalized a particularly classed way of seeing nature.[34] His point of view defined nature in relation to a pristine sublimity dependent upon both an absence of human habitation and a tourist form of spectatorship.[35] In yoking the sublime to the beautiful, Watkins's photography produces a domesticated sublime in two ways. Most obviously, he produced photographs that can be viewed in art galleries or gazed at as private possessions in one's own home, far from the terrors of sublime wilderness. Equally significant within the photographs themselves Watkins created a safe space for the spectator—the beautiful place—from which to view the sublime spectacle. This dynamic, at work in many of the photographs, is particularly evident in the photograph "Yosemite Falls."

In "Yosemite Falls," the beautiful literally frames the sublime. An idyllic meadow occupies over a third of the photograph. In the immediate foreground is a flat space ringed by flowering plants, grasses, and four trees. The area resembles nothing so much as a picnic site. The trees occupy entirely the left and right sides of the frame, creating a frame within the frame. Within this treed frame, positioned in the upper center of the photograph, is the spectacular sight of Yosemite Falls cascading down the cliffs of the canyon. The cascading plume rivens the canyon walls and links to the washed-out sky.

Two compositional elements sharpen the sublimity of Watkins's "Yosemite Falls." First, the foregrounded scene draws viewers to the summit. The meadow and framing trees position the canyon walls and waterfalls as the image's vanishing point. The second is the contrast in lighting. While the foreground is shot in a familiar black and white, the background is a distinctly brighter shade, giving the canyon walls a decidedly celestial hue. The twice-enframed sublime is domesticated and commodified, a view for the taking, the common currency of the tourist trade. In the union of the sublime and the beautiful is born the tourist gaze. The beautiful foreground gives the tourist a pleasing place from which to view the spectacular spectacle of the sublime. Positioned in the meadow, viewers experience the scene at ground level. From the picnic site, the viewer gazes across a wide expanse of meadow to the cliffs and Yosemite Falls. Apprehending the scene from this plane envelops viewers within a garden rather than positioning them at the precipice. Watkins anticipates and constructs a sublime experience in which comfort displaces risk as the spectator replaces the participant. The distanced position of the spectator obviates the emotional experience of the sublime. In a sense, Watkins's images blaze a trail for the tourist at the expense of the adventurer and hollow out the sublime, leaving only spectacle.

In addition to the classed way of seeing the American wilderness naturalized by Watkins's photography, the discursive framing of Yosemite through the language of the sublime and Gothic frames nature from a Eurocentric perspective by naming a granite formation Cathedral Rock. Named for what Josiah Whitney describes as "isolated columns of granite, at least 500 feet high, standing out from but connected at the base with, the walls of the valley," the Spires of Cathedral Rock appear from some angles as if to be twin towers of a Gothic cathedral. The deep reverence for nature induced by the soaring sublime of Watkins's images assumes an even greater significance as a result of this naming and the site's subsequent appropriation as a holy symbol and divine space.

Compositionally, Watkins creates a sense of an almighty dominion that resonates with visitors' divine experience of the site: "It seemed indeed, that I was making a pilgrimage to some vast cathedral shrine of Nature. The stately trees were as columns through which one finds their way along the vast colonnade, when approaching St. Peters. There is a hush in the air—you feel you are in a mighty presence. This wondrous valley—Nature's own great cathedral, where her votaries come from all lands to wonder and admire—whose cathedral spires point to heaven, whose domes have withstood the storms and tempest of all the ages, seems set apart from all the world to show forth the mighty works of Omnipotent Power."

Samuel Bowles developed the analogy suggested by the name, writing that Cathedral Rocks replicated "the great impressiveness, the beauty and the fantastic form of the Gothic architecture. From their shape and color alike, it is easy to imagine, in looking upon them, that you are under the ruins of an old Gothic cathedral, to which those of Cologne and Milanare but baby-houses."[36]

Watkins and his peers, in creating a way of viewing wilderness, in a very material sense created wilderness. Although the U.S. wilderness parks are now understood as natural and national sacred sites, that cultural perception had to be learned. Before the 1870s, whites perceived Yellowstone as a hell on earth. When mountain man James Ohio Pattie became one of the first whites to stumble upon the Grand Canyon, he found the view "horrid." As de Tocqueville observed of Americans in 1832, "In Europe people talk a great deal of the wilds of America, but the Americans themselves never think about them; they are insensible to the wonders of inanimate nature and they may be said not to perceive the mighty forests that surround them till they fall beneath the hatchet."[37] Watkins's images played a pivotal role in making Americans perceive inanimate nature as sublime wilderness.

SAVING WHITE WILDERNESS: FRAMING MUIR'S VIEW

In popular accounts, John Muir stands as the savior of Yosemite and the mythical founder of environmentalism. Arriving in Yosemite four years after it had been designated the world's first wilderness park, Muir was soon popularizing the area and trying to insure its preservation through his writings. His popularity garnered him audiences with such notables as Ralph Waldo Emerson and President Theodore Roosevelt. By 1889, Muir's reputation was such that Robert Underwood Johnson, an associate editor of *Century,* the nation's leading literary monthly, urged Muir to write on behalf of making Yosemite a national park. Muir's two essays, "The Treasures of the Yosemite" and "Features of the Proposed Yosemite National Park," appeared in the August and September 1890 issues of *Century.* On September 30, 1890, a park bill resembling John Muir's proposal passed both houses of Congress and was signed into law the next day by President Harrison.[38]

Our preceding account of the effects of Watkins's photographs suggests that the Muir myth is not the whole story. The railroads as well as landscape photographers and painters were crucial players in the cultivating of a wilderness sensibility and in the founding of national parks. Still, Muir remains a pivotal figure. His writings certainly contributed to the shaping of the social imaginary with respect to wilderness and preservation politics. Besides these two articles functioning as a blueprint for establishing Yosemite National Park, they provide a vision of wilderness, white wilderness, that many environmental groups still embrace today.

Muir's *Century* essays are framed and interlaced by the values of whiteness. This is seen most obviously and yet most unquestioningly in the choice of Yosemite as an object worthy of being saved. Muir considers Yosemite a "temple lighted from above" that presents the "most striking and sublime features on the grandest scale." Regarded on aural as well as visual dimensions, Yosemite Falls is a "sublime psalm," "pure wildness." The values of white wilderness are also evident in Muir's descriptions of Yosemite. Muir translates the pictorial conventions of Watkins's photographs into word-pictures. Most telling, Muir's descriptions at times echo the sublime/beautiful dynamic at work in Watkins's photographs: "the main canyons widen into spacious valleys or parks of charming beauty, level and flowery and diversified like landscape gardens with meadows and groves and thickets of blooming bushes, while the lofty walls, infinitely varied in form, are fringed with ferns, flowering plants, shrubs of many species, and tall evergreens and oaks." Muir also translates the conventions of Watkins's landscapes into word-pictures through the repeated deployment of key terms: sublime, massive, immense, imposing, grandeur, serene majesty, temple, psalm, and so on. Both essays are filled with such word-pictures of the various treasures of the Yosemite region and threats to that wilderness. In addition, the two essays include twenty-two pictures attesting to the sublime wilderness beauty of Yosemite. These pictures reproduce most of Watkins's famous images: the view of Yosemite Valley from Inspirational Point, Cathedral Rocks, Yosemite Falls, El Capitan, Three Brothers, and Sentinel Rock. In other writings, Muir's view of Yosemite is framed by Watkins's vision of sublime wilderness. In his book *The Yosemite,* Muir remarks on the pristine quality of the region: "In general views no mark of man is visible upon it." Many of the descriptions in the book of the iconic treasures of Yosemite are copied verbatim from the earlier essays.[39]

Muir's use of religious imagery is crucial in distinguishing Yosemite as a sacred place that deserves comparison not to the exotic spaces of Africa or South America but the sacred places of white civilization. By comparing American wilderness to cathedrals, Muir transforms it from a potentially corrupting place into a divine place that reaffirms America's connection to European civilization. Muir and other popularizers of American wilderness understood themselves to be in a commercial competition for tourist dollars as well as a nationalist competition for cultural capital. By viewing Yosemite as a white wilderness of "mountain temples," it becomes a sign of the blessing of the white man's god and maintains Americans' connection to Europe even in the New World. In arguing for the preservation of Hetch Hetchy, Muir makes the connection in no uncertain terms: "Dam Hetch Hetchy! As well dam for water tanks the people's cathedrals and churches, for no holier temple has ever been consecrated by the heart of man."[40] The association of wilderness with European civilization has continued in contemporary environmentalism. In successfully preventing the damming of the Colorado River in the Grand Canyon during the 1960s, the Sierra Club ran an ad that asked, "SHOULD WE ALSO FLOOD

THE SISTINE CHAPEL SO TOURISTS CAN GET NEARER THE CEILING?"[41]

The white wilderness actualized in Muir's writing is a result of his focus on Yosemite, his mode of description, and his ability to evoke a sublime response. Despite Watkins's ability to depict the sublime, Muir more aptly translated a domesticated sublime for his middle-class, eastern, urban readers.[42] Watkins offered a beautiful vantage point from which to view the sublime sights, while Muir positioned readers to vicariously experience the sublime by living through his adventures. In many of his writings, Muir constructs his persona as a knowledgeable but reckless wilderness guide. In the first of the two *Century* essays, after conducting an extensive tour of the sublime sights Muir closes with accounts of numerous dangerous adventures that risked his life, including rock and ice-climbing escapades, traipsing through a spring-time deluge at night, and Muir's most fear-inspiring specialty, defying the fates at Yosemite Falls. Muir gets to experience the sublime and the readers enjoy a domesticated sublime in the comfort of their homes, perched on the ledge with John Muir, but only in their imagination.

BANNING CLASS

A white notion of civilization as not only separate from but also the antithesis of wilderness permeates the two essays as well. The wilderness of Yosemite must be saved from the encroachments of such a civilization. The civilization/wilderness dichotomy is imbued with the values of whiteness, so it is laced with class distinctions. This is clear in Muir's discussion. Muir warns that "all that is perishable is vanishing apace . . . every kind of destruction is moving on with accelerated speed."[43] The "ravages of man" include logging, farming, and grazing (sheep and cattle). To save Yosemite "it is proposed to reserve [it] out of the public domain for the use and recreation of the people."[44] The universalized "people" masks race, ethnicity, and class dimensions that influence for whom and from whom Yosemite is being saved. Who counts as "people" and what counts as "use and recreation" is determined by the prerequisites of white wilderness. Muir makes these distinctions more explicitly in other parts of the essay. "The Yosemite Valley, in the heart of the Sierra Nevada, is a noble mark for the traveler: he writes, whether tourist, botanist, geologist, or lover of wilderness pure and simple."[45] White wilderness permits only visitors, but not all visitors. The recreations that Muir mentions are notably upper class. Noticeable by its absence is hunting. Though hunting traverses classes, in the United States hunting for food becomes associated with rural lower classes, not the elite, eastern urban dwellers that Muir is appealing to. This is evident a few decades later in the conservationist tract *Our Vanishing Wildlife,* wherein author William Hornaday condemns the crimes of "Italians, negroes and others who shoot song-birds as food" and concludes that "all members of the lower classes of southern Europe are a dangerous menace to our wild life."[46] In this sentiment, the mix of class and race condemn those who do not share the white values of WASP culture. Another key distinction in the "use and recreation" phrase is that between leisure and work. Aside from scientists, Muir limits "use and recreation" to tourists, explicitly arguing against those who do manual work in nature. Muir laments the presence of loggers and shepherds. In other places, Muir mocks those who work in the wilderness. Much of Muir's first summer in the Sierras was spent working with Billy, a shepherd. Muir mocks Billy's clothes: "his wonderful everlasting clothing on . . . These precious overalls are never taken off, and nobody knows how old they are"; his fear of bears "seems afraid that he may be mistaken for a sheep"; and his blindness to the sublime beauty of the wilderness.[47] This last point irks Muir. "I pressed Yosemite upon him like a missionary offering the gospel, but he would have none of it. 'Tourists that spend their money to see rocks and falls are fools, that's all. You can't humbug me. I've been in this country too long for that.' Such souls, I suppose, are asleep, or smothered and befogged beneath mean pleasures and cares."[48]

Muir's vitriolic wit and class bias foretells the split between labor and environmentalism that has haunted the latter in the twentieth century. Muir is unable to take Billy seriously. He ridicules his work clothes and fears. He dismisses Billy's pleasures and cares as "mean." Since his tone is one of derision and condescension, it is no wonder his missionary work fails. Though Muir dismisses Billy as asleep or befogged, the passage actually contains a dialogue and if we pay attention to Billy's words, it is clear that his is an insightful position. For Billy, Yosemite is not an idle place of sightseeing, but a workplace. His knowledge of Yosemite is not that of the sublime aesthetic transposed from Europe but an intimate knowledge derived from on-the-ground practices in Yosemite. Muir and Billy's encounter is not a meeting of wisdom and ignorance but a clash of different knowledges and worldviews. In this dialogue, Muir models an attitude that environmentalists have imitated to great detriment. This attitude implicitly includes a classed notion of work. Blue-collar labor in the woods is bad, white-collar labor in city skyscrapers escapes notice. This position is problematic, for while the latter activity isolates humans from the natural world, physical labor in nature is one of the primary ways that humans come to know and connect with the natural world.[49] The irony of Muir's position becomes evident when he recognizes that the work of some makes possible the play of others in the wilderness—that play is founded on work, both in civilization and in the wilderness. Muir's acknowledgment that the shepherd's work enables him to be "free to rove and revel in the wilderness all the big immortal days" adds specificity to his white perspective.[50] Further, Muir did advocate tourism, which requires a work force of cooks, maids, and guides to enable the tourist to consume the wilderness experience. As Karl Jacoby explains in his analysis of the designation of the Adirondack Park and the class tensions played out in constructing that wilderness, "The paradox of the touristic understanding of the Adirondacks—as a place apart, free from the corrupting influences of contemporary life—was that it was achieved by transforming the Adirondacks into a workplace . . . a scenic vista for visiting tourists, whose arrival had helped to solidify the presence of a wage-based service industry in the region."[51] Muir's position is the precedent for today's ecotourists, for whom wilderness is not a place to be worked but a recreational fantasy to be consumed.[52]

In addition to his work as a shepherd, in order to support his wilderness wanderings, Muir operated a sawmill for the owner of the largest hotel in the valley. In a sense, Muir embodies the interactions of aesthetic and use values but celebrates only the former. In short, although Muir presents his perspective as universal and ideal, he is advocating a white wilderness that is a social construction with roots in culture, class, and race and which works to mark social distinctions and affirm hierarchies.

RACE IN THE WOODS

White wilderness certainly forbids inhabitants. Muir only briefly mentions the Native Americans of the region, when, in discussing Hetch Hetchy, Muir observes, "Furthermore it was a home and stronghold of the Tuolumne Indians, as Ahwahne was of the grizzlies."[53] Muir's association of the Tuolumne with the grizzlies is telling, for often in the civilization/wilderness dichotomy, those not part of white civilization get coded as nature. This example is not atypical of Muir's writings. In *My First Summer in the Sierra,* Muir writes admiringly of the vague Indian "instinct" of "walking unseen": "All Indians seem to have learned this wonderful way of walking unseen,—making themselves invisible like certain spiders I have been observing here . . . Indians walk softly and hurt the landscape hardly more than the birds and squirrels, and their brush and oak huts last hardly longer than those of wood rats."[54] Again, Muir clearly associates Native Americans with nature—spiders, birds, squirrels, and wood rats. They are constructed as a natural part of wilderness. To Muir's credit, during the concluding stages of the military campaign against Native Americans, Muir clearly admires and respects them.[55] In the passage immediately following the above quote, Muir decries the destructive marks of the white man. Yet Muir's perspective is still more complicated, and it reflects both discourses in his own time and foreshadows the United States's continuing simultaneous demonization and idealization of Native Americans. In *The Yosemite,* Muir's account of the removal of Yosemite's Native Americans is quite conventional, and he sees nothing wrong with the military campaign.[56] Muir describes Native American resistance to the encroachments of miners as "their usual murdering, plundering style" and describes the "Yosemite Indians" as "this warlike tribe."[57]

Even Muir's admiration is problematic. Muir is clearly operating from the perspective that Indians are a part of nature and not human agents that transform nature. He praises them for having no more impact than birds, squirrels, and wood rats. Such an understanding requires Muir to ignore the Native American practice of setting fire to the meadows, the very practice that creates the pristine, Edenic garden that he celebrates. Muir's attitude of "demeaning exoticization"[58] holds sway today: "We are pious toward Indian peoples, but we don't take them seriously; we don't credit them with the capacity to make changes . . . This is why our flattery (for it is usually intended to be such) of 'simpler' peoples is an act of such immense condescension. For in a modern world defined by change, whites are portrayed as the only beings who make a difference."[59] The point is that within the context of whiteness, those not part of white civilization are, at best, seen as part of nature. At worst they are often expelled from wilderness and forcibly "civilized." The myth of pristine wilderness is founded on the erasure of the humanity, presence, and history of Native Americans.[60] Muir's "wilderness blinders" materially impacted Yosemite Valley. Most of those with long experience of Yosemite Valley and its state managers argued for continuing the Ahwahneechee practice of annual burning and noted the effects of not burning. In 1894, Galen Clark, an early state guardian of Yosemite, recalled how on his first visit to Yosemite in 1855 clear open meadow land was at least four times as large.[61] The *Biennial Report of the Commissioners to Manage Yosemite Valley and the Mariposa Big Tree Grove for the Years 1891–1892* described Yosemite originally as a forest park, dotted with open meadows and maintained by Indians through annual burning and weeding. Such interventions by Native Americans made visible the sublime vistas that earned the valley its renown as a tourist wilderness destination. The *Report* noted that leaving nature alone had negatively impacted the beauty of the valley as underbrush and scrub trees obscured the views.[62] Muir, however, saw burning as antithetical to wilderness and adamantly opposed the practice. His position carried the day, so that in 1929 when Totuya, the granddaughter of Chief Tenaya, visited her native Yosemite Valley for the first time since being forced out in 1851 she remarked, "Too dirty; too much bushy."[63]

While Muir perceived the practice of burning as a violation of the wilderness, he did not see tourism as a practice that threatened the wilderness. Although Muir had doubts about tourists, he advocated building tourist accommodations in Yosemite Valley, declaring that "A large first-class hotel is very much needed."[64] By 1895, besides hotels, the infrastructure of tourism cluttered Yosemite Valley: "warehouses, barns, livery stables and a blacksmith; ice house; vegetable stalls, a general store, a bakery, a butcher, and a laundry; a lumberyard, a cabinet shop; saloons; express, telegraph, and post offices; a chapel; two art galleries, and two photographic studios. There were also countless private dwellings with attendant cow sheds and chicken coops."[65]

IN RETROSPECT

The rhetoric of white wilderness practiced by Watkins, Muir, their peers, and their corporate and political backers established the arc of environmentalism for its first century. The successes of wilderness advocacy have been great and its cultural impact impressive. Yet in taking as their task wilderness preservation, mainstream environmental groups rescued themselves from the responsibility of protecting urban and inhabited rural areas and of critiquing industrial consumer society in general. Starting with the emergence of pollution as an "environmental" issue in the 1960s, and now with the growth of environmental justice activism, environmental groups, to their credit, have begun to engage issues outside the bounds of their traditional focus. Environmental groups' move beyond wilderness will always be haphazard as long as the concept of wilderness is left undisturbed. In this move, scholarly efforts to deconstruct and historicize wilderness can be of vital significance.

Admittedly, the stakes for environmental politics are great, for our deconstruction of wilderness challenges the essential object of much of environmentalism. Indeed, for many groups, wilderness functions as both the reason for their existence and the goal of their activities. So why deconstruct wilderness? Clearly, such a process is fraught with danger. What makes sense theoretically can be harmful politically. Still, the deconstruction of wilderness potentially benefits environmental politics. The origin myth of environmental politics is that John Muir and others of his ilk came across wilderness and were so inspired that they dedicated their lives to saving it. The construction of wilderness in the works of Watkins and Muir belies the belief in the mythic origins of environmentalism anchored in the formation of the first wilderness parks.

Quite clearly, wilderness parks are the products of multiple discourses and serve a role within the paradigm of industrial progress. The parks had no clear environmental mission and were not even ends in themselves but means to attract tourist dollars and extend industrial development. Further, the sublime feeling produced by wilderness is not an innate, universal feeling but a culturally conditioned response. Within the discourses of the late nineteenth century, the sublime becomes not so much a feeling but a commodity produced through specific techniques. In the world of politics, where pristine wilderness has become an effective wedge issue, the benefit of the deconstruction of wilderness is clear. Most obviously, this shatters the belief that wilderness is a natural object that people will "naturally" respond to. Wilderness is not a natural fact, but a political achievement. When environmentalists keep this in mind, they will never take for granted that others will necessarily share their feelings about wilderness once they are exposed to it. Preserving wilderness always requires political struggle. It also requires cultural education. Too often when certain types of people, loggers or urban dwellers, do not revere wilderness, environmentalists have dismissed or ignored them. This tendency has earned environmentalism a reputation as an elitist movement and has isolated it from potential allies in the struggle against the depredations of industrialism. Instead, environmentalists need to accept that wilderness does not have value in and of itself but instead has social value that must be communicated and fought for. Wilderness is a social construction—one worth preserving. The deconstruction of wilderness as a founding concept, the revealing of wilderness to be unnatural, is not an argument for the abandonment of wilderness and preservation politics. It is to realize that an unquestioning embrace of pristine wilderness has political and social costs as well as benefits. The proponents of industrial progress have deployed wilderness to divide environmental groups and workers. Yet the industrial juggernaut threatens the environment, the working class, and minorities. To preserve wilderness as anything more than tattered relics symbolizing the failure of environmentalism requires recognizing the social character of wilderness and the role of people in it. The struggle to preserve wilderness must not center on issuing proclamations of divine revelations of wilderness as sacred spaces and denouncements of the unimpressed as maleficent or ignorant. Instead, preservation must rest on the recognition that wilderness is not a divine text but a significant social achievement. The preservation and expansion of that achievement depends on making arguments about the worth of wilderness to the social and biological worlds and on forging uncommon alliances. Maybe Thoreau was right, that in wildness is the preservation of the world. Conversely, in society is the preservation of wilderness.

NOTES

1. Southwest Organizing Project, "The Letter that Shook a Movement," *Sierra* (May/June 1993): 54.

2. "A Place at the Table: A Sierra Roundtable on Race, Justice, and the Environment," *Sierra* (May/June 1993): 28, 57.

3. "A Place at the Table," 51.

4. Gender, too, is etched and elided in these early texts. An analysis of the role of gender in the construction of wilderness will be the focus of another essay.

5. Richard Dyer, *The Matter of Images: Essays on Representations* (New York: Routledge, 1993), 142–43.

6. Karl Marx, "The German Ideology," in *The Marx Reader,* ed. Robert C. Tucker (New York: W. W. Norton, 1978), 146–200; Raymond Williams, *Problems in Materialism and Culture Keywords* (Great Britain: Redwood burn, 1980), 71.

7. This process has been particularly intense in feminism. See Carolyn Merchant, *Death of Nature: Women, Ecology, and the Scientific Revolution* (San Francisco, Calif.: Harper & Row, 1980); Donna Haraway, *Simians, Cyborgs, and Women: The Reinvention of Nature* (New York: Routledge, 1991); and Londa Schiebinger, *Nature Body: Gender in the Making of Modem Science* (Boston: Beacon Press, 1993).

8. Roderick Nash, *Wilderness and the American Imagination* (Binghamton, N.Y.: Vail Ballou, 1973), 1, 273; William Cronon, "The Trouble with Wilderness, or Getting Back to the Wrong Nature," in *Uncommon Ground: Toward Reinventing Nature,* ed. William Cronon (New York: W. W. Norton, 1995), 69.

9. William Cronon, "The Trouble with Wilderness; or, Getting Back to the Wrong Nature," *Environmental History* 1 (1996): 7–28; Samuel P. Hays, "Comment: The Trouble with Bill Cronon's Wilderness," *Environmental History* 1 (1996): 29–32; Michael P. Cohen, "Comment: Resistance to Wilderness," *Environmental History* 1 (1996): 33–42; Thomas R. Dunlop, "Comment: But What Did You Go Out into the Wilderness to See?" *Environmental History* 1

(1996): 43–46; William Cronon, "The Trouble with Wilderness: A Response," *Environmental History* 1 (1996): 47–55; J. Baird Callicott and Michael P. Nelson, eds., *The Great New Wilderness Debate* (Athens: The University of Georgia Press, 1998); Michael Soulé and Gary Lease, *Reinventing Nature? Responses to Postmodern Deconstruction* (Washington, D.C.: Island Press, 1995); Flays, "Comment: The Trouble with Bill Cronon's Wilderness," 31; Cohen, "Comment: Resistance to Wilderness," 33.

10. See the collection of essays in *Uncommon Ground: Rethinking the Human Place in Nature* (1995), especially the essays by Anne Whiston Spirn, Candace Slater, Caroline Merchant, and Karl Jacoby, "Class and Environmental History: Lessons from 'The War in the Adirondacks,'" *Environmental History* 2 (1997): 324–42.

11. For further discussion of the relationship between Eurocentricism, wilderness, and Native Americans, see Mark Spence, "Dispossessing the Wilderness: Yosemite Indians and the National Park Ideal, 1864–1930," *Pacific Historical Review* (1996): 27–59; Rebecca Solnit, *Savage Dreams: A Journey into the Hidden Wars of the American West* (San Francisco, Calif.: Sierra Club Books, 1994); Marcy Darnovsky, "Stories Less Told: Histories of U.S. Environmentalism," *Socialist Review* 22 (1991): 11–54.

12. Joe Freeman, *The Politics of Women's Liberation: A Case Study of an Emerging Social Movement and Its Relation to the Policy Process* (New York: Longman, 1975).

13. Quoted in Mark David Spence, *Dispossessing the Wilderness: Indian Removal and the Making of the National Parks* (New York: Oxford University Press, 1999), 27.

14. Ibid., 10.

15. Cronon aptly summarizes this position when writing: "Wilderness suddenly emerged as the landscape of choice for elite tourists, who brought with them striking urban ideas of the countryside through which they traveled. For them, wild land was not a site for productive labor and not a permanent home; rather, it was a place of recreation. One went to the wilderness not as a producer but as a consumer, hiring guides and other backcountry residents who could serve as romantic surrogates for the rough riders and hunters of the frontier if one was willing to overlook their new status as employees and servants of the rich." Cronon, "The Trouble with Wilderness," in *Uncommon Ground,* 78.

16. Editorial, "Abolish the White Race," *Race Traitors,* ed. Noel Ignatiev and John Garvey (New York: Routledge, 1996), 9. Similarly, Audrey Smedley characterizes race as "a 'knowledge system'; a way of knowing and looking at the world and of rationalizing its contents (in this case, other human beings) in terms that are derived from previous cultural-historical experience and re-

flective of contemporary social values, relationships, and conditions." Audrey Smedley, *Race in North America: Origin and Evolution of a Worldview* (Boulder, Colo.: Westview Press, 1998), 15.

17. Noel Ignatiev, *How the Irish Became White* (New York: Routledge, 1995), 41; also, Noel Ignatiev, "Immigrants and Whites," in *Race Traitor,* ed. Noel Ignatiev and John Garvey (New York: Routledge, 1996).

18. Alden Vaughan, "From White Man to Redskin: Changing Anglo-American Perceptions of the American Indian," *American Historical Review* 8 (October 1982): 918.

19. A number of scholars detail the various descriptors used to mark immigrant groups as unassimilable. See Ignatiev, *How the Irish Became White,* 96; David Roediger, *Abolition of Whiteness: Race and the Making of the American Working Class* (London: Verso, 1991), 141.

20. Ignatiev, *How the Irish Became White,* 96; Roediger, *Abolition of Whiteness,* 68.

21. See Theodore Allen, *The Invention of the White Race* (New York: Verso, 1994); David Roediger, *Towards the Abolition of Whiteness: Essays on Race, Politics, and Working Class History* (New York: Verso, 1994); Richard Brookhiser, *The Way of the WASP: How It Made America, and How It Can Save It; So to Speak* (New York: Free Press, 1991); and Joe Kincheloe, ed., *White Reign: Deploying Whiteness in America* (New York: St. Martin's Press, 1998).

22. Quoted in Robert Gottlieb, *Forcing the Spring: The Transformation of the American Environmental Movement* (Washington, D.C.: Island Press, 1993), 235.

23. For histories of the status of photography in the nineteenth century, see Robert Trachtenberg, *Reading American Photographs: Images as History: Mathew Brady to Walker Evans* (New York: Hill and Wang, 1989); Richard Masteller, "Western Views in Eastern Parlors: The Contribution of the Stereograph Photographer to the Conquest of the West," in *Prospects: The Annual of American Cultural Studies,* 6, ed. Jack Salzman (New York: Burt Franklin and Company, Inc., 1980).

24. Quoted in Peter Bacon Hale, *American Views and the Romance of Modernism in Photography in Nineteenth-Century America,* ed. Martha Sandweiss (New York: Harry Abrams, Inc., Publishers, 1991), 206.

25. Reverend H. J. Morton, "Yosemite Valley," *Philadelphia Photographer* 3 (December 1866): 107.

26. Quoted in Hale, *American Views and the Romance of Modernism in Photography in Nineteenth-Century America,* 207.

27. Watkins, deposition of 27 August 1858 in the case *United States vs. Charles Fossat,* quoted in Palmquist and Sandweiss, *Photographer of the American West,* 9.

28. Quoted in Han Huths, "Yosemite: The Story of an Idea," *Sierra Club Bulletin* 33 (1948): 47–48.

29. Mary Warner Marien, "Imaging the Corporate Sublime," *Carleton Watkins: Selected Texts and Bibliography,* ed. Amy Rule (Boston: C. K. Hall & Co., 1993), also Albert Boime, *The Magisterial Gaze: Manifest Destiny and American Landscape Painting, ca. 1830–1865* (Washington, D.C.: Smithsonian Institution Press, 1991).

30. Robert Harriman, "Terrible Beauty and Mundane Detail: Aesthetic Knowledge in the Practice of Everyday Life," *Argumentation and Advocacy* 35 (1998): 10–18.

31. Edward Wilson, "Views in the Yosemite Valley," *Philadelphia Photographer* 3 (April 1866): 106–7.

32. Although artists associated with the Hudson River school also celebrated the nation's majestic natural resources, artists such as Thomas Cole and Alfred Agate also dramatized the transformation of the American landscape from wilderness to pastoral scene. See: William Cronon, "Telling Tales on Canvas: Landscapes of Frontier Change," *Discovered Lands, Invented Pasts* (New Haven, Conn.: Yale University Press, 1992), 37–87. Weston Naef, David Robertson, David Featherstone, Tome Fels, Peter Palmquist, Amy Rule, "Looking West: The Photographs of Carleton Watkins," *In Focus: Carleton Watkins* (Los Angeles, Calif.: The J. Paul Getty Museum, 1997), 99–100.

33. William [John] Keats quoted in Diego Saglia, "Looking at the Other: Cultural Difference and the Traveler's Gaze in *The Italian,*" *Studies in the Novel* 28 (spring 1996): 12; Lieutenant Colonel A. V. Kautz and William B. Brewer quoted in Alfred Runte, *National Parks: The American Experience,* 3rd ed. (Lincoln: University of Nebraska Press, 1997), 17–19.

34. Frederick Law Olmsted, the first commissioner for the organization established to protect the landscape, illustrates how a particular relationship to nature functions as index of civility and class. In his 1865 commission report on Yosemite Valley and Mariposa Grove, Olmsted justifies the protection of Yosemite on the following grounds: "The power of scenery to affect men is in a large way proportionate to the degree of their civilization and to the degree in which their taste has been cultivated." Frederick Law Olmsted, "Preliminary Report upon the Yosemite and Big Tree Grove," *The Papers of Frederick Law Olmsted: The California Frontier 1863–1865*, vol. 5, ed. P. F. Ranney (Baltimore, Md.: Johns Hopkins University Press, 1985), 505.

35. Although this may have been an artistic imperative (and Watkins's naming of his studio the Yosemite Art Gallery speaks of his artistic aspirations for photography), it was also a commercial imperative as Watkins left the financial security and comforts of portrait work for the uncertainties and hardships of landscape photography.

36. Josiah Letchworth, Letter to Mrs. Delia Skinner, Yosemite Valley, May 22, 1880, Yosemite Collection, 31; Alfred Runte, *Yosemite: The Embattled Wilderness* (Lincoln: University of Nebraska Press, 1990), 15.

37. Quoted in François Leydet, *Time and the River Flowing: Grand Canyon,* ed. David Brower (San Francisco: Sierra Club, 1964), 37 (1945 p. 78).

38. Roderick Nash, *Wilderness and the American Mind: Revised Edition* (Binghamton: Vail Ballou, 1973), 130. Johnson's urging was Muir's doing. Johnson had arrived in San Francisco looking for someone to write about the romance of gold-hunting. Muir lured him into a camping trip in Yosemite. When Johnson saw the destruction caused by the hoofed locusts (sheep), he exclaimed a need for a park and proceeded to try to sell Muir on an idea that Muir had proposed in 1881.

39. John Muir, "The Treasures of the Yosemite," *The Century Magazine* (August 1890): 484–85, 492, 493.

40. Muir, *The Yosemite,* 197.

41. Quoted in Nash, *Wilderness and the American Mind,* 231.

42. Christine Oravec, "John Muir, Yosemite, and the Sublime Response: A Study in the Rhetoric of Preservationism," *The Quarterly Journal of Speech* 67 (1981): 247, 257.

43. Muir, "The Treasures of the Yosemite," 483.

44. John Muir, "Features of the Proposed Yosemite National Park," *The Century Magazine* (September 1890): 667.

45. Muir, "The Treasures of the Yosemite," 483.

46. Quoted in Darnovsky, "Stories Less Told," 25.

47. John Muir, *My First Summer in the Sierra* (New York: Houghton Mifflin, 1911), 171, 173, 259.

48. Ibid., 197–98.

49. Richard White, *The Organic Machine: The Remaking of the Columbia River* (New York: Hill and Wang, 1995).

50. Muir, *My First Summer in the Sierra,* 174.

51. Karl Jacoby, "A Class and Environmental History: Lessons from 'The War in the Adirondacks,'" *Environmental History* 2 (July 1997): 336.

52. Cronon, *Uncommon Ground,* 78–79.

53. Muir, "Features of the Proposed Yosemite National Park," 66.

54. Muir, *My First Summer in the Sierra,* 71–73.

55. This is not true in all of Muir's writings. Sometimes Muir describes the Yosemite Indians as "dirty" and "lazy."

56. John Muir, *The Yosemite* (San Francisco, Calif.: Sierra Club Books, 1914 [1988]), 168–74.

57. Ibid., 168.

58. Darnovsky, "Stories Less Told," 21.

59. White, *The Organic Machine,* 175.

60. Cronon, *Uncommon Ground;* Spence, *Dispossessing the Wilderness;* Robert H. Keller and Michael F. Turek, *American Indians and National Parks* (Tucson: University of Arizona Press, 1998).

61. Quoted in Runte, *Yosemite,* 3.

62. Ibid., 6.

63. Quoted in Margaret Sanborn, *Yosemite: Its Discovery, Its Wonders, and Its People* (New York: Random House, 1981), 238.

64. Ibid., 116; Muir, *The Yosemite,* 175.

65. Sanborn, *Yosemite,* 229.

Lynn Maria Laitala

Jackfish Pete (2001)

Pete LaPrairie's Story

I WAS LIVING IN WINTON in 1918 when the government agents removed the people on Jackfish Bay, scattering them among reservations in Minnesota and Ontario, as far as Lac La Croix, Grand Portage and White Earth. I didn't know who had died in the flu epidemic and who had been taken away by the agents. I traveled to the three reservations before I found my cousin Mary at Lake Vermilion.

"Your stepmother died," Mary told me. "All your brothers died except Jake. Josie was still alive, too. The agents took them both, I don't know where."

I found my little brother and sister staying with an old woman at Fond Du Lac and brought them back to Winton.

I was working at the Swallow and Hopkins mill planing lumber when the boss came over with a couple of men dressed in silly looking hats and vests with wicker creels slung over their shoulders. "Hey Pete," the boss said. "Take my friends out to a real good fishing hole. Some place where they can pull in one right after the other. Preferably bass. I'll pay your regular wages."

It sounded like a good deal so I took the greenhorns out, cooked for them and showed them where to fish. They caught enough fish to feed a village for a week but they still wanted more—for the sport of fighting the fish, they said.

There's nothing I like better than a meal of fresh fish—but fight fish for sport? If you look at it one way, it's torturing creatures for fun. Look at it another way, you're playing with your food.

Anyhow, those guys were real happy, and they gave me extra money when they left. In a month or two some of their friends from Chicago came up, wanting to catch fish like that too. I was the first real tourist guide out of Winton.

After the mills shut down, I earned enough money from guiding in the summer so I could stay in Winton and keep house for Jake and Josie during the school months. We did okay. I picked up an odd job here and there. Josie sewed most of our clothes, Jake and I hunted, and every year we went over to Nett Lake to make rice. We ate a lot of fish, wild rice, and venison.

I took Sig Olson out on his first trip. That was the year after the government destroyed the village on Jackfish. I told him my name was Pete LaPrairie.

"LaPrairie. That's French. Pete must be short for Pierre," Sig said. "Pierre LaPrairie."

From that time on he called me Pierre.

"Pierre," he said, "you have the blood of romance flowing in your veins. Yours is the legacy of the noble voyageur."

We were paddling across Basswood on a rare calm day, green canvas canoe gliding easily on smooth water.

"Can you sing for me, Pierre? Do you know a *chanson*?"

I sang one of my grandfather's songs.

"*Goutons boire qui le vin est bon, goutons boire, oui, oui, oui. . . .*"

"That's it, that's it! A *chanson,*" Sig cried.

Antti and some of the other lumberjacks had started guiding, but I was Sig's favorite. I took him out many times. He was likable enough, but he was so eager that we couldn't resist having a little harmless fun with him, like the time Antti and some of the guys took him to Laitinen's public sauna.

"Too hot for you in here?" Antti asked. "Well, if you want to cool it down, throw a little cold water on the rocks." Sig did. I've never seen a man so pink move so fast out the door. I don't think he ever went back to the public sauna.

Sig had a knack for asking stupid questions. When I tried to explain he interrupted and told me the answers he wanted to hear. Like how I was a Frenchman, and what it must have been like to be a hearty voyageur or a simple primitive native. You couldn't take offense at his ignorance. He liked us colorful people.

But when he asked a practical question, none of us could manage a straight answer.

"What's the best firewood to cook with?" he asked.

"Spruce," Antti said.

Next time I was out with him, Sig gathered up armloads of dry spruce branches for the cook fire. The fire roared up, showering sparks into the brush. We built up good appetites stamping out fires around the campsite.

"*Sacre Dieu,*" he cried when I brought him over a portage to a lake he hadn't seen before. "Oh the life of the voyageur! The adventure of exploring new lands is like no other feeling in the world!"

For Sig, an explorer was someone who didn't know where he was going.

He didn't understand the difference between an explorer and a voyageur, either. My grandfather was a voyageur. Voyageurs worked for the big fur companies, men so desperate for work they allowed themselves to be used as pack animals. The company worked them like hell, treated them like shit and expected them to die young. A voyageur got a pension if he got to be thirty years old, and the company didn't spend a lot of money on pensions.

The worst part of working as a voyageur was living without women, Grandfather said. Like a lot of the young men from Quebec he ran off from the company, married a Chippewa woman, "went native." We were already pretty mixed by then—the trade had been going on for a long time, hundreds of years.

It wasn't blood that made you what you were in this country, it was how you lived. Grandfather was despised in Quebec because his father was poor. Rich people can be very cruel, he said, but the Indians took him as he was, as long as he followed their rules.

The old Chippewa rules made sense to me. When you followed them you didn't own much, but you lived well, but the government wasn't going to let us live like that anymore. Indian agents had come around hassling us for years. It was when the agent came to Jackfish to force me back to boarding school that I ran away and went work in the lumber mill with Antti.

The mill was noisy and dangerous but it was better than boarding school. Once in awhile I'd go home to Jackfish with presents for everybody, stay up there for a while, and feel all right again. Then the government agents destroyed the village. A few people straggled back, but the real life of the Chippewas was gone from Jackfish Bay. I planned to take Jake and Josie to live on the reservation at Vermilion, but when I visited my cousin Mary and her husband Joe, they talked me out of it.

"The Bureau of Indian Affairs hires agents as firewood inspectors," Mary told me. "All they do is check our woodpiles. If our woodpile is small, they tell us we're lazy. If we have a big woodpile, they accuse us of commercializing in wood."

"We fill out forms and forms about our allotments, but we don't have any control over the land," Joe said. "My allotment was logged off. The government keeps the stumpage fees in an account for us, because they say we can't handle money. But when I wrote to ask how much money was in my account, the government claimed that my allotment hadn't been logged, and there was no money in my account."

"You can't cut and sell your own wood?" I asked.

"No."

"How do they expect you to make a living?"

Joe laughed.

"We're supposed to farm. The Indian agent says that farming will make us like white people," he said. "I work in the camps in the winter, do some trapping. Like everybody else."

"It's against the law for us to have guns," Mary said. "I guess they're afraid that we'll go on the warpath." She laughed. The Chippewa had never fought a war with the Americans. They'd preferred negotiation, knowing they'd lose the land either way, but less life if there wasn't an excuse to kill them off.

"How do you hunt?"

"Oh, we have guns. We couldn't survive if we had to obey all their laws. We're even supposed to get a permit whenever we leave the reservation."

The Chippewas had never needed prisons. Now the government put them all in prison. If Sig wanted to think I was French, I'd be French.

"Stay in Winton, Pete," Joe told me. "You know, even Nanaboujou's given up on the Chippewa. When he saw how things were going, he got himself an easy government job working for the BIA."

"Are you going to go ricing at Nett Lake this year?" Mary asked when I left.

"Of course," I said.

Sig and I were up on Kawnipi one time, and he was going on as usual about the glories of living the real man's life out in the wilderness. Usually I didn't pay much attention when he rambled on like that, but all of a sudden I had a vision of my mother kneeling on the rock, dipping water from the lake into her iron pot. My family had often camped on the same site Sig and I were camping on.

"You don't want women here?" I asked.

Sig tipped back his hat and took a puff from his pipe. "Women aren't made for the rigors of the wilderness," he said.

"No children?"

"Boys are okay on short trips."

"No old people?" I thought of my white-haired grandfather playing his fiddle, my father teaching the children to step dance on the cliff.

"The wilderness is a place for men who hunger for action, distance and solitude," Sig said.

Not long ago this country was full of people, a village or a fur post no more than a day's paddle apart. Sig wanted solitude. He could not see the beauty my French grandfather had found in this place, the beauty of people who knew how to get along with each other, who found joy working together, living from creation and protecting each other from its terrors.

"There's more to living up here than paddling and portaging," I said.

"It takes skill for a man to provide for others. It's not as simple as paddling through, catching a few fish, maybe shooting some ducks. A man gets his honor by taking care of other people, being generous. That was the Chippewa way."

"What stage of progress did Indians attain? By conquering frontiers, white men created the greatest civilization the world has known," Sig said. He walked to the edge of the cliff and looked out over the lake. "Now we need places like this to reinvigorate ourselves to meet the challenges of civilized life, live like the frontiersman of yesteryear—so we can continue to progress."

I didn't argue. What harm was there in letting a man keep his fantasies?

"Anyway," Sig said, "the Indians couldn't see the beauty here like I can."

One of the first times I guided Sig, we took refuge in a trapping shack during a thunderstorm. There was one like it on every lake, the kind that was thrown together by a couple of guys in a few days, with a little stove and a bed made out of poplar poles and spruce boughs.

"Look," said Sig, "the owner must have left in a hurry. He left his matches and food."

"No," I said. "That's what you do in this country. You leave food and wood for anyone who needs them."

Sig's imagination quickened. "Think of the life this man lived, far from the responsibilities of modern living," he said.

I couldn't tell Sig anything, but I tried to make sure he understood that it was his responsibility to replace the firewood we'd used before we left.

One time Sig wanted me to take him to a lake off the map. I took him up the Maligne River and followed a creek to a little lake.

"Does this lake have a name?" he asked.

"Not on the map," I said.

As Sig stood there, pretending he was an explorer laying eyes on land no white man had seen before, he spotted the cabin.

"This lake's inhabited," he said, with real disappointment.

"Not for a long time," I said.

"Let's go take a look," he said.

"No. Leave it be."

"How come?"

"It's a place of tragedy," I said, "the scene of death."

"Ah, one of your superstitions," Sig said. Since he was in the stern of the canoe, I had little choice when he paddled us over to the cabin.

It wasn't a shack, but a well-made log cabin, though the roof had collapsed and a sapling was growing in the middle. It had been the home of Claude Bouchay.

Claude was married to my mother's sister Nellie. They lived in the cabin during the winter, close to his trapline. My mother worried about them, living by themselves like that.

Claude left Nellie alone with the children while he was checking his traps. Eva was seven, Charlie was five, and Nellie was pregnant with a third child. Something went wrong. The baby came early and Nellie didn't stop bleeding.

Mother said that Grandma could have saved her if she'd been at home in the village.

Eva left Charlie to feed the fire and went to look for her father.

Charlie kept the fire going while his mother died and waited there alone with her body. It was many days before his father came back.

Eva wasn't with him.

They never did find Eva.

Claude brought Charlie back to the village. Claude married again and had other children. All of them died. People said he was being punished. He went to the medicine man, and did what he was told. His next child lived. That was my cousin Mary.

I'd come here often with Charlie and Ira. Charlie said prayers for his mother and sister. Aunt Nellie was a Catholic, so first he prayed for her on her rosary, and then he prayed a Chippewa prayer for the dead.

The door had fallen free of the leather hinges.

"Don't go in there," I told Sig, but he didn't listen.

He emerged after a time with a small wooden box and opened it in the sunlight, revealing a tiny pair of soft white rabbit skin moccasins, fur side in, each decorated with three beads.

Nellie had made them for her new baby.

"I'd say that a Frenchman kept a squaw up here," Sig said. "What do you think?"

"I think you should put the moccasins back," I said.

"They'll just rot anyway," he said, putting them in his pocket.

We were paddling home, going around a bend in the Maligne, when I spotted a moose cow with her calf standing near the river bank. I gestured to Sig. He stopped paddling and we drifted by. The cow caught wind of us and lifted her head, a clump of rushes hanging from her mouth. She turned and galloped off into the brush.

"God, there has to be a way to make money off this country," Sig said.

He started an outfitting business in Winton in an old horse barn left over from logging days—Border Lakes Outfitting. Tourists were supplied with grub, packs, canoes, and a guide. That was me or Antti or one of the other guys who'd worked out in the camps.

One day Antti and I took the morning train to Ely to pick up some supplies. We walked past the barber shop.

"I need a haircut," I told Antti, turning back. "You could use one, too."

Antti sat in the chair first. I leafed through a *Field and Stream* magazine while I waited.

"Hey, Antti, there's an article by Sig in this magazine," I said. "About guides. Listen to this. 'As a breed, they are blessed of men, for they live a life more appealing to them than any other occupation on the face of the earth.'"

Antti snorted. "My favorite work," he said. "Wiping the asses of whining tourists."

I read on down the page. "The longer a man lives away from civilization, the more natural he becomes. Gone is the smooth veneer that makes him acceptable in society, and he is at last an individual with the God-given right to exercise his own free will."

"Yah," Antti said. "That's freedom. Go on a camping trip."

"There's more," I said. "To the true woodsman, the wilderness is always at its best . . . the motto of the guides in the canoe country is, 'No matter how wet and cold you are, you're always warm and dry.'"

Antti was laughing.

"Jesus, Antti, did you tell him that?"

"It was one miserable trip, rained for days. He started complaining on the second day. So I told him a real woodsman is always warm and dry. I meant a real woodsman knows how to keep himself warm and dry. But you haven't heard him complain about the weather since then, have you?"

Whenever Sig published something after that, the guides passed it around Border Lakes for a laugh. Sig must have thought we couldn't read when he wrote those things—or maybe he didn't care what we thought.

"This is a good one. Listen to this one," Big Art said. "'It was high noon of a breathless day in August when Joe Mafreau, the old half-breed, spotted the magnificent silver black fox and her three pups. In all his years of effort to bring back to the fur trading post a single specimen of the silver black, Joe had been frustrated . . .' Old half-breed. Must have been talking about you, Pete."

"What were you doing trapping in August, Pete?" Swedstrom asked. Big Art laughed so hard he had to stop reading. Swedstrom took the magazine and read on.

"'He revealed his plans to no one. Jackfish Pete, a renegade Indian, just released from prison, guessed Joe's motives, however.'"

"That sounds more like Pete," Swedstrom said. "Renegade Indian. When were you in prison, Pete?"

I took the magazine.

"'De fines' pup een de worl',"' I read. "'Eet ees a shame to put you een a cage wen de woods ees all around. Mebbe who knows, some day you come back to Lac La Croix.'"

"Yup. That's how you talk," Big Art said, and started laughing again.

After Sig put an article in *Field and Stream* about how a person could come up here and find lakes no man had ever seen, we had a new challenge.

"There is one thrill that never grows old," Sig had written, "the thrill of seeing for the first time new land or water. I do not mean water that is new to you only, but new to everyone; a spot of blue that has never been on a fisherman's map, something untried and untouched."

Lots of people read that article, and our customers started demanding to go somewhere no one had been before.

We'd get three or four parties going out at once, each for a week, each wanting to get to a place no one had ever seen.

We consulted each other to make sure that we weren't going to take two parties to enjoy the same primitive view at once. We renamed a few of the lakes.

"I'm going to take these folks up to Unseen Lake," I'd tell Antti. "You better steer your swampies over to Untouched."

Often, when I took a party through Basswood, Old Man Muskrat and his niece Bessie paddled over from the village to give me the news. They enjoyed talking to the tourists.

Sig took me aside after one of those trips.

"I'm getting complaints from customers. They don't like being bothered by Indians."

"It's just Old Man Muskrat and Bessie. We smoke our pipes, talk a little, then they go home. Old Man Muskrat is being hospitable, making visitors feel welcome."

"Well, it annoys the guests. They come a long way to find solitude up here, not be pestered by natives."

I guess you need solitude if you don't know how to be friendly.

Not long after that, government agents came to clear out Old Man Muskrat and the few Indians who were left on Jackfish Bay.

People said Sig was behind it because he was always talking to the government. I asked him, but he wouldn't give me a straight answer.

"It was bound to happen sooner or later," he said. "The Indians are a dying race. They're already corrupted by the white man, and they make people nervous when they hang around. Besides, they net too many fish. They're going to spoil the fishing for sportsmen."

I tried to follow Sig's logic. According to him, Indians were all right if they were stuck in time while everybody else changed, but how could they do that when the land changed, the world changed? If they changed with the times like everyone else, then they were corrupted.

The way Sig had it, Indians were doomed no matter what they did.

It looked like everybody was doomed when the Depression hit. Banks failed, mines closed and lumber camps shut down. Railroad cars rusted on sidings, loaded with logs that rotted away. But there were still plenty of rich people in America who wanted to come up north for a canoe trip. Border Lakes Outfitting did a booming business and Sig built himself a fine house in Ely on Snob Hill, in the darkest days of the Depression. Swedstrom went there once.

"Nice shack," he told us.

By then Jake and Josie were nearly grown, going to school in Ely. They always started school late every fall, after the wild rice harvest, but now that they were in high school I wondered if they would still want to come with me when I went to Nett Lake to make rice.

"You're joking," said Josie. "We wouldn't miss ricing for anything."

We took the train to Tower on a Saturday and waited in front of the movie theater. When the movie was over, out came the Indians.

"Hey, Pete! Josie! Jake! Boujou!" Ira shouted out when he saw us.

"Must be ricing time. Pete shows up," Bessie Waboose teased.

"Good thing, too. We need more pretty girls like Josie over at Nett Lake. You should have come early and seen the show. It was a western," Ben Geshick said.

We walked down to the river where they'd left their canoes pulled up on shore, paddled down to Lake Vermilion and over to the village.

I sat big shot with Ira and Ben in Ben's leaky old birch bark canoe. "We're trying to build a road from the reservation to town, but the government says we can't have one," Ben said. "It's okay to paddle in the summer, and it's not a bad walk over the ice in the winter. But during freeze up and break up, you can't get supplies, can't get to a doctor."

"Why can't you have a road?" I asked.

Ben shrugged. "The government says that isolation protects us."

I laughed. The Indians were always on the go, down to Wisconsin, up into Canada, visiting relatives. Chippewa territory was big country, with cousins all over the place.

Mary and Joe were busy getting their gear together for the trip to Nett Lake. Next day we started out with fifteen canoes. I paddled with

Bessie and Old Man Muskrat. Jake and Josie went with the Keewatin kids.

The canoes were packed with babies, old people, teenagers. Dogs raced along the shore. We paddled, shouting and joking across the water, to the place where Chief Wakemup's village used to be on the other end of Vermilion and made the portage to the Little Fork. We followed the Little Fork due west for twenty-five miles and camped overnight before walking the trail to Nett Lake.

Nett Lake was a village of log cabins and tarpaper shacks, like Vermilion. The people who came from Vermilion camped together under the elms and maples.

I stood on the shore, looking out over the vast lake. Across the water, the great bed of wild rice rippled in the breeze, glinting with gold. Maude Waboose and members of the rice committee were out marking ripe patches, sticking tall poles with white flags into the mud. The flags beckoned, fluttering above the gold-green of the rice leaves. From a time beyond history, Chippewa people had made rice here each autumn.

We stayed up late that night, talking around campfires. Little kids sat quietly, drinking in every word of the talk, Jake and Josie laughed with their friends beyond the circle of firelight.

The morning dawned crisp and bright, a perfect day to make rice.

Bessie and I let our canoe drift among the others near the landing, waiting for the dew to dry off the rice, waiting for the signal for the harvest to begin.

I scattered tobacco on the lake. "Migwitch," I said, thanking God for the harvest. Maude waved a scarf. A great roar rose from our throats as we headed out across the lake and into the field of wild rice.

I knocked rice into the canoe while Bessie paddled, thinking of Uncle One-Eye who'd taught me to knock rice. I was careful not to break the stalks, not to shake off the green rice before it ripened, not to knock leaves in with the kernels. I remembered some of his jokes and told them to Bessie, who remembered others. Our spirits were so high we didn't feel our muscles tire.

We paddled back in the early afternoon, to roast the rice we'd harvested. The fragrance of warm, parched rice wafted from the village.

Old women stirred the rice in great iron pots. Old men put on new moccasins and danced in the pots, jigging off the husks. Josie worked with Mary and Bessie, tossing rice from birch trays, letting the chaff blow away.

Fires glowed in every yard. We worked steadily through the twilight into the night, murmuring to each other and singing to ourselves while we finished the rice.

Ira came and put his arm around my shoulders.

"It's a good harvest, Pete," he said, and winked. "Looks like we'll be around another year."

Sahotra Sarkar

Wilderness Preservation and Biodiversity Conservation (1999)

Keeping Divergent Goals Distinct

Conservation biology, as developed and practiced in the United States, has the explicit aim of maintaining and encouraging biodiversity. The term "biodiversity" was introduced in 1986 by Walter Rosen as a shorthand for "biological diversity." Although Rosen's original intention was quite precise, biodiversity, according to a survey of US conservation biologists, has become a fashionable scientific—but no more precise—substitute for the undeniably vague term "nature" (Takacs 1996). These conceptions of biodiversity are actually quite different, and the differences matter when strategies for biodiversity conservation have to be devised.

"Biological diversity" may be hard to define, but its intended meaning is not hard to fathom: It refers to diversity at all levels of biological organization, from alleles, to populations, to species, to communities, to ecosystems. "Nature," by contrast, is a much more vague term: In the United States, at least, it seems mostly to refer to "wilderness" (Cronon 1996b). Meanwhile, "wilderness," according to the 1964 US Wilderness Act, is a place "where man himself is a visitor and does not remain." Humans are sometimes admitted as being part of a wilderness, especially if they are members of indigenous groups already resident in that "wilderness." But from this Eurocentric point of view, these humans are not much different from other animals: Bereft of "civilized" culture, they do not destroy the sanctity of a pristine wilderness. Another aspect of "wilderness" is that the wilder a place, the more natural it is. An Antarctic landscape is more of a wilderness than the interior of an Amazonian rainforest—the latter has a higher density of human inhabitants.

Biodiversity conservation, therefore, cannot be identical with wilderness preservation (see also Haila 1997). In this article, I explore the differences—that is, examine exactly how the two goals differ and what that difference entails, particularly for biologists. The goals differ not only with respect to their explicit and implicit long-term objectives, but also with respect to their justifications, their immediate targets and obstacles, and the strategies that are likely to achieve these targets (Table 1). In some instances, the tasks of biodiversity conservation and wilderness preservation converge, but at least as often they do not. When they do not, the conservation of biodiversity is often more feasible when that goal is not conflated with that of wilderness preservation. This point is important because there is a third factor that is often critical to conservation efforts: social interests, whether those of social justice movements (whose immediate goals often coincide with the interests of conservationists) or aspirations for economic improvement (which may or may not conflict with biodiversity conservation). If wilderness preservationism is cast aside as a predetermined goal, it can become easier for biodiversity conservationists to negotiate and, often, to achieve consensus with these social interests.

WILDERNESS AND ITS PRESERVATION

"Wilderness" as a category of positive concern—as opposed to "waste" lands to be tamed and used efficiently by humans—is of recent and highly localized vintage. As Nash (1973) put it: "Friends of wilderness should remember that in terms of the entire history of man's relationships to nature, they are riding the crest of a very, very recent wave." Although the origins of this concept of wilderness are usually traced back to eighteenth-century European romanticism, its relevant use in today's

Table 1 Wilderness preservation and biodiversity conservation—summary of issues.

Issue	Wilderness preservation	Biodiversity conservation
Objective	Landscapes without humans	Biological diversity at all levels of organization
Justifications	Aesthetic	Intellectual interest; present and future utility
Targets	National Parks; wilderness preserves	High-biodiversity regions; representative sample of biodiversity
Obstacles[a]	Economic interests; over-consumption; human encroachment; invasive technologies	Economic interests; over-consumption; human encroachment; invasive technologies; habitat fragmentation; human exclusion, in some cases; diversion of scarce resources from conversation to wilderness
Strategies	Legislation; habitat purchase	Diverse methods

[a]Obstacles are of particular importance because the results of wilderness preservation, such as the creation of small national parks and complete exclusion of human use, may generate problems for biodiversity conservation. In such a situation, the two goals are in conflict. See text for further discussion.

context emerges only in the late nineteenth and early twentieth centuries, primarily in the United States (Oelschlaeger 1991, Denevan 1992, Cronon 1996a, 1996b). Wildernesses as uninhabited areas in the United States were generally created through the exclusion of the human residents (i.e., the First Nations) and an erasure of their history. Erasure refers to the systematic, if unconscious, reconstruction of memory to recast as uninhabited "wildernesses" the lands from which the original inhabitants were forcibly expelled. The final stage of exclusion was achieved at the end of the last "Indian" wars, when the remnants of the First Nations were herded into reservations and their traditional lands were declared to have been unoccupied by humans from the beginning of time (Cronon 1996b).

John Muir, the founding figure of the wilderness myth in the United

States, accepted the presence of the First Nations in such putative wildernesses as Yosemite (which was set aside as a park in 1864) and claimed that "Indians walked softly and hurt the landscape hardly more than the birds and squirrels" (Nabhan 1995). He also recognized the importance of the use of fire by the First Nations in creating the landscapes that he hoped to preserve as wildernesses. Subsequent wilderness advocates generally forgot this caveat. But they all agreed that wilderness was what the First Nations—for instance, the Miwok of Yosemite—had called home. What was universally ignored by these later advocates was that Yosemite and other wildernesses had been intensively but relatively stably modulated by their First Nation residents for centuries.

During the first decade of this century, Muir and his followers fought the first modern environmentalist battle, over whether the city of San Francisco had a right to dam the Tuolomne River (in the Hetch Hetchy Valley inside Yosemite National Park) to augment its water supply (Cronon 1996b). They lost that battle but ultimately won the war to designate and preserve wildernesses throughout the United States.

Wilderness preservationism spread. The immediate target was the creation of national parks for recreational use, primarily by short-term visitors. The strategy of choice—and, in retrospect, a very effective one—was federal intervention eventually imposed with the force of law (Nash 1973, Graber 1995). Concern for biological diversity played no role in the selection of US national parks in the early decades of this century: The first swamp was so designated (the Everglades National Park) only in the 1940s, and there is still no national park dedicated to preserving grasslands (Cronon 1996b). Rather, the national parks were "sublime" landscapes: mountains, waterfalls, and other landforms of exquisite and deep aesthetic appeal to transient visitors, who usually came from an urban elite rather than from the surrounding rural population. Nevertheless, throughout the world, especially since 1950, the creation of national parks has emerged as the predominant strategy of biodiversity conservation.

Arguably, the concept of national parks is a US export to the rest of the world. At the very least, the United States had priority in the formulation of official policies to create and maintain national parks. For example, it was not until 1930 that Canada passed a National Parks

Act (earlier parks had been aimed at the preservation of historic sites rather than landscapes; Doern and Conway 1994); a system of wildlife reserves, including parks, was established in Kenya after 1945 (Olindo 1991); and although the British created some kinds of forest reserves (mainly multiple use) in India in the nineteenth century and in what is now Malaysia at around 1900, it was not until around 1930 that identifiable parks were created in those countries (Gadgil and Guha 1992, Aiken and Leigh 1995). Wilderness preservationism reached most of Latin America even later, with Mexico being an unusual early adopter of parks.

Because the concentration of economic power in the so-called First World has made it the source of calls for nature preservation during the last 25 years, the creation of national parks in the US mold to preserve wilderness has emerged as a part of almost every conservationist and preservationist strategy (Cronon 1996b). The Convention on Biological Diversity, formulated at the 1992 Rio de Janeiro United Nations Conference on Environment and Development (UNCED), even included wilderness as a type of ecosystem and habitat to be targeted for conservation.

CONSERVATION OF BIOLOGICAL DIVERSITY

During the 1980s, biodiversity became the focus of concern of biologists alarmed by the increasing pace of anthropogenic extinction, particularly in the neotropics (e.g., Wilson 1988). Concern for biodiversity came to include the traditional focus on the potential extinction of charismatic (and culturally symbolic) species as well as those recognized as endangered by the 1973 Convention on International Trade in Endangered Species of Wild Fauna and Flora (CITES). "Biodiversity" includes allelic diversity within populations, structural differences between populations, diversity of species, and diversity at higher levels of phylogenetic and ecological organization (e.g., ecosystems). In recent years, the targets of biodiversity conservation have expanded from the traditional concern with species and even ecosystems. For instance, they now include "endangered biological phenomena" such as the migration of the monarch butterfly (*Danaus plexippus*) in the United States and Mexico, each cycle of which takes several generations (Brower and Malcolm 1991).

A serious problem with the concept of biodiversity is that there is no fully satisfactory quantitative measure for it. Three commonly used indexes, α-, β-, and γ-richness, refer, respectively, to the number of species within a homogeneous habitat, the rate of change of species composition between habitats, and the rate of change across larger units. These indexes do not, however, capture the value of rarity, for instance, of endangered biological phenomena or of unique species in otherwise unexceptional or biologically impoverished habitats. Identifying conservation targets often remains dependent on educated intuition. Given the diversity of the appropriate foci for conservation, these targets can vary widely, ranging from deserts and forests to farms and horticultural gardens (e.g., if these two "unnatural" systems contain unique species). The important point is that, except perhaps for a few species, such as large, wide-ranging predators, there is no *a priori* reason to suppose that conservation of biodiversity requires wilderness, such as national parks that exclude humans. Whether it turns out that this is so will depend on empirical data, as will be discussed below.

The intellectual rationale for biodiversity conservation was forcefully enunciated by Janzen (1986) in the context of a plea for the conservation of tropical forest systems: He asserted that the unknown biological systems of the tropics present research problems of unparalleled biological interest. Moreover, both the species and the ecosystems are unique and threatened with impending permanent extinction. Janzen added that past experience has also shown that tropical forests are a potential source of many practically valuable commodities. The last assertion is not a biological reason, but in a late twentieth century context, when the sciences often have to justify expenditure and investment in them in terms of economic value, it was an important point to make.

Ehrenfeld (1976) and some other conservation biologists have presented other, generally less utilitarian rationales for biodiversity conservation (for discussions of the various positions see Norton 1986, 1987, Rolston 1994). If the importance of intellectual interest is acknowledged, along with the potential for new resources, then these other reasons (which have generally been controversial) are not strictly necessary for an adequate defense of biodiversity conservation. Nevertheless, in the political arena, these other reasons—including, for instance, consonance with certain cultural and religious traditions—may well have more persuasive power than the recognition of the undeniable intellectual interest of biodiversity.

CONFLUENCES

In spite of the radical differences in the long-term objectives and usual targets of biodiversity conservation and wilderness preservation, and in the justifications offered for these two goals, there are points of confluence. The existence of such overlaps explains why the two goals have so often been pursued in tandem and why some biologists who recognize the differences between them still downplay the significance of these differences (e.g., see Graber 1995). There are both positive and negative reasons for the confluence between biodiversity conservation and wilderness preservation. The positive reason is that they share one type of target habitat: fragile ecosystems or those with rare or endangered species that show little or no human influence. Such habitats exist but are probably rare. None of the 18 global hotspots of biodiversity identified by Myers (1988, 1990) satisfies the criterion of either minimal human presence or influence.

More important are the negative reasons for the confluence: Many issues of concern for wilderness advocates are equally important to biodiversity conservationists. These issues include at least four anthropogenic factors that contribute to the extinction of populations, species, and ecosystems:

- *Economic cornucopianism.* Whether it be attempts to drill for oil within the Arctic National Wildlife Refuge in Alaska, cattle ranching in Amazonia, or logging in Borneo, unbridled market forces have been inimical to the interests of both wilderness preservationists and biodiversity conservationists. For both camps, a potential solution is to assign economic value to environmental goods or services. There have been significant attempts in this direction (e.g., Pearce 1993), but in practice there is little reason for optimism that such a strategy will be effective in preventing destruction of nonrenewable natural resources for short-term profits.

- *Overconsumption.* Wilderness and biodiversity advocates generally agree that current patterns of natural resource use, if accompanied by current rates of population increase, cannot be sustained in the long run. Overconsumption arises from excessive per capita consumption in most industrialized countries and high population densities in many developing countries. It endangers wilderness preservation by inevitably requiring the development of wild lands, and biodiversity conservation either by direct use of relevant resources or by reduction of habitat. How much population a region or the entire world can sustain depends on choices about patterns of living; contrary to the often emotional debates about this issue, there are no known relevant absolute limits (Cohen 1995).
- *Human encroachment.* By definition, human encroachment would destroy a wilderness. Human encroachment can—although it need not—also deplete biodiversity, for instance, by destroying biotically fragile habitats. Encroachment may be, but is not necessarily, a result of overconsumption. Even necessary human activities far from a region can lead to encroachment, for instance, through the emission of atmospheric pollutants that affect the appearance of landscapes or the viability of populations.
- *Invasive technologies.* Even if wildernesses admit only temporary visitors, it matters for preservation whether these visitors come on foot or in motorized vehicles. Even people who argue for the compatibility of biological conservation and human use admit that modern technologies, such as mechanized forestry, pose special problems. Gómez-Pompa and Kaus (1992) make a useful distinction between technologies that are internal to a local environment and those that are external. The former emerge and evolve by trial and error within communities, usually those with a stake in the long-term renewable management of resources. The latter often introduce irreversible changes that are generally

detrimental to the entire habitat, including the biological diversity that it contains.

CONFLICTS

Given how common it is that wilderness preservation and biodiversity conservation are conflated, it may be surprising that the area of confluence is so minute. The reason for the small overlap is that, contrary to a pervasive implicit belief, the presence of humans *per se* is not necessarily detrimental to biodiversity. Human groups, particularly those that do not use invasive technologies and have lived in a region for many generations, are often integral parts of ecosystems and may have little or no negative impact on biodiversity. Moreover, even intrusive human use may not always be detrimental to biodiversity.

Two examples appear to be typical of such benign, although intrusive, use. First, Latin American tropical rainforests, particularly those along the Amazon, are ecosystems of extraordinarily high biodiversity. Hecht and Cockburn (1990) have systematically documented not only high densities of past human populations in many parts of the Amazon, but also evidence of intensive but stable modulation of forests (see also Posey and Balée 1989). Second, in Costa Rica, the La Selva Biological Station straddles a transition between tropical premontane and wet forest life zones (Hammel 1990). It borders the Braulio Carrillo National Park and is a major repository of biodiversity (Barry 1990, Clark 1990). Until recently, it was universally believed to be devoid of human influence. But, in recent years, pottery shards and crop residues have been discovered in La Selva (Yoon 1993).

There are many more examples of humans forming integral parts of ecosystems that have shown no recorded biodiversity decline. However, the fact that there is no recorded decline from their use of natural resources does not mean that there was no general biodiversity decline or that it increased. No study to date adequately distinguishes the effects of indigenous groups on general biodiversity from their effects on the abundance of key resources (Nabhan 1995).

Three examples show how traditional resource management systems have been used to maintain or increase an abundance of desired species, which are the key resources in this context (for other examples, see Anderson 1996). First, the Nass River watershed in northern British Columbia (Canada) was the traditional home of the Nishga First Nation. Each Nishga community controlled its own part of the watershed, and a hierarchical system of controls allocated specific fishing sites to individuals (who acted on behalf of groups). This allocation system prevented overharvesting (Gadgil and Berkes 1991). Second, the *dina* system in Mali manages biotic resources by ensuring that different groups specialize in different resources. For instance, the Bozo people specialize in shallow-water fishing, whereas the Somono specialize in net fishing. Four different groups specialize in farming, and one, the Fulani, specializes in herding. Once again, the system, which was formalized in the nineteenth century, seems to act to prevent the disappearance of any single resource (Gadgil and Berkes 1991). A third example comes from Mexico, where traditional shifting (i.e., slash-and-burn) agriculture involved processes of clearing, planting, and fallowing, which resulted in a mosaic of forest patches at different growth stages, including a significant amount of mature forest. This mosaic prevented forest fires from spreading (Gómez-Pompa and Kaus 1992). For the farmers, it also ensured adequate amounts of sufficiently renewed soil for agriculture. The prevention of fires almost certainly helped to preserve biodiversity in the surrounding forest.

Unless the targeted resources act as keystone species in their ecosystems, it is impossible to conclude from these examples that general biodiversity was maintained. Nevertheless, it seems likely. Indeed, circumstantial evidence suggests that, in many cases, human intervention has been critical to the maintenance of biodiversity. Although most such reports are anecdotal, two cases have been extensively documented.

First, in Keoladeo National Park in Rajasthan, India, a 450 ha artificial wetland with shallow bodies of water was created in the eighteenth century. This wetland attracts tens of thousands of wintering waterfowl and also supports large numbers of bird species that breed during monsoons (Gadgil and Guha 1995). Before Indian independence (1947), the area was a hunting reserve that also served as a grazing ground for cattle from the surrounding villages and as a water source for irrigation during the dry postmonsoon period. After 1947, it was set aside as a national park. On the advice of Indian and US ecologists, who had not, however, carried out detailed field studies, grazing was banned in the early 1980s in an effort to promote bird diversity. When villagers protested the loss of fodder, the Indian state responded with violence; the police killed several protesters. The ban on grazing has, despite its intent, devastated Keoladeo as a bird habitat, especially for wintering geese, ducks, and teals. Paspalum grass, which had been kept in check by grazing, has now established a stranglehold on the wetland, choking the shallow bodies of water (Vijayan 1987).

A second example comes from the Sonoran Desert, where two oases, one on either side of the United States–Mexico border, were subject to different management regimes. On the US side, the protection of an oasis by its inclusion in the Organ Pipe Cactus National Monument led to a significant decline of species diversity over a 25-year period. On the Mexican side, continued traditional land use by Papago farmers at the Quitovac oasis 54 km to the southeast led to no such decline (Nabhan et al. 1982).

These examples can be easily discounted by skeptics on the grounds that the available comparisons are not proper controls. Unfortunately, it is virtually impossible to find adequate controls for actual conservation targets. Experimental work similar to studies of the effect of habitat fragmentation on biodiversity should also be conducted to study human impacts. The examples given here suggest that many forms of human use need not have an adverse effect, and may indeed have a positive effect, on biodiversity.

POLITICS

Biological conservation does not proceed in a sociopolitical vacuum. The wilderness preservation strategy of setting up national parks largely to exclude human resource use while encouraging transient visitors, often to finance park maintenance, leads in at least two ways to a contentious political terrain that is inimical to biodiversity conservation. These problems typically occur together. First, setting up a national park that excludes humans from traditional habitats and prevents them from using resources denies them perceived cultural, economic, or social enhancement. The result is a political conflict that can significantly hurt biodiversity conservation efforts. Second, the strategy of concentrating on the creation of national parks in which all human resource use is excluded may result in compromises in which regions outside the reserves are entirely unprotected. The reserves then become isolated habitats, in many ways similar to islands, with all the attendant negative consequences for biodiversity.

Two cases illustrate the first problem. One is that of Tortuguero, on Costa Rica's Caribbean coast, where turtle conservation efforts began in the 1950s. In the 1960s, decrees were promulgated to regulate hunting, and legislation passed in 1970 and 1975 converted the protected area into the Tortuguero National Park and banned hunting completely (Boza and Mendoza 1981, Lefever 1992). Between 1970 and 1995, the park became a major ecotourism destination. The facilities created to house the visitors range from expensive rustic lodges to cheap rooms in Tortuguero village. Access to the village is by a 6-hour motorized boat ride. Regular boat service existed for a while, but it petered out in the late 1980s. The villagers now have to rely on the goodwill of tour operators for all transportation. Although boating was the traditional mode of transportation before the creation of the park, the villagers feel that they are being denied the material advancement common elsewhere in Costa Rica. This situation, along with the absence of adequate educational and medical facilities in the village, had generated significant local resentment by 1995 (Sahotra Sarkar, unpublished local interviews, August 1995). The residents perceived a state policy of preference for foreign tourism over local needs. Their solution was to propose a road to Tortuguero that would pass through parts of fragile forests near the Tortuguero reserve. The road would also probably lead to illegal logging and destruction of biologically important habitat. Although the road has yet to be built, it has significant political support in all surrounding communities.

The social problems at Tortuguero pale in comparison to the scale of those created by the establishment of national parks and reserves in India. This process gained momentum around 1960, promoted by a coalition of Indian hunters-turned-preservationists (from a declining colonial elite) and international organizations such as the World Wildlife

Fund and the International Union for the Conservation of Nature and Natural Resources (Guha 1989). By 1989, India had 65 national parks and 380 wildlife sanctuaries (Kothari et al. 1989, Agarwal 1992). The parks were usually based on the habitats of large charismatic mammals, such as the Asian elephant (*Elephas maximas*), the greater one-horned rhinoceros (*Rhinoceros unicornsis*), and the Bengal tiger (*Panthera tigris tigris*). The establishment of the parks led to massive dislocation of villages, often without any adequate provisions for the displaced (Guha 1989, Agarwal 1992). No official figures for the numbers of displaced people are available, but the following example from a region with almost no large parks is probably representative: In the Jharkhand region of east-central India, thousands of people were displaced to create parks and reserves. Because there is no official policy of culling excess animals, villages surrounding the parks suffer from depredations (Sukumar 1989, 1994, Agarwal 1992). Tension between state officials and villagers is unusually high (Agarwal and Narain 1993, Gadgil and Guha 1995). Poachers sometimes hire villagers and successfully evade anti-poaching efforts using local knowledge; there have even been acts of arson against parks and reserves by villagers adversely affected by their establishment (Gadgil and Guha 1995).

Two examples are characteristic of the second political problem—that is, that protected reserves can become "islands" in a sea of unprotected regions. One example comes from Costa Rica, which is usually touted as a success story of biodiversity conservation through what, in this article, is called wilderness preservation. Approximately 29% of the land area of Costa Rica is set aside on paper as national parks and reserves (Meffe and Carroll 1994). Yet Faber (1993) estimates that, whereas 55% of Costa Rica was forested in 1961, only 22% remained forested in 1991. Only approximately half of this forested area was in parks and reserves. This fact makes the 29% land area in reserves irrelevant because much of this area is neither forested nor otherwise important for biodiversity. In addition, Repetto (1992) estimates that approximately 28% of Costa Rica was deforested between 1966 and 1989. Even if Faber's lower, 33% estimate of total deforestation is used, the average annual deforestation rate was 6.9%, by far the highest in Latin America. This high deforestation rate apparently resulted because the establishment of official conservation areas led to almost total deforestation outside these areas, usually to satisfy economic interests. There was no premium on establishing simultaneous human use and biodiversity preservation through the 1980s, although this attitude is slowly changing. Meanwhile, the national parks have become isolated wildernesses of varying size. Many of them are also ravaged by excessive ecotourism (Wallace 1992). Thus, the strategy of wilderness preservation has not been particularly successful in conserving biodiversity.

A second example of the second problem is provided by Sabah, in Malaysian Borneo. Sabah has 386,375 ha in national parks and reserves (Aiken and Leigh 1995), which form more than 5% of the land area. What this figure hides is that some parks, such as the Tawau Hills Park, were extensively logged (approximately 40% of Tawau Hills) before their designation as national parks or reserves (Aiken and Leigh 1995). Over two-thirds of the reserve area is occupied by commercial forests (Cleary and Eaton 1995). Outside of a few major national parks, such as Mt. Kinabalu, Sabah is almost entirely logged. The major purpose the designation of parks seems to have served is to deflect environmental criticism. More important, the parks are isolated and separated by extensively logged regions, thus restricting many species to small and isolated habitats.

The two types of problems with national parks—political conflict and the promotion of destruction outside the parks—are often compounded in the same situation. An example is the situation created by the Gunung Mulu National Park, a 52,890 ha park in Sarawak in Malaysian Borneo that was created in 1974. The indigenous nomadic Penans who had lived in the area that became the park retained their right to use it. Although the park has developed into a major tourist attraction, the area outside it is being extensively logged, despite much-publicized protests by Penans and other indigenous groups living in these areas. Logging has entirely isolated the forest in the park (Cleary and Eaton 1995). Newly dislocated Penans from the other areas are not permitted to use the park (Sahotra Sarkar, unpublished local interviews, July 1996). Indigenous Berawans, who also traditionally lived within the area of the park, now live outside, with no right to use resources inside (unlike the local Penans). For a while, they catered to visitors by providing tourist facilities. However, their houses are now slated for demolition and their land is being expropriated to create a golf course outside the park intended primarily to attract Japanese tourists. Local resentment is severe.

It is important to note that in every one of these cases, there is no essential contradiction between social interests and biodiversity conservation. In Tortuguero, for example, conservationists could have helped to establish regular transportation by boat, using the leverage that they have with the Costa Rican government, and attempted to ensure local medical services and educational facilities. These strategies would have simultaneously helped conservation and improved conditions for local people. In India, the prevention of village dislocation, and the incorporation of local inhabitants into conservation, would have helped to avert poaching. (Moreover, in this case, whatever biodiversity that existed had done so in association with human presence for millennia, albeit with significantly lower populations—so there was probably no need for the parks in the first place.) All except a tiny minority of ranchers would have benefited from the partial maintenance of forests throughout Costa Rica. The beneficiaries would have included agriculturists, who probably form approximately 80% of Costa Rican farmers (Caufield 1991). Finally, in Sabah and Sarawak, logging benefited a tiny fraction of the population while destroying several indigenous cultures. Isolated national parks, while being at best questionable safeguards for biodiversity, served mainly to deflect environmentalist criticism. Biodiversity conservation would have been better served had traditional land usage been allowed to continue.

Because of the political problems caused by the creation of national parks and reserves, social ecologists (i.e., those who see ecological health as a social justice issue) have routinely criticized the creation of national parks and reserves, especially when these are created by decree of distant national governments acting in concert with international agencies and ignoring local needs. The arguments and evidence presented in this article show that this strategy is actually a result of biodiversity conservationists using a strategy borrowed inappropriately from wilderness preservationists. The real dispute should be between social ecologists and wilderness preservationists, not between social ecologists and biodiversity conservationists. Some of these issues have been implicitly acknowledged in the last decade, for instance, through UNESCO's "biosphere reserve" projects and integrated conservation and development projects (ICDPs), which were initially floated in the early 1970s but popularized only during the last decade (see Alpert 1996). In contrast to national parks, biosphere reserves and ICDPs allow human habitation and use of the designated areas, as long as that use does not cause irreversible degradation of the habitats.

CONCLUSIONS

Wilderness preservation cannot be used as a surrogate for biodiversity conservation. Indeed, wilderness preservation and biodiversity conservation may be in conflict. Biodiversity conservation can avoid contentious political issues while simultaneously promoting long-term conservation by maintaining independence and distance from wilderness preservation. In the conflict between wilderness preservation and biodiversity conservation, the biologist's professional concerns are (with the exception of situations such as that of ecologists desiring specifically to study ecosystems bereft of humans) restricted to biodiversity conservation. Janzen's (1986) plea for activist conservation serves as a manifesto for the latter. The biologist must ensure biodiversity conservation to prevent systems of outstanding intellectual interest from disappearing forever through extinction. And to the extent that biologists must justify their demand on societal funds by pointing toward utility, then, extrapolating from past experience, it is likely that unexplored biological systems will provide resources of significant value.

This is not to say that wilderness preservation (i.e., the maintenance of currently uninhabited or sparsely inhabited landscapes) may not have value in its own right. Preservation should, however, involve negotiation with other interests, such as those of developers. There is also the important question of how much preservation is affordable. Affordable preservation would probably include the national parks and many other areas in the United States and other wealthy nation-states. Appropriate negotiation and settlement would also probably lead to the wilderness designation of areas such as forests in the Pacific Northwest (which are contested between loggers and preservationists) without having to rely on the politically vulnerable fate of single species, such as the Northern spotted owl (*Strix occidentalis caurina*). By following a strategy of negotiation and settlement, the preservationist acts on deeply held aesthetic values and, in contested situations, should expect to pay to uphold them.

Social concerns are not concerns of the biologist as biologist. As a member of a civil society, a biologist may have a social conscience. There may be cases in which biological conservation necessarily conflicts with social justice, and there may be no tangible process through which such conflicts can be avoided. For instance, hunting an endangered turtle may be an eastern Indian community's only method of avoiding starvation. Because of socio-political constraints, the economic changes necessary to defuse such a situation may not be possible in the time frame during which the turtle becomes extinct. In such contexts, social justice must take precedence over the biologist's desire for conservation. Fortunately, situations of this kind are vanishingly rare—no clear example is available.

Similarly, a biologist may also value wilderness. But when wilderness preservation conflicts with biodiversity conservation, the biologist's professional judgment must favor biodiversity. Once again, such a choice is rare at present (although not unimaginable, as the examples of Keoladeo and the Organ Pipe Cactus National Monument show). Moreover, choices of this sort are likely to become more frequent because biodiversity protected in wildernesses is often a biased, nonrepresentative sample of biodiversity (e.g., see Pressey 1994). Therefore, wilderness preservation may conflict with biodiversity conservation when the former restricts options for the latter by diverting resources from biodiversity to wilderness.

A more usual choice occurs when wilderness preservation conflicts with social justice. Ordinary ethics dictates the precedence of the latter, but it is important to note that in these conflicts, biodiversity conservation usually gives no reason for preferring wilderness preservation over social justice. In such a circumstance, the issues must be kept separate, and biodiversity concerns—and professional judgments—must not be allowed to be used to hide the real choice: wilderness or justice. The stark choice must be faced by the individual as aesthete and citizen. That reality, perhaps, is the major—and most troubling—point of this article.

ACKNOWLEDGMENTS

Thanks are due to Rebecca Chasan, Ramachandra Guha, Yrjö Haila, Kurt Jax, Chris Margules, Bryan G. Norton, Kent Peacock, and four helpful referees for comments and criticisms of earlier drafts of this paper. Needless to say, they should not be held responsible for any unpalatable views in this piece. Thanks are also due to the Wissenschaftskolleg zu Berlin and the Max-Planck-Institut für Wissenschaftsgeschichte in Berlin for their support of this work.

REFERENCES CITED

Agarwal A. 1992. Sociological and political constraints to biodiversity conservation: A case study from India. Pages 293–302 in Sandlund OT, Hindar K, Brown AHD, eds. Conservation of Biodiversity for Sustainable Development. Oslo (Norway): Scandinavian University Press.

Agarwal A, Narain S. 1993. Towards green villages. Pages 242–256 in Sachs W, ed. Global Ecology: A New Arena of Political Conflict. London: Zed Books.

Aiken SR, Leigh CH. 1995. Vanishing Rain Forests: The Ecological Transition in Malaysia. Oxford: Clarendon Press.

Alpert P. 1996. Integrated conservation and development projects. BioScience 46:845–855.

Anderson EN. 1996. Ecologies of the Heart: Emotion, Belief, and the Environment. New York: Oxford University Press.

Barry H. 1990. The distribution of diversity among families, genera and habitat types in La Selva. Pages 75–84 in Gentry AH, ed. Four Neotropical Rainforests. New Haven (CT): Yale University Press.

Boza MA, Mendoza R. 1981. The National Parks of Costa Rica. Madrid: Incafo.

Brower LP, Malcolm SB. 1991. Animal migrations: Endangered phenomena. American Zoologist 31: 265–276.

Caufield C. 1991. In the Rainforest: Report from a Strange, Beautiful, Imperiled World. Chicago: University of Chicago Press.

Clark DB. 1990. La Selva biological station: A blueprint for stimulating tropi-

cal research. Pages 9–27 in Gentry AH, ed. Four Neotropical Rainforests. New Haven (CT): Yale University Press.
Cleary M, Eaton P. 1995. Borneo: Change and Development. Kuala Lumpur (Malaysia): Oxford University Press.
Cohen JE. 1995. How Many People Can the Earth Support? New York: W. W. Norton.
Cronon W. 1996a. Introduction: In search of nature. Pages 23–56 in Cronon W, ed. Uncommon Ground: Rethinking the Human Place in Nature. New York: W. W. Norton.
———. 1996b. The trouble with wilderness; or, Getting back to the wrong nature. Pages 69–90 in Cronon W, ed. Uncommon Ground: Rethinking the Human Place in Nature. New York: W. W. Norton.
Denevan WM. 1992. The pristine myth: The landscape of the Americas in 1492. Annals of the Association of American Geographers 82: 369–385.
Doern GB, Conway T. 1994. The Greening of Canada: Federal Institutions and Decisions. Toronto (Canada): University of Toronto Press.
Ehrenfeld DW. 1976. The conservation of non-resources. American Scientist 64: 648–656.
Faber D. 1993. Environment Under Fire: Imperialism and the Ecological Crisis in Central America. New York: Monthly Review Press.
Gadgil M, Berkes F. 1991. Traditional resource management systems. Resource Management and Optimization 8: 127–141.
Gadgil M, Guha R. 1992. This Fissured Land: An Ecological History of India. New Delhi (India): Oxford University Press.
———. 1995. Ecology and Equity: The Use and Abuse of Nature in Contemporary India. New Delhi (India): Penguin Books India.
Gómez-Pompa A, Kaus A. 1992. Taming the wilderness myth. BioScience 42:271–279.
Graber DM. 1995. Resolute biocentrism: The dilemma of wilderness in national parks. Pages 123–135 in Soulé ME, Lease G, eds. Reinventing Nature? Responses to Postmodern Deconstruction. Washington (DC): Island Press.
Guha R. 1989. Radical American environmentalism and wilderness preservation: A Third World critique. Environmental Ethics 11: 71–83.
Haila Y. 1997. 'Wilderness' and the multiple layers of environmental thought. Environment and History 3: 129–147.
Hammel B. 1990. The distribution of diversity among families, genera, and habitat types in the La Selva flora. Pages 75–84 in Gentry AH, ed. Four Neotropical Forests. New Haven (CT): Yale University Press.
Hecht S, Cockburn A. 1990. The Fate of the Forest: Developers, Destroyers and Defenders of the Amazon. New York: Harper Perennial.

Janzen DH. 1986. The future of tropical ecology. Annual Review of Ecology and Systematics 17: 305–324.
Kothari A, Pande P, Singh S, Variava D. 1989. Management of National Parks and Sanctuaries in India: A Status Report. New Delhi (India): Indian Institute of Public Administration.
Lefever HG. 1992. Turtle Bogue: Afro-Caribbean Life and Culture in a Costa Rican Village. Selinsgrove (PA): Susquehanna University Press.
Meffe GK, Carroll CR. 1994. Principles of Conservation Biology. Sunderland (MA): Sinauer Associates.
Myers N. 1988. Threatened biotas: 'Hot spots' in tropical forests. The Environmentalist 8: 187–208.
———. 1990. The biodiversity challenge: Expanded hot-spots analysis. The Environmentalist 10: 243–256.
Nabhan GP. 1995. Cultural parallax in viewing North American habitats. Pages 87–101 in Soulé ME, Lease G, eds. Reinventing Nature? Responses to Postmodern Deconstruction. Washington (DC): Island Press.
Nabhan GP, Rea AM, Reichardt KL, Mellink E, Hutchinson CF. 1982. Papago influences on habitat and biotic diversity: Quitovac oasis ethnoecology. Journal of Ethnobiology 2: 124–143.
Nash R. 1973. Wilderness and the American Mind. 2nd ed. New Haven (CT): Yale University Press.
Norton BG, ed. 1986. The Preservation of Species: The Value of Biological Diversity. Princeton (NJ): Princeton University Press.
———. 1987. Why Preserve Natural Variety? Princeton (NJ): Princeton University Press.
Oelschlaeger M. 1991. The Idea of Wilderness: From Prehistory to the Age of Ecology. New Haven (CT): Yale University Press.
Olindo P. 1991. The old man of nature tourism: Kenya. Pages 23–38 in Whelan T, ed. Nature Tourism: Managing for the Environment. Washington (DC): Island Press.
Pearce DW. 1993. Economic Value and the Natural World. Cambridge (MA): MIT Press.
Posey DA, Balée WC, eds. 1989. Resource Management in Amazonia: Indigenous and Folk Strategies. New York: New York Botanical Garden.
Pressey RL. 1994. Ad hoc reservations: Forward or backward steps in developing representative reserve systems. Conservation Biology 8: 662–668.
Repetto R. 1992. Accounting for environmental assets. Scientific American 266: 94–100.
Rolston H III. 1994. Conserving Natural Value. New York: Columbia University Press.
Sukumar R. 1989. The Asian Elephant: Ecology and Management. Cambridge (UK): Cambridge University Press.

———. 1994. Elephant Days and Nights: Ten Years with the Asian Elephant. New Delhi (India): Oxford University Press.
Takacs D. 1996. The Idea of Biodiversity: Philosophies of Paradise. Baltimore: Johns Hopkins University Press.
Vijayan VS. 1987. Keoladeo National Park Ecology Study. Bombay (India): Bombay Natural History Society.
Wallace DR. 1992. The Quetzal and the Macaw: The Story of Costa Rica's National Parks. San Francisco: Sierra Club Books.
Wilson EO, ed. 1988. BioDiversity. Washington (DC): National Academy Press.
Yoon CK. 1993. Rain forests seen as shaped by human hand. New York Times (27 July): C1, C10.

Feng Han

Cross-Cultural Confusion (2008)

Application of World Heritage Concepts in Scenic and Historic Interest Areas in China

Scenic and historic interest areas form a designated national park system in China. These areas are characterized by outstanding cultural as well as natural qualities and are the most attractive and popular tourism destinations. They are also significant components of the global park system. Fifteen of China's thirty-one World Heritage Sites are also Scenic and Historic Interest Areas or are located in these areas.

Recently, the management of Scenic and Historic Interest Areas, especially in the World Heritage Sites, appears to be caught up in the general pattern of globalization—which is also Westernization or even Americanization. This has raised hot debates in China (Huang and Wu; ChinaYouth 2002; L. Wang 2003; W. Wang 2003). These debates focus on Chinese national government policies derived from criteria of the Natural Heritage in the Operational Guidelines of the World Heritage Convention: the removal of local inhabitants and the demolition and strict restriction of man-made structures within the properties (Guo 2003; Mingxing 2003; CNWH 2004). These policies are strongly opposed by local communities and local governments because local people are uprooted and traditional lifeways and subsistence economies are ruined. They are also opposed by urban Chinese interested in visiting Scenic and Historic Interest Areas as tourists because such policies are not consistent with the traditional Chinese attitudes toward and values regarding nature.

THE TRADITIONAL CHINESE VIEW OF NATURE

The traditional Chinese view of nature has its philosophical origins in Confucianism and Taoism and has continued to develop historically. The Chinese have maintained a philosophical, humanist, and holistic attitude to the human-nature relationship that is distinguished from the traditional Western human detachment from nature (Moore 1967; Wang 1990; Li 1996; Zhou 1999). From the Chinese point of view, nature has never excluded human activities; instead, it is a place that always embraces humans. Scenic and Historic Interest Areas are the places where natural beauty and cultural artifacts are at "perfect oneness" and present the Chinese ideal of nature as beautiful, peaceful, spiritually charged, and gracefully and proportionately inhabited by human beings.

Historically, wild nature is not within the scope of Chinese appreciation. Traditionally, the Chinese have valued nature that has been aesthetically and morally enhanced by cultural refinement. Indeed, one might go so far as to say that historically the Chinese have valued nature that imitates art more than nature imitated by art. The Chinese developed a unique culture of landscape poems, landscape gardens, and landscape paintings a millennium or more before these artistic traditions emerged in the West. Thus, landscape has evolved its specific meaning over time in China. In Mandarin it is called *shanshui* (mountain and water), referring to those "great" or "scenic" mountains and waters expressing equally great moral and aesthetic ideals. Traditionally, landscape is morally and aesthetically centered in China. Loving and traveling in morally and aesthetically idealized nature has been the prime virtue of the Good Man ever since the Jin Dynasty (265–420 CE). (It is widely accepted that landscape, as an isolated object without connection to a unified scene, emerged from the Jin Dynasty in China.)

Distinguished from the West, the traditional Chinese view of nature is marked by the following characteristics (Lin 1935, 1937; Zhang 1992; Wang 1998; Yang, Zhang, and Sun 2001; Yu 2001; Shen 2002; Han 2003):

1. It is humanistic rather than religious.
2. It is aesthetic rather than scientific.
3. There is great value and beauty expressed by nature.
4. Nature is consistent with human culture.
5. Nature is the extension of home; it is an enjoyable and inspiring place.
6. Artistic representations of nature are more beautiful than their originals.
7. Nature that is managed to imitate art is more beautiful than uncultured nature.
8. Natural aesthetics is highly developed in China.
9. Traveling in nature aims to be companionable and enjoyable instead of solitary and physically daunting.

WORLD HERITAGE CATEGORIES AND CROSS-CULTURAL CONFUSION

Thus, it is not difficult to imagine that the World Heritage classifications of natural heritage, cultural heritage, mixed heritage, and cultural landscape heritage are confusing for the Chinese to apply in Scenic and Historic Interest Areas where nature and culture are highly integrated. Typically, many of these sites have associative cultural significance. They are landscapes associated with various artistic, cultural, moral, or ethical values that are often not obvious to the outsider. Thus, the Chinese consider it arbitrary to sort them into these different categories. Managers are usually poorly guided by the policies of the convention because of this cross-cultural confusion. Allowing themselves to be guided by these classifications, managers are liable to a heavy-handed separation of nature from culture on these sites.

There is not one single site that could be considered natural by Western (American) standards. On the contrary, the cultural inventory in the most natural-appearing areas is always apparent upon closer inspection. They always manifest either material culture or associative culture. A common natural-looking stone in the deep mountains may have been where Li Bai (one of the greatest poets of the Tong Dynasty, 701–62 CE) lay drunk and composed his famous landscape poem. A flat platform with a beautiful view could be the place where Bai Juyi (another great poet of the Tong Dynasty, 772–846 CE) constructed his straw hut and lived spiritually with nature. When we Chinese travel in nature, we are not alone. History and our ancestors are our companions. The Chinese have woven so much emotion and energy into nature through thousands of years that it is impossible to separate nature from culture in such a tightly knit net (Lin 1937; Feng 1990).

CULTURAL LANDSCAPE: A PROBLEMATIC CONCEPT

Somewhat ironically, even the term "cultural landscape" is problematic for the Chinese. As noted, landscape has its specific meanings in China. The Chinese would take it for granted that landscape is cultural, so it is redundant to put "cultural" in front of "landscape" even as "a useful tautology" (O'Hare 1997:47). The concept of cultural landscape was coined to broaden the concept of landscape to include human habitation and other investments of humans in nature. But the concept of cultural landscape is difficult for the Chinese to understand because they lack the contrasting contrary concept of a culture-free, purely natural landscape—that is, the concept of wilderness.

Although economically productive cultural landscapes, such as mountains terraced for rice fields, were not within the mainstream of traditional natural aesthetics, they were still an organic part of nature as "pastoral landscapes." But now that puts Scenic and Historic Interest Areas in conflict with another ecological movement in China called "return the terraces back to the forest." What is happening in China is similar to what happened in Australia in the 1970s, when national parks authorities sought to create "pristine" wilderness areas by erasing the traces of Aboriginal habitation (O'Hare 1997:29). Battles between social and cultural constructions of nature are by no means unique to China.

CULTURAL TRADITION VERSUS SCIENCE AND GLOBALIZATION IN CHINA

The influence of cultural tradition on today's Chinese is profound, and culture is usually inherited unconsciously. For example, Chinese housewives routinely slaughter fresh animals for cooking as they have for century upon century. Children in kindergartens recite Tang poems loudly without realizing they are one thousand years old. And the Chinese consider it appropriate to travel in natural areas and build houses there just to enjoy the beautiful scenery as did their ancestors. Such cultural traditions are hard for the Chinese to abandon. The Chinese expect three traditional things from Scenic and Historic Interest Areas: first, beautiful scenery; second, cultural enhancement of the scenery; third, convenient access to remote scenic spots and tourist facilities, such as hotels, to provide comfort and enjoyment. These demands have caused huge environmental impacts in these areas (BRN 2002).

Such impacts are frequently attributed to the great pressure of a large population and growing tourism market instead of to cultural traditions. Certainly, mass tourism is a big part of the problem. But the impact of sheer numbers of nature tourists is compounded by cultural expectations of what nature tourism entails. Traditionally, Chinese people gathered in beautiful places, had parties, and composed poems while drinking wine and gazing at beautiful scenery (265–420 CE) (Kubin 1990; Wang 1990). To repair to nature in order to experience solitude, taking only pictures and leaving only footprints, as Westerners wish to do, is foreign to the Chinese. Nature is an open-air theater. This perspective, while maintaining the philosophical spirit of the Chinese view of nature as being in harmony and unity with humanity, however, has been vulgarized as well as commercialized and democratized, with dire environmental consequences.

The great artistic achievement of natural aesthetics also has a profound influence. The Chinese love humanizing nature more than any other people, for they believe that humanized nature is more beautiful than pristine nature. In one survey, 92 percent responded that they feel that a site lacks spirit if it is purely or wildly natural without any cultural artifacts (Han 2004). This overwhelming preference on the part of Chinese tourists has resulted in today's many (and some tasteless) man-made structures and altered landscapes in Scenic and Historic Interest Areas.

CONTEMPORARY ENVIRONMENTAL SCIENCE AND GLOBALIZATION

Since China reopened to the world in 1978, contemporary environmental-science ideas have spread rapidly in China. Sustainability is one of the most important concepts developed since the mid-1970s in the West. The World Heritage movement encourages the sustainability of the valuable properties that the Chinese have inherited from their rich history.

However, the concept of sustainability is rooted in contemporary Western philosophical foundations. The rethinking of the relationship of human beings with the natural environment in the West over the last thirty years reflects a widespread perception in the 1960s that the late twentieth century faced a serious environmental crisis. This process of rethinking has engendered lively theoretical debates—everything from challenging traditional Western anthropocentrism and resourcism to challenging the concepts of wilderness and ecological restoration. The set of World Heritage classifications—natural heritage, cultural heritage, mixed heritage, and now cultural landscape heritage—is consistent with current trends in environmental philosophy toward integrating nature and culture, based on cultural diversity around the world, instead of the old-school dichotomizing of nature and culture. But in China the practice of World Heritage Sites in Scenic and Historic Interest Areas appears to be having the opposite effect. Nature is beginning to be disentangled from culture. The influence of globalization is obvious, especially in the spread of the concepts of wilderness and "pristine nature."

CONFLICTS

However, the global influence is mainly limited to management authorities and governments. It is certainly not manifest in local communities and tourists. Some management policies, which are misapplied from the convention by management authorities due to cross-cultural confusion, have resulted in two unfortunate consequences in Scenic and Historic Interest Areas in China. One is the removal of local inhabitants from these areas, which results in the rapid disappearance of the living traditional culture; the other is the restriction of new man-made structures in these areas, which is strongly against traditional Chinese cultural values. Both of these policies are implemented in the name of natural heritage preservation, and they are leading to cultural stasis in these sites. They are creating moral and cultural crises while dealing with ecological crises. Battles between government and local communities, management authorities, and visitors are essentially battles between international universal values and traditional Chinese values.

CASE 1: DEMOLITION IN THE WULINGYUAN SCENIC AND HISTORIC INTEREST AREA

The case of the Wulingyuan Scenic and Historic Interest Area is a typical example. It is a hotly debated case and calls for deep thought. Wulingyuan is one of the most popular Scenic and Historic Interest Areas in China, with a large annual visitation of more than 5 million (XHN 2003). It was designated as a World Natural Heritage Site by UNESCO in 1992. In 1998 it was severely criticized by the Centre/IUCN mission in its state of conservation report because it was "overrun with tourist facilities, [which have] a considerable impact on the aesthetic qualities of the site" (UNESCO 1998). The mission was also sharply critical of increasing indigenous agricultural activity as well as urbanization caused by rapidly developing exogenous tourism (XHN 2003). It seemed that Wulingyuan was in danger of being delisted as a World Heritage Site. In order to meet UNESCO's requirement of World Natural Heritage designation, the Central and Provincial Governments of China decided to demolish 340,000 square meters of recently built tourist facilities and artificial scenic spots to respond to the committee's critics. It was to accomplish this restoration over five years beginning in 2001. In addition, from 2001 to 2003 1,791 indigenous people from 546 families were to be resettled in order to make the site natural, wild, and pristine (XHN 2003).

This multifaceted ecological restoration project is strongly resisted by local governments and communities. Besides its huge financial cost (about 1 billion yuan), it is also criticized for "erasing history"—albeit, admittedly, recent history. There is also deep confusion among the local farmers, which can also be seen in their children's eyes. They cannot understand why they should move off the land where they have lived for generations and why their existence is a deleterious "ecological and visual impact on nature." They are also worried about how to survive in a strange new world outside their ancestral mountain demesne with limited financial compensation from the government. But all this is happening in the name of "World Heritage."

CASE 2: JIUZHAIGOU VALLEY SCENIC AND HISTORIC INTEREST AREA: AN ARTIFICIAL NATURAL "EARTHLY FAIRYLAND"

Jiuzhaigou Valley Scenic and Historic Interest Area is another World Natural Heritage Site designated in 1992 that is especially famous for the beautiful colors of its waters. Jiuzhaigou is doing its best to create an "earthly fairyland" or "fairy-tale world" with beautiful natural scenery that seems never to have been touched by humans. It is a new Chinese interpretation of the Western "wilderness" concept.

Jiuzhaigou was once polluted due to deforestation, exacerbated later by large numbers of tourists (more than a million annually). The ecorestoration involved complete removal of tourist accommodations in the valley to be replaced by new hotels restricted to areas outside of the property. The management effort to restore the ecosystem and the model of partnership between authorities and the local people were commended by the Bureau of World Heritage Committee (UNESCO 1998). Now this model is strongly recommended by the central government of China, and all other sites are requested to learn from Jiuzhaigou's experience.

However, the price of the removal of all tourism facilities and the prohibition of grazing by the local Tibetan population is the disappearance of a unique culture. Traditional local culture, the origins of which go back five thousand years, has been totally changed. Until 1975 there were nine Tibetan villages in the Jiuzhaigou Valley living according to their own customs, grazing their flocks, and raising their crops generation after generation. Then Han Chinese logging operations stripped the vegetation, polluted the waters, and disrupted the traditional Tibetan way of life. Now the local people still live in the valley, but since 1984, when the area was opened to Chinese visitors, their existence has revolved around tourism. They have become simulacra—the tourists' image of Tibetan herdsmen. They were forced to abandon their traditional way of living in nature to be replaced by joining the national economy, more particularly, the tourist industry. Tourism has eliminated their need to "exploit" the valley's natural resources on which they formerly depended for their livelihood, but at the price of their culture.

COMMENTS AND ARGUMENTS

Both cases are driven by the laudable effort to effect ecological restoration in World Natural Heritage Sites. Restoration is mandated when nature is threatened. But if we think of the environmental impacts of tourism just in ecological terms, our interpretation will be too shallow (Naess 1973). There are deep philosophical issues underlying the ecological phenomena. Essentially, the conflict is not between preservers and developers; rather, it is between the different meanings of nature. I am arguing for no one meaning in conflict with others. Rather, I am arguing that the issues are complex and that we need to think as we act. We must reflect about why we do what we do before we do what we do.

While the local people are losing their homeland, we are losing our living culture. In its place we are creating "dead culture" (outdoor museums). The poor indigenous people are being dispossessed, possibly because they are regarded as low, "uncivilized" people. Their existence interferes with the aesthetic experience that some self-congratulating "civilized" and "nature-seeking" tourists want to enjoy. The Wulingyuan indigenes have been removed or resettled to new areas that are "out of the view of tourists" (XHN 2003; Zhang 2003). If the Jiuzhaigou people were not Tibetans protected by special policies and did not have tourism value, they would probably be removed as well. Compounding the conflict between nature tourism, Chinese style, and the continuation of indigenous lifeways and livelihoods, the Western wilderness idea (Callicott 2000) is rapidly enthralling the management of World Heritage Sites in China. Now whenever discussion about the management of these properties comes up, the first reaction of local authorities is clearing the local people out. This is a form of ethnic cleansing, and it is tragic.

The restriction on human buildings in natural areas is also against Chinese traditional cultural activities in nature. Nature is culturally and socially constructed, and there is no correct or incorrect way that it is constructed. Culture can be guided but should not be suppressed as such in the process of preserving nature. For example, in Wulingyuan a huge elevator has just been built that breaks three world records—biggest, fastest, and highest—at the same time as the big tourist facilities demolition project is moving forward. This elevator rivals the Great Wall as a man-made structure in a natural area. The contradiction in this situation is as ridiculous as it is mysterious to outsiders, but it is perfectly accepted by common Chinese people.

The existence of World Heritage sites in China is allied with commercial and political change in China. Almost all successful applications for World Heritage designation are accompanied by huge demolition and relocation projects to meet the convention's criteria. Then the properties are commercially exploited as "golden tourism bait" to make the investment back once the application is approved. While the government actively applies for the World Heritage status, local communities, such as those living in Zhouzhuang near Shanghai, live in fear of being cleared out and sacrificing their way of life—and sometimes their very lives—for the World Heritage status. They claim that "World Heritage application is pursued for political advantage . . . but for us, . . . we just want to live better lives." In response the politicians complained, "We are struggling with the local communities" (ChinaYouth 2002).

Many policies applied in China are against the central spirit of the convention. Heritage is a living concept, as is cultural landscape, in which history and meanings can be read and interpreted as texts (O'Hare 1997;

Armstrong 2001). While we pass on yesterday to the present generation, we have to think what of today can and should be passed on to future generations. When culture and ethics encounter universal science and globalization, there emerge many difficult issues. We need to be "more sensitive about who counts and why" (Light and Rolston 2003). There is no conclusion in this essay. Rather, it calls urgent attention to the dialectical dynamic generated at the interface of cultural diversity and ecological sustainability, and it makes a plea for thoughtful and cautious action to keep the vitality in sustainable cultural landscapes.

REFERENCES

Armstrong, H. 2001. *A New Model for Cultural Landscape Interpretation.* Brisbane: Cultural Landscape Research Unit, Queensland University of Technology.

BRN. 2002. When There Is Conflict between Ecotourism and Ecosystem . . . Botany Research Network. http://botany.szu.edu.cn/magazine/2002.11.15/assets/0204305.htm.

Callicott, J. B. 2000. *Contemporary Criticisms of the Received Wilderness Idea.* "Wilderness Science in a Time of Change" Conference, Ogden, PMRS.

ChinaYouth. 2002. Closing Shops or Applying for the World Heritage: Zhouzhuang Comes to Its Crossroad. http://www.china.com.cn/chinese/TR-c/145623.htm.

CNWH. 2004. The Earthly Fairyland: Preservation and Management of Huang Long Scenic and Historic Interest Area. http://www.cnwh.org/news/news.asp?news=295.

Feng, J. 1990. Man and Nature. *Journal of Architecture* 1.

Guo, S. 2003. The Management Investigation of Taishan Scenic and Historic Interest Area. http://www.cnwh.org/news/news.asp?news=25.

Han, F. 2003. *The Chinese View of Nature: Tourism in Scenic and Historic Interest Areas.* Brisbane: Queensland University of Technology.

———. 2004. *Motives of Traveling in Nature.* Shanghai: Tongji University.

Huang, J., and Y. Wu. How to Pass the Heritages to the Generations. http://www.cnwh.org/news/news.asp?news=38.

Kubin, W. 1990. *Nature Perspective in Chinese Literature.* Shanghai: Shanghai People's Press.

Li, W., ed. 1996. *Propagating Chinese Humanities.* Beijing: Beijing Broadcast and Television Press.

Light, A., and H. Rolston III. 2003. Introduction to *Ethics and Environmental Ethics.* Oxford: Blackwell Publishing.

Lin, Y. 1935. *My Country and My People.* Beijing: Foreign Language Teaching and Research Press.

———. 1937. *The Importance of Living.* Beijing: Foreign Language Teaching and Research Press.

Mingxing. 2003. Fights between World Heritage Conservation and Benefit of Development. http://www.huaxia.com/wh/whsd/00132274.html.

Moore, C. A., ed. 1967. *The Chinese Mind.* Honolulu: East-West Center Press, University of Hawaii Press.

Naess, A. 1973. The Shallow and the Deep, Long-Range Ecology Movement. *Inquiry* 16:95–100.

O'Hare, D. 1997. *Tourism and Small Coastal Settlements: A Cultural Landscape Approach for Urban Design.* Oxford: Oxford Brooks University.

Shen, Z., ed. 2002. *Tourism and Chinese Culture.* Beijing: Tourism Education Press.

UNESCO. 1998. *State of Conservation Report of the Bureau of the World Heritage Committee.*

Wang, L. 2003. Thinking about the World Heritage. http://www.cnwh.org/news/news.asp?news=8.

Wang, S. 1998. *Chinese Tourism History.* Beijing: Tourism Education Press.

Wang, W. 2003. Don't Make Heritage Application a Movement. *Liberation Daily* (Shanghai).

Wang, Y. 1990. *Gardens and Chinese Culture.* Shanghai: Shanghai People's Press.

XHN. 2003. After the Yellow Warning Card. Xin Hua Network. http://www.hn.xinhuanet.com/ptfx/2003-06/05/content_571612.htm.

Yang, L., Z. Zhang, and J. Sun. 2001. *The Graceful Chinese.* Shanghai: Xuelin Press.

Yu, H. 2001. *The Romantic Chinese.* Shanghai: Xuelin Press.

Zhang, B. 1992. *Chinese Tourism History.* Yunnan: Yunnan People Publication.

Zhang, J. 2003. Struggling between Natural and Man-Made World. http://www.hn.xinhuanet.com/ptfx/2003-06/05/content_571803.htm.

Zhou, M. 1999. *Exceeding and Transcending.* Chengdu: Sichuan People Publication.

Antonio Carlos Diegues

Recycled Rain Forest Myths (1998)

THE CREATION OF PROTECTED areas has been one of the principal strategies adopted for the conservation of nature, in particular in the countries of the Third World. The establishment of these areas increased substantially in the 1970s and 1980s, when 2,098 protected areas were created around the world, encompassing more than 3.1 million square kilometers. According to a 1996 report of the World Conservation Center, today about 5 percent of the earth's surface is legally protected under twenty thousand different categories, not only at federal levels but also at provincial, state, and municipal levels, covering an area the size of Canada and spread throughout 130 countries.

In 1990, Brazil had 34 national parks, 23 biological reserves, 21 ecological stations, 38 national forests, 14 environmentally protected areas, and 4 extractive reserves, totaling 31,294,911 hectares, or almost 4 percent of the territory. Ninety percent of this area is located in the Amazon region. The increased interest in creating protected areas in Brazil could be explained by a combination of factors: the rapid devastation of the Amazonian rain forests and the Mata Atlântica (Atlantic rain forest); the loss of biodiversity; the availability of international funding for conservation efforts; the possibility of revenue generation from tourism in the parks; and, above all, the pressure on the World Bank to create new protected areas to counterbalance the development projects it is funding in fragile areas such as the Amazon. The establishment of protected areas is also a powerful political weapon for the dominant elite of many countries of the Third World, who continue to obtain external financing for large projects that have an impact on fragile ecosystems.

Already, there is more protected area in Brazil than in many European countries. If the proposal of the U.N. Environment Programme (UNEP) is achieved, in which approximately 10 percent of national territory would be put under some form of protection, around eight hundred thousand square kilometers of Brazilian territory will become parks and reserves, a surface area equivalent to France and Germany combined. Apparently, most environmental agencies maintain that the greater the area that is put under some form of protection the better it is for conservation. Today, about 18 percent of the Amazonian region is protected. UNEP's goal has in fact already been achieved in seven countries in Asia, fourteen countries in Africa, and six countries in Latin America. But in the United States, one of the proponents of this idea, less than 2 percent of the protected territory is designated as national parks, and in Europe, it is less than 7 percent. Judging from this, it would seem that UNEP deems the idea of national parks to be more appropriate for the Third World than for industrialized countries—and this in spite of the fact that many Third World countries are experiencing food shortage crises, which are in part due to insufficient agricultural land and inequitable land distribution.

A North American model of conservationism, which dichotomizes "people" and "parks," has spread rapidly throughout the world. Because this approach has been adopted rather uncritically by the countries of the Third World, its effects have been devastating for the traditional populations—extractivists, fisherfolk, and indigenous peoples. This model was transposed from industrialized countries with temperate climates to the Third World, whose remaining forests have been, and continue to be, inhabited by traditional populations.

The United Nations has estimated indigenous populations at three hundred million in seventy countries and throughout various ecosystems, ranging from savannas and forests to polar regions. According to Jeffrey McNeely, the people known as tribals, natives, traditionals, and other cultural minorities occupy about 19 percent of the land surface, living in isolated regions with fragile ecosystems.[1] It is most often these ecosystems that are labeled as "natural" and transformed into protected areas from which the residents are expelled. With this authoritarian action, the state contributes to the loss of a wide range of ethnoknowledge and ethnoscience—of ingenious systems for managing natural resources—and of cultural diversity itself. The expulsion of inhabitants has contributed to even more degradation of park areas because, due to insufficient monitoring—despite the fact that the majority of the budget for these protected areas is allocated for monitoring and enforcement—they are invaded by logging industries and miners who illegally exploit the natural resources. Inhabitants also illegally extract their means of subsistence from these protected areas.

Governments rarely assess the environmental and social impact that the creation of parks will have on the local inhabitants, whose land-use practices often have preserved these natural areas over the years. They are transferred from regions where their ancestors lived to regions that are ecologically and culturally different. The hunters, fisherfolk, and other resource users who have developed a symbiosis with the forests, rivers, and coastal areas have great difficulty surviving, once relocated to other areas, due to the accompanying prohibition of their traditional activities.

These populations have difficulty comprehending how their traditional activities could be considered detrimental to nature when hotels and tourism infrastructures are created for the use of people from outside the area. Very little of the budget for protected areas is allocated for improving the living conditions of the traditional people, who, if encouraged, could make a positive contribution. When they have organized and become vocal about defending their historical right to remain on ancestral land, they are accused of being against conservation. In most cases, these are people who are illiterate, lack political power or legal ownership of the land, and are therefore not compensated when their land is expropriated. But, as has occurred in the Mata Atlântica in Brazil, when land is expropriated from the large landowners, who often have obtained their land by usurping the rights of the traditional residents, they are royally compensated because they can prove legal ownership.

The authorities who are responsible for the preserved areas perceive the traditional inhabitants as destroyers of wildlife, which eliminates any real opportunity for their inclusion in the conservation project. In many cases, and especially in the Amazon, the so-called participation of traditional populations in the establishment of parks and reserves does not go beyond well-intentioned words that are offered to assuage international demands from such large institutions as the World Bank, the International Union for the Conservation of Nature, and the World Wide Fund for Nature.

This model of preserving wilderness has been criticized both inside and outside the United States, and part of this opposition has come from the American "pure preservationists." John Rodman holds that the idea of parks subscribes to an anthropocentric view, that the creation of parks principally values the aesthetic, religious, and cultural motivations of humans, demonstrating that it isn't wilderness in and of itself that is considered valuable and worthy of being protected. Yet Rodman considers this mode of preservation, based on the model of parks and natural reserves, to be unjustly selective because it privileges natural areas that appeal to a Western aesthetic—such as forests, large rivers, and canyons—and discriminates against natural areas that are considered less noble—swamps, bogs, and marshes.[2] Arturo Gomez-Pompa and Andrea Kaus have also criticized this notion of a "natural world" that privileges an urban perspective:

> The concept of wilderness as untouched or domesticated is fundamentally an urban perception, a view of people who live far from the natural environment on which they depend for raw material. The inhabitants of rural areas have different perceptions of the areas that the urbanites designate as wilderness, and base their use of the land on alternate views.[3]

More recently, a socio-environmental focus has been adopted in the critique of "the Yellowstone model." This new approach to conservation arose out of the collaboration between the social movements that fight for the continued access of peasants, fisherfolk, and forest people to land and natural resources and the Third World environmentalists who see the environmental crisis in their countries as being linked to the existing model of development. This movement, which Eduardo Viola and Hector Leis have called "peasant ecology,"[4] critiques the imported environmentalism for its lack of consideration of the traditional communities that depend on the forests for their livelihood.

In North America, the myth of "wilderness" as an uninhabited space has fueled the move to create protected restricted-use areas. By the end of the nineteenth century, after the conquest and widespread massacre of the native peoples and the westward expansion of the frontier by European settlers, the land was perceived to be uninhabited. With the movement of human settlements to the West, the mid-nineteenth century saw natural areas being degraded by mining and forestry companies. This raised protests from the nature lovers who had been influenced by the ideas of Henry David Thoreau and George Perkins Marsh. In 1864, in his widely read book *Man and Nature,* Marsh argued that the preservation of virgin areas was justified as much for artistic and poetic reasons as it was for economic reasons, and he held that the destruction of the natural world threatened the very existence of humans on earth.

In the early nineteenth century, artist George Catlin traveled throughout the American West. He cautioned that the Indians as well as the bison were threatened with extinction and suggested that the native peoples, the bison, and the virgin areas could be equally protected if the government were to establish a national park that incorporated humans and animals "in all their primitive and natural beauty."[5] This idea was not implemented, however, and the notion of wilderness as a virgin, uninhabited area prevailed. On March 1, 1872, when the decision was made to create Yellowstone National Park, the U.S. Congress decided that the region could not be colonized, occupied, or sold, but would be separated as a public park or recreation area for the benefit and enjoyment of the people. Any person who occupied any part of this park would be breaking the law and would be removed.

In the 1960s, after much of the "wilderness" had been "tamed" and even destroyed in most of the northern countries, environmental preservationists, in search of this lost, untouched nature, turned their attention to the vast rain forests and savannas in tropical countries, particularly in Africa and South America. In Brazil, the Amazonian rain forest became the focus for the construction of a new myth. Called the "lungs of the earth," this tropical forest was considered to be "empty space," only sparsely inhabited by the remaining indigenous tribes—although it is now estimated that at the beginning of the sixteenth century, five to seven million Amerindians were living in the region, largely concentrated in the river floodplains (*várzeas*), an even higher density than today.

The Brazilian military group in power in the 1960s and 1970s exploited this neo-myth in order to occupy the region, which led to the rapid transformation of vast rain forest areas into large cattle-raising and agricultural farms. It is no coincidence that most of the protected areas also began to be established during this period in order to counterbalance the widespread forest destruction. Neither the preservationists nor the military acknowledged the presence of the people living in those areas. Indians were confined in special reserves, and non-Indian local inhabitants were resettled outside the borders of the newly created national parks and other strictly protected reserves.

Due mainly to a lack of support for this type of conservation within southern countries—particularly among the communities that live inside and adjacent to protected areas—there have been frequent failures in the implementation of protected areas. Consequently, nature-conservation practices and the underlying ideas that have guided the creation of protected areas are changing in many countries around the world, including Brazil. There is a growing awareness that the reason for this lack of social support is the unsuitability of this conservation model to local realities rather than, as some preservationists argue, the lack of appreciation for the importance of protected areas. National parks and other strictly protected areas cannot simply be considered "islands" created to conserve biodiversity, since biological diversity also lies beyond the parks.

In southern countries, environmental movements are emerging that are different from those in northern countries in that they are attempting to harmonize nature conservation with the need to improve the living conditions of the inhabitants of national parks and adjacent regions. These new socio-environmental movements recognize the importance of the knowledge and management practices of traditional populations.

In many of these countries, the process of decolonization and democratization has also led to the challenging of the imported model of nature conservation. People living inside protected areas have mounted spontaneous and increasingly organized resistance against resettlement.

There are basically two representations of nature—and particularly of forests and wood lands—that coexist in modern mythology. By "mythology" I mean the symbolic representations of the natural world that are a cultural and historical product of the various forms and moments of the relations between diverse societies and their physical surroundings.

On the one hand is the naturalist myth of an untouched nature or wilderness in a "pure" state, prior to the appearance of humans. This myth presupposes the incompatibility between the actions of any human group and the conservation of nature. Regardless of their culture, humans are, in this equation, destroyers of the natural world. The idea of a "paradise lost" informed the creation of the first North American national parks in the second half of the nineteenth century, where portions of territories that were considered "untouched" were closed off to human habitation. These "wild" areas were created for the benefit of urban North Americans who could visit them and appreciate their "natural beauty." This "modern" model of conservation and its underlying ideology have spread to the rest of the world in cultural contexts distinct from those in which it was created, generating serious consequences.

On the other hand is the representation of forests as a natural resource to be traded. According to this view, nature has value only when it is transformed into commodities for human use. The ideal would be to transform the tropical forest, with its great variety of tree species, into a homogeneous forest, like those of the temperate climates, which would be more easily managed (cut) and used industrially. This view has fueled the extensive transformation of the rich Atlantic forest of Brazil into plantations of pines and eucalyptus through the fiscal incentives that the Instituto Brasileiro de Desenvolvimento Florestal (Brazilian Institute of Forestry Development) has granted to the timber companies since the 1960s.

Paradoxically, both of these approaches see the forest as uninhabited, negating the existence of innumerable societies who live in the forest and make use of it within a sociocultural framework very different from urban-industrial societies. The human communities that live in the forests would at most be identified as a "species of fauna" or "threatened species"—one more component of the natural world—the local culture and its myths and complex relationships with nature deemed "savage" and "uncivilized."

In their re-creation of the myth of a "wild" nature, preservationists from North America and other countries ignored the myths that guided and interpreted the relations between the North American indigenous populations and nature. For these peoples, the world referred to as "wild" by whites did not exist. But these myths—which Edgar Morin has called "bio-anthropomorphic"—are not exclusive to the indigenous populations in North America. They also exist among populations of hunters, extractivists, fisherfolk, and peasants in the Third World, who still live somewhat apart from the market economy of the urban-industrial world.

In many traditional societies, "wilderness" and the "natural world" are understood contextually in terms of myths in which humans might assume natural features and plants and animals might present humanized characteristics and behavior. According to Morin, in this mythological universe, the fundamental features of animate beings are encountered in inanimate things. This unity/duality of humans is also reflected in the ways that reality is perceived. One is empirical, technical, and rational, by which complex botanical, zoological, ecological, and technological knowledge is accumulated (today the subject of ethnoscience); the other is symbolic, mythological, and magical.[6] However, these forms of knowledge, although quite distinct, do not live in two separate universes; they are practiced in the same (although dual) universe. According to Mircea Eliade, in this dual universe, space and time are both the same and different—mythical time, the time past, is also always present, returning in regenerative ceremonies.[7]

This symbolic representation of the cycles in which all of creation is born, dies, and is reborn is strong among the indigenous societies of Brazil, but it is also present in the communities of peasants, fisherfolk, and gatherers that continue to live according to nature's cycles and a complex agricultural or fishing calendar. There is a time for *coivara* (burning of vegetation that has grown after the first burning), to prepare the land, to sow, to weed, and to harvest; and there is also a time to wait for species of migratory fish, such as mullet (*tainha*). Upon completing one cycle, the next cycle is begun. These activities are often ordered by signs—such as a particular phase of the moon, the appearance of rain, and so on—that are celebrated in festivities that mark the planting or harvesting of a specific crop.

According to Morin, contemporary history, while dissolving old mythologies, creates others, regenerating symbolic/mythological thought in a modern form. He holds not only that mythological thought persists in remote rural regions, but that there is also a resurgence of myths in the urban world. And Eliade suggests that myths related to nature endure and resist the incursions of science, surviving as "pseudo-religions" or "degraded mythologies." He goes on to say that in modern societies that declare themselves atheist, religion and myths are buried in the unconscious, periodically returning to the surface as new mythologies. P. Thuillier states that in hundreds of texts inspired by ecological concerns, the old myths reappear with an almost religious enthusiasm and apocalyptic vigor.[8]

In Brazil, the first inspiration for the creation of national parks came from the abolitionist André Rebouças in 1876 and was based on the model of North American parks. In defending the creation of the National Park of Itatiaia, the geographer Hubmayer, as early as 1911, stated that this national park was

> without equal in the world[;] it will be at the doorstep of our beautiful Capital [at that time Rio de Janeiro] offering scientists and researchers immeasurable potential for the most diverse research, as well as offering the ideal retreat for physical and psychological renewal after the exhausting work in the cities. Also, it will provide a source of satisfaction for travelers and visitors interested in the attractions of nature in the area.[9]

The first national park was created in Itatiaia in 1937, upon an initial proposal by the botanist Alfredo Loftgren in 1913, with the objective of encouraging scientific research and offering leisure to urban populations. Little thought was given to the indigenous populations and the fishing and gathering populations that were already there.

The concern for traditional populations that live in conservation areas is relatively recent in Brazil, and until a short time ago (and still today for classical preservationists), this was considered "a police matter," since they were to be expelled from their traditional lands to make way for the creation of parks and reserves. The positions of the environmental movements in Brazil vary regarding the presence of traditional communities in conservation areas. The "preservationists" dominate the older and classical conservation groups—such as the FBCN (Brazilian Foundation for the Conservation of Nature), created in 1958, and the more recent ones, such as the Fundação Biodiversitas, Funatura, and Pronatura, which are more linked to international preservation organizations. Their influence continues to prevail in many of the institutions that have been responsible for the creation and administration of parks, such as IBAMA (Brazilian Institute for the Environment) and the Forest Institute of São Paulo. These groups have generally been formed by professionals in the natural sciences who consider any human interference in nature to be negative. Ideologically, they were, and continue to be, influenced by the U.S. preservationist view: they consider wild nature to be untouched and untouchable.

Working in difficult circumstances, these preservationists very often have dedicated their lives to protecting endangered flora and fauna, and probably, without their devotion, many unique habitats and species would have disappeared. In some cases, the protected areas they helped to create prevented the expulsion and resettlement of the traditional populations by outside logging and tourist industries. However, despite their accomplishments and goodwill, their approach to conservation has led to conflicts with local populations, and they have contributed less and less to finding a real solution to existing problems. Many of these preservationists are still very influential in Brazilian government conservation institutions, and they resist any attempt to change their imported model of environmental protection. Rather than attributing the failure of this model to its inappropriateness, they have usually blamed its failure on inadequate funding and enforcement.

Beginning in the 1970s, an "ecologism" of denunciation emerged in Brazil, represented by AGAPAN (Gaúcha Association for the Protection of the Natural Environment), Ecological Resistance, Catarinian

Association for the Preservation of Nature, and APPN (São Paulo Association for the Protection of Nature). The military regime in power at that time was more tolerant of non-leftist movements, such as environmental nongovernmental organizations (NGOs), and repressed social protest movements. The 1970s was a time of rapid growth for the Brazilian economy, particularly through megaprojects that resulted in serious impacts on nature. Most of these, such as chemical and petrochemical plants, were established or expanded in coastal zones, the most populous areas of the country, and brought levels of degradation never before seen in Brazil. At the same time, the agricultural industry was growing considerably, resulting in a massive increase in the use of biocides and insecticides. Millions of rural workers were forced to move to the cities, which led to the growth of *favelas* (slums).

This extensive environmental degradation and social pauperization was masked by the ideology of the so-called economic miracle, in which the Brazilian government's objective was to attract industries of the industrialized countries. It is in this context that the *Brazilian Ecological Manifesto: The End of the Future* emerged in 1976, headed by ecologist José Lutzemberger and representing ten ecological organizations. Written at the height of the repressive military regime, the document was indeed a courageous act. The manifesto advocated the human-nature relations of traditional societies—the indigenous people and small-scale subsistence farmers—as an alternative to the predatory use of natural resources. The *Ecological Manifesto* played an important role in the ecological struggles of the 1970s and 1980s, denouncing environmental degradation, construction of nuclear power plants, and militarism.

In the mid-1980s, another type of environmentalism, more linked to social questions, began to emerge. This new movement developed along with the beginnings of redemocratization after decades of military dictatorship and constituted a critique of the model of economic development whose inequitable concentration of wealth and destruction of nature had had its apogee during that period. The widespread destruction of the Amazon and Atlantic forests led to the beginning of what has been called "social ecologism," a movement that struggles to maintain access to territories with natural resources and places a high value on systems of production that are based on traditional technologies. The National Council of Rubber-Tappers, the Movement of People Affected by Dams, the Movement of Artisanal Fishermen, and the Indigenous Movement are all part of this movement, which reached one of its highest points in 1989 in Altamira, with the Meeting of the Indigenous People of Xingu. These movements acknowledge the necessity to rethink the role of national parks and reserves as well as the role of the traditional inhabitants within the parks. The final declaration of the Altamira meeting counseled: "Do not destroy the forests, the rivers, that are our brothers, since these territories are sacred sites of our people, Home of the Creator, that cannot be violated."[10]

Some of the local movements, which have no direct links to national movements, fight against the curtailment of their traditional activities in conservation areas. Other local organizations have pressured park administrations to begin negotiating alternative uses of natural resources. But they are incipient and fragile and are still subordinate to the local movements that are under state control. More spontaneous local resistance movements—the small-scale local extractivist producers defending their traditional territory against outsiders—are struggling to gain control over access to natural resources. For example, in response to their reduced access to local fishing sites because of fences that were erected by large landowners and to the threat posed by incoming commercial fishers who use predatory fishing equipment, one such action was the "closing of the lakes" in the Amazon region and the establishment of lake reserves by many *vargeiros* (riverine communities) of Amazonia, who themselves assumed control over the territories they have traditionally occupied.

The traditional populations that lived in the areas that were made into parks have been ignored by the state authorities for decades. When the State Park of Ilha do Cardoso was created in 1962 on the land along the south coast of São Paulo, a sophisticated and detailed management plan was developed for the flora and fauna and for support structures for tourism and research. This top-down plan, developed by the Forest Institute with the assistance of two "specialists" from the Food and Agriculture Organization (FAO), did not even mention the existence of the hundreds of families who lived there. Fortunately, the plan was shelved. Nevertheless, many of the families left their birth place because of persecution by the park wardens.

Some local movements in isolated regions—such as the Movement of the Riverine Population of Mamirauá, Amazonas—are supported by NGOs and research institutes, although they are not linked to any major social movement at the national level. The incorporation of traditional populations in restricted conservation areas is a project of the Mamirauá Ecological Station (EEM), administered by the Mamirauá Civil Society and supported by several international environmental NGOs, among them the World Wide Fund for Nature. The EEM was created to protect a large part of the floodplain between the Japurá and Solimões Rivers. Forty-five hundred vargeiros live in this huge area, spread over fifty small communities, with an average of fourteen households in each. They live from fishing and from hunting and gathering forest products. However, logging takes place along with these traditional activities, and the wood is sold to the sawmills in the cities. Rather than expelling the vargeiros, as was legislatively mandated, the project administrators decided to allow them to remain in the territory.

During the floods, water covers millions of hectares, making law enforcement, carried out exclusively by government officials, an impossible task. The management team, belonging to a local NGO, believed that the biodiversity and culture of the region could be protected only through community participation. This type of management, which differs substantially from the management plans established and imposed by scientists and bureaucrats, takes longer to develop since it depends on constant dialogue and consultation with local populations, the inclusion of social research teams, and more flexibility in planning. It places more value on the process of decision making than on the establishment of rigid conservation objectives. This project demonstrated that once a decision is made by the local population, it has a much greater chance of being followed. In the consensus that was reached by the local population regarding the conservation and sustainable use of lakes—which was extremely important, both biologically and socioeconomically—the communities decided to define six categories of lakes. These included lakes for reproduction of fish (untouchable, with the shoreline included in the area of total preservation); "subsistence lakes" (for exclusive use of the community for subsistence fishing); "market-oriented lakes" (for exclusive use of the community, with the fish to be sold); and "lakes for use of the nearby urban centres" (where fishing is permitted to satisfy the needs of municipalities).

The rubber-tappers' extractive reserves are one of the outcomes of the rubber-tappers' movement, which was created in the 1970s during the height of the conflict over land in Acre. This movement organized the first *empate* (blockade), in which the organized rubber-tappers confronted the machines that were cutting down the forest and threatening their way of life. In 1975, when the first rural union was created in Basiléia in Acre—an area with a high density of rubber trees—the reaction of the landowners was violent, and in many cases the houses of the rubber-tappers were burned and the leaders were assassinated.

The National Council of Rubber-Tappers, established in 1985, pursued the creation of "extractive reserves." The extractive reserves gained international notoriety in 1988 when the rubber-tappers' leader, Chico Mendes, was assassinated. The first official extractive reserve was created in 1988, and in 1990 the extractive reserves became part of the protected areas system. The extractive reserves are administered communally. Although not allocated in individual lots, families have the right to exploit the resources along their traditional extractivist *colocações* (tapping routes). The land cannot be sold or transformed into non-forest uses, except for small areas that are allowed to be cleared for subsistence agriculture (approximately 1 to 2 percent of the area of the reserve).

Despite the organized opposition through the UDR (Democratic Rural Union) of large landowners, the rubber-tappers' movement expanded not only into Acre, where already by 1980 around 60 percent of the municipalities had rubber-tappers' organizations, but also into other states, such as Amapá, Rondônia, and Amazonas, which include ten extractivist settlements and four extractivist reserves, covering 3,052,527 hectares and benefiting about nine thousand families. In 1992, IBAMA created the National Council of Traditional Populations (CNPT) to lend technical support for the reserves in Amazonia and to disseminate the idea to other regions of the country. There are also extractivist reserves outside of the region, based on *babassu* found in the *cerrado* (savanna vegetation in semiarid areas) and on fishing resources in the state of Santa Catarina.

The movement to establish extractivist reserves is an effort to defend, reinforce, and re-create threatened ways of life. Furthermore, in Amazonia it is an alternative that can enable a sustainable use of natural resources that respects both biological diversity and traditional ways of life. Official and public recognition of these reserves was made possible only through the collaboration and solidarity that grew between the strong social movement and the National Council of Rubber-Tappers. Together they seek national as well as international legitimacy, especially in their struggle against other forms of ownership, particularly large landholdings. The frequent meetings of the leaders of the National Council with the rubber-tappers in many regions of Amazonia have helped them to organize additional associations that will propose new reserves.

One of the preservationists' arguments against the existence of traditional populations in "restrictive" protected natural areas is the assumed incompatibility between their presence and the protection of biodiversity. The establishment of protected areas for the preservation of biodiversity is, however, a relatively recent objective, promoted by international environmental organizations in response to the disappearance of species and ecosystems. The earlier parks were created primarily for environmental education, research, and the recreation and enchantment of urbanites.

Recent studies have shown that the maintenance and even the enhancement of biological diversity in tropical forests is intimately related to the shifting agriculture practiced by traditional communities. The use of small areas of land for agriculture and their abandonment after the decline of agricultural production (shifting agriculture) has an effect similar to that produced by the occasional destruction of the forests by natural causes. Shifting agriculture has been a natural means of using the regenerative properties of the rain forest for the benefit of humans. Arturo Gomez-Pompa suggests that tropical ecologists have recognized that "a large part of the primary vegetation of many zones, seen as virgin, actually contain vestiges of human disturbances, and there is more and more difficulty in finding zones that are totally virgin."[11] Many dominant species of the primary forests of Mexico and Central America were actually protected by humans in the past, and their current abundance is related to this fact. In the case of tropical forests, it is very difficult to distinguish "virgin" forests from "disturbed" forests, especially in areas where itinerant agriculture is practiced. The establishment of protected natural areas that respect these traditional practices can contribute to sociocultural diversity as well as to conservation of the natural world, whether it be "virgin" or already altered by traditional populations.

Protected areas, especially those with very restricted use, are more than a government strategy of conservation; they are emblematic of a particular relation between humans and nature. The spread of the mid-nineteenth-century U.S. idea of uninhabited national parks is based, first, on the myth of an untouched natural paradise, an image of Eden from which Adam and Eve were expelled, and, second, on what Serge Moscovici has called "reactive conservationism." This reactive conservationism of the nineteenth century, in which the natural world is attributed all the virtues and society all the vices, was a reaction to culturalism, which sees in nature the infirmity of man, a threat of return to savagery to which culture must be opposed.[12]

Even when urban-industrial society and the advance of science has desacralized the world and weakened the power of myths, the image of national parks and other protected areas as a paradise in which "virgin nature" is expressed in all its beauty—transformed into an object of reverence by urban humanity—confirms the idea that mythologies continue and can be reborn under the shadow of rationality. This myth of an untouched and untouchable nature not only reshapes old creeds but also incorporates elements of modern science—such as the notion of biodiversity and ecosystem function—in a symbiosis expressed by the alliance between particular currents of natural science and preservationist ecology. The persistence of the idea of a wild and untouched natural world has considerable force, especially with urban and industrial populations that no longer have daily contact with the rural environment. This occurs despite growing scientific evidence that for thousands of years, humans have, in one way or another, interfered with many terrestrial ecosystems, so that today very little untouched virgin nature remains.

In tropical countries, the historical realization of the myth of an untouched nature in the creation of national parks and reserves continues unabated. The conflict between the views of the so-called traditional populations and the preservationist and state conservationist institutions cannot be analyzed simply in terms of the oppositions between different mythologies and symbolisms. The conflict also revolves around a political ecology to the extent that the state imposes new spaces that are "modern and public" upon territories where traditional populations live—the parks and reserves from which, by law, inhabitants must be expelled. To those with power, these social actors are invisible. The knowledge of their existence and their importance to the conservation and maintenance of biological diversity is a recent phenomenon—the result of the socio-environmental ecologism that has developed in Third World countries.

This new ecologism has been translated into social movements that propose a new alliance between humans and nature, the need for democratic participation in nature conservation, and a respect for cultural diversity as the basis for the maintenance of biological diversity. Park inhabitants became more visible as a result of the conflicts that arose when landless populations occupied park areas that were not effectively administered by the government. Traditional populations and newcomers have recently begun to organize against the enforcement actions of the state, which, in most cases, impede the social and cultural reproduction of these human communities.

In Brazil, at the federal level as well as in some NGOs, the question of the presence of traditional inhabitants in national parks and other conservation areas has been dealt with from a conservative point of view, one that is still influenced by urban perceptions of the natural world and wilderness. In underdeveloped countries, conservation could be better achieved through the real integration and participation of the traditional populations, which to a great extent have been responsible for maintaining the biological diversity that today we are trying to rescue.

However, there is also a need to guard against a simplistic view of the "ecologically noble savage."[13] Not all inhabitants are "born conservationists," but among them there exist traditional people with a vast store of empirical knowledge of the workings of the natural world in which they live. We need to better understand the relations between the maintenance of biological diversity and the conservation of cultural diversity. An interdisciplinary view is urgently needed, whereby biologists, forestry engineers, sociologists, anthropologists, and political scientists, among others, work in an integrated way in cooperation with traditional populations. As Gomez-Pompa and Kaus have said, we are discussing and establishing policies on a subject that we know little about; and traditional populations, who know their environment better than we do, rarely participate in debates and decisions about conservation management.

NOTES

1. Jeffrey McNeely, "Afterword to People and Protected Areas: Partners in Prosperity," in E. Kemf, ed., *The Law of the Mother: Protecting Indigenous Peoples in Protected Areas* (San Francisco: Sierra Club Books, 1993), p. 90.

2. John Rodman, "What Is Living and What Is Dead in the Political Philosophy of T. H. Green," *Western Political Quarterly* 26 (1973): 566–86.

3. Arturo Gomez-Pompa and Andrea Kaus, "Taming the Wilderness Myth," *Bioscience* 42/4 (1992): 273.

4. Eduardo Viola and Hector Leis, "Desordem global da biosfera e a nova ordem international: O papel das organizações do ecologismo," in H. Leis, ed., *Ecologia e política mundial* (Rio de Janeiro: Vozes/Fase, 1991).

5. J. McCormick, *Rumo ao puraíso* (Rio de Janeiro: Ed. Relume-Dumará, 1992).

6. Edgar Morin, *La Méthode 4, Les idées, leur habitat, leur vie, leurs moeurs, leur organization* (Paris: Seuil, 1991).

7. Mircea Eliade, *Imagens e símbolos* (São Paulo: Martins Fontes, 1991).

8. P. Thuillier, "Les mythes de l'éau," *La Recherche,* special issue (May 1990): 221.

9. In Maria Pádua and Adelmar F. C. Filho, *Os parques nacionais do Brasil* (São Paulo: Edit. José Olympio, 1979), p. 122.

10. M. Waldman, *Ecologia e lutas sociais no Brasil* (São Paulo: Contexto, 1992), p. 90.

11. Arturo Gomez-Pompa, C. Vasquez-Yanes, and C. Guevara, "The Tropical Rainforest: A Non-Renewable Resource," *Science* 177/4051 (1972): 762–65.

12. Serge Moscovici, *Hommes domestiques, hommes sauvages* (Paris: Unión Généralle d'Editions, 1974).

13. K. Redford, "The Ecologically Noble Savage," *Orion* 9/3 (1990): 25–29.

G. W. Burnett, Regine Joulié-Küttner, and
Kamuyu wa Kang'ethe

A Willing Benefactor (1996)

An Essay on Wilderness in Nilotic and Bantu Culture

Too many conservationists in Africa assume that, because Africans are not privy to the romantic and transcendental notions prerequisite to wilderness conservation, they cannot be trusted with the task and are irrelevant to it (Adams & McShane, 1992; Burnett & Conover, 1989). European imperialism required assertion of brutal superiority over the cultures it sought to dominate. Africans suffered more than most, and their religion-philosophy was viewed with particular contempt. An early scholar of the African experience in America, Ulrich Phillips, summarizes this scorn: "No people is without its philosophy and religion . . . [but] of all regions of extensive habitation equatorial Africa is the worst" (quoted in Joyner, 1940/1986, p. xvi); whereas Bohannan and Curtin (1988) found that rational consideration of African religion was always criticized as "see no evil" optimism. Recent reevaluations of African philosophy and religion by African and Western scholars have allowed us to challenge, using common historiographic sources—early ethnographic reports—the idea that traditional African thought is wanting in substance with respect to wilderness, at least among the highland Bantu (Burnett & Kamuyu, 1994; Spears, 1981).

Here, we extend the discussion to Kenya's Nilotes as defined by Murdock (1959), with the specific objective of comparing Nilotic and Bantu thought about wilderness. Our approach is ethnophilosophical (Deren, 1953/1970; Dieterlen, 1951; Griaule, 1950; Jahn, 1961; Kagame, 1976; Mulago, 1968; Temples, 1949). For Jahn (1961), a new African culture is conceived as being born of "native-foreign" contact perceived at three levels. The first, the *déjà le,* is the language, moral codes, beliefs, and rituals of the people, available in this case in the ethnographic reports from the period of early colonization. The second level consists of explications of the *déjà le,* which reveal interconnections of foundational beliefs and practices through examination of their form and unity. In this essay, this second level is provided by Burnett and Kamuyu (1994), who depended on the explications of Temples (1949) and Mbiti (1969, 1975). The third level consists of critical discourse on the first two levels.

The ethnophilosophical approach is highly relativistic. Cultural relativism has been criticized (e.g., Hollis, 1977; Hollis & Lukes, 1982; Horton, 1967) largely for too lightly masking the West's achievement of having constructed the only system for acquiring objective insights into nature. A relativistic approach seems appropriate in this case, however, because wilderness, and its desirability or undesirability, is normative and nonscientific. For this reason, we have also avoided any rigorous *a priori* definition of wilderness. In any society, the preservation or destruction of wilderness can proceed scientifically only after a fundamental attitude toward wilderness is established. A relativistic approach permits a sympathetic but objective view of the people's outlook from what remains of it in available texts.

Our approach assumes that in the popular culture, the political level where much policy is given its fundamental formulation in democratic societies, Western traditions will not entirely replace but rather meld with African traditions, resulting in a synthesis useful to the modern African situation. Our purpose, therefore, is to contribute to the search for an indigenous philosophy of wilderness and related topics that will serve Africans and their resources better than exclusive reliance on exotic Western ideas, specifically romanticism and transcendentalism.

THE PEOPLE AND THE SOURCES

Eastern Sudanic Nilotes originated west of the Ethiopian Highlands (Maxon, 1989; Murdock, 1959). By the end of the first millennium B.C., three groups were apparent, from west to east: the river-lake Nilotes, the plains Nilotes, and the highland Nilotes. In Kenya, a highland group, the Kalenjins, divided into several distinct communities: Elgon, Pokot, Nandi, Elgeyo, Marakwet, Tugen, and Kipsigi. Occupying much of the same geographic area but different niches as the highland Nilotes, the plains Nilotes dispersed as two groups, the Turkana and the Masai. The river-lake Nilotes developed as a single agricultural group, the Luo, settling in the Nyanza's lake-plains.

Nilotes were pastoralists, but as they migrated, their economies changed with environmental opportunities and the lessons learned from other peoples. Agriculture came to predominate in the Kalenjin communities of the humid highlands. The Turkana supplemented pastoralism with hunting and fishing. The supreme pastoralists, the Masai, were not a single, united people, and, like others, they varied their economies according to circumstances (Maxon, 1989). The Ilmasai relied exclusively on their herds and disdained vegetable and game food. The Iloikop Masai, however, grew crops and the Njemps around Lake Baringo developed intensive, irrigated agriculture.

Available substantial ethnographic texts and a novel practically limit our discussion to plains and highland Nilotes. We make no attempt to discuss Kenya's river-lake Nilotes, the Luo. Nilotic sources must be approached warily. Unbiased observations are rare and descriptions superficial compared with writings about the sedentary Bantu. Europeans did not live closely with the seminomadic Nilotes (Lado, 1993). Home as a livestock kraal repulsed many observers, while Nilotic military organization appealed to others. Also, Nilotic peoples have produced few scholars who have undertaken interpretation of Nilotic philosophy and religion.

BANTU WILDERNESS

The most fundamental questions that can be asked about creation concern its purposes, because purpose defines nature's relevance (Glacken, 1967). Answers to questions about creation are found in mythical, religious, and philosophical ideas that were common to both Nilotes and Bantu. Late precolonial Nilotes and Bantu enjoyed considerable violent and peaceful interplay. Raiding for stock and brides was common, as were trading, inter-marriage, adoption, and "pawning" children to prosperous groups during difficult times. These actions facilitated exchange of vocabularies and customs and, presumably, a shared world view, including an attitude toward wilderness. Consequently, we begin investigating Nilotic wilderness with the hypothesis that Nilotic and Bantu views are substantially the same. The hypothesis admits the theory that at least the major language groups of tropical Africans share a common intellectual culture and world view (Bohannan & Curtin, 1988; Davidson, 1969; Masolo, 1994; Turnbull, 1976). The major Bantu ideas relevant to wilderness (Burnett & Kamuyu, 1994) are

1. A benevolent God created the world for humans. The order of creation was generally understood to be a single integrated social order rather than by species.
2. The wilderness is not a source of spiritual inspiration or self-realization for individuals or groups.
3. With rare exceptions, wilderness is conceived as an extension of human living space or location.
4. Wilderness will be, or at least can be, dominated by humans through concerted social action.
5. Wildlife, a portion of God's creation, is alienated from human society.
6. Wildlife is an objective manifestation of a wider variety of beings that inhabit wilderness and compete with humankind for control of wilderness.

7. Consequently, individual experience within wilderness is likely to be harmful to the individual and to the group the individual belongs to.

WILDERNESS AND NILOTIC RELIGION AND ONTOLOGY

The highland Bantu's world was rigorously monotheistic and ontologically complex (Burnett & Kamuyu, 1994). God, creator and maintainer, was all-benevolent and oversaw a lively world in which various ontological levels reflect "vital force" (Temples, 1949). God was more existential than ancestors and spirits, both of which were more vital than humans, who were above animals, plants, and nonbiological things (Mbiti, 1975). Ontologically, wilderness was human environment, existential and subjective (Burnett & Kamuyu, 1994).

In comparison with Bantu religion, Nilotic religion was described by early observers as maddeningly vague (Beech, 1911; Emiley, 1927; Gulliver, 1952; Hobley, 1902). This apparent ambiguity has several explanations. Early observers did not live close to the Nilotes so religious viewpoints, always private and difficult to express to strangers, would not have been readily accessible. Furthermore, Nilotic religious leaders, identified with activities such as cattle raiding and warfare condemned by colonial authorities, made even casual conversation about religion dangerous (Gulliver, 1951). Also, environmental caprice required adaptability, which hindered development of systematic theology. Strongly militaristic groups living in harsh environments valued success, dependent on flexibility and individuality, even among women and children, more than rigid orthodoxy (Temples, 1965). Neither circumstances nor personality would have encouraged structured belief.

For Europeans, religious ambiguity marked Nilotes as "primitive." Gulliver (1951), however, appreciated the absence of systematic theology as ecologically advantageous, encouraging practical responses to unpredictable circumstances. His assessment of the Turkana was that they sought naturalistic explanations, turned to magic and mysticism only as a last resort, and were not surprised when either failed. In any event, that early observers found no systematic theology in Nilotic religion does not mean that the Nilotes had no theology. Systematic theology arises from the idea that God is subject to rational analyses, an idea not generally entertained in African religion (Kamuyu, 1981).

Nilotes, like Bantu, unequivocally recognized a supreme being associated with up, the sky, or the sun. Nilotic mythology also associated God with other celestial activities and personalities (Barton, 1921; Beech, 1911; Hollis, 1905, 1909), and many observers have attempted to parallel Nilotic cosmology with that of Europeans (e.g., Merker, 1910). Consequently, portrayals of Nilotic cosmologies must be treated cautiously. In creation stories (Hollis, 1905, 1909; Huntingford, 1927), God is not alone in the universe, but rather brings a margin of order to a scene already filled with being. A creation where God begets the moon only to appear on an Earth where various creatures are already interacting confounds Western logic.

The Nilotic God is both a universal creator and the creator of a particular people. This implies an obvious paradox: God is an all-powerful, benevolent creator who can, but need not, control everything. There is God's way of doing things and the way that things happen. Among the Turkana, for example, God controls the rains and might reasonably be approached to provide them in their appointed season, but God does not change the natural order of things by, for example, converting deserts into gardens (Gulliver, 1951). God exists in a creation that is good and complete which He does not routinely will to alter. God is paradoxically both above and entirely within, rather than extrinsical to, nature. God assists in natural processes such as pregnancy of women or livestock, the germination of plants, and the falling of rains in their appointed season, and God might even be expected to violate his own laws; but, because creation is final and all-good, it is too much to expect God to do the unnatural just for man's convenience, though no harm is done in asking. Consequently, extreme sacrificial practices, such as human sacrifice, are unknown among Kenya's Nilotes as well as Bantu. Bantu, however, were more inclined to see God as nature's direct sovereign and to account for nature's whimsy by understanding creation as more ontologically active. God remains for both groups, however, a court of last resort capable of altering nature.

For both Nilotes and Bantu, creation is ontologically complex, but for

Nilotes this complexity is commonplace. For example, among Nilotes, at least some spirits of the dead were temporarily active and potentially pesky; yet, ancestor worship did not exist. Most groups believed that the souls of great men might enter the bodies of snakes, where, in the kraal, they might guard children or cattle; but the Pokot killed snakes because they thought the act might take an ancestor to final rest (Beech, 1911). In order to free the spirit, Nilotes disposed of their dead in the bush, where hyenas ate everything but the skulls. The Pokot provided the skulls of the good with a little food or tobacco, whereas those of the bad were neglected so that their death might be all the more unbearable (Beech, 1911). Death was relative, although absolute and irrevocable.

Nature, because it was ontologically complex, was lively and spirit filled, but these spirits were of little concern. In Turkanaland, evil spirits dwelt in mountains and could cause insanity, but they seldom were close to where people might be. They did not prevent Turkana from visiting mountains after performing a minimum ritual of throwing a rock on a pile along the trail (Gulliver, 1951). The Masai lived in a world of benevolent spirits that required little attention. They thoroughly disliked forests, which they believed were populated with a variety of monsters. The Nandi hid their cattle there when the Masai were raiding. But Ole Kulet's fictional Masai made frequent use of the forest (Huntingford, 1950; Ole Kulet, 1972). Practical materialism was the cardinal trait of the Nilotes' world view. A people dominated by cattle but without a cattle cult would reasonably view rocks, trees, and animals as such, and little else.

Early observers were quick to find evidence of profound superstition among all Africans. Superstition indicated "primitiveness," which proved dependence and justified imperialism. That superstition existed, just as it existed in Europe, is certain, but interpreting the extent to which superstition, as distinct from ontology, reflected a world view is a difficult task, the more so when observation is sometimes 70 years removed from its interpretation or reinterpretation. There is little to suggest that Nilotes or Bantu innately feared an ontologically complex world and even less to suggest how superstition and magic affected broader world views.

LINEAGE AND LAND

Among pastoralists a concept of personal land ownership was unnecessary except to protect private development (Lado, 1993). A house site and garden were respected as private, and this was permanent and inheritable for the nonnomadic Nandi (Huntingford, 1950). Although privately owned water was unheard of, the right to improved watering sites was recognized; but those rights were terminated naturally when floods destroyed the sites. The traditional grazing range, the collective gift from the ancestors, was special, although the ancestors' spirits did not inhere in the gift. The Nandi observed tradition inflexibly in the homeland, including its wilderness and frontier margins, but their observances diminished declivitously beyond Nandiland (Huntingford, 1969). Nilotes defended their frontiers stubbornly.

In the homeland were several classes of grazing land, but "wilderness" was both wildland and political boundary. Among the Nandi, this frontier, or *kaptich* (Snell, 1954), was crucial because it provided seasonal grazing and potential areas for settlement. There the young initiated men herded the stock and thus in the wilderness learned leadership, discipline, and the ways of nature. These soldiers were accompanied by the older uninitiated girls, who kept house, cooked, and acted as runners and messengers. Therefore, virtually all adult Nandi were experienced in the bush.

In the homeland, some places were more special than others. As among the Bantu, some trees were involved in Nandi ritual and were the sites for council meetings and important ceremonies (Snell, 1954). To harm such a tree was to invite wrath, though a gift of beer to the elders generally resolved any offense. Trees near homesteads belonged to the homesteaders and some trees were, without explanation, otherwise special (Huntingford, 1950; Snell, 1954). The Nandi avoided unnecessary tree cutting, did not fell forests, and generally supported British forest reservations (Huntingford, 1950). Known special places in Nandiland were the hill, Chepeloi, where the first ancestor was disposed, and Chepeloi, a hill where ghosts burned the grass each year. The extinct volcano, Menengai, in Masailand, was also regarded as holy

(Huntingford, 1927). Among the Turkana, Mt. Kailongol or "a willing benefactor" was respected because of its always lush pastures (Emiley, 1927). All pastoralists considered salt licks and hot springs as particularly special, although not sacred.

WILDLIFE RELATIONS

Among Nilotes, hunting for food was disdained as a mark of poverty but necessary among those living in extreme environments. The Turkana ate everything except hyena, jackal, dog, and snake, and highland Pokot used elephant as their main source of meat (Beech, 1911; Emiley, 1927; Gulliver, 1951; Hobley, 1902; Huntingford, 1969). Most Masai avoided game meat entirely, and Huntingford (1950) described the Nandi as uninterested in wildlife. Still, most Nilotes hunted if only to provide training for soldiers (Huntingford, 1950; Snell, 1954). Even the Masai hunted—elephants for their tusks, buffaloes and elands for their hides, ostriches for their feathers, and lions and leopards to protect stock and for the thrill (Merker, 1910; Ole Kulet, 1972). Some hunting supported ritual (Gulliver, 1951, 1952; Ole Kulet, 1972). The Turkana used live-trapped gazelle in rainmaking and Masai boys used bird wings in their initiation. The Turkana, who valued ostrich feathers, trapped, plucked, and released the birds. Hunting was social and could win prestige but it was not a spiritual exercise. Observers were impressed with the Nilotic hunting practices and avoidance of wanton slaughter.

Certain animals stood in special relationship to humans. Hildebrandt (1878) speaks of a hyena cult among the Masai. The hyena, humanity's graveyard, was obviously regarded as exceptional; however, other literature does not support the hyena's cult status. Snakes were peculiar. As possible recipients of recently dead souls, they were respected when they appeared in the kraal or house. Rats and other peri-human rodents might be treated as snakes. For all Nilotes, frogs were vile (Hollis, 1905).

Some wildlife was identified with clans, age sets, and other social groupings. A totem is a hereditary emblem of a group, an animal, or less commonly a plant or other natural object to which the group is fraternally related. Nilotes were totemistic. But totemism was found in only one highland Bantu group, the Akamba, who were more given to pastoralism and hunting than other Bantu, and even there it was almost unnoticeable (Lindblom, 1920/1969). Totemism can take many forms (Levi-Strauss, 1962). Some require physical trials intended to reconcile an individual with nature and thereby establish a deep, personal relationship with the totem. For this reason, totemism is often assumed to reflect an "accommodating relationship with the nonhuman order of things" (Murray, 1993, p. 17). The Nilotic totem was sufficiently removed from this concept to leave doubt about the term's propriety (Barton, 1921, 1923). Nilotic totems were generally described as simply indifferent objects commanding loyalties little deeper than those to football teams (Barton, 1921). Hollis (1909) insisted that Nandi totems were serious, but Huntingford (1927) held that the totem was only a form of polite address among people shy of formal names. Barton (1921) thought that Suk totems were once more serious but offered no evidence.

The lists of totemic objects do, however, suggest aspects of Nilotic ontology (Barton, 1921, 1923; Beech, 1911; Hobley, 1902; Hollis, 1909; Huntingford, 1969). Included are the expected animals and a few plants but also some universal food items, such as white ants, and such things as hyenas, frogs, and snakes, which had other, often unpleasant, relations with humans. They also include natural phenomena such as rain, thunder, and lightning associated with God or other aspects of mythology. In addition, there are a few definitely human processes, such as hut burning, and human objects, such as unmudded hair. Although Nilotic totems did not provoke a mystic relationship with nature, they suggest an other-than-modern classification of objective categories and processes. Animals are more existential than plants, but this need not imply an absolute distinction. Natural processes are easily understood to be animal-like, and human objects may, at least occasionally, be included in classes that embrace nature. Furthermore, things may have more than one aspect at a time. Frogs may be both vile, because God made them that way, and special in the sense that a clan totem is special.

However, both Nilotes and Bantu drew a distinction between social things and wild things. For the Bantu, all animals were initially part of the human order. Some animals alienated themselves from humankind and became wildlife in order to escape painful slaughter by stone and wooden tools, a problem that was solved by the invention of iron (Kenyatta, 1965). Nilotes tell a similar story, but, in theirs, alienation resulted from a negligent child herdsman, and Nilotic stories generally assume a time when humankind and wildlife shared society (Merker, 1910). Domestic livestock in the wild became like wildlife, a free good without owner and thus there for the taking.

Throughout Nilotic folklore humans are portrayed as nature's cutthroats. Terrorizing monsters inhabit Bantu folk wilderness, but Nilotic wilderness monsters are craven even before women and children. In a particularly rakehell story (Hollis, 1905), a neglected brother and sister wander into a monster's kraal. They drive him to suicide and, while taking his cattle home, outwit lions, leopards, and hyenas. At home, they murder their heedless parents and, after their initiation, marry. Admired by the narrator, the children defeat both nature and all social convention except initiation. In another story (Hollis, 1905), three young women meet a warrior at a dance and accompany him to his wilderness home. Discovering he is a monster, they escape, but one returns to retrieve her beads, is captured, and is forced to become his wife. She eventually connives his murder and the murder of their son and returns home, rich and respected. In Ole Kulet's (1972) novel, when Merresho learns that his archenemy has secured witchcraft intended to banish him to hopeless wandering in the wilderness, he bombastically disregards the threat. Children exposed to stories such as these learned to fear very little, real or imagined, in creation.

Nilotic folklore, unlike Bantu, is rich in beast fables. Many, demonstrating real understanding of complex ecological relations, explain wildlife behavior naturalistically (Hollis, 1909; Massam, 1927). In allegorical beast fables, the little and wily, obviously including humans, are the most admired for their ability to fool, often cruelly, the powerful. The stories present humans as vicious and feared competitors, not as helpless pawns of nature. The agricultural Bantu had no reason to become involved with wilderness, and their stories taught utter fear of the wild; but, for the pastoral Nilotes, the wild was an inescapable reality. Their folklore domesticated wilderness, socialized wildlife, and illustrated humankind's singularity in nature.

TRANSCENDENT WILDERNESS

For the agricultural Bantu, a frontier people socially disposed to settle wilderness, individual wilderness experiences were a threat to the God-ordained social order. Folklore taught the Bantu to fear wild places and to approach them cautiously. Nilotes, herding and raiding, lived in the wild. The Nilotic education began with folk stories and proceeded to experience, often for both boys and girls, in the wilderness. There is little to suggest that Nilotic culture encouraged an emotional or transcendent experience with wilderness, but there are several intriguing hints. Wilderness beckoned young Turkana men to the religious life. Touched, the chosen would wander alone in the mountains sometimes for months (Gulliver, 1951). Returning always in good health, the inspired youth would tell tales of a fabulous journey. Elsewhere, in Ole Kulet's (1972) novel, Merresho, horrified by his son's aspirations, wanders in the forest in search of solace.

Like the Bantu's, Nilotic society confirmed the Nilotes' humanity, distinguishing it from nature by elaborate social rituals and commitments (Davidson, 1969). Unlike Bantu, the Nilotes' experience with the wild must have defined the individual's being, identity, and place in nature. Consequently, all Nilotes had a transcendent experience with wilderness to the extent that wilderness was integral to and inseparable from the individual's being.

DISCUSSION

The hypothesis that highland Bantu and Kenya's plains Nilotes traditionally viewed wilderness in much the same way can be cautiously accepted. Both Nilotes and Bantu believed they could dominate wilderness through concerted social action, but Nilotes would likely have added, "Not only can we, but we do." East Africans agreed that God intended creation to be a single, integral, human-centered social order and looked back to a time when wildlife and human society were a unity. Wildlife became alienated from human society, the natural order, as a consequence of the human's pugnacious, irascible competitiveness. Wilderness taught that if society is not maintained, humans face the hyena's fate. Human society was considered the centerpiece and point of creation. Human centrality demanded elaborate social and kinship groupings as well as rituals marking life's events because these defined society's relationship to nature. The social, not the biological, distinguished man and hyena. Humans must work at being social or become just another beast in the wilderness.

Accepting the hypothesis, however, should not veil several differences that separated the two peoples. The most important differences arise from the role wilderness played in their economies. Surrounded by so much wilderness, both cultures had to deal with it as fundamental. Bantu culture minimized access to wilderness while maintaining a social organization allowing its systematic exploitation and settlement. Nilotic culture existed directly in wilderness and was more comfortable and confident with it. This Nilotic alliance with nature should not be romanticized, however. Nilotes were materialists, centermost in creation, and superior to wild things and wilderness. Nilotes agreed with Bantu that uncontrolled individual experience with wilderness results in chaos. Because wilderness was unavoidable, it was dealt with through an elaborate educational system that began with folklore and culminated in an imposing military organization. The extent to which Nilotes allowed wilderness to be a source of spiritual inspiration was, at most, slight. Because nature was more proximate, Nilotic ecological observations, demonstrated in their beast fables, alone distinguished Nilotes from the Bantu; however, Nilotes certainly did not encourage private wilderness experiences. Nilotes would further agree that wilderness embraced a variety of visible and invisible states of being; however, the invisible was minimized. Concurrence with the Bantu idea of wilderness as an extension of human space is ambiguous because, for Nilotes, wilderness and living space were essentially congruent.

IMPLICATIONS

Ethnophilosophy has been criticized for only being able to reconstruct a static world view (Hountondji, 1970). Capturing how Africans felt about wilderness as colonization began hardly reveals their current thought, so reconstruction is useful only as long as many East Africans still adhere significantly to tradition. Our experiences lead us to believe that tradition is still prevalent. Among East Africans, only a remnant live traditionally while a similar minority are purely westernized. The vast majority, including much of the electorate, the bureaucracy, and the intelligentsia, who debate and set natural resource policies, live in a coalescence of tradition and modernity. We believe, too, that many Westerners agree with us, as is evidenced by the number of environmental education programs intended to instill Western conservation values in lieu of tradition in East African youth.

Whereas science usefully describes and draws inferences about objective things, conservation is not so much objective as the application of science to accomplishing objectives generated by other world views, be they romantic, transcendental, or Christian dictates about stewardship. Because the romantic and transcendental are unappealing to Africans, it is reasonable to found conservation programs on the people's actual thoughts about nature. In fact, little is known about how contemporary Africans view wilderness and associated conservation issues, but if traditional ideas are significant in popular culture, they must be confronted and used to the advantage of the resources and society. Because the ruthless subjugation of ideas employed during the colonial period must be dismissed as having failed, the alternative is to seek an accommodation, including an accommodation of ideas, between conservation objectives and the needs of affected people. This proposition is central to "sustainable development" projects.

A place to begin, and the rationale of this essay, is to determine, to the extent possible, how people used to think about wilderness. There is little in traditional thought to suggest that East Africans were innately hostile to the natural environment or that they would object to its scientific conservation and much to suggest that contemporary hostility toward conservation arises from failed economic systems (Mbaku, 1994) and differences in opinion over the methods, rather than objectives, of conservation (Akama et al., 1995). The major implication of having established something of the traditional attitude toward wilderness, is to make apparent the need to examine the extent that traditional ideas prevail in contemporary society and how these might be made more useful to national conservation objectives. One approach is "ethnoclassification" (Berlin, 1992), which reveals much about how traditional people close to nature order their environment. Another approach might be to view current African concepts of nature through the theories of developmental psychology, as did Hallpike (1979) in his analysis of preliterate thought. However, because literacy is now common in East Africa, it should be possible to measure environmental perceptions directly, particularly among schoolchildren and their teachers. This would reveal the extent to which the next generation accepts traditional world views and internalizes traditional and Western viewpoints about the environment. If this seems extravagant in a Third World situation, consider that many resource administration agencies in the West expect exactly this information about their client groups. More immediately, because Kenya, among other African nations, requires primary school study of both religion and environmental science, the curriculum could be amended to include instruction at the interface of these two subjects.

Ultimately, however, the articulation of a world view accommodating African thought toward wildlands management is the responsibility of Africa's intellectuals in a process that should differ little from that described by Nash (1982) for the United States. Western involvement in this discourse is desirable if only for its benefit to the West. Despite Bergson's (1911) discussions, the idea that creation has vital force eludes most Westerners; however, in the form of the Gaia hypothesis, the idea of a living Earth is being reevaluated among Western intellectuals (Asimov, 1988; Murchie, 1978; Myers, 1991; Schneider & Boston, 1991). Among its most determined proponents, Myers (1991) obtained much of his basic ecological orientation among East Africans. African ontology and the Gaia hypothesis may, consequently, provide Africans and Westerners with a common ground for the discussion of wilderness. The tragedy is that Africans must conclude that discourse in a matter of decades, if not years, whereas Americans had the luxury of generations to conduct their conversations.

REFERENCES

Adams, J. S., and T. O. McShane. 1992. *The myth of wild Africa: Conservation without illusion.* New York: W. W. Norton.

Akama, J. S., C. L. Lant, and G. W. Burnett. 1995. Conflicting attitudes toward state wildlife conservation programs in Kenya. *Society and Natural Resources* 8:133–144.

Asimov, I. 1988. *Prelude to foundation.* Garden City, NY: Doubleday.

Barton, J. 1921. Notes on the Suk Tribe of the Kenya Colony. *Journal of the Royal Anthropological Institute of Great Britain and Ireland* 51:82–99.

Barton, J. 1923. Notes on the Kipsikis Lumbwa Tribe of the Kenya Colony. *Journal of the Royal Anthropological Institute of Great Britain and Ireland* 53:42–78.

Beech, M. W. H. 1911. *The Suk. Their language and folklore.* Intro. by Sir Charles Eliot. Oxford: Clarendon Press. 1969 reprint, New York: Negro Universities Press, Greenwood Publishing Co.

Bergson, H. 1911. *Creative evolution,* ed. Arthur Mitchell. New York: Henry Holt.

Berlin, B. 1992. *Ethnobiological classification: Principles of categorization of plants and animals in traditional societies.* Princeton, NJ: Princeton University Press.

Bohannan, P., and P. Curtin. 1988. *Africa and Africans,* 3rd ed. Prospect Heights, IL: Waveland Press.

Burnett, G. W., and R. Conover. 1989. The efficacy of Africa's national parks: An evaluation of Julius Nyerere's Arusha Manifesto of 1961. *Society and Natural Resources* 2:251–259.

Burnett, G. W., and Kamuyu wa Kang'ethe. 1994. Wilderness and the Bantu mind. *Environmental Ethics* 16:145–160.

Davidson, B. 1969. *The African genius: An introduction to African social and cultural history.* Boston: Little, Brown.

Deren, M. 1953/1970. *Divine horsemen: The living gods of Haiti.* New York: Delta.

Dieterlen, G. 1951. *Essai sur la religion Bambara.* Paris: University Press of France.

Emiley, E. D. 1927. The Turkana of Kolosia District. *Journal of the Royal Anthropological Institute of Great Britain and Ireland* 57:157–201.

Glacken, C. J. 1967. *Traces on the Rhodian shore: Nature and culture in Western thought from ancient times to the end of the eighteenth century.* Berkeley: University of California Press.

Griaule, M. 1950. Philosophie et religion des Noirs. *Présence Africaine* 8–9: 307–312.

Gulliver, P. H. 1951. *A preliminary survey of the Turkana.* A report compiled for the Government of Kenya. Communications from the School of African Studies, University of Capetown, 26 (n.s.): 1–281.

Gulliver, P. H. 1952. The Karamajong cluster. *Africa* 22:1–21.

Hallpike, C. R. 1979. *The foundations of primitive thought.* Oxford: Clarendon Press.

Hildebrandt, J. M. 1878. Ethnographische Notizen über Wakamba und ihre Nachbaren. *Zeitschrift für Ethnologie,* pp. 347–406. Berlin: Hildebrandt.

Hobley, C. W. 1902. *Eastern Uganda: An ethnological survey.* Occasional Paper No. 1. London: Anthropological Institute of Great Britain and Ireland.

Hollis, A. C. 1905. *The Masai: Their language and folklore.* Intro. by Sir Charles Eliot. New York: Books for Libraries Press. 1971 reprint, Freeport, NY: Black Heritage Library.

Hollis, A. C. 1909. *The Nandi: Their language and folklore.* Intro. by Sir Charles Eliot. Oxford: Clarendon Press.

Hollis, M. 1977. *Models of man.* Cambridge: Cambridge University Press.

Hollis, M., and S. Lukes, eds. 1982. *Rationality and relativism.* Cambridge, MA: MIT Press.

Horton, R. 1967. African traditional religion and Western science. *Africa* 37:50–71, 155–187.

Hountondji, P. 1970. Remarques sur la philosophic Africaine contemporaine. *Diogène* 71:120–140.

Huntingford, G. W. B. 1927. Miscellaneous records relating to the Nadi and Kony tribes. *Journal of the Royal Anthropological Institute* 57:417–461.

Huntingford, G. W. B. 1950. *Nandi work and culture.* London: His Majesty's Stationery Office for the Colonial Office.

Huntingford, G. W. B. 1969. *The Southern Nilo-Hamites* (Ethnographic Survey of Africa, ed. D. Ford, East Central Africa, Part VIII). London: International African Institute.

Jahn, J. 1961. *Muntu: An outline of the new African culture.* New York: Grove Press.

Joyner, C. 1940/1986. Introduction to the Brown Thrasher Edition. Savannah Unit, Georgia Writers' Project, WPA. *Drums and shadows: Survival studies among the Georgia coastal negroes,* pp. ix–xxviii. Athens: University of Georgia Press.

Kagame, A. 1976. *La philosophie Bantu comparée.* Paris: Présence Africaine.

Kamuyu wa Kang'ethe. 1981. The role of the Agikuyu culture and religion in the development of Karing'a nationalist movement. Ph.D. dissertation, University of Nairobi, Nairobi.

Kenyatta, J. 1965. *Facing Mount Kenya: The tribal life of the Gikuyu.* Intro. by B. Malinowski. New York: Vintage Books.

Lado, C. 1993. The perceptions of pastoralism and rural change in Masailand, Kenya. *Journal of Third World Studies* 10:148–183.

Levi-Strauss, C. 1962. *Totemism.* Trans. R. Needham. Boston: Beacon Press.

Lindblom, G. 1920/1969. *The Akamba in British East Africa: An ethnological monograph,* 2nd ed. New York: Negro University Press.

Masolo, D. A. 1994. *African philosophy in search of identity.* Bloomington: Indiana University Press.

Massam, J. A. 1927. *The cliff dwellers of Kenya.* London: Seeley, Service and Co.
Maxon, R. M. 1989. *East Africa: An introductory history.* Nairobi: Heinemann, Kenya.
Mbaku, J. M. 1994. Africa after more than thirty years of independence: Still poor and deprived. *Journal of Third World Studies* 11(2): 13–58.
Mbiti, J. S. 1969. *African religions and philosophy.* London: Heinemann.
Mbiti, J. S. 1975. *An introduction to African religion.* London: Heinemann.
Merker, M. 1910. *Die Masai: Ethnographische Monographie eines östafrikanischen Semirenvolkes.* Berlin: Verlagsbuchhandlung Dietrich Remimer. 1968 reprint, *Landmarks in anthropology,* ed. W. La Barre. New York: Johnson Reprint Co.
Mulago, V. 1968. Le Dieu des Baritu. *Cahiers des Religions Africaines* 2:23–24.
Murchie, G. 1978. *The seven mysteries of life: An exploration in science and philosophy.* Boston: Houghton Mifflin.
Murdock, G. P. 1959. *Africa: Its peoples and their culture history.* New York: McGraw-Hill.
Murray, J. A. 1993. *Wild Africa: Three centuries of nature writing from Africa.* New York: Oxford University Press.
Myers, N. 1991. *Gaia atlas of future worlds.* New York: Doubleday.
Nash, R. 1982. *Wilderness and the American mind,* 3rd ed. New Haven, CT: Yale University Press.
Ole Kulet, H. R. 1972. *To become a man.* Nairobi: Longman.
Schneider, S. H., and P. J. Boston, eds. 1991. *Scientists on Gaia.* Cambridge, MA: MIT Press.
Snell, G. S. 1954. *Nandi customary law.* London: Macmillan.
Spears, T. 1981. *Kenya's past: An introduction to historical method in Africa.* London: Longman.
Temples, P. 1949. *Philosophie Bantoue.* Paris: Présence Africaine.
Temples, P. 1965. *Bantu philosophy.* Paris: Présence Africaine.
Turnbull, C. M. 1976. *Man in Africa.* London: David and Charles.

Kimberly K. Smith

What Is Africa to Me? (2005)

Wilderness in Black Thought, 1860–1930

> *Herein the longing of black men must have respect: . . . the strange renderings of nature they have seen may give the world new points of view.*
> W. E. B. Du Bois[1]

When environmentalists refer to wilderness they usually have in mind a concept defined by the canonical texts of the wilderness preservation movement including works by such familiar figures as Thoreau, Muir, and Leopold. There's room within this tradition for disagreement of course (is Walden Pond a wilderness?). But we can nonetheless identify a broad consensus on the meaning of wilderness to those who provided key ideological support for the preservation movement. For the nineteenth-century romantics and twentieth-century preservationists, wilderness typically meant the part of the landscape that is unaffected by human impacts, a pristine region where "earth and its community of life are untrammeled by man, where man is a visitor who does not remain."[2] Wilderness reflects a natural as opposed to human order, and can therefore serve as a place of moral regeneration and spiritual renewal.

It's a useful concept, to be sure, but it does have limitations. According to its critics, it reinforces a problematic opposition between what is human and what is natural and provides little guidance on how to integrate wilderness into human society. Moreover, it is outdated; there are no longer any places we can sensibly view as truly pristine.[3] Like all concepts, *wilderness* bears the imprint of the historical and social location of the actors who formulated it. It undoubtedly captures an important dimension of the experience of some nineteenth- and early twentieth-century middle-class white American men. John Muir apparently *was* able to experience Yosemite Valley as pristine and radically different from human civilization—but differently situated actors might not, and the traditional concept of wilderness may not carry much meaning for them. To discover such alternative understandings of wilderness we will need to look beyond the canonical nature writers.

One promising place to look is the black intellectual tradition. Conventional wisdom has it that the concept of wilderness was invented by and for white men,[4] but in fact wilderness also plays an important role in the political and literary works of black Americans. These works deserve greater attention from environmental theorists. The black intellectual tradition reflects both blacks' unique experiences and the broader American culture; it is, as Henry Louis Gates put it, "two-toned," both distinctly black and recognizably American. The concept of wilderness at play in this tradition thus bears some relation to the preservationist concept, but it carries a different inflection, reflecting the distinctive experiences and concerns of the black American community.[5] Exploring this concept can illuminate alternative meanings of wilderness, giving us a critical perspective on the traditional concept and making available a different understanding of humans' relationship to the undeveloped landscape.

The traditional concept imagines wilderness as radically opposed to human society, a physical region lying beyond society and outside of human history.[6] Indeed, under this view the value of wilderness lies in its separation from human society and detachment from human history. The black tradition, in contrast, is centrally concerned with the relationship between identity and landscape, and particularly the historical relationship between a community and the land as that relationship is mediated by memory. Those concerns give shape to a distinctive concept of wilderness that (at the risk of oversimplifying this complex tradition) I call "the black concept." Importantly, wilderness in this tradition is not always confined to the external landscape; there is also a wilderness within, an untamed vital energy that derives from and connects one to the external wilderness in which the race originated. Thus, wilderness is not radically differentiated from human society; it is the origin and foundation of culture, and intimately connected to one's cultural (and particularly racial) identity. Preserving wilderness means preserving not merely the physical landscape but the community's cultural forms and consciousness—its collective memories of the community's aboriginal environment.

This concept of wilderness is quite rich, but it has also been troublesome to black writers. Both the black concept and the traditional concept were influenced during the early twentieth century by scientific racism and primitivism. By the 1920s, black writers were raising concerns about the racial essentialism that infuses both concepts. Those concerns, I would argue, do not entirely undermine the value of the black concept of wilderness, but they do provide an important critical perspective on the wilderness concept in general.

This study is necessarily limited in scope. I focus on the writings of black elites, such as Alexander Crummell and W. E. B. Du Bois, because their works are comparable in richness and sophistication to those of the canonical preservationists. However, I begin with slave culture, which was an important source for later black writers. I continue with the nineteenth-century discourse on African colonization, and conclude by examining how scientific racism and primitivism influenced and complicated the black concept of wilderness.

WILDERNESS IN SLAVE CULTURE

White immigrants to the New World in the seventeenth century arrived already equipped with a concept of wilderness. "Wilderness" referred to the part of the landscape that was not under agricultural cultivation or human settlement of the European pattern. It was the abode of savages and beasts, a place of danger—but also a place governed by a natural order, in opposition to the human order reflected in rural and urban landscapes. It could therefore be a spiritual or political refuge.[7]

The Africans enslaved and forcibly transported to the New World brought a different set of conceptual resources. Although religiously diverse, most enslaved Africans embraced an animistic belief system that imbued the natural world with spiritual meaning. In some respects, however, their views of the landscape were similar to those of their white captors, and they had a category for the part of the landscape that Europeans called wilderness. In many West African cultures, "the bush"—the region beyond the human settlement—was an important spiritual resource, the location of religious rituals and spiritual transformation.[8] These African religious beliefs persisted among American slaves well into the nineteenth century, particularly in regions such as South Carolina where there were large concentrations of slaves. However, intermittent efforts at Christianization also had effect, as did the influence of various white folk traditions. By the nineteenth century, these influences had resulted in a distinctive slave culture—not uniform or universal, to be sure, but shared by the majority of slaves who worked on large plantations. We should not overstate the differences between slave culture and white folk culture, of course; there was much interaction between them, and white folk culture in the South was itself a mosaic of beliefs that included religious, magical and scientific perspectives on nature. Nevertheless, slave culture did have some distinctive emphases.[9] In particular, the slaves' symbolic culture imbued the natural landscape with moral and spiritual meaning, emphasizing the close relationship between human morality and the condition of the landscape. This interdependence between the community and the land would become central to the concept of wilderness as it developed in the black intellectual tradition.

Two aspects of nineteenth-century black culture were particularly relevant to the evolution of the black concept of wilderness: the spirituals and the fugitive slave narratives. The spirituals, a rich body of religious vocal music, drew heavily on Christian imagery to imbue the landscape with religious significance—a tendency probably reinforced by the persistence of African spiritual beliefs, which also infused the natural landscape with spiritual meaning. Slaves sang "O Canaan, sweet Canaan, / I am abound for the land of Canaan"; called on the river to "roll, Jordan, roll"; and asked "Did yo ever / Stan' on mountain / Wash yo' han's / In a cloud?" Black troops marched into war singing "Go in de wilderness, / Jesus call you. Go in de wilderness / To wait upon de Lord." According to Lawrence Levine, these songs illustrate that "the sacred world of the slaves was able to fuse the precedents of the past" with "the conditions of the present" so that the material world and the spiritual world merged.[10] The result was a sacred geography that could be imposed on the natural landscape: a river could become a type of the river Jordan, the North a type of Canaan, and wild spaces could become the desert in which the children of Israel wandered or a place to which one retreats to seek spiritual transformation.[11]

We find a striking example of this rhetorical practice in Martin Delany's serial novel, *Blake* (1861–1863). Henry, the hero, has escaped from a plantation with the intent of leading a slave revolt. As he enters the wilderness and reaches the Red River, he is overwhelmed by the task in front of him:

> Standing upon a high bank of the stream, contemplating his mission, a feeling of humbleness and a sensibility of unworthiness impressed him. . . . Henry raised in solemn tones amidst the lonely wilderness:
>
> Could I but climb where Moses stood,
> And view the landscape o'er;
> Not Jordan's streams, nor death's cold flood,
> Could drive me from the shore!

The modern reader may be struck by the sudden emergence of an Old Testament landscape in the American wilderness, but Henry is thoroughly at home in this biblical world. He climbs his mountain and crosses the alligator-infested Jordan, his faith "now fully established."[12]

The imposition of foreign geographies onto the American landscape persists in black literature, as do biblical imagery and the depiction of the wilderness as a place of spiritual trial. But black literature drew on other traditions as well, including the fugitive slave narratives published by abolitionists.[13] These narratives added another dimension to the black concept of wilderness. Although the narrators did frequently echo the spirituals (describing the South as "Egypt" and the North as "Canaan," for example), they also drew on conventions familiar to the white audiences to which they were directed—particularly the literary conventions of pastoralism. Fugitive Henry Bibb echoed a common pastoral theme in describing the beauties of nature "on free soil": "the green trees and wild flowers of the forest; the ripening harvest fields waving with the gentle breezes of Heaven." Charles Ball's narrative contrasts the pastoral North with a South so corrupt that agriculture was failing and wilderness was overtaking the plantations: "In some places, the cedar thickets . . . continued for three or four miles together," a melancholy "deserted wilderness."[14]

In general, these stories depict wilderness in a negative light; it is a perilous place teeming with dangerous beasts and savage men (sometimes native Americans, but more often violent, uncivilized white men). Moses Roper escapes his plantation only to "wander through the wilderness for several days without any food," and encounter alligators and wolves in the forest. William Wells Brown nearly freezes to death during his wilderness journey. Josiah Henson also confronts hunger and wolves.[15] To slaves, the wilderness was hostile and frightening—but no more so than the tyrannical plantations, where a worse kind of wildness reigns. This emphasis on the frightening aspect of the wilderness may be surprising, since most plantation slaves were probably familiar with the forests and swamps surrounding the plantation, where they often hunted and gathered roots and herbs.[16] But the narratives remind us that slaves were severely restricted in their freedom of movement; although their immediate locale might be familiar, a fugitive would quickly find him- or herself in an unknown and dangerous landscape. The political purpose of the narratives is relevant here: narrators probably emphasized the dangers of escape in order to win the audience's sympathy and admiration.[17]

The narratives' political purpose also accounts for the fact that they were less likely than the spirituals to give the landscape spiritual meaning; narrators typically described the wilderness not as a place of spiritual transformation but as a temporary—and dangerous—refuge from tyranny.[18] Nevertheless, the slave narratives like the spirituals invest the landscape with moral significance. Plantations turn into desolate wilderness, in Charles Ball's narrative, because the planters lack the virtues necessary to maintain the land's fertility; the South, in Frederick Douglass' words, was "cursed with a burning sense of injustice."[19] This relationship between the natural environment and human morality rehearses an old theme in the Christian tradition: the notion that nature as well as humanity was corrupted by Adam's transgression, and is similarly in need of redemption. These themes of interdependence between the landscape and human morality and the land's need for redemption shape black discourse about wilderness through the nineteenth century, becoming intertwined with a related theme connecting wilderness and black identity.

WILDERNESS AND DESTINY

Wilderness for nineteenth-century white Americans typically meant the western frontier, and attitudes about the wilderness were linked to the ideology of manifest destiny: the belief that it was the duty of white Americans to civilize the West. The ideology is a nineteenth-century gloss on a long standing theme in American thought, that God intends for the wilderness to be subdued and that this intention justifies colonization of the New World.[20] Slaves and freedmen were important agents in this project, of course. But some black Americans, both before and after Emancipation, believed that the frontier they were destined for lay in Africa.

Africa was central to both black racial identity and to black discourse about wilderness; indeed, it constituted the primary conceptual connection between identity and nature in black thought. It was a troubled connection, however. While many American blacks apparently preserved a sense of African identity into the nineteenth century, others strongly objected to being called African and worried that claiming their African heritage would make it more difficult to win acceptance into American society. To most white Americans, after all, Africa was a wild region inhabited by savages unfit for participation in American civilization.[21] The problem of black identity was therefore intimately connected to the meaning and value of the African wilderness.

One influential approach to this problematic connection is represented by the nineteenth-century African colonization movement. Throughout the nineteenth century, black leaders such as Martin Delany, Edward Blyden, and Alexander Crummell urged black Americans to colonize the African wilderness—essentially, to elevate the black race by transforming the African landscape. Colonizationists claimed divine sanction for their project, citing the Ethiopian Prophecy (Psalm 68). "Princes shall come out of Egypt; Ethiopia shall soon stretch out her hands unto God," declared Martin Delany in his call for the colonization of East Africa. The land belongs to the black race, and "all that is left for us to do, is to make ourselves the 'lords of terrestrial creation'" by possessing it.[22] It is more than a right; it is the duty of black Americans to "civilize" Africa: "Africa is our fatherland and we its legitimate descendants." Delany recognized that much of Africa was already settled, but he nevertheless evoked the familiar image of Africa as a tropical jungle: undeveloped, unpossessed, and inhabited by savages. His plan, he declared, was "the first voluntary step that has ever been taken for her regeneration."[23]

Edward Blyden also relied on the Old Testament—Deuteronomy 1:21—in making clear God's intent that black Americans "possess the land" (in this case, Liberia). Africa is "ours as a gift from the Almighty," who preserved it for centuries from European domination.[24] He described what they might accomplish: the visitor to Liberia today, encountering a lonely, unbroken forest, would wonder "when and how are those vast wildernesses to be made the scene of human activity and to contribute to human wants and happiness?" A few years later, however, the visitor might return to find roads and bridges. As Blyden imagined the transformation, "The gigantic trees have disappeared, houses have sprung up on every side. . . . The waving corn and rice and sugar-cane . . . have taken the place of the former sturdy denizens of the forest." These "wonderful revolutions" were all to be accomplished by the American Negro colonist, who would cause "the wilderness and the solitary place to be glad—the desert to bloom and blossom as the rose—and the whole land to be converted into a garden of the Lord." Blyden went on to compare Liberia favorably to the iconic American wildernesses, the Rocky Mountains and Yosemite Valley. But he emphasized that he was not advocating the settling of a new country by strangers; he was calling for the "repatriation" of Africa.[25]

African colonization rhetoric thus characterized Africa as wilderness, and like the rhetoric of manifest destiny it assumed that wilderness must be subdued and given value through human labor. In this respect it echoed the rhetoric of white Americans intent on settling the western frontier. But it lacked the ambivalence often expressed by white elites—an ambivalence stemming from a competing set of ideas, rooted in eighteenth-century romanticism, that informed American preservationism. In this tradition, wilderness is a place to escape the corruptions of civilization; savages even have a certain nobility by virtue of their natural innocence. For writers such as Thoreau and Muir, the wilderness was a place of spiritual and moral regeneration, to be valued for its own sake.[26]

The rhetoric of black colonizationists contrasts with this preservationist tradition in three respects. First, as noted above, in this preservationist tradition wilderness is typically conceived as outside of human history; it is "detached from all temporal relationship." Because humans have no history with wilderness areas, they are places where we can escape our own bad history and start over. To preserve wilderness is to preserve this possibility.[27] For the black colonizationists, in contrast, the African wilderness has an important historical dimension. True, nineteenth-century blacks often shared white historians' view that Africa itself had no political history. Nevertheless, interest in Africa among black elites in the late nineteenth century was aimed at claiming their history, a history that began in the African wilderness.[28] For them, the African wilderness was located in the historical past; going to Africa meant recovering one's own history. In fact, the colonizationists urged black Americans to return to Africa *because of* their historical ties to it; they argued that black Americans had obligations to Africa based on that history. As Alexander Crummell put it, "The land of our fathers is in great spiritual need, and . . . those of her sons who haply have the ability to aid in her restoration will show mercy to her, and perform an act of filial love and tenderness which is but their 'reasonable service.'"[29] In this tradition, black Americans' relationship to the wilderness is fundamentally historical: while white Americans "discover" the West, black Americans "return" to Africa. That historical relationship is the basis for black Americans' obligations to redeem the African landscape.

Second, in the preservationist tradition, wilderness is typically a place of natural innocence, where natural law rules.[30] Journeying into the wilderness can therefore lead to moral regeneration, as the corrupting influences of society are replaced by the positive influences of nature. This idea is largely absent from nineteenth-century black colonization rhetoric. Blyden and Crummell usually described Africa as spiritually dark, at least amoral and possibly corrupt. The wildness of the landscape is evidence of this corruption: enlightened, civilized people would cultivate the land. "Regenerating" the land is therefore both a moral and a physical task; it involves enlightening the "benighted heathens" as much as transforming the landscape.[31] Of course, after the turn of the century, black elites would question this judgment. Du Bois, for example, had a more favorable view of traditional African culture, and also argued that the moral corruption that Crummell condemned was the result of white imperialism and the slave trade.[32] Other twentieth-century developments, discussed below, would undermine the negative judgments of the African wilderness further. Nevertheless, this element of the colonization rhetoric had lasting impact on black thought. Although it implies that wilderness is without intrinsic moral value, it also makes regenerating the land a moral duty—a duty rooted in history and requiring positive action by black Americans.

Finally, some American preservationists went beyond the rhetoric of manifest destiny, suggesting that conquering nature was an end in itself, a way to exert one's independence and express one's masculinity. This version of rugged individualism celebrates the domination of nature, but not through transforming the landscape. Instead, it aims at preserving the wilderness in order to preserve the opportunity to cultivate the "physical independence," "individuality and competence" achieved by adventuring in the wild.[33] Such rhetoric of rugged individualism was sometimes echoed by Booker T. Washington, who saw struggle as the primary means of racial development. But even Washington recognized that blacks did not have to test themselves against mountains and forests; the social environment provided enough challenges. And Washington notwithstanding, black thought is generally less individualistic than other varieties of American liberalism. It highlights the individual's dependence on the community, and typically defines freedom as the freedom to create a home and community.[34] This tradition therefore focuses on the relationship between the wilderness and the *community,* not just the wilderness and the individual. The goal of

African colonization was to create homes and communities in Africa, and it was an obligation rooted in group history and solidarity, not an expression of individuality.

In sum, African colonization rhetoric casts the wilderness in a different light: the African wilderness is intimately connected to black Americans' identity; it is part of the group's history that must be reclaimed and restored through positive action. Unlike preservationism, this tradition focuses on regenerating a degraded landscape rather than preserving a pristine one, it calls for action based on the group's historical ties to an ancestral homeland, rather than mere contemplation of a newly discovered landscape. These themes of historical connection and obligation continue to shape the black concept of wilderness into the twentieth century, as the relationship between wilderness and culture becomes a central focus of black thought.

BLACK PRIMITIVISM

Albert Barnes announced in 1925 that the art of the New Negro would make us feel "the majesty of Nature, the ineffable peace of the woods and the great open spaces." The Negro has "kept nearer to the ideal of man's harmony with nature"—not just Africans but American Negroes, who enjoy "a primitive nature upon which a white man's education has never been harnessed."[35] Strange things to say about the highly educated, city-dwelling artists of the Harlem Renaissance, perhaps, but Barnes' language was not out of place in Alain Locke's *The New Negro.* It reflects an important evolution in black thought in the early twentieth century: the connections between nature, human culture and racial identity remained central to the concept of wilderness, but they took on a new dimension as black writers came to conceptualize wilderness as a reservoir of creative energy, a source of culture that lay not only in the historic past but in the consciousness of the race itself.

To understand this evolution, we have to explore its roots in two nineteenth-century intellectual trends: scientific racism and artistic primitivism. Racists and artistic primitivists held opposing views on the value of non-Western cultures, but they shared the late Victorian anxiety about the loss of cultural vitality. To many social critics, Western culture in the late nineteenth century seemed to be stagnant. Preservationism itself reflects this fear; one rationale for preserving wilderness was to revitalize Americans by giving them more opportunities for contact with nature.[36] Black Americans, drawing on scientific race theory and artistic primitivism, suggested another source of cultural revitalization: the creative energy of a race that, through centuries of oppression and dislocation, had never lost its primitive connection to nature.

The idea that contact with nature is the primary source of cultural vitality draws on the vitalist concept of nature associated with nineteenth-century romanticism. According to vitalists, nature is animated by a vital force that drives and governs all biological growth. Under this view, nature is a reservoir of creative energy, and because humans are also part of nature, they also share in its vital power—a power that a sensitive soul can draw on in order to create great works of art. For the romantics, contact with external nature was the most obvious way to tap into its vital force. Thus, individuals or groups who interacted with the natural world on a practical, spiritual and emotional level were the most capable of creating a rich, vital culture.[37] Hence, the romantic interest in "primitives," rural folk and others who lived "close to nature."

These ideas were particularly influential among white supremacists and the artists who developed the intellectual foundations of modern art, both of which groups would have profound influence on the development of black thought. White supremacists, for example, remained obsessed with vitality long after the rest of the scientific community had abandoned vitalism in favor of a mechanical model of nature. American race theorists such as Nathaniel Shaler, Joseph Le Conte, William McDougall, Madison Grant, and Lothrop Stoddard contended that human races evolved out of a dynamic interaction between the race's inner vital force and the external environment.[38] Le Conte, for example, contended that there was a vital principle in nature, and that "resident or inherent forces" in a race determine how successfully it will respond to the natural environment. Grant similarly argued that each race had an "inherent capacity for development and growth"—a capacity that cannot be altered by environmental influences. The "vigor and power" of the

Nordic race explains its dominance, which should continue as long as the "energy" of the race does not dissipate in its increasingly urbanized, less demanding environment.[39]

Scientific racists argued that the inner vitality of the race, in reaction to the natural environment, determines cultural as well as physical development. Indeed, they posited a close connection between natural environment and culture: mental processes, they claimed, are directly determined by racially differentiated brain structures. Therefore, biological adaptation produces a mental life and culture peculiarly adapted to the natural environment.[40] Shaler elaborated this relationship between nature and culture through the concept of the cradle land: humans, he argued, are like all organisms affected by environmental conditions such as climate and the condition of the soil. Although evolving originally in one place, humans spread out over the earth and settled into environmentally diverse niches: the "cradle lands." As they adapted to these different environmental conditions over the course of many centuries, they developed into distinctive races. For Shaler (as for virtually all the scientific racists), the environmental conditions of northern Europe made it the cradle of "strong peoples," because its climate was harsh enough to require vigor and industry, but not so harsh as to discourage effort. In contrast, Africa did not produce great peoples because (Shaler imagined) tropical climates are easy to live in. Thus, the races cradled in these lands never developed civilization; their native vitality and capacity for culture dissipated in their undemanding environment.[41]

Of course, white supremacists had to account for the achievements of tropical and subtropical cultures, such as the Mediterranean and Indian civilizations. Defenders of Nordic civilization argued that such peoples were *limited* to artistic and spiritual achievements and would never make substantial progress in science, philosophy or technology. The influential English historian Henry Thomas Buckle, for example, argued that peoples living in severe environments—among the mountains and great rivers of India, or endless deserts of the Mideast—are stimulated by the scenery so that they become more imaginative and less analytical, and are also rendered passive by the sheer enormity of the natural forces they face.[42] Northern Europe, and particularly England, lacked such sublime landscapes; as a result, the Nordic race was more analytic, rational and active than southern races. Under this reasoning (endorsed, for example, by William McDougall) the hardy, vigorous races that originated in the forests of northern Europe could be expected to produce more and better cultural products than weak and passive races produced by tropical wildernesses.

Unless, that is, the Nordic race degenerated in its new environment. Not all race theorists believed that environmental pressures were still driving racial development, and American theorists tended to worry about miscegenation more than environmental degradation. Nevertheless, some race theorists expressed anxiety that the Aryan race, cradled in a rugged northern European climate, would languish in an industrialized and urbanized environment—a fear that led many (notably Grant, Le Conte and George Bird Grinnell) to support wilderness preservation. A careful reader of Buckle might have had some concerns about preserving the sublime landscapes of the American West, but American racists tended to view the American wilderness as precisely the right kind of environment to maintain the vitality of Nordic civilization.[43]

Artistic primitivists, on the other hand, weren't so sure. They were also concerned about the vitality of Western civilization, but they had a different perspective on tropical cultures than did the scientific racists. Artistic primitivism was an aesthetic movement that arose in opposition to the nineteenth-century academic school of painting. It drew on the more diffuse literary and philosophical primitivism that had long permeated Western culture, but artistic primitivism was a distinctive tradition, and highly influential among black theorists.[44] Artistic primitivists complained that European art was constrained by convention; it was too civilized, and therefore too far removed from the vital force of nature. To remedy this artistic ennui, they suggested that European artists look for inspiration to the art of the tribal cultures of Africa and Oceana. Following the scientific racists, primitivists characterized these peoples as a kind of evolutionary cul-de-sac—a people without culture and therefore free of "the conventionalities and unnaturalness of the civilized." But primitivists thought this lack of culture and analytic abilities an advantage: primitive arts (particularly those nurtured in fertile, vibrant landscapes) drew more directly on the vital force of nature.[45] "Are not savages artists who have forms of their own as powerful as the form of thunder?" asked August Mack, of the Blau Reiter school. Under this view (developed, for example, by Wassily Kandinsky), artistic creativity is a matter of not merely imitating the forms of nature but producing in the audience the *feeling* that the artist experiences when communing with nature.[46] What these artists hoped for from primitive art was a direct if crude expression of that feeling, which could inspire Western artists' more sophisticated and technically superior art.

Primitivism stimulated interest in African art. According to one enthusiast, "These Africans being primitive, uncomplex, uncultured, can express their thoughts by a direct appeal to the instinct. Their carvings are informed with emotion."[47] Elie Faure declared that "even when transported in great numbers to places like North America that have reached the most original . . . degree of civilization . . . , the black man remains, after centuries, what he was—an impulsive child, ingenuously good, and ingenuously cruel." He sought in Negro art "that still unreasoned feeling which merely obeys the most elementary demands of rhythm and symmetry. . . . Brute nature circulates in them, and burning sap and black blood."[48] To be fair, not all primitivists were so blinded by racial stereotypes. But the underlying motive of the movement was to discover an art produced through an untutored, almost instinctual process free of culture, history and tradition: an art that was virtually organic, the product of nature's vital energy.

Artistic primitivism became a major influence on black thought in the early decades of the twentieth century. Drawing on primitivism and scientific racism, black Americans could claim to be a young race, still in touch with the aboriginal wilderness, and therefore a source of creative energy that could regenerate American culture. Du Bois' famous essay from 1897, "The Conservation of Races," reflects this synthesis of racial and primitivist ideas. Although resisting a strictly biological concept of race, Du Bois insisted that

> there are differences . . . which have silently but definitely separated men into groups. While these subtle forces have generally followed the natural cleavage of common blood, descent and physical peculiarities, they have at other times swept across and ignored these. At all times, however, they have divided human beings into races, which, while they perhaps

transcend scientific definition, nevertheless, are clearly defined to the eye of the historian and sociologist.

According to Du Bois, the African race had as much to contribute to world civilization as the white race. In particular, it would contribute its gift of artistry and spirituality: Negroes "are that people whose subtle sense of song has given America its only music, its only American fairy tales, its only touch of pathos and humor amid its mad money-getting plutocracy."[49] This is primitivism underwritten by race: black Americans are members of a race characterized by the creativity and vitality of a primitive people—specifically, a tropical people, with the imaginative and spiritual gifts produced by a warm, fertile environment. They can regenerate American culture because their racial characteristics are those of true primitives. In short, black Americans retained a racial connection to the tropical African wilderness, and that connection is the source of their cultural vitality.

Primitivism and scientific racism thus gave new meaning to the concept of wilderness in black thought. Du Bois explored this meaning in his novel from 1911, *The Quest of the Silver Fleece.* The story follows Zora, a black girl born in a swamp in Georgia, as she educates herself, travels to New York, becomes politically active, and returns to Georgia to help the black community. The swamp is a central image in the novel: Du Bois describes it as wild and savage, a dark, "sinister and sullen" place of "strange power." Zora, daughter of the swamp's only inhabitant, the witch Elspeth, is "a heathen hoyden"; when we first encounter her she is dancing in the firelight to "wondrous savage music." Du Bois underscores her primitive vitality and spirit; she glows "with vigor and life."[50]

But Zora does not remain a primitive. Under the tutelage of her school teacher, Zora's savagery diminishes; she becomes "a revelation of grace and womanliness." Without losing her energy and spirit, she develops into "a brilliant, sumptuous womanhood; proud, conquering, full-blooded, and deep bosomed—a passionate mother of men." She also begins to speak "better English," drifting into "an upper world of dress and language and deportment." She continues her education in politics and economics in the North, but eventually returns to Georgia, intent on improving the economic and social condition of the black farmers. Her plan is "a bold regeneration of the land": she encourages the farmers to clear the swamp and establish a collective farm. The result is a transformation akin to Zora's own evolution from wild to civilized: the farmers create a thriving communal cotton plantation on the site of the swamp. The swamp does not wholly disappear, however. Its vital energy persists in the fertility of the soil, and the swamp itself remains in Zora's memory—not "cold and still" but "living, vibrant, tremulous," the origin of her strength and love.[51]

Du Bois' story draws on primitivist themes, but there are also important continuities with the nineteenth-century discourse on wilderness. The swamp, although located in the United States, seems to be contiguous with the African jungle by virtue of cultural continuity between American blacks and their African forbears (represented by the witch Elspeth). Zora's return to Georgia is, in a sense, a return to Africa. Moreover, the wilderness is located in history; it is a part of the heroine's personal history and the history of the black community. In another respect, too, Du Bois holds to the older concept of wilderness: although the swamp is a reservoir of nature's creative energy, that energy is essentially amoral—or even immoral. Zora's mother, for example, is no noble savage; she prostitutes her daughters to the wealthy white landowners (a reference perhaps to Africa's exploitation by Europeans as well as Americans' exploitation of black slaves). Thus, the swamp is not a place of spiritual regeneration. In order to provide the basis for a well-ordered human community, the land itself must be regenerated—transformed through human labor into higher cultural forms.

Nevertheless, for Du Bois, the point of transforming the wilderness is to *preserve* it—to maintain its vital energy and make that energy more widely available to the community. What is valuable in wilderness—its vital power and natural beauty—is still present after its transformation, in the fertility of the soil and in cultural representations of the swamp (the community's collective memory of this original wilderness). The swamp also remains within Zora herself, in her memory and racial consciousness. Du Bois' heroine relates to the wild by remaining in contact with the inner vital force that is part of her racial heritage (the wilderness within) as well as with the culture that develops out of the group's interaction with nature (the wilderness without). Thus, the novel's message is that the black artist must stay in touch with her roots in the wild earth; her task is to preserve that primitive energy in her own memory and to embody it in cultural products that reflect the race's highest ideals.

To a traditional preservationist, this sort of preservation may seem inadequate; surely we need to preserve the swamp itself in order to maintain the land's fertility, not to mention biodiversity. But Du Bois points out that if a goal of wilderness preservation is to forge a closer tie between wilderness and human culture, then some sort of transformation may be necessary to make what is valuable about wilderness *accessible* to the human community. White, middle-class preservationists often elided this question of access, but for black Americans—who had provided much of the labor of transforming the wilderness and who had been subject to segregation and exclusion throughout their tenure on the American continent—access to nature's gifts could not be taken for granted. Further, Du Bois suggests that the individual's memory of the wilderness—his or her understanding of its place in the group's history—is also critical to the connection between wilderness and human culture that ensures cultural vitality. Thus, to preserve and make available the creative power of the wilderness experience, we must attend to maintaining both the individual's sense of membership in his or her community (in this case, his or her racial identity) and the community's collective memory of its aboriginal wilderness.

MODERN MAN-MADE JUNGLES

Primitivism remains a ubiquitous theme in black literature during the Harlem Renaissance period, as does the belief that the source of artistic creativity is the culture that one's community developed during the period when it still had a close connection to nature.[52] Under this view, the vital elements of black culture were those shaped by the race's history in Africa and in the American South. Black artists should therefore look back and within, in order to make contact with the vital creative energy of nature. As Jean Toomer said of his inspiration for *Cane:*

> Georgia opened me. And it may well be said that I received my initial impulse to an individual art from my experience there. For no other section of the country has so stirred me. There one finds soil, soil in the sense the Russians know it,—the soil every art and literature that is to live must be imbedded in.[53]

But black artists, under this view, did not actually have to go to Georgia to find artistic inspiration. Because their wilderness lies within, they can get in touch with it in the city, through an inner journey aimed at becoming aware of one's history and racial identity. The city can therefore be a "modern man-made jungle." Claude McKay described a Harlem nightclub as "a real throbbing little Africa"; for Jessie Redman Fauset, people in Harlem lived "at a sharper pitch of intensity"—"coloured life" was "so thick, so varied, so complete."[54] Just as Martin Delany had imposed an Old Testament landscape on the American wilderness, twentieth-century writers imposed a tropical landscape on New York. In Countee Cullen's poem, the tropics appear suddenly in a New York market: "Bananas ripe and green, and ginger root / Cocoa in pods and alligator pears" bringing "memories / Of fruit-trees laden by low-singing rills."[55] Harlem was imagined as a place of primitive vitality where even white Americans could get in touch with nature; as Albert Barnes suggested, the artistry of black Americans could regenerate American culture by showing us the "ineffable peace of the woods." But it can accomplish this goal only if black urban culture retains its consciousness of its roots in Africa and the American South—the places where the race formed through constant interaction with wild nature. The cultural vitality of black America depends on preserving the African wilderness and the America South in the community's consciousness and cultural forms.

This program, however, rests on a racial essentialism that many Harlem Renaissance artists found troubling. As Cullen pointedly asked, "What is Africa to me?" His poem seems to reflect the primitivist conceit that American blacks still hear the "great drums throbbing through the air," but it also questions that conceit.[56] Could one "three centuries removed" still remember Africa's copper sun or scarlet sea? Alain Locke thought not; the culture of black Americans, he insisted, was *not* essentially African: "What we have thought primitive in the American

Negro . . . are neither characteristically African nor to be explained as an ancestral heritage." He explained them as the result of blacks' "peculiar experience in America and the emotional upheaval of its trials and ordeals." For Locke, the American Negro's temperament was a product of social and historical forces "rather than the outcropping of a race psychology."[57]

Locke's concerns point to the basic difficulty with primitivism: it seems to be inescapably essentializing. It assumes the existence of a racial "essence"—the wilderness within—that was preserved in black Americans through centuries of acculturation and dislocation. Of course, an essentializing racial ideology could be politically useful for a people seeking to gain acceptance in American society. But racial essentialism was a major pillar of early twentieth-century white supremacy, and by the 1920s black artists experimenting with primitive themes were under attack from activists (like Du Bois himself) for supporting harmful racial stereotypes.[58] Moreover, artists began to find the conventions of primitivism constraining. As they explored its limits and examined the conflict between racial solidarity and individualism, they suggested that escaping from the past and community—and from white images of black identity—may be as important to artistic creativity as remaining in touch with the "wilderness within."[59]

Did this growing distrust of primitivism and essentialism affect blacks' attitudes toward the wilderness preservation movement? It's hard to say. The Harlem Renaissance artists did pose Harlem as an alternative to the American wilderness as a source of cultural vitality, but they offered little criticism of the preservationists' agenda. Du Bois' guarded remarks in *Darkwater* may represent a broader attitude: when asked by his white friends why he doesn't enjoy the wilderness more often, he lectures them on the indignities of traveling in the era of "Jim Crow."[60] Du Bois probably did not oppose preservationism; his prose exhibits an extraordinary sensitivity to natural beauty. But he had to be suspicious of a movement that claimed Grant and Le Conte among its supporters, and that showed little interest in challenging the segregationist policies that made the nation's parks inaccessible to many blacks.

By the end of the Harlem Renaissance period, the black intellectual community increasingly embraced a cultural rather than essentialist concept of race. The full implications of that transition are beyond the scope of this article. But rejecting racial essentialism does not require us to reject the insights derived from this exploration of the black concept of wilderness—the idea, for example, that a historical connection to the group's aboriginal wilderness, preserved in the group's collective memory, may be a source of inspiration, identity, and sense of obligation to the land. This concept of wilderness as a source of cultural creativity, a fatherland to which the artist must return (at least in memory) in order to unlock his or her own creative power, remains an important part of the black literary heritage. It could become even more relevant as the Romantics' pristine wilderness recedes into history, leaving in its place a degraded landscape burdened by a history of injustice.

On the other hand, black writers' critiques of primitivism require us to explore further the connection between landscape and racial identity. Indeed, the black concept of wilderness is useful to us precisely because it foregrounds questions about one's relationship to the community and to the land, and about the obligations of history—questions that the traditional concept seems designed to avoid. In sum, wilderness for black writers is not a receding frontier but an increasingly complex and difficult terrain; it is a place not to escape from but to confront one's history, community, and identity.

NOTES

1. W. E. B. Du Bois, *Souls of Black Folk* (New York: Penguin Books, 1989), p. 90.

2. J. Baird Callicott and Michael P. Nelson, eds., *The Great New Wilderness Debate* (Athens: University of Georgia Press, 1998), pp. 3–4; Wilderness Act, U.S. Code, vol. 16, sec. 1131(c) (1964).

3. J. Baird Callicott, "The Wilderness Idea Revisited," in Callicott and Nelson, *The Great New Wilderness Debate,* pp. 348–55; William Cronon, "The Trouble with Wilderness, or, Getting Back to the Wrong Nature," in ibid., pp. 471–99.

4. See, e.g., Bruce Braun, "On the Raggedy Edge of Risk," in *Race, Nature, and the Politics of Difference,* ed. Donald S. Moore, Jake Kosek, and Anand Pandian (Durham: Duke University Press, 2003), pp. 175–203.

5. "Criticism in the Jungle," in *Black Literature and Literary Theory,* ed. Henry Louis Gates (New York: Methuen, 1984), p. 4. There is a small but

growing literature on the history of black environmental thought. See Melvin Dixon, *Ride Out the Wilderness* (Urbana: University of Illinois Press, 1987); Elizabeth Blum, "Power, Danger, and Control: Slave Women's Perceptions of Wilderness in the Nineteenth Century," *Women's Studies* 31 (2002): 247–66; Dianne Glave, "A Garden So Brilliant with Colors, So Original in Its Design," *Environmental History* 8 (2003): 395–411; Cassandra Johnson and J. M. Bowker, "African-American Wildland Memories," *Environmental Ethics* 26 (2004): 57–75; Kimberly Smith, "Black Agrarianism and the Foundations of Black Environmental Thought," *Environmental Ethics* 26 (2004): 267–86; Christine Gerhardt, "The Greening of African-American Landscapes," *Mississippi Quarterly* 55 (2002): 515.

6. Cronon, "The Trouble with Wilderness," pp. 483, 484.

7. Roderick Nash, *Wilderness and the American Mind,* 3rd ed. (New Haven: Yale University Press, 1982), pp. 8–22, 44–66.

8. Margaret Creel, *"A Peculiar People": Slave Religion and Community-Culture among the Gullahs* (New York: New York University Press, 1988), pp. 47–51.

9. Charles Joyner, *Shared Traditions* (Urbana: University of Illinois Press, 1999), pp. 35–36.

10. Lawrence Levine, *Black Culture and Black Consciousness* (New York: Oxford University Press, 1977), pp. 39, 51–53.

11. John Blassingame, *The Slave Community,* 2d ed. (Oxford: Oxford University Press, 1979), p. 145; Eugene Genovese, *Roll, Jordan, Roll* (New York: Vintage Books, 1972), pp. 242–55; Dixon, *Ride Out the Wilderness,* pp. 3–4, 13–14; Nell Painter, *Exodusters* (New York: W. W. Norton and Co., 1970), pp. 191, 195.

12. Martin Delany, *Blake, or the Huts of America* (Boston: Beacon Press, 1970), pp. 69–70.

13. Charles T. Davis and Henry Louis Gates, Jr., *The Slave's Narrative* (Oxford: Oxford University Press, 1985); Williams Andrews, *To Tell a Free Story* (Urbana: University of Illinois Press, 1986); Mary Wilson Starling, *The Slave Narrative* (Boston: G. K. Hall, 1981).

14. "Narrative of the Life and Adventures of Henry Bibb" [1849] in Yuval Taylor, *I Was Born a Slave* (Chicago: Lawrence Hill Books, 1999), 2:36; Charles Ball, "Slavery in the United States" [1836] in ibid., 1:279.

15. "A Narrative of the Adventures and Escape of Moses Roper" [1838] in ibid., 1:495, 511–12; William Wells Brown, "Narrative" in ibid., 1:712; "The Life of Josiah Hensois" [1847] in ibid., 1:746.

16. Stuart Marks, *Southern Hunting in Black and White* (Princeton: Princeton University Press, 1991), pp. 27–28; Mart Stewart, *What Nature Suffers to Groe* (Athens: University of Georgia Press, 1996), pp. 134–36; Sharla Felt, *Working Cures* (Chapel Hill: University of North Carolina Press, 2002), p. 68.

17. Kimberly Smith, *The Dominion of Voice* (Lawrence: University Press

of Kansas, 1999), pp. 175–76, 231. On slaves' fear of wilderness, see Blum, "Power, Danger, and Control," pp. 251–56.

18. Smith, *Dominion of Voice,* pp. 221–23, for further discussion of the narratives' spiritual meaning.

19. "Address before the Tennessee Colored Agricultural and Mechanical Association," in *African-American Social and Political Thought,* ed. Howard Brotz (New Brunswick: Transaction Publications, 1992), p. 291.

20. Henry Nash Smith, *Virgin Land* (Cambridge: Harvard University Press, 1950), pp. 35–43; Nash, *Wilderness and the American Mind,* pp. 24–25.

21. Mechal Sobel, *Trabelin' On* (Westport, Conn.: Greenwood Press, 1979), pp. 40, 72, 227; Levine, *Black Culture and Black Consciousness,* pp. 86–87.

22. "The Condition, Elevation, Emigration, and Destiny of the Colored People of the United States" [1852] in *African-American Social and Political Thought, 1850–1920,* ed. Howard Brotz (New Brunswick: Transaction Publications, 1992), pp. 111, 101, 103.

23. Ibid., p. 103.

24. "The Call of Providence to the Descendants of Africa in America" [1862] in Brotz, *African-American Social and Political Thought,* pp. 115, 117.

25. "The African Problem and Its Method of Solution" [1890] in ibid., p. 136; "The Call of Providence," in ibid., pp. 125, 137.

26. The classic studies of this tradition include Smith, *Virgin Land;* Nash, *Wilderness and the American Mind;* Leo Marx, *The Machine in the Garden* (Oxford: Oxford University Press, 1964).

27. Cronon, "The Trouble with Wilderness," pp. 483–84.

28. August Meier, *Negro Thought in America, 1880–1915* (Ann Arbor: University of Michigan Press, 1963), pp. 50–51.

29. "The Relations and Duties of Free Colored Men in America to Africa" [1860] in Brotz, *African-American Social and Political Thought,* p. 174.

30. Nash, *Wilderness and the American Mind,* pp. 46–48.

31. Crummell, "Relations and Duties," in Brotz, *African-American Social and Political Thought,* p. 173.

32. W. E. B. Du Bois, *The Negro* (1915; reprint ed., Oxford: Oxford University Press, 1970), pp. 62–85, 93–94.

33. Nash, *Wilderness and the American Mind,* pp. 526–27; Robert Marshall, "The Problem of Wilderness," *Scientific Monthly* 30 (February 1930): 143; Simon Schama, *Landscape and Memory* (New York: Vintage Books, 1995), pp. 392–97, 490–98.

34. Michael Dawson, *Black Visions* (Chicago: University of Chicago Press, 2001), p. 11.

35. "Negro Art and America," in *The New Negro,* ed. Alain Locke (1925; reprint ed., New York: Touchstone, 1992), pp. 24, 19, 20.

36. Roderick Nash, "The American Cult of the Primitive," *American Quarterly* 18 (Fall 1966): 520–37.

37. Joseph Beach, *The Concept of Nature in Nineteenth-Century English Poetry* (New York: Macmillan, 1939), pp. 17–20, 45–109; Donald Worster, *Nature's Economy,* 2d ed. (Cambridge: Cambridge University Press, 1994), pp. 17–18.

38. This notion reflects the continuing influence of the Lamarckian view that acquired characteristics could be inherited. Therefore, characteristics acquired as a response to the environment could be inherited by one's offspring and become a racial trait. J. B. Lamarck, *Zoological Philosophy* (1809; reprint ed., Chicago: University of Chicago Press, 1984), p. 2.

39. Le Conte, *Evolution: Its Nature, Its Evidences, and Its Relation to Religious Thought,* 2d rev. ed. (New York: D. Appleton and Co., 1902), pp. 28, 328; Madison Grant, *The Passing of the Great Race,* 4th rev. ed. (New York: Charles Scribner's Sons, 1924), pp. 97–98, 169–70, 215.

40. See, e.g., Herbert Spencer, "Social Statics" [1851], in *On Social Evolution,* ed. J. D. Y. Peel (Chicago: University of Chicago Press, 1972), pp. 8–9, 33; Nathaniel Shaler, *The Neighbor* (Boston: Houghton Mifflin and Co., 1904); Le Conte, *Evolution;* William McDougall, *The Group Mind,* 2d rev. ed. (New York: G. P. Putnam's Sons, 1920); Grant, *The Passing of the Great Race;* Stoddard, *Rising Tide of Color.*

41. Nathaniel Shaler, "Nature and Man in America," *Scribner's Magazine* 8 (1890): 360, 361–63, 365.

42. Thomas Buckle, *History of Civilization in England,* 2 vols. (New York: D. Appleton and Co., 1880), 1:85–86. Buckle, we should note, did not claim such mental characteristics were heritable.

43. Stephen Fox, *The American Conservation Movement* (Madison: University of Wisconsin Press, 1981), pp. 115, 118, 345–51.

44. Colin Rhodes, *Primitivism and Modern Art* (London: Thames and Hudson, 1994).

45. Ibid., pp. 16, 131.

46. "Masks," in *Blau Reiter Almanac* [1912], ed. Wassily Kandinsky and Franz Marc (Munich: R. Piper and Co., 1965), p. 85; Wassily Kandinsky, *Concerning the Spiritual in Art and Painting in Particular* [1912] trans. Michael Sadleir, *Documents of Modern Art* (New York: Wittenborn Art Books, 1947), 5:23.

47. Quoted in Gelett Burgess, "The Wild Men of Paris" [1910] in *Primitivism and Twentieth-Century Art,* ed. Jack Flam and Miriam Deutch (Berkeley: University of California Press, 2003), p. 39.

48. Elie Faure, "L'Histoire de l'art," in ibid., pp. 55–56.

49. *The Oxford W. E. B. Du Bois Reader,* ed. Eric Sundquist (Oxford: Oxford University Press, 1996), pp. 40, 41, 44.

50. W. E. B. Du Bois, *The Quest of the Silver Fleece* (1911; reprint ed., New York: Negro Universities Press, 1969), pp. 13, 16.

51. Ibid., pp. 125, 127–28, 400, 433.

52. See, e.g., James W. Johnson, *Autobiography of an Ex-Coloured Man* (1912; reprint ed., New York: Hill and Wang, 1960); Carl Van Vechten, *Nigger Heaven* (1926; reprint ed., Urbana: University of Illinois Press, 2000); Claude McKay, *Home to Harlem* (1928; reprint ed., Boston: Northeastern University Press, 1987); Jessie Redman Fauset, *Plum Bun* (1928; reprint ed., London: Pandora Press, 1985).

53. Quoted in Locke, "The Negro Youth Speaks," in Locke, *The New Negro,* p. 51.

54. Joel Rogers, "Jazz at Home," in ibid., p. 218; McKay, *Home to Harlem,* p. 29; Fauset, *Plum Bun,* pp. 57, 96.

55. Countee Cullen, "The Tropics in New York," in Locke, *The New Negro,* p. 135.

56. Countee Cullen, "Heritage," in *The Portable Harlem Renaissance Reader,* ed. David Levering Lewis (New York: Penguin Books, 1994), pp. 244–45.

57. Alain Locke, "The Legacy of the Ancestral Arts," in Locke, *The New Negro,* pp. 254, 255.

58. George Hutchinson, *The Harlem Renaissance in Black and White* (Cambridge: Harvard University Press, 1995), pp. 65, 182–208.

59. Langston Hughes, "Slave on the Block" and "Rejuvenation Through Joy," in *The Collected Works of Langston Hughes,* ed. R. Baxter Miller (1932; reprint ed., Columbia: University of Missouri Press, 2002), 15:30–36, 56–71; Nella Larsen, "Quicksand," in *Quicksand and Passing* (1928; reprint ed., New Brunswick: Rutgers University Press, 1986).

60. *Darkwater: Voices from within the Veil* (1920; reprint ed., New York: Humanity Books, 2003), pp. 229–31.

Cassandra Y. Johnson and J. M. Bowker

African-American Wildland Memories (2004)

. . . acknowledging the ambiguous legacy of nature myths does at least require us to recognize that landscapes will not always be simple "places of delight" — scenery as sedative, topography so arranged as to feast the eye. For those eyes, as we will discover, are seldom clarified of the promptings of memory. And the memories are not all of pastoral picnics.

Simon Schama[1]

INTRODUCTION

MAURICE HALBWACHS ARGUES that memory is only retained in groups or communities of people.[2] For events to withstand the test of time, there must be a mutual sharing of information about such events; otherwise memories die. Certain memories do not exist apart from social milieus and in fact are particular or especially salient only to those who are members of a given group or mnemonic community. Because most of humankind lives as social creatures and not isolated individuals, we cannot escape the shadow of the past which continually reminds us of the glory, pain, or shame experienced by the various collectives or communities to which we belong.[3] A ready example of this type of "collective memory" is the experience of Jews during the Holocaust and the "memory" of the Holocaust by successive generations of Jews. Although younger generations of Jews in contemporary Germany and elsewhere have no personal memories of the Holocaust, they may recount vividly stories related to them by parents, older relatives, and others who lived these experiences.[4]

As noted by May and also as indicated above, much of the scholarship on collective memory has focused on events of political or cultural significance or national figures—for instance, Marten-Finnis' study of German and Polish national identities or Schwartz's study of Abraham Lincoln.[5] In contrast to these national-level sociocultural and political memories, there have been few theoretical or empirical considerations of collective memories about wildlands for particular racial or ethnic groups in U.S. society.[6] An exception is the extensive documentation of Native American land disfranchisement and Native American recollections of the same.[7] But relatively few studies have explored the historical relationship of African Americans to wildlands,[8] and how the collective experience and memories that derived from the land may influence a contemporary black wildland view.[9] We focus on black impressions of wildlands because these landscapes are considered to form the core of American national identity. Yet, empirical studies show some of the greatest black/white differences in outdoor recreation participation occur with wildland interaction.[10]

We propose that collective memory can be used as a conceptual tool to consider African-American perceptions of wildlands and black interaction with such areas. As indicated, collective memory involves the relaying or handing down of cultural history from generation to generation. Successive generations can be influenced by events that impact a nation, ethnic/racial group, or gender even though subsequent generations have no direct memory of such events. For example, current generations of African Americans have no personal memories of slavery; and the proportion of blacks who labored in southern turpentine and lumber camps or who were alive when lynchings routinely occurred is small relative to the entire African-American population. Still, similar to Jews, blacks may recollect stories told to them by family members or other public information about hardships that occurred in wildland, backcountry type areas. These "memories" may contribute to ethnic identity formation for African Americans.

Stokols refers to the historical symbolism of places or landscapes as social imageability or the perceived social field of a milieu.[11] This is the "capacity of a place or type of place to evoke vivid and collectively held social meanings among its occupants or potential occupants." Social imageability can arise through environmental symbolism, which is a gradual process of assigning meaning to a place or landscape based on past experiences. According to Stokols, this symbolism can come about even for people who have no direct contact with an area.[12] For example, the history of a place can be passed down to successive generations via word of mouth or some other medium; and the place can come to symbolize a certain atmosphere or mood although no direct personal contact has been established.

The white, middle-American view of wildlands and wilderness constructs these areas as benign places—spiritual, sanctified refuges distinct from the profanity of human modification (see Schama's description of Yosemite valley).[13] Middle America imagines wildlands as mystical places having the power to transform one's essence because wild nature is perceived as sacred. One enters a wilderness but emerges somehow changed. One of American wilderness' earliest advocates, John Muir, located the divine in nature. He describes wild nature as a "window opening into heaven, a mirror reflecting the Creator."[14]

The idea of wildlands as benevolent entities also abounds in the secular realm. Turner proposed that wilderness, the actual physical aspects of primeval forests and unchartered territory, helped to establish American democracy.[15] Tyranny, contrasted with democracy, is contained in civilized society, not in the wild. Turner evokes an environmental determinism, the "frontier thesis," which posits that the behavior and ideas of a people are strongly influenced by the physical environment. Because early white Americans had the unique experience of vast, open lands, and the freedom to explore these places and appropriate them, this unrestricted freedom resulted in American "individualism, independence, and confidence," ideal characteristics which distinguish Americans from others.[16] Similarly, Hammond argues that wilderness is good because it contributes to a uniquely American character and is symbolic of American national heritage.[17] But to whose heritage or character? We agree that there is a homogeneity across the varying racial and ethnic subgroups in America, and that there is something that can be described generally as the American character. However, we disagree that the exact same set of values, ideals, or social movements, for instance, resonate to the same extent among all sub-populations in American society. Perhaps many African Americans do not consider wilderness as a heritage value.

We take the position of Greider and Garkovich that the *perception* of the natural environment is a social construction;[18] and we would add that this interpretation is largely the result of the observer's imaginings—images which are influenced, in part, by a group's past relationships with particular environments. Wild places are not objective entities which hold the same value, meaning, or symbolism for all who behold them. Wildlands do not exist in the minds of all Americans as beneficent or uncontaminated places, detached from society's ills. Mainstream environmentalists frame wildlands, singularly, as healing, revitalizing "therapeutic landscapes" or "fields of care,"[19] having the power to recreate the human spirit. However, for African Americans, these same terrains may be what cultural geographers refer to as "sick places" which evoke horrible memories of toil, torture, and death.[20]

We consider the contradictory relationship blacks have demonstrated towards wildlands. On the one hand, we review evidence suggesting that black collective memories of labor involving work on forests, slavery/plantation agriculture, sharecropping, and lynching have contributed to a black adversarial relationship with wildlands. The institutions of slavery, forest work camps, and sharecropping exploited black labor, and lynchings were essentially terrorist acts perpetuated against blacks in wildland areas.

Despite these experiences, however, there are also indications that there has been a connectedness between blacks and wildlands. African Americans have always fished and hunted in wildland settings and are returning to the rural South to establish homesteads adjacent to the wild. To better understand these apparent discrepancies and to set the discussion of black land memories in context, the following section reviews the literature on black interaction with wildlands, including

African-inspired ontology, black concern for the environment, and the progression of American environmentalism. Black working relationships with both wildlands and other natural areas are discussed and set in the context of the evolution of mainstream environmentalism.

LITERATURE REVIEW

AFRICAN AMERICANS AND WILDLANDS

National parks and forests, federally designated wilderness areas, and other wild, primitive outdoor settings are esteemed national treasures. These areas have come to be held in such regard in large part because of the conservation and preservationist ideas advanced by early environmentalists such as Ralph Waldo Emerson, Henry David Thoreau, Frederick Law Olmsted, and George Perkins Marsh, among others. These conservationists and preservationists were, in turn, influenced by romantic conceptions of nature which originated among European intellectuals during the Enlightenment.

Because of its foundations in European intellectualism and romanticism, Taylor[21] argues that the American environmental, wilderness, and wildland recreation movements have largely been a concern of white, middle-class males.[22] Indeed, surveys of visitors to federally designated wilderness areas show that the overwhelming majority of recreationists are white, college educated, middle to upper income wage earners.[23] More general research on wildland recreation also shows a relative lack of black participation in outdoor activities that occur in such areas. Goldsmith reports that national park visitors are also mostly white, despite the park service's intent to attract a more ethnically and racially diverse visitor base.[24] Also, regional-level household surveys show blacks are significantly less likely than whites to interact with wildlands, even when blacks live adjacent to such areas and socioeconomic factors are held constant.[25]

In a review of ethnic and racial research in outdoor recreation, Gramann also found African Americans were less likely than whites to engage in wildland recreation, the exception being fishing and hunting.[26] But even for a wildland based activity like hunting, Marks' multi-ethnic investigation of rural, male hunters in North Carolina showed that blacks, compared to whites and Native Americans, were much less likely to report that they enjoyed the aesthetics of nature when hunting.[27]

In contrast, recent empirical analyses of national level data show few significant differences between blacks and whites for wilderness concern.[28] Moreover, Mohai, Jones, Arp and Kenny, and Parker and McDonough question the assumption of black apathy for the environment.[29] Though blacks may be less active than whites in joining mainstream environmental organizations or voting for environmental agendas, this difference should not be taken to mean that blacks are not concerned about nature. Rather, African-American interest may be demonstrated in non-conventional environmental forms such as concern for community integrity (i.e., clean, crime-free neighborhoods and work place conditions) rather than in traditional concern for wildlands and wildlife habitats.

Again, to better understand the gap between positive black wildland and environmental sentiments, on the one hand, and a relative lack of black wildland interaction, on the other, we believe it useful to consider history. What were the chief concerns of black Americans during the formation of American environmentalism? In particular, what was the black relationship to land during these periods? We focus exclusively on the relationship of southern blacks to the land because the majority of blacks have lived in the South. Traditional southern culture provides the basis for much of contemporary black American culture, although the majority of blacks now live in non-rural areas.

AMERICAN ENVIRONMENTALISM

Dorceta Taylor chronicles the rise of the American environmental movement and also considers the parallel and particularistic histories of non-whites and women during eras when environmentalism was being defined.[30] She identifies four periods of environmental thought. This comparison of mainstream white interests and the concerns of periphery groups shows how issues of basic civil liberties such as voting and worker's rights, immigration, and protection against domestic terrorism were chief concerns of marginalized groups during the time

Anglo-Americans were concerned with environmental protection. The first environmental period is described as the pre-environmental movement era, between 1820 and 1913. The related environmental paradigm or environmental philosophy (exploitive capitalist paradigm) was based on exploitation and intense extraction of natural resources.

The next phase is the early environmental movement, which commenced in 1914 after the Hetch Hetchy dam dispute between San Francisco and preservationists. During this phase, between 1914 and 1959, environmental issues were embraced more by the wider populace; whereas prior to this time, environmental protection was advocated most often by artists, physical scientists, and other intellectuals. Passage of the 1897 Forest Management Act and later the Hetch Hetchy controversy, in some ways, marked a turning point for the environmental movement because of differences in goals between more biocentric environmental advocates (preservationists) and those who favored more utilitarian natural resource use (utilitarian conservationists).[31] The exploitive capitalist paradigm still prevailed as an environmental ideology, although the romantic environmental paradigm began to gain prominence. Romantic views of the environment framed the natural world in idealistic, hyper-real terms. As discussed earlier, romanticism endows wild areas with a mysticism which exceeds the actual physics of the resource.

The third environmental phase began in the early 1960s with the publication of Rachel Carson's *Silent Spring.* Even greater attention was focused on environmental concerns such as air and water contamination. Membership in mainstream environmental organizations increased appreciatively. During this time, the romanticized view of nature was replaced by the new environmental paradigm. This new paradigm and its advocates challenged the dominating, positivistic assumptions of technology and rationalism as providing the sole answers to social ills. The new environmental paradigm was personified in student environmental activists who protested against nuclear energy, the Vietnam war, and the general environmental exploitation of non-industrial peoples and their lands by Western nations.

The latest environmental phase is the post–Three Mile Island/Love

Canal era from about 1980 to the present. In this present phase, like the preceding one, the new environmental paradigm is still the leading environmental frame of reference among most mainstream environmentalists. However, alternative environmental thought such as the environmental justice paradigm and ecofeminism, as well as more radical environmental activism (e.g., Earth First!) are emerging as competing environmental discourses.

PRE-MOVEMENT ERA: SLAVERY, SHARECROPPING, LYNCHING

When the pre-movement environmental era began around 1820, there were approximately 1.5 million slaves in the United States.[32] There have been numerous accounts written of slavery and the "nightmare of drudgery" under which most blacks lived.[33] Some skilled slaves worked in southern cities as domestic servants, artisans, or factory workers, but the majority worked on various sized plantations where their primary task was toiling on the land.[34] An ex-slave is quoted in Bennett, *Before the Mayflower:* "it seems the fields stretched 'from one end of the earth to the other.'"[35]

Though slaves lived close to nature like other racial/ethnic groups of the period and extracted sustenance from the land (when permitted) they could not explore the wider environment. The very condition of being a slave dictated a life of extreme restrictions. The slave stood as antonym to the American myth of unrestricted wilderness exploration. Taylor remarks that while free, white men had the privilege of discovering wildlands, slaves (and other oppressed people, e.g., women, Asian immigrants, the poor) were severely circumscribed in their movement by a white, male dominated society that enacted slave codes in each slaveholding state.[36] Some of these laws prohibited blacks from assembling in large groups away from their plantations or forbade any slave from leaving the plantation without written permission. Even free blacks were subject to the circumscriptions contained in these laws.[37]

The ambiguity blacks appear to have with wildland environments may have begun with the slave experience. Blum writes that slaves assigned multiple meanings to wilderness.[38] Though black movement was severely restricted, slaves still managed to access wildlands covertly.

Wildlands provided a place of escape, either temporarily or permanently, from the oppression of plantation life. Blum remarks:

> Slavery affected how blacks thought about the wilderness, altering and melding African beliefs. Slaves maintained some elements of their aesthetic, adapting and changing others to fit their new environment. In many African religions, for example, the wilderness or "bush," far from a place to be feared or avoided, actually was seen as a place of refuge and transformation. . . . Interestingly, this concept held by slaves mirrored in some ways the view held by white women and transcendentalists of nature as a place of refuge and spirituality. For the slaves, however, the concept of wilderness as a refuge remained a palpable reality, rather than a poet's or scholar's rhetoric.[39]

Blum's interpretation of early black American wildland interaction is more consistent with the growing preservationist sentiments described by Taylor during this era. At the same time, however, Blum also stresses that slaves perceived both fear and danger in wilderness. Slaves especially dreaded wild animals (poisonous snakes, panthers) and to a lesser extent other humans (both black and white), and supernatural forces believed to inhabit wildlands. The wilderness was a place to be avoided for many, and some slave parents were concerned that their children understand the potential danger wilderness contained. A slave quoted in Blum from the slave narratives talks about the defenselessness of humans in the wild: "De poisonous snakes strike wicked fangs into bare heels, danger hides everywhere in de streams too so we much know how to escape form [*sic*] hit. De wild animals have nimble feet and wings to save dem from de ones dat kill dem but de nigger had to save hisself."[40]

Whites also sought to discourage blacks from venturing into wilderness areas by telling blacks of the horrors that awaited them in the woods. Blum's description of slave wildland interaction suggests, again, that blacks did not view these terrains as romanticized landscapes removed from human influence but rather in more practical terms. Wilderness was both perceived and used as a haven but was also kept at bay.

In an article on black women's relationship to wilderness areas, Evelyn C. White also writes how the slave experience may have negatively influenced black impressions of such landscapes:

> The timidity African American women feel about the outdoors is colored, I believe, by our experiences of racism and sexism in this nation. It is steeped in the physical and psychic damage we have suffered as a result of being forcefully removed from Africa and enslaved on southern plantations. Ask yourself why a black woman would find solace under the sun knowing that her foremothers had toiled in the brutal, blistering heat for slavemasters [*sic*].[41]

As stated, we also argue that the actual work of plantation agriculture and related tasks in forested wildlands served to create a negative imagery of such places among blacks. While descriptions of black labor on cotton and tobacco plantations are plentiful, relatively few accounts exist of slave work specifically in wildland environments, such as naval stores operations (especially turpentining).[42] Yet, Starobin writes that during the antebellum period, turpentine extraction "was entirely dependent on slave labor."[43] Gay Goodman Wright's ethnohistorical account of turpentining in southern pine forests notes that black labor in the naval stores industry has been overlooked by historians because of confusion in the nineteenth century as to whether naval stores products should have been classified as industrial or agricultural products.[44]

During the colonial era, the British used pine derived naval stores products such as tar, pitch, gum, turpentine, and rosin in shipbuilding. When supplies in northern Europe became threatened, the British turned to the extensive pine forests in the American colonies for supplies.[45] Initially, small farmers in North Carolina produced naval stores, but as demand for these products increased, the planter class moved into the industry in the 1830s and 1840s, and naval stores operations became part of the plantation economy. By 1850, naval stores ranked third in export products from the South behind cotton and tobacco[46] and by 1860, roughly 15,000 slaves labored in the naval stores industry.[47]

Turpentining took place in remote pine forests. Workers used hatchets or hacks to make incisions into trees. Carved receptacles or "boxes" (later attached cups) collected the crude gum or oleoresin that flowed from the opened spots in the trees. Periodically, workers would empty collected crude gum into a larger storage bin for later distillation. The turpentine extraction period ran from about March or April until November. Until the 1930s, turpentining was a purely extractive operation. The longleaf pine forests were viewed much like minerals to be extracted rather than as renewable resources.[48]

In addition to turpentine extraction in forested wildlands, we also submit that the exploitation and subjugation associated with plantation agriculture during slavery influenced the way blacks perceive wild areas. This point is crucial because we believe that not only direct work in wildlands but work associations with land, generally, have informed black land memories. As Cronon remarks:

> Ever since the nineteenth century, celebrating wilderness has been an activity mainly for well-to-do city folks. Country people generally know far too much about working the land to regard *un*worked land as their ideal.[49]
>
> The dream of an unworked natural landscape is very much the fantasy of people who have never themselves had to work the land to make a living. . . . Only people whose relation to the land was already alienated could hold up wilderness as a model for human life in nature. . . .[50]

Cronon suggests class differences in wildland appreciation: that the nineteenth century agrarian working class did not romanticize or idealize wildlands because their livelihood was more directly dependent upon land resources. Such was the case not only for African Americans involved in plantation agriculture, but also for poorer whites and other groups living close to the land. But what further distinguishes the historical black wildland relationship from that of white America is the intersection of class oppression with racism. African Americans have not belonged to the elite group that appropriated wild spaces as cultural ideal because the subjugated black position with respect to cultivated lands would not allow this association. But just as impressive upon the collective black land memory is European racism which placed blacks on par with the wild and uncivilized. In contrast to the image of the overcivilized, middle-class white American,[51] blacks were believed to have retained a great measure of primitivism. As DeLuca remarks, ". . . within the context of whiteness, those not part of the white civilization are, at best, seen as part of nature."[52]

After the Civil War, blacks continued to be kept in a virtual system of "involuntary servitude" by both legal (e.g., Black Codes) and extra-legal means (e.g., fear tactics used by vigilante groups such as the Ku Klux Klan).[53] In the post-bellum South, planters used debt incurred by blacks through sharecropping and tenancy arrangements to restrict black mobility. In effect, blacks were rendered immobile and compulsively tied to the land through planter-backed ordinances such as "enticement laws, emigrant agent restrictions, contract laws, vagrancy statutes, the criminal-surety system, and convict labor laws. . . ."[54]

Agrarianism remained the dominant economic system in the United States until the last decade of the nineteenth century. When slavery ended, blacks felt they had a right to land they had helped cultivate during slavery. They reasoned that their work had contributed substantially to both the southern and northern economies. Blacks realized that land ownership would be crucial in uncoupling them from the exploitative plantation economy. Harding writes of the newly freed slaves: "Of course the search for land, the need to hold on to land, was still central to the black hope for a new life in America."[55] Vernon Wharton, quoted in Mandle also states: "their [ex-slaves'] very lives were entwined with the land and its cultivation; they lived in a society where respectability was based on ownership of the soil; and to them to be free was to farm their own ground."[56]

However, the freedmen were disappointed when the federal government did not redistribute seized southern land to former slaves but returned it to former owners. According to Mandle, such land redistribution would have been a "radical act" which would have left white plantation owners landless. This denial of land to blacks left the overwhelming majority of blacks landless and effectively perpetuated the antebellum plantation economy, and with it continued black disfranchisement.[57]

Most blacks could not obtain land independent of federal redistributions because they had no means to buy land; also, many whites refused to sell land to blacks even when the latter could afford the asking price. Landownership after the Civil War became concentrated in the hands of fewer people, affecting both blacks and smaller white landowners.[58] The inability to acquire land, in addition to lack of industrialization in the South, left blacks with few opportunities for gainful employment, save sharecropping.

The 1890s marked the end of rural agrarianism as the dominant economic system in the U.S. By the early part of the twentieth century, industrialization had replaced agrarianism as the primary economy. The final decade of the nineteenth century was significant for wider black participation in American democracy because of Booker T. Washington's 1895 "Atlanta Compromise," which was an official black acquiescence of Jim Crow and social inequality.[59] The decade was also significant because of the Supreme Court's 1896 *Plessy v. Ferguson* decision that ruled that blacks had no rights which whites were bound to respect. Thus, the twentieth century started with blacks solidified in a subordinate position.

EARLY ENVIRONMENTAL MOVEMENT: BLACK MIGRATION, TURPENTINING LUMBER CAMPS, SHARECROPPING, LYNCHING

In 1914, blacks were still, according to Mandle mostly "southern, rural, and poor,"[60] although the "Great Black Northern Migration" had already commenced and its net effect would continue over the next four decades. Blacks remaining in the rural South before World War II were still mostly employed in agriculture and domestic or service positions because opportunities for other viable employment in the region remained limited.[61] The 1910 through 1940 censuses show that farm labor accounted for more than fifty percent of all black labor in the deep South states of Alabama, Arkansas, Georgia, Louisiana, Mississippi, North Carolina, South Carolina, and Texas.[62]

After the Civil War, naval stores operations moved from plantations to camps established by "producers" who secured financing for the operations. The industry also moved steadily southward as pine forests in the Carolinas became depleted.[63] Blacks continued to make up the overwhelming proportion of workers in the industry. In both 1910 and 1920, blacks accounted for at least eighty percent of turpentine laborers in twelve southern states.[64] The descriptor "Turpentine Negro" was a common term applied to blacks employed in this industry. In a 1971 report on contemporary turpentining, Tze I. Chiang, W. H. Burrows, William C. Howard, and G. D. Woodard, Jr. remark: ". . . it is thought by some that the nonwhites [blacks] are the only ones who can harvest gum because of their superior ability to withstand the heat during the summer months when production is at its peak."[65]

These woods workers lived either in turpentine camps and were shuttled to the woods for work; or they lived in backwood shanties near the work site. Sometimes the camps were located within the boundaries of national forests, such as the New Home community on the Choctawhatchee National Forest in Florida (circa early 1900s).[66] Todes describes typical conditions in the work camps:

> Negroes predominate in the turpentine camps of Georgia and Florida where exploitation of the workers is notorious. Mexican and Negro workers only are employed in the insect ridden cypress swamps. To cut cypress, the workers must wade in humid swamps, often up to their hips in water, and must live with their families in house boats built over the swamps. Living quarters for Negro workers are "match-box" shacks or box cars, segregated from white workers.[67]

Aside from the dangerous working conditions, daily living conditions in the camps were also exploitative. Like plantations before them, turpentine camps continued to operate as micro-societies with a distinct set of morals, social norms, and economic guidelines.[68] Turpentine producers controlled all aspects of a camp's social and economic life. Producers established housing, schools, churches, and recreation in the camp, and a system of debt peonage bound the laborers to their work. Because of low wages, laborers had to borrow money from producers to pay off debts incurred at company stores. Again however, because of low wages, workers were usually unable to repay loans, so the worker was indebted to his employer and his labor was controlled by the same. Principally in Florida, a largely black convict labor force was also exploited in the industry.[69] By the 1960s, traditional turpentining had ceased to be a significant industry in the South, due mainly to the lack of an available work force.[70] Black respondents in Chiang et al.'s investigation cite exacting working conditions such as "foul weather, insects, loneliness, snakes, underbrush, and rough terrain" as reasons for seeking employment outside the forests.

In addition to turpentining, blacks also labored in the southern lumber industry. As stated, the forestry-related industries moved southward in the latter half of the 1800s because of depleted resources in other parts of the country. In both 1909 and 1918, the southern region accounted for the largest percentage of lumber cut in the U.S., thirty-three and thirty-five percent, respectively.[71] Black males accounted for one-half of all southern forest laborers from 1910 to 1940.[72] During this period, forest workers were often migrants who moved across the South following timber jobs. More permanent lumber and mill towns were also erected adjacent to lumbering operations to house timber workers and their families. Conditions reported in these camps are similar to those described in turpentine camps with the same closed system of indebtedness and company dictated mores. According to Mayor, blacks performed the most dangerous work in timber processing. Typically, they loaded cut logs onto railroad cars that transported the timber to the sawmill or they comprised the "rail gang" which laid tracks for the makeshift rail line into the timber stands.

Todes also writes about working conditions for blacks in lumbering:

> To work at the heaviest jobs is the lot of the Negroes. In the woods, they fell and buck the trees, handle the hooks or tongs, form the labor gang in the skidder crew, work on railroad construction and do the heavy work in the loading process. In the mills, they ride the carriage or haul and stack lumber while the white workers handle the machines. In the Great Southern Lumber Company's camps where the white sawyers have an 8-hour shift, Negro workers riding the carriage or "rig" must work 10 hours a day. White workers get paid for two holidays a year but the Negroes get no vacations at all.[73]

Descriptions of blacks in post-bellum timber-related occupations provide a more direct account of black interactions with wildlands. These accounts should be contrasted with those presented by early wildland preservationists during the same period. Black working experiences in wildlands suggest blacks were in a more marginal position with respect to wild nature.

The collective memory of turpentining and lumbering has been relayed to successive generations by word of mouth. For instance, Chiang et al.'s study reports that older blacks who had been turpentiners strongly discouraged younger family members from becoming involved in the work.[74] Older blacks recounted the hardships involved in turpentining, and it was viewed as a low class occupation. Official knowledge of turpentining and lumbering has been kept alive in official memory sites such as the anthropological databases contained in the Florida Folk Life Collection. Songs and folk tales recorded by southern turpentine workers in the 1930s are included in this database compiled by the Work Project Administration in the 1930s.[75]

We also hypothesize that lynchings served to further alienate blacks from wildlands because some of these acts occurred in isolated woodland areas.[76] For instance, eight of eleven black lynchings that took place in Florida in the 1930s, happened in "open country" or wooded areas. The incidences of lynching lessened towards the middle of the twentieth century, but the threat of such violence remained (the greatest number of lynchings occurred in the last decade of the nineteenth century).[77] According to an NAACP report, 4,743 people were lynched in the United States between 1882 and 1968; of these, 72.7 percent were black.[78]

Lynchings did not always take place in forested wildlands. However, the fact that some of these murders occurred in wooded areas sufficed to influence black perceptions of wildlands. Because isolated wildlands are not familiar, everyday landscapes for most people, the backdrop or environment for such events become impressed in collective memorization, not just the act of terror itself. In such cases, the isolated rural landscape is not distinguished from large forested settings because for many African Americans, these settings represent the unknown "Other." For instance, White comments about fears black women, in particular, have of venturing into the wilderness: "Some black women shun the wilderness because we cannot erase the memory of Emmett Till. . . . As a child, my feelings about being outdoors were skewed by the powerful pictures of Emmett's beaten and bloated body displayed in the media."[79]

The murders of Emmett Till and Mack Charles Parker, both in Mississippi in the 1950s are two of the more recent collective recollections of lynchings. Neither of these lynchings took place in large, forested land tracts, still these murders are coupled with "the woods" and are especially salient to African Americans and influence the way some perceive wildlands.

MODERN ENVIRONMENTAL MOVEMENT: ENVIRONMENTAL JUSTICE AND BLACK RETURN TO THE LAND

In 1982 the environmental justice movement began in Warren County, North Carolina when African Americans protested the proposed siting of a hazardous waste landfill in their county.[80] The environmental justice movement focused attention on what some charge as the inequitable distribution of hazardous and toxic waste sites in lower income and minority communities.[81] Chief concerns of environmental justice advocates are issues relating to pollutants and environmental toxins which threaten the integrity of local neighborhoods and workplaces, both in urban and rural areas. These concerns of primarily female and lower-income groups have been contrasted with the goals of mainstream (majority white) environmental groups which seek to preserve federally designated wilderness areas and wildlife and fish habitats. In contrast, the environmental justice movement represents an effort by marginalized groups to preserve the "natural environment" of home and community.

The environmental justice movement coincided with another mobilization among African Americans in the 1970s and 1980s. This other movement involved the return of blacks to the South, including the rural South. In the decade from 1970 to 1980, more than one million African Americans migrated to the South from the Northeast, Midwest, and West. This migration compares with an out-migration from the South of 950,000 blacks during the same time period.[82] This trend remained throughout the 1990s and is expected to continue into the twenty-first century. Blacks, like other race/ethnic groups returned to the South because of better job prospects in the urban areas of the region, such as metropolitan Atlanta and Charlotte and Raleigh-Durham, North Carolina.

In *A Call to Home: African Americans Reclaim the Rural South,* Stack notes that blacks are also returning to rural regions of the South, despite the lack of viable economic opportunities or improved racial relations.[83] According to Stack, social conditions in some rural areas have not improved appreciatively since blacks left en masse a half century ago. Still, some blacks are moving back to the rural South because of the need to reconnect with family and to reclaim the land. As Stack notes:

> ... by the end of the 1970s, the Great Migration had turned back on itself, and the old southern homeplaces were welcoming the prodigals. How could things change so quickly? What forces on earth could reverse such precipitous decline? The appeal of God's little acre crosses all bounds of race and time, but the urgency could seem shrill for African Americans. If security and liberty were to be found anywhere, wouldn't it be under one's own roof, safe on one's own land?[84]

The black return to rural landscapes again highlights the paradoxical relation of blacks towards wildland environments. Blacks are returning to rural areas despite the hardships encountered in these places by earlier generations. The back-to-the-land migration and an environmentally centered social justice movement suggest a black desire to engage wildlands rather than an avoidance of such landscapes. Certainly, rural residence would provide more opportunities for blacks to interact with wildland places, less inhibited now by the constraints which accompanied their fore parents.

This return migration indicates that there may be factors which mitigate black land memories, for instance increased urbanization and affluence among African Americans. Because the mostly urban, black population is farther removed from the land than its rural predecessors, present generations of African Americans may also be farther removed from negative images of wildlands. There is also a larger black middle-class compared to fifty years ago with greater access to information about wildland recreation resources and official data concerning environmental degradation. If there has been a rupture in black collective land memory, blacks may now hold wildland attitudes more like middle-class whites, even though blacks interact with these resources less than whites. Precisely how these two movements, return migration and environmental justice, might translate into black interaction with wildlands is not known. Empirical investigations are needed to assess the degree to which environmental memories inform contemporary black views of wildlands.

NOTES

1. Simon Schama, *Landscape and Memory* (New York: Alfred A. Knopf, 1995), p. 18.

2. Maurice Halbwachs, *The Collective Memory* (New York: Harper and Row, 1980).

3. Eric Zerubavel, "Social Memories: Steps to a Sociology of the Past," *Qualitative Sociology* 19 (1996): 283–99.

4. Lynn Rapaport, *Jews in Germany after the Holocaust: Memory, Identity, and Jew Relations* (Cambridge: Cambridge University Press, 1997).

5. Reuben B. May, "Race Talk and Local Collective Memory among African American Men in a Neighborhood Tavern," *Qualitative Sociology* 23 (2000): 201–14; Susanne Marten-Finnis, "Collective Memory and National Identities: German and Polish Memory Cultures: The Forms of Collective Memory," *Communist and Post-Communist Studies* 28 (1995): 255–61; Barry Schwartz, "Collective Memory and History: How Abraham Lincoln Became a Symbol of Racial Equality," *Sociological Quarterly* 38 (1997): 469–96.

6. We define wildlands as primitive, backcountry areas. These are undeveloped or roadless areas that retain pristine attributes or those which otherwise have the character of "the woods" or wild nature. Federally designated wilderness areas are included in this definition, but "wilderness" used herein does not necessarily refer to lands in the National Wilderness preservation system.

7. John G. Neihardt, *Black Elk Speaks* (New York: Simon and Schuster, 1972); Dee Alexander Brown, *Bury My Heart at Wounded Knee* (New York: H. Holt, 2001).

8. African American and black are used interchangeably.

9. Dorceta E. Taylor, "The Rise of the Environmental Justice Paradigm Injustice Framing and the Social Construction of Environmental Discourses," *American Behavioral Scientist* 43 (2000): 508–80. See also Schama's extensive analysis of the widely accepted historical myths and memories European and North American societies hold towards nature in *Landscape and Memory.* Dorceta E. Taylor, "Meeting the Challenge of Wild Land Recreation Management: Demographic Shifts and Social Inequality," *Journal of Leisure Research* 32 (2000): 171–79.

10. James H. Gramann, *Ethnicity, Race, and Outdoor Recreation: A Review of Trends, Policy, and Research,* Miscellaneous Paper R-96-l, March 1996, Washington, D.C.: U.S. Army Corps of Engineers.

11. Daniel Stokols, "Group X Place Transactions: Some Neglected Issues in

Psychological Research on Settings," in David Magnusson, ed., *Toward a Psychology of Situations. An Interactional Perspective* (Hillsdale: Lawrence Erlbaum Associates, 1980), pp. 393–415.

12. Daniel Stokols, "Instrumental and Spiritual Views of People-Environment Relations," *American Psychologist* 45 (1990): 641–46.

13. Schama, *Landscape and Memory,* pp. 7–9.

14. Roderick Nash, *Wilderness and the American Mind* (New Haven and London: Yale University Press, 1967), p. 125.

15. Frederick Jackson Turner, *The Frontier in American History* (New York: Henry Holt and Co., 1953).

16. Roderick Nash, "Qualitative Landscape Values: The Historical Perspective," in Ervin H. Zube, R. O. Brush, and J. G. Fabos, eds., *Landscape Assessment: Value, Perceptions, and Resources* (Stroudsburg: Dowden, Hutchinson, and Ross, 1975), pp. 10–17.

17. John L. Hammond, "Wilderness and Heritage Values," *Environmental Ethics* 7 (1985): 165–70.

18. Thomas Greider and Lorraine Garkovich, "Landscapes: The Social Construction of Nature and the Environment," *Rural Sociology* 59 (1994): 1–24.

19. Wilbert M. Gesler, "Therapeutic Landscapes: Medical Issues in Light of the New Cultural Geography," *Social Science Medicine* 34 (1992): 735–46.

20. Robert D. Bixler and Myron F. Floyd, "Nature Is Scary, Disgusting, and Uncomfortable," *Environment and Behavior* 29 (1997): 443–67; Dorceta E. Taylor, "Blacks and the Environment: Toward an Explanation of the Concern and Action Gap," *Environment and Behavior* 21 (1989): 175–205; Yi-fu Tuan, *Landscapes of Fear* (New York: Pantheon Books, 1979).

21. Taylor, "Meeting the Challenge of Wild Land Recreation Management"; Nash, *Wilderness and the American Mind.*

22. White women have also contributed significantly to environmental thought and action. Carolyn Merchant, *Earthcare: Women and the Environment* (Routledge, 1995), approaches ecology from a feminine perspective.

23. Robert C. Lucas, "A Look at Wilderness Use and Users in Transition," *Natural Resources Journal* 29 (1989): 41–55; Alan F. Watson, Daniel R. Williams, Joseph W. Roggenbuck, and J. J. Daigle, *Visitor Characteristics and Preferences for Three National Forest Wildernesses in the South,* Resource Paper INT-455 (Ogden: U.S. Department of Agriculture, Forest Service, Intermountain Research Station, 1992).

24. Goldsmith, "Designing for Diversity," *National Parks* 68 (1994): 20–21; Myron F. Floyd, "Race, Ethnicity and Use of the National Park System," *Social Science Research Review* 1, no. 2 (1999): 1–24.

25. Cassandra Y. Johnson, Patrick M. Horan, and William Pepper, "Race, Rural Residence, and Wildland Visitation: Examining the Influence of Sociocultural Meaning," *Rural Sociology* 62 (1997): 89–110.

26. Gramann, *Ethnicity, Race, and Outdoor Recreation.*

27. Stuart A. Marks, *Southern Hunting in Black and White* (Princeton: Princeton University Press, 1991).

28. Ken Cordell, J. M. Bowker, Carter Betz, and Cassandra Johnson, "Wilderness Awareness and Participation: A Comparison Across Race and Ethnicity: Preliminary Results: NSRE 2000," presentation at the National Wilderness Workshop, Washington, D.C., U.S.D.A. Forest Service, July 2000.

29. Paul Mohai, "Black Environmentalism," *Social Science Quarterly* 71 (1990): 744–65; Robert Emmet Jones, "Black Concern for the Environment Myth Versus Reality," *Society and Natural Resources* 11 (1998): 209–28; William Arp and Christopher Kenny, "Black Environmentalism in the Local Community Context," *Environment and Behavior* 28 (1996): 267–82; Julia Dawn Parker and Maureen H. McDonough, "Environmentalism of African Americans: An Analysis of the Subculture and Barriers Thesis," *Environment and Behavior* 31 (1999): 155–77; Robert Emmet Jones, J. Mark Fly, and H. Ken Cordell, "How Green Is My Valley? Tracking Rural and Urban Environmentalism in the Southern Appalachian Ecoregion," *Rural Sociology* 64 (1999): 482–99.

30. Dorceta E. Taylor, "American Environmentalism: The Roles of Race, Class and Gender in Shaping Activism, 1820–1995," *Race, Gender, and Class* 5 (1997): 16–62; Dorceta E. Taylor, "The Rise of the Environmental Justice Paradigm: Injustice Framing and the Social Construction of Environmental Discourses," *American Behavioral Scientist* 43 (2000): 508–80.

31. J. Douglas Wellman, *Wildland Recreation Policy: An Introduction* (New York: John Wiley and Sons, 1987), pp. 106–12.

32. U.S. Department of State, *Census for 1820* (Washington, D.C.: Gales and Seaton, 1821).

33. Leronne Bennett, Jr., *Before the Mayflower* (New York: Doubleday, 1968), p. 87.

34. Another exception is house slaves.

35. Bennett, *Before the Mayflower,* p. 87.

36. Taylor, "The Rise of the Environmental Justice Paradigm."

37. Eugene D. Genovese, *Roll, Jordan, Roll: The World the Slaves Made* (New York: Pantheon, 1972).

38. Elizabeth D. Blum, "Power, Danger, and Control: Slave Women's Perceptions of Wilderness in the Nineteenth Century," *Women's Studies* 31 (2002): 247–67.

39. Ibid., p. 250.

40. George Rawick, *The American Slave: A Composite Autobiography* (Westport: Greenwood Publishing, 1972), vol. 3.

41. Evelyn C. White, "Black Women and Wilderness," *Outdoor Woman* (1991).

42. James Battle Averit, "Turpentining with Slaves in the 30's and 40's," in

Thomas Gamble, ed., *Naval Stores: History Production, Distribution and Consumption* (Savannah: Review Publishing and Printing Co., 1921), pp. 25–27; Donnie D. Bellamy, "Slavery in the Microcosm: Onslow County, North Carolina," *Journal of Negro History* 62 (1977): 339–50.

43. Robert S. Starobin, *Industrial Slavery in the Old South* (New York: Oxford University Press, 1970), p. 26.

44. Gay Goodman Wright, "Turpentining: An Ethnohistorical Study of a Southern Industry and Way of Life" (Master's thesis, University of Georgia, 1979).

45. Thomas Gamble, *Naval Stores: History, Production, Distribution and Consumption* (Savannah, Ga.: Review Publishing and Printing Co., 1921), p. 17; Eldon Van Romaine, "Naval Stores, 1919–1939," *Naval Stores Review* 100 (1990): 6–16.

46. Wright, "Turpentining," pp. 33–35.

47. Starobin, *Industrial Slavery in the Old South,* p. 26.

48. Thomas C. Croker, Jr., "The Longleaf Pine Story," *Southern Lumberman* 239 (December 1979): 69–74.

49. William Cronon, "The Trouble with Wilderness or, Getting Back to the Wrong Nature," *Environmental History* 1 (1996): 7–28, p. 15 (emphasis in the original).

50. Ibid., 16–17.

51. Frederick Jackson Turner, *The Frontier in American History* (New York: Henry Holt and Co., 1953), pp. 2–4, encouraged European settlers to establish homesteads in wild nature so that they might reclaim their simpler, more natural character.

52. Kevin DeLuca, "In the Shadow of Whiteness," in Thomas K. Nakayama and Judith N. Martin, eds., *Whiteness: The Construction of Social Identity* (Thousand Oaks: Sage Publications, 1991), pp. 217–46. In the eighteenth century, European philosophers reasoned that the African and European were different types of humans—blacks being closer to primates than whites. Blacks were seen as the ignoble savage. Hegel, in particular, held a decidedly negative view of blacks, likening Africans to children and describing blacks as a *Kindernation* or an infantile people who needed to be educated by whites. Later, both Schopenhauer and Nietzsche interpreted these perceived differences in a more positive light, but Nietzsche also reasoned that blacks existed in a more primitive state than whites. This image of blacks continues to some extent and informs black wildland memories. It has not been a black project to revert, however temporarily, to a more primitive state because in the European eye, blacks have never left this state. Blacks by contrast, have sought to dissociate from the primitive.

53. William Cohen, "Negro Involuntary Servitude in the South, 1865–1940: A Preliminary Analysis," *Journal of Southern History* 42 (1976): 33–35; W. Fitz-

hugh Brundage, *Lynching in the New South: Georgia and Virginia, 1880–1930* (Urbana and Chicago: University of Illinois Press, 1993).

54. Pete Daniel, *Breaking the Land: The Transformation of Cotton, Tobacco, and Rice Cultures since 1880* (Urbana: University of Illinois Press, 1985), p. 6.

55. Vincent Harding, *There Is a River: The Black Struggle for Freedom in America* (New York: Harcourt Brace Jovanovich, 1981), p. 315.

56. J. Mandle, *The Roots of Black Poverty: The Southern Plantation Economy after the Civil War* (Durham: Duke University Press, 1978), p. 106.

57. Ibid., pp. 105–07.

58. Daniel, *Breaking the Land,* p. 4.

59. Louis R. Harlan, *The Booker T. Washington Papers* (Urbana: University of Illinois Press, 1974), vol. 3, pp. 583–87.

60. Mandle, *The Roots of Black Poverty,* p. 84.

61. Ibid., p. 84.

62. U.S. Department of Commerce, *Thirteenth Census of the United States, Population-Occupation Statistics,* vol. 4 (1914); U.S. Department of Commerce, *Fourteenth Census of the United States, Population-Occupations,* vol. 4 (1923); U.S. Department of Commerce, *Fifteenth Census of the United States, Population-Occupations by States,* vol. 4 (1933); U.S. Department of Commerce, *Sixteenth Census of the United States, Population,* vol. 3 (1943).

63. Archer H. Mayor, *Southern Timberman* (Athens and London: University of Georgia Press, 1988).

64. U.S. Department of Commerce, *Thirteenth Census of the United States, Population-Occupation Statistics,* vol. 4 (1914); U.S. Department of Commerce, *Fourteenth Census of the United States, Population-Occupations,* vol. 4 (1923); U.S. Department of Commerce, *Fifteenth Census of the United States, Population-Occupations by States,* vol. 4 (1933); U.S. Department of Commerce, *Sixteenth Census of the United States, Population,* vol. 3 (1943).

65. Tze I. Chiang, W. H. Burrows, William C. Howard, and G. D. Woodard, Jr., *A Study of the Problems and Potentials of the Gum Naval Stores Industry* (Atlanta: Georgia Institute of Technology, 1971), p. 82. Zora Neal Hurston, *Mules and Men* (New York: Negro Universities Press, 1969), also provides an anthropological look at southern turpentines. She focuses on the folk tales relayed to her by black turpentine and lumber workers. David C. Nichells, "Migrant Labor, Folklore, and Resistance in Hurston's Polk County: Refraining Mules and Men," *African American Review* 33 (1999): 467–70.

66. Carroll B. Butler, *Treasures of the Longleaf Pines: Naval Stores,* 2nd ed. rev. (Shalimar: Tarkel Publishing, 1998), pp. 127–39.

67. Charlotte Todes, *Labor and Lumber* (New York: International Publishers, 1931), pp. 83–84.

68. Wright, "Turpentining."

69. Jeffrey A. Drobney, "Where Palm and Pine Are Blowing: Convict La-

bor in the North Florida Turpentine Industry, 1877–1923," *Florida Historical Quarterly* 72 (1994): 411–34.

70. Chiang et al., *A Study of the Problems and Potentials of the Gum Naval Stores Industry.*

71. Industrial Workers of the World, *The Lumber Industry and Its Workers* (Chicago: Industrial Workers of the World), p. 45.

72. U.S. Department of Commerce, *Thirteenth Census of the United States, Population-Occupation Statistics,* vol. 4 (1914); U.S. Department of Commerce, *Fourteenth Census of the United States, Population-Occupations,* vol. 4 (1923); U.S. Department of Commerce, *Fifteenth Census of the United States, Population-Occupations by States,* vol. 4 (1933); U.S. Department of Commerce, *Sixteenth Census of the United States, Population,* vol. 3 (1943).

73. Todes, *Labor and Lumber,* p. 83.

74. Chiang et al., *A Study of the Problems and Potentials of the Gum Naval Store Industry.*

75. http://memory.loc.gov/ammem/flwpahtml/flwpahome.html.

76. Arthur F. Raper, *The Tragedy of Lynching* (Chapel Hill: University of North Carolina Press, 1933), p. 6.

77. Robert L. Zangrando, *The NAACP Crusade against Lynching. 1909–1950* (Philadelphia: Temple University Press, 1980).

78. Ibid., pp. 4, 6–7.

79. White, "Black Women and Wilderness."

80. Andrew Szasz and Michael Meuser, "Environmental Inequalities: Literature Review and Proposals for New Directions in Research and Theory," *Current Sociology* 45 (1997): 99–120.

81. Robert D. Bullard, *Dumping in Dixie: Race, Class and Environmental Quality* (Boulder: Westview, 1990).

82. K. E. McHugh, "Black Migration Reversal in the United States," *Geographical Review* 77 (1987): 171–82.

83. Carol Stack, *A Call to Home: African Americans Reclaim the Rural South* (New York: Basic Books, 1996).

84. Ibid., p. 42.

PART THREE

The Wilderness Idea Roundly Criticized and Defended . . . Again

Gary Snyder

Is Nature Real? (2000)

I'M GETTING GRUMPY ABOUT the slippery arguments being put forth by high-paid intellectuals trying to knock nature and knock the people who value nature and still come out smelling smart and progressive.

The idea of nature as a "social construction"—a shared cultural projection seen and shaped in the light of social values and priorities—if carried out to the full bright light of philosophy, would look like a subset of the world view best developed in Mahayana Buddhism or Advaita Vedanta, which declares (as just one part of its strategy) the universe to be *maya,* or illusion. In doing so the Asian philosophers are not saying that the universe is ontologically without some kind of reality. They are arguing that, across the board, our seeing of the world is biological (based on the particular qualities of our species' body-mind), psychological (reflecting subjective projections), and cultural construc-

A slightly different version of this essay appeared under the title "Nature as Seen from the Kitkitdizze Is No 'Social Construction'" in *Wild Earth* (Winter 1996–97): 8–9.—Eds.

tion. And they go on to suggest how to examine one's own seeing, so as to see the one who sees and thus make seeing more true.

The current use of the "social construction" terminology, however, cannot go deeper, because it is based on the logic of European science and the "enlightenment." This thought-pod, in pursuing some new kind of meta-narrative, has failed to cop to its own story—which is the same old occidental view of nature as a realm of resources that has been handed over to humanity for its own use. As a spiritually (politically) fallen realm, this socially constructed nature finally has no reality other than the quantification provided by economists and resource managers. This is indeed the ultimate commodification of nature, done by supposedly advanced theorists, who prove to be simply the high end of the "wise use" movement. Deconstruction, done with a compassionate heart and the intention of gaining wisdom, becomes the Mahayana Buddhist logical and philosophical exercise that plumbs to the bottom of deconstructing and comes back with compassion for all beings. Deconstruction without compassion is self-aggrandizement.

So we understand the point about wilderness being in one sense a cultural construct, for what isn't? What's more to the point, and what I fail to find in the writings of the anti-nature crowd, is the awareness that wilderness is the locus of big rich ecosystems, and is thus (among other things) a living place for beings who can survive in no other sort of habitat. Recreation, spirituality, aesthetics—good for people—also make wilderness valuable, but these are secondary to the importance of biodiversity. The protection of natural variety is essential to planetary health for all.

Some of these critical scholars set up, then attack, the notion of "pristine wilderness" and this again is beating a dead horse. It is well known that humans and proto-humans have lived virtually every where for hundreds of millennia. "Pristine" is only a relative term, but humanly used as the landscape may have been, up until ninety years ago the planet still had huge territories of wild terrain that are now woefully shrunken. Much of the wild land was also the territory of indigenous cultures that fit well into what were inhabited wildernesses.

The attacks on nature and wilderness from the ivory towers come at just the right time to bolster global developers, the resurgent timber companies (here in California the Charles Hurwitz Suits at Pacific Lumber) and those who would trash the Endangered Species Act. It looks like an unholy alliance of Capitalist Materialist and Marxist Idealists in an attack on the rural world that Marx reputedly found idiotic and boring.

Heraclitus, the Stoics, the Buddhists, scientists, and your average alert older person all know that everything in this world is ephemeral and unpredictable. Even the earlier ecologists who worked with Clementsian succession theory knew that! Yet now a generation of resource biologists, inspired by the thin milk of Daniel Botkin's theorizing, are promoting what they think is a new paradigm that relegates the concept of climax to the dustheap of ideas. Surely none of the earlier scientific ecologists ever doubted that disturbances come and go. It looks like this particular bit of bullying also comes just in time to support the corporate clear-cutters and land-developers. (Despite blow-downs, bugs, fires, drought, and landslides, vast plant communities lasted in essence for multimillions of years prior to human times.)

It's a real pity that many in the humanities and social sciences are finding it so difficult to handle the rise of "nature" as an intellectually serious territory. For all the talk of "the other" in everybody's theory these days, when confronted with a genuine Other, the non-human realm, the response of the come-lately anti-nature intellectuals is to circle the wagons and declare that nature is really part of culture. Which maybe is just a strategy to keep the budget within their specialties.

A lot of this rhetoric, if translated into human politics, would be like saying "African-American people are the social construction of whites." And then they might as well declare that South Central Los Angeles is a problematic realm that has been exaggerated by some white liberals, a realm whose apparent moral issues are also illusory, and that the real exercise in regard to African Americans is a better understanding of how white writers and readers made them up. But liberal critical theorists don't talk this way when it comes to fellow human beings because they know what kind of heat they'd get. In the case of nature, because they are still under the illusion that it isn't seriously there, they indulge themselves in this moral and political shallowness.

Conservationists and environmentalists have brought some of this on themselves. We still have not communicated the importance of biodiversity. Many if not most citizens are genuinely confused over why such importance appears to be placed on hitherto unheard-of owls or fish. Scientists have to be heard from, but the writers and philosophers among us (myself included) should speak our deep feelings for the value of the nonhuman with greater clarity. We need to stay fresh, write clean prose, reject obscurity, and not intentionally exaggerate. And we need to comprehend the pain and distress of working people everywhere.

A *Wilderness* is always a specific place, because it is there for the local critters that live in it. In some cases a few humans will be living in it too. Such places are scarce and must be rigorously defended. *Wild* is the process that surrounds us all, self-organizing nature: creating plant-zones, humans and their societies, all ultimately resilient beyond our wildest imagination. Human societies create a variety of dreams, notions, and images about the nature of nature. But it is not impossible to get a pretty accurate picture of nature with a little first-hand application—no big deal, I'd take these doubting professors out for a walk, show them a bit of the passing ecosystem show, and maybe get them to help clean up a creek.

J. Baird Callicott

Contemporary Criticisms of the Received Wilderness Idea (2000, 2008)

In one of the most ancient and venerable sources of Chinese philosophy, the *Analects,* his disciple asks Confucius what he would do first were he to become the prime minister of the State of Wei. Without question, Confucius replies, first I would rectify names (Hall and Ames 1987). His disciple was puzzled by this saying; and for a long time so was I. No more, for here my project is precisely to rectify one domain of names—the wild domain.

The answer to Juliet's question, "What's in a name?" in Shakespeare's play, is "Really, quite a lot." Consider, by way of analogy, a different domain of names: various names for women—chicks, babes, broads, ladies. The feminist movement has made us keenly aware that what we call someone or something—what we name him, her or it—is important. A name frames, colors and makes someone or something available for certain kinds of uses . . . or abuses. The feminist project in the domain of names for women also makes us keenly aware that someone who criticizes a name is not necessarily critical of what the name refers to. Indeed, often quite the contrary. Women themselves have, of course, taken the lead in purging polite and respectful discourse of such names as "chicks," "babes" and "broads." Even the name "lady" is freighted with so much baggage that it is not worn comfortably by many women.

Just as the women who criticize some of the names they are called do not intend to criticize themselves or other women, I want to note here at the outset, in the most direct and emphatic way I can, that I do not criticize the *places* we call "wilderness." Quite the contrary. Rather, I criticize a *name,* a concept, the *received* wilderness idea. I am as passionately solicitous of the places called wilderness as any of the defenders of the classic wilderness idea. However, in my opinion, the name "wilderness" improperly colors them, frames them, and makes them available for inappropriate uses and abuses.

As Michael Nelson and I brashly write in the introduction to our 1998 anthology, *The Great New Wilderness Debate,* the wilderness idea is "alleged to be ethnocentric, androcentric, phallogocentric, unscientific, unphilosophic, impolitic, outmoded, even genocidal" (Callicott and Nelson 1998). I hasten to say that *we* are not necessarily the ones who allege that the received wilderness idea is *all* these bad things, just that such things have been alleged—some of them by him or me, some by other writers included in the book. Here, I take up each of these indictments, try to explain why they have been filed and expose the evidence on which they are based.

But first, what is meant by the "received" or "classic" or "traditional" wilderness idea? The idea of wilderness we have inherited—received—from its framers, going back now at least several centuries, but shaped most fully during the first half of the 20th century. The *received* wilderness idea is eloquently conveyed in the definition of "wilderness" in the oft-quoted Wilderness Act of 1964: "A wilderness, in contrast to those areas where man and his own works dominate the landscape, is hereby recognized as an area where the earth and its community of life are untrammeled by man, where man himself is a visitor who does not remain" (Public Law 88-577).

First, the received wilderness idea is ethnocentric. My friend and fellow environmental philosopher, Holmes Rolston, III—a staunch defender of the classic wilderness idea—told me awhile back that when he was lecturing in Japan, his translators could come up with no Japanese word for the English word "wilderness." Recently, I asked another friend, Roger T. Ames, who translates and interprets Chinese philosophy, if there is a word for wilderness in Chinese. There is for wild man, he said, and wild woods, but no word for wilderness. Even for most Europeans, wilderness is a foreign concept. The notable exceptions are the Norwegians and other Scandinavians, who, significantly, have an arctic frontier inhabited by indigenous peoples, formerly called Lapps and now Sami.

The wilderness idea is most familiar in American and Australian discourse. The United States and Australia have colonial histories, both beginning as English colonies. As opposed to the French, Spanish and Portuguese, who seem to have been more interested in extracting and appropriating resources and leaving behind their genes, the English colonial enterprise was focused on land to live on and to make over into a landscape like the one they left behind. The English colonists called the new lands of North America and Australia "wilderness," an idea originally taken from the English translation of the Bible—about which more below. This designation enabled them to see the American and Australian continents as essentially empty of human beings, and thus available for immediate occupancy. The Australian bureaucratic term for wilderness, *terra nullius,* a Latin phrase meaning "empty land," says it all quite explicitly (Bayet 1998). So does the U.S. Wilderness Act of 1964, "an area of undeveloped Federal land retaining its primeval character and influence, without permanent improvements *or human habitation*" (Public Law 88-577, emphasis added).

The Australian and American continents were not, however, empty lands when "discovered" by Europeans and settled by English colonists. Until recently, the voices of American Indians and Australian Aborigines were either ignored or silenced (Plumwood 1998). In the third edition of *Wilderness and the American Mind,* Roderick Nash (1982) belatedly noted that American Indians disputed the idea that European colonists found North America in a "wilderness condition," and quoted Luther Standing Bear to that effect. Nelson and I could find no earlier American Indian protest against the wilderness idea, so we republished the chapter of Standing Bear's 1933 book, *Land of the Spotted Eagle,* in which it occurs. As many readers of Nash's preface to the third edition will recall, Standing Bear (1998) said,

> We did not think of the great open plains, the beautiful rolling hills, and winding streams with tangled growth, as "wild." Only to the white man was nature a "wilderness" and only to him was the land "infested" with "wild" animals and "savage" people. To us it was tame. Earth was bountiful and we were surrounded with the blessings of the Great Mystery. Not until the hairy man from the east came and with brutal frenzy heaped injustices upon us and the families we loved was it "wild" for us. When the very animals of the forest began fleeing from his approach, then it was that for us the "Wild West" began.

This is as clear a piece of evidence as one could want to show that the wilderness idea is ethnocentric; it was an idea not entertained by a representative member of and spokesperson for a non-European ethnic group. Shakespeare's Juliet goes on to say "a rose by any other name is still a rose." Yes. And in almost every language, there are different but mutually translatable names for common plants, such as roses, and animals, such as rabbits, and all the other general features—rivers, mountains, valleys—of the landscape. But not every language has a name for wilderness—not Lakota, not Japanese, not Chinese, not Dutch, probably not most. Unlike the names "mountain" or "river," which are just the English labels for actual topological features of the landscape, the name wilderness *socially constructs,* as we now say, the landscape, in a way not shared by all social groups. It is therefore an ethnocentric idea.

Note that I am here making a more limited claim than that recently ridiculed by Gary Snyder (1996)—"the idea of Nature as being a 'social construction.'" Because "liberal critical theorists," according to Snyder ". . . are still under the illusion that it isn't seriously *there,* they indulge themselves in this moral and political shallowness." While other philosophers, East and West, past and present, have doubted the robust reality of the whole of *nature*—mountains and rivers, plants and animals, sun and moon, stars and planets—I do not. In my opinion, nature and all its components and processes are incontestably real, "there" as Snyder would have it. But to call certain areas of the natural world "wilderness"—just as to call certain aspects of nature "natural resources"—is to put a spin on them; it is to socially construct them, not as objective, autonomous nature, but nature in relationship to us human beings. The name "natural resources" socially constructs nature as a self-renewing larder existing for our consumption; "wilderness" socially constructs it in a variety of ways, many of which are the subject of this discussion. Most feminists believe that women too are seriously *there,* but to call them "babes" or "ladies" is to put a spin on them, it is to socially construct them in relationship to men, not neutrally as autonomous female persons. The name "babes" socially constructs women as sexual objects; "ladies" socially constructs women as paragons of virtue standing on a pedestal.

The quote from Standing Bear provides a transition to the ugliest allegation against the wilderness idea—that it is genocidal. More precisely, it was and is a *tool* of genocide. Suppose you come to a place already inhabited by people and declare it to be a wilderness, that is, "an area . . . where man . . . is a visitor who does not remain," an area that is devoid of "human habitation," to quote once more the essential characteristic of wilderness in the Wilderness Act of 1964. Then you have "erased," as Australian environmental philosopher Val Plumwood (1998) puts it, the indigenous inhabitants from the landscape as you and your group socially construct it. If you don't acknowledge their existence in the first place, it makes it easier to dispossess and delete them.

There is another way of defining wilderness, which is not in contrast to man and his works, but in contrast to a certain kind of man and certain kinds of his works—to civilized man and to the works of civilization (Duerr 1985). By that definition, a wilderness may be inhabited by wild people without invalidating its wilderness condition. Which means, in effect, that the noncivilized *Homo sapiens* living in the wilderness are just another form of wildlife. So it is of as little moral consequence to hunt them as it is to hunt other kinds of wildlife. We know that in North America, indigenous peoples were often regarded as "vermin" to be shot on sight, man, woman, or child—as indicated in the infamous frontier phrase, "the only good Indian is a dead Indian"—while in Australia, Aborigines were actually hunted for sport (Berkhofer 1978; Whitelock 1985).

But that was then, and this is now. In Central America—Guatemala most notoriously—genocidal campaigns against indigenous peoples have been conducted, with the complicity of the United States government, right into the 1990s, although not under cover of the wilderness idea (Broder 1999). In Africa and India, however, wilderness areas have been created, quite recently, by clearing out inconvenient human inhabitants. A quarter century ago, Colin Turnbull (1972) shocked the world with an account of an African tribe called the Ik who seemed to have lost their humanity, living sullenly in their huts with only minimum human contact and turning their children out at three years of age to fend for themselves. What had happened to these people to turn them into a travesty of humanity? They were victims of the wilderness idea. They had the misfortune to live in the Kidepo Valley, where, a decade earlier, the dictator of Uganda, Milton "Apollo" Abote, decided to establish an American-style national park in which "man himself is a visitor who does not remain" (Harmon 1998). Thus, he couldn't have the Ik remaining there, where they had happily, successfully, sustainably and humanly lived as gatherer-hunters from time immemorial. The Ik seem literally to have abandoned their humanity, in their abject despair over having been evicted from their homeland and forced to live in sedentary villages. A similar fate, though with less dramatic human consequences, befell the Juwasi San, among the famous Kalahari bushmen, when Etosha National Park was created in Namibia (Thomas 1990). These are not isolated cases. They only loom large because they were publicized in popular media, such as trade books, magazines, and film. All over Africa, according to Raymond Bonner (1993), similar cases abound.

The most bitter critique of the wilderness idea in our anthology was written by Ramachandra Guha, who documents several Indian examples of ethnic cleansing in recently declared "wilderness areas." Here is his characterization of one case.

> The Nagarhole National Park in Southern Karnataka has an estimated forty tigers . . . [and] is also home to about 6,000 tribals, who have been in the area longer than anyone can remember, perhaps as long as the tigers themselves. The state forest department wants the tribals out, claiming they destroy the forest and kill wild game. The tribals answer that their demands are modest, consisting in the main of fuel wood, fruit, honey,

> and the odd quail or partridge. They do not own guns, although coffee planters on the edge of the forest do. . . . [T]hey ask the officials, if the forest is only for tigers, why have you invited India's biggest hotel chain to build a hotel inside it while you plan to throw us out? . . . [T]he Nagarhole case is not atypical. All over India, the management of parks has sharply pitted the interests of poor tribals who have traditionally lived in them [so far, apparently, without destroying them] against wilderness lovers and urban pleasure seekers who wish to keep parks "free of human interference"—that is, free of *other* humans. These conflicts are being played out in the Rajaji sanctuary in Uttar Pradesh, in Simlipal, in Orissa, in Kanha in Madhya Pradesh, and in Melghat in Maharashtra (Guha 1998).

I now take up the allegation that the wilderness idea is politically suspect. As mention of these African and Indian cases demonstrates—along with a thousand other things, from Coca Cola, Kentucky Fried Chicken, and Marlboro, to *Bay Watch,* blue jeans, and Michael Jordan—the influence of American culture is global. The politics of wilderness preservation in the rest of the world, however, is not what it is in the United States. In the U.S. a beleaguered minority of mostly liberal and progressive people heroically battle the industry-funded Wise Use Movement; the industry-owned congressional delegations of Alaska, Idaho, and Utah; and a variety of ideologically driven right-wing, private-property zealots on behalf of some roadless fragments of backcountry with a precious modicum of remaining ecological integrity. In many other parts of the world, the politics of wilderness preservation are not so unambiguously respectable. There, preservationists often find themselves bed-fellowed with wealthy urban elites, state-sponsored paramilitary terrorists, and undemocratic regimes against remnant populations of Fourth World peoples—tribals as Guha calls them—living by traditional modes of subsistence in scattered pockets of Third World nation states.

Take the androcentric and phallogocentric charges against the received wilderness idea next. 'Androcentric' means male-centered. I'm not sure what 'phallogocentric' means. Evidently, it is a feminist neologism combining *phallus* with *logos* and adding the suffix "centric"—a self-centered rational prick, I guess. In any case, the bottom line here is that the wilderness idea is macho.

I am not sure if it still is, but it certainly once was. One hundred years ago, Theodore Roosevelt was advocating wilderness preservation for reasons based on historian Frederick Jackson Turner's then new and convincing "frontier thesis." After generations of contact with the wild frontier, the transplanted Northwestern European had become a new kind of human being on the face of the earth, an American. Or so it was thought. However, by the end of the 19th century—with the construction of the transcontinental railroad, the slaughter of the bison, and the final solution of the wild Indian problem—the frontier closed. To conserve the American character, Roosevelt advocated conserving simulacra of the frontier, to which future Americans could repair for rugged, character-shaping adventure. One might call this the "American-character" rationale for wilderness preservation.

By the way, we forgot to mention in the introduction to the anthology that, ancillary to the American-character rationale, the wilderness idea is liable to the allegation of racism, not to be confused with ethnocentrism. Aldo Leopold (1998a, emphasis added), early in his career a warm supporter of the American-character rationale for wilderness preservation, expresses the racist undercurrent in this argument with his characteristic flair for elegant succinctness: "For three centuries [wilderness] has determined the character of our development; it may in fact be said that coupled with the character of our *racial* stocks, it is the very stuff America is made of. Shall we now exterminate this thing that made us Americans?" Leopold (1998a) defined the distinctive American character as follows: "a certain vigorous individualism combined with the ability to organize, a certain intellectual curiosity bent to practical ends, a lack of subservience to stiff social forms, and an intolerance of drones." Leopold (1991), incidentally, was fully aware that designated wilderness areas provide only illusions—simulacra—of the bygone frontier:

> [T]he loss of adventure into the unknown . . . causes the hundreds of thousands to sally forth each year upon little expeditions, afoot, by pack train, or by canoe, into the odd bits of wilderness which commerce and "development" have regretfully and temporarily left us here and there. Modest adventurers to be sure, compared with Hanno, or Lewis and Clark. But so is the sportsman with his setter dog in pursuit of partridges, a modest adventurer compared with his neolithic ancestor in single com-

bat with the Auroch bull. The point is that along with the necessity for expression of racial instincts there happily goes that capacity for illusion which enables little boys to fish happily in wash-tubs. That capacity is a precious thing, if not overworked.

But, getting back to androcentrism, according to Roosevelt, the distinctive American character was decidedly machismo: "wilderness promoted 'that vigorous manliness for the lack of which in a nation as in an individual, the possession of no other qualities can atone,'" while its absence risked a future breed of American "'who has lost the great fighting masterful virtues'" (Nash 1982).

The androcentrist aspect of the wilderness idea is also prominent in the work of two wilderness-movement giants in the first half of the 20th century, Robert Marshall and Sigurd Olson. First, according to Marshall (1998), some men,

> become so choked by the monotony of their lives that they are readily amenable to the suggestion of lurid diversion. Especially in battle, they imagine, will be found the glorious romance of futile dreams. William James has said that "militarism is the great preserver of ideals of hardihood, and human life with no use for hardihood would be contemptible." The problem, as he points out, is to find a "moral equivalent of war." . . . This equivalent may be realized if we make available to every one the harmless excitement of the wilderness.

The androcentric brassiness of Olson's prose, like Marshall's written in the 1930s, is a little shocking in the 1990s. His essay, "Why Wilderness?" opens with the following sentence: "In some men, the need of unbroken country, primitive conditions, and intimate contact with the earth is a deeply rooted cancer gnawing forever at the illusion of contentment with things as they are" (Olson 1998). And it continues through to the end in the same vein:

> I have seen them come to the "jumping off places" of the North, these men whereof I speak. I have seen the hunger in their eyes, the torturing hunger for action, distance and solitude, and a chance to live as they will. I know these men and the craving that is theirs; I know also that in the world today there are only two types of experience which can put their minds at peace, the way of wilderness and the way of war. . . . The idea of wilderness enjoyment is not new. Through our literature we find abun-

> dant reference to it, but seldom of the virile, masculine type of experience men need today (Olson 1998).

In the first third of the 20th century, the dominant argument for wilderness preservation was recreation of a primitive and unconfined sort. Now, ironically, wilderness recreation has become one of the most gadget-laden and rule-bound forms of sport available—what with the freeze-dried food, Swiss Army knives, nylon tents, permits, designated camp sites, open-flame restrictions, packing-out-your-garbage-and-feces requirements and all. But back then, wilderness recreation was imagined to be the crudest and freest form available, suitable for the men about whom Olson waxes poetic. They take off their clothes, "laugh as they haven't laughed in years and bellow old songs in the teeth of a gale. . . . I can honestly say, that I have heard more laughter in a week out there than in any month in town. Men laugh and sing as naturally as breathing once the strain is gone" (Olson 1998).

After Roosevelt, so entangled with the notion of an unconstrained, virile, masculine type of recreation was the wilderness idea that in the 1920s Leopold routinely characterized the minimum size of an area to qualify for wilderness designation in terms of such recreation—"big enough to absorb a two weeks pack trip." And by "pack trip," he didn't mean back pack; he referred to a donkey train.

Next, the wilderness idea is unscientific. One scientific problem with the wilderness idea has already been indicated in connection with the ethnocentric allegation. The fact is, there were people here in the Americas before Columbus's landfall in the late 15th century and in Australia before the European discovery of that continent by James Cook in the mid-18th century. That not only renders the received wilderness idea ethically and politically problematic, it also creates an ecological conundrum for "wilderness science" based on that idea. When the existence of such peoples was acknowledged at all by the framers of the wilderness idea, their ecological impact was minimized. Read Marshall (1998) on the matter:

> When Columbus effected his immortal debarkation, he touched upon a wilderness which embraced virtually a hemisphere. The philosophy

> that progress is proportional to the amount of alteration imposed upon nature never seemed to have occurred to the Indians. Even such tribes as the Incas, the Aztecs, and Pueblos made few changes in the environment in which they were born. "The land and all that it bore they treated with consideration; not attempting to improve it, they never desecrated it." Consequently, over billions of acres the aboriginal wanderers still spun out their peripatetic careers, the wild animals still browsed unmolested in the meadows and the forests still grew and moldered and grew again precisely as they had done for undeterminable centuries. It was not until the settlement of Jamestown in 1607 that there appeared the germ for that unabated disruption of natural conditions which has characterized all subsequent American history.

But the ecological impact of American Indians, over the 13,000 years or more since America was originally discovered, and of Australian Aborigines, over the 40,000 years or more since Australia was originally discovered has probably far exceeded the ecological impact of the rediscovery and resettlement of those continents by European peoples (Denevan 1998; Pyne 1997). For starters, there is a disturbing coincidence, and very probably a causal connection, between the extinction of some 30 genera of megafauna in the Americas and the arrival of Siberian big game hunters, the infamous Clovis spearmen (Martin 1973). Paul S. Martin (1984) traces the same coincidence of a spreading wave of *Homo sapiens* out of Africa and a wave of megafaunal extinctions all over the planet.

After the ecological spasm that followed initial invasion of the Western Hemisphere, Homo sapiens became naturalized in the Nearctic and Neotropics. The species became an ecological keystone, structuring biotic communities by horticulture, irrigation, cultural fire and unremitting predation on grazers and browsers (Doolittle 1992; Kay 1994; Krech 1999; Pyne 1982). Therefore, the classic wilderness idea as defined by the Wilderness Act of 1964, serves only to befuddle the science of wilderness management. The goal of wilderness management in the United States—to preserve "the condition that prevailed when the area was first visited by the white man"—was set by the enormously influential Leopold Report in 1963 (Leopold and others 1998). But that condition was heavily influenced by the red man; it was not a condition "affected primarily by the forces of nature, with the imprint of man's work substantially unnoticeable" as specified in the Wilderness Act of 1964. Knowing what we now know about the ecological impact of indigenous peoples in the Americas, preserving these "vignettes of primitive America" requires a continuous trammeling of the kind historically imposed by their original human inhabitants on areas designated as wilderness, "where the earth and its community of life are untrammeled by man, where man himself is a visitor who does not remain." You see how confusing this is.

Adding to the confusion and paradox is the post-contact demographic history of the Western Hemisphere. Today's demographers estimate the 15th-century New World human population to be ten times greater than demographers estimated it to be in Marshall's day, the 1930s (Denevan 1998). Earlier estimates did not take account of Old World diseases that reduced the pre-Columbian population of the Western Hemisphere by up to ninety percent. In 1750—the half-way point between Columbus's "immortal debarkation" and the present—subtracting all the Indians that died, and adding all the Europeans and Africans that immigrated and multiplied, the human population of the Western Hemisphere was thirty percent *less* than it was in 1492, according to William Denevan (1998). Old World diseases stalked the New World, transmitted from Indian to Indian, well in advance of the leading edge of European conquest and settlement. Thus, the condition that prevailed in areas of the North American west—where most designated wilderness is located—when first visited by the white man was one of rapid ecological transition. Denevan (1998) refers to it tendentiously as in a process of "recovery." In any case, between 1492 and 1750, the keystone species that had for centuries, if not millennia, structured the biotic communities of the New World was reduced by epidemic disease and warfare with the white (and black) invaders from the Old World. Thus, whatever we call it, the biotic communities in the Western Hemisphere were headed toward new ecological domains of attraction. So here's a nice scientific mind bender. The upshot of Denevan's thesis is this: When the white man first visited the interior of North America, he did, after all, find a wilderness condition as defined by the Wilderness Act of 1964—"an area where the earth and its community of life are untrammeled by man, where man himself is a visitor who does not remain"—but it was an *artificial* wilderness condition, an anthropogenic ecological effect, created by the depopulation of the country following the devastating demographic consequences of contact.

As the quotation from Marshall shows, the received wilderness idea is entangled with another unscientific assumption, the now discredited balance-of-nature paradigm in ecology. Stripped of its fancy language, this is the picture Marshall paints: The Western Hemisphere was virgin territory when discovered and then raped by the white man. Well, maybe there were some Indians here, but there were not very many of them, and they were so technologically backward and environmentally ethical that they either couldn't or wouldn't change the Nature of which they were a part. Thus—and here's the new point—for centuries, if not millennia, Nature remained in a dynamic equilibrium. In his own words, again, Marshall believed that "the wild animals still browsed unmolested in the meadows and the forests still grew and moldered and grew again precisely as they had done for undeterminable centuries." In the absence of a robust human presence in the Western Hemisphere, Nature remained in a steady state.

By now, however, we have all read and absorbed the work of Margaret Davis (1969) and other palynologists who paint a very different picture of the constantly shifting ecological mosaic of North America over time. By now, we have all read and absorbed the work of Pickett and White (1985) and other landscape ecologists who emphasize the normalcy, not the abnormalcy, of ecological disturbance, whether anthropogenic or nonanthropogenic. Daniel Botkin (1991) has popularized the new "shifting paradigm" in ecology in his well-read book, *Discordant Harmonies.*

Now, next to last, I discuss the allegation that the received wilderness idea is unphilosophic. As Nash (1982) points out in his classic *Wilderness and the American Mind,* the word "wilderness" hails from Old English and was used in English translations of the Bible. In its biblical context, it stood opposed not only to civilization, but to the Garden of Eden, and often referred, more especially, to arid, desert regions such as the Sinai. And as Nash (1982) also points out, it was applied to North America by the Puritans especially, for whom at first it had a wholly negative connotation. North America in the perfervid Puritan imagination was the stronghold of Satan, and the indigenous population—what was left of it anyway—was composed of the devil's minions. In the words of historian Perry Miller (1964), the Puritans believed themselves sent by God on an "errand into the wilderness," to convert the continent to Calvinism and replace the wilderness with fair English-style, small-hold farms and shining European-style cities on hills.

After several generations had succeeded well in this enterprise, later Puritan fervor found sin in those very cities. The country having been ethnically cleansed and pacified, Nature began to appear to be God's pristine, undefiled creation. A central doctrine of Puritan theology was original sin and human depravity. Our 1998 wilderness anthology opens with selections from Jonathan Edwards, who finds "images or shadows of divine things" not in his fellow men or in human works, but in Nature. To make a long story short, the received wilderness idea is ultimately a legacy of Puritan theology. At the heart of that theology is a dualism of man and nature. To the first generations of Puritans in America, man was created in the image of God and, if not good, at least the Elect among men were put in the service of a good God to enlighten a benighted, dismal, and howling wilderness continent. To later generations of Puritans, the positive and negative poles of the dualism were reversed. The Fall and man's consequent evil were stressed. In the minds of some influential Puritan thinkers like Edwards, Nature became a foil for man's sin and depravity. It was transformed into the embodiment of goodness. After all, Nature was created by God and declared to be good, as you may read for yourself in the first pages of Genesis. And as Muir (1916) astutely observed in the mid-19th century, Nature and its nonhuman denizens remain, in sharp contrast to man, "unfallen" and "undepraved."

In many of the most passionate framers of the received wilderness idea, strains of what might be called neo-Puritan Nature theology run strong. Man and his works are sinful, Nature is pure and divine. It is there in Thoreau, there most vividly in Muir, there in Marshall, there in Rolston (1998)—an ordained Presbyterian minister, I might add—and there in Dave Foreman (1998a), who likes to boast about his pugnacious Scotch-Irish temperament. I don't see it in Leopold, and, curiously, the

Scandinavian Olson complains that this meme in the memetic makeup of the received wilderness idea gets in the way of the expression of the macho meme: According to Olson (1998),

> Typical of this tone of interpretation is Thoreau with his "tonic of wildness," but to the men I have come to know his was an understanding that does not begin to cover what they feel. To him the wild meant the pastoral meadows of Concord and Walden Pond, and the joy he had, though unmistakably genuine, did not approach the fierce unquenchable desire of my men of today. For them the out-of-doors is not enough; nor are the delights of meditation.

Olson's men had to be out there yukking it up and kicking butt in the real wilderness, not silently sniffing the ladyslippers and communing with the Oversoul in a rural wetland. Muir, on the other hand, being the most orthodox neo-Puritan framer of the received wilderness idea, opposed the macho meme in it for which Roosevelt is most responsible. According to Nash (1982), Muir took Roosevelt on a hike and sleepover in Yosemite and there said to him, "'Mr. President, when are you going to get beyond the boyishness of killing things . . . are you not getting far enough along to leave that off?'"

Finally, I come to the outmoded allegation. Recreation may be an important purpose served by designated wilderness areas, but the most adamantine apologists for the wilderness idea and I agree that wilderness areas have a higher calling in these desperate days. According to Snyder (1996), "we are not into saving relatively uninhabited wild landscapes for the purpose of recreation or spirituality even, but to preserve home space for nonhuman beings." Historically, according to Foreman (1998b), "The most common argument for designating wilderness areas . . . touted their recreational values." But now, "core wilderness areas [should] be managed to protect and, where necessary, to restore native biological diversity and natural processes. Traditional wilderness recreation is entirely compatible, so long as ecological considerations come first." The most important *raison d'être* of designated wilderness areas is habitat for species that do not coexist well in close proximity with Homo sapiens—for whatever reasons; brown bears, wolves, and mountain lions are the most frequently cited examples (Grumbine 1992,

1998). They do not coexist well with people because of direct conflict of interest. Usually people, out of fear or ignorance, attack them; or, more rarely, they attack people or pets or livestock. Less often mentioned, many species of birds, with which people have no direct conflict, need interior forest habitat, which is increasingly fragmented, and thus ruined for them, by suburban and exurban real estate development (Robinson and others 1995).

What to do now? We're talking about a name, an idea, a concept. We could redefine it. We could purge it of its macho baggage, its neo-Puritan theological freight, its connotation of a resource for either virile or meditative recreation, its penumbra of Arcadian ecology, its undercurrents of ethnocentrism and racism, and sanitize or disavow its colonial origins and functions. Or we could start with a fresh concept for the purposes of wilderness science. I recommend the latter alternative; except in that case, of course, we wouldn't call it "wilderness science."

If we call the habitat of wolves, bears, lions, lynx, and warblers "wilderness areas," inevitably our minds are flooded with all the hogwash the wilderness idea is steeped in. Visions well up in our imaginations of virile, unconfined recreation, or of reverential pilgrimages in holy sanctuaries unsullied by the presence of profane people and preserved forever just as they always existed, in splendid harmony and balance from time immemorial. We get confused by the wilderness name and think that these places should remain "natural" and not be actively managed by, say, prescribed burns or therapeutic hunts to make them fit habitat for threatened species, or that they mainly exist for us to recreate in, just as the Wilderness Act of 1964 says. But that should not be their primary purpose. They should exist, primarily, for the animals, whose homes they are, and for us to manage with their needs exclusively in view or, in some cases, to stay the hell out of altogether. Imagine insisting on calling a Battered Women's Shelter, the House of Babes or the Home for Uppity Ladies. It might attract the wrong kind of attention.

Well, how do I propose we rename the places misnamed wilderness areas? The scientific community seems to be settling on "biological reserve" or "biodiversity reserve" as the most straightforward name (Scott 1999). That idea, however, is not new (Grumbine 1998; Scott 1999). A similar concept was proposed in the first quarter of the 20th century, just as the wilderness movement was gathering steam, most notably by Victor Shelford (1920, 1933), Francis Sumner (1920, 1921), G. A. Pearsons (1922), and George Wright and others (1933). It was reasserted again at mid-century by S. C. Kendeigh and others (1950) and Ray Dasmann (1972). But the proposal for a representative biological reserve system for the United States, as opposed to a system of "wilderness playgrounds" (Leopold 1991), was eclipsed by the human-experience-oriented wilderness idea and, until now revived, was consigned to the dust bin of American conservation history (Grumbine 1998; Scott 1999). The names of Shelford, Sumner, Pearsons and Wright became but footnote fodder in the scholarly tracts of conservation historians, while the names of Thoreau, Muir, Roosevelt, Leopold, Marshall, and Olson live on in wilderness legend.

We are in the midst of a global conservation crisis. The present generation is witness to and the cause of only the sixth abrupt mass extinction event in the 3.5 billion–year biography of the Earth. To address this gargantuan problem a new transdisciplinary science called conservation biology has taken shape. Its *summum bonum*—or "greatest good"—is biodiversity. As one of its architects, Michael Soulé (1985), put it in a field-defining paper, "Diversity of organisms is good." Period. The only way to save species populations from extinction is to provide them with habitat. *Ex situ* conservation without the hope of reintroduction into fit habitat is a kind of living extinction for species, just as confinement in a hospital with no hope of going home is a kind of living death for sick people. Human activities, such as agriculture and suburban and exurban development, provide some organisms—such as raccoons and white-tailed deer—with excellent habitat. But the habitats of many other organisms are severely degraded by the cultural modifications of landscapes that characterize contemporary industrial civilization. These organisms need places that are otherwise suitable for them where modifications of that kind are prohibited. Such places are biodiversity reserves, and a major focus of the science of conservation biology is reserve selection, design, and management (Meffe and Carroll 1997).

Designated wilderness areas now serve as biodiversity reserves, but only as an afterthought. As Dave Foreman (1998b) points out, they were selected for purposes other than providing threatened and endangered species with habitat. They were selected, rather, because of their potential for a virile and unconfined type of recreation, or because they contained spiritually uplifting "monumental" scenery, or often because they simply had no other foreseeable utility (Foreman 1998b). They may not be the best places for biodiversity reserves, but they are a start.

However, to continue to call these places and future reserves "wilderness areas" not only confuses the public about what their most important function is, it also confuses their scientific management. What are the goals of wilderness science? To preserve vignettes of primitive America? To preserve pristine Nature in perfect balance? To provide opportunities for meditative solitude, or for travel by canoe and pack animal? To monitor what happens to a biotic community when its keystone species has been prohibited from remaining and can only visit as a tourist or scientist? To provide habitat for threatened and endangered species? Some of these? All of these? Or only one of these? If we had already decided to take my suggestion and rename wilderness areas "biodiversity reserves," then "wilderness science" would be more narrowly focused, as I believe it ought to be, on the science of reserve selection, design, and management.

With a clearer focus on goals, we can achieve not only a clearer focus on the scientific task, but also a better criterion for determining what is and what is not an acceptable human presence in biodiversity reserves. In some places in the Amazon, for example, traditional extractive activities, such as rubber tapping and nut gathering, may be consistent with many if not all the biodiversity conservation goals in those areas (Peters and others 1989). In which case conservation may not be at odds with the economic activities of indigenous peoples. Such peoples and conservationists may then form political alliances against capital-intensive development schemes, hatched in distant capitals, that would destroy the economies of both tribal peoples living by traditional means and the habitat of the species with which they have traditionally coexisted. On the other hand, in the First World, bourgeois recreational use of wilderness areas may come in conflict with species recovery plans. If renamed, "biodiversity reserves," the priority question—the species recovery plan versus the bourgeois recreational use—is not even up for debate. We allow recreational and other human uses of biodiversity reserves, certainly, but only up to the point that the maintenance of threatened species populations is not compromised.

As Grumbine (1998) notes, "the concept of 'biodiversity' has become a central rallying cry for a growing portion of the US environmental movement." It may not be, however, so central on the radar screens of most laypersons. While the term "biodiversity reserve" may serve to better focus the energies of the scientific subset of the environmental movement, it may be a nonstarter politically. But "wilderness" too, as already noted, is a politically loaded term. For purposes of public relations, other more or less equivalent names might be deployed. We might resurrect, for example, the term employed by Kendeigh and others (1950) at mid-century, "nature sanctuaries." Or if "nature" is too vague and inclusive a term, we might call them "wildlife sanctuaries." A task for future social science research might be to discover the most appealing name for what future erstwhile wilderness scientists will refer to as "biodiversity reserves," if my advice is heeded.

In conclusion, I have criticized a name, "wilderness," not the places—wilderness areas—that bear the name. I might add that I am not criticizing the framers of the received wilderness idea either: Edwards, Emerson, Thoreau, Muir, Roosevelt, Aldo Leopold, Marshall, Olson (well, maybe Olson), Starker Leopold, and Howard Zahniser, the ghostwriter of the Wilderness Act of 1964. Their times were different from ours today. Now we are in the midst of a global biodiversity crisis that, with one exception, they knew nothing about. That exception was Aldo Leopold (1998b) who, although he continued until the end to call them wilderness areas, recast his arguments for their preservation in the same terms as I am here, namely habitat for "threatened species," as had Shelford and a few other of his now all-but-forgotten contemporaries. The baggage that freights the received wilderness idea, in my opinion, makes it an unsuitable conceptual tool to meet the challenge of the biodiversity crisis. We need a new name that will better focus our contemporary conservation goals and, therefore, our conservation policy and science.

I suggest that new name, at least within the scientific sector of the conservation community, should be "biodiversity reserves."

REFERENCES

Bayet, F. 1998. Overturning the doctrine: Indigenous people and wilderness—being Aboriginal in the environmental movement. Pages 314–324 in Callicott and Nelson, editors. The great new wilderness debate. University of Georgia Press, Athens.

Berkhofer, R. F. 1978. The white man's Indian: Images of the American Indian from Columbus to the present. Alfred A. Knopf, New York.

Bonner, R. 1993. At the hand of man: Peril and hope for Africa's wildlife. Alfred A. Knopf, New York.

Botkin, D. 1990. Discordant harmonies: A new ecology for the twenty-first century. Oxford University Press, New York.

Broder, J. M. 1999. Clinton offers his apologies to Guatemala. New York Times 148 [March 11]: A1, A12.

Callicott, J. B. and M. P. Nelson. 1998. The great new wilderness debate. University of Georgia Press, Athens.

Dasmann, R. E. 1972. Towards a system for classifying natural regions of the world and their representation by national parks and reserves. Biological Conservation 4: 247–255.

Davis, M. 1969. Climatic changes in southern Connecticut recorded by pollen deposition at Rogers Lake. Ecology 50: 409–422.

Denevan, W. M. 1998. The pristine myth: The landscape of the Americas in 1492. Pages 414–442 in Callicott and Nelson, editors. The great new wilderness debate. University of Georgia Press, Athens.

Doolittle, W. E. 1992. Agriculture in North America on the eve of contact: A reassessment. Annals of the Association of American Geographers 82: 386–401.

Duerr, H. P. 1985. Dreamtime: Concerning the boundary between the wilderness and civilization. Basil Blackwell, Oxford.

Foreman, D. 1998a. Wilderness areas for real. Pages 395–407 in Callicott and Nelson, editors. The great new wilderness debate. University of Georgia Press, Athens.

Foreman, D. 1998b. Wilderness: From scenery to nature. Pages 568–584 in Callicott and Nelson, editors. The great new wilderness debate. University of Georgia Press, Athens.

Grumbine, R. E. 1992. Ghost bears: Exploring the biodiversity crisis. Island Press, Washington.

Grumbine, R. E. 1998. Using biodiversity as a justification for nature preservation in the US. Pages 595–616 in Callicott and Nelson, editors. The great new wilderness debate. University of Georgia Press, Athens.

Guha, R. 1998. Deep ecology revisited. Pages 271–279 in Callicott and Nelson, editors. The great new wilderness debate. University of Georgia Press, Athens.

Hall, D. L. and R. T. Ames. 1987. Thinking through Confucius. State University of New York Press, Albany.

Harmon, D. 1998. Cultural diversity, human subsistence, and the national park ideal. Pages 217–230 in Callicott and Nelson, editors. The great new wilderness debate. University of Georgia Press, Athens.

Kay, C. E. 1994. Aboriginal overkill: The role of Native Americans in structuring western ecosystems. Human Nature 5: 359–358.

Kendeigh, S. C., H. I. Baldwin, V. H. Cahalane, C. H. D. Clarke, C. Cottam, W. P. Cottam, I. M. Cowan, P. Dansereau, J. H. Davis, F. W. Emerson, I. T. Craig, A. Hayden, C. L. Hayward, J. M. Linsdale, J. A. McNab, and J. E. Potzger. 1950. Nature sanctuaries in the United States and Canada: A preliminary inventory. Living Wilderness 15: 1–45.

Krech, S. 1999. The ecological Indian: Myth and history. W. W. Norton, New York.

Leopold, A. 1991. The river of the mother of god. Pages 123–127 in S. L. Flader and J. B. Callicott, editors. The river of the mother of god and other essays by Aldo Leopold. University of Wisconsin Press, Madison.

Leopold, A. 1998a. Wilderness as a form of land use. Pages 75–84 in Callicott and Nelson, editors. The great new wilderness debate. University of Georgia Press, Athens.

Leopold, A. 1998b. Threatened species: A proposal to the wildlife conference for an inventory of the needs of near-extinct birds and animals. Pages 513–520 in Callicott and Nelson, editors. The great new wilderness debate. University of Georgia Press, Athens.

Leopold, A. S., S. A. Cain, C. M. Cotham, I. N. Gabrielson, and T. L. Kimball. 1998. Wildlife management in the national parks. Pages 103–119 in Callicott and Nelson, editors. The great new wilderness debate. University of Georgia Press, Athens.

Marshall, R. 1998. The problem of the wilderness. Pages 85–96 in Callicott and Nelson, editors. The great new wilderness debate. University of Georgia Press, Athens.

Martin, P. S. 1973. The discovery of America. Science 179: 967–974.

Martin, P. S. 1984. Prehistoric overkill: The global model. Pages 354–403 in P. S. Martin and R. G. Klein, editors. Quaternary extinctions: A prehistoric revolution. University of Arizona Press, Tucson.

Meffe, G. K. and C. R. Carroll et al. 1997. Principles of conservation biology. Second edition. Sinauer Associates, Sunderland, Mass.

Miller, P. 1964. Errand into the wilderness. Harper and Row, New York.
Muir, J. 1916. A thousand mile walk to the Gulf. Houghton Mifflin, Boston.
Nash, R. 1982. Wilderness and the American mind. Third edition. Yale University Press, New Haven, Conn.
Olson, S. 1998. Why wilderness? Pages 97–102 in Callicott and Nelson, editors. The great new wilderness debate. University of Georgia Press, Athens.
Pearsons, G. A. 1922. The preservation of natural areas in the national forest. Ecology 3: 284–287.
Peters, C. M., A. H. Gentry, and R. O. Mendelsohn. 1989. Valuation of an Amazonian rainforest. Nature 339: 656–657.
Pickett, S. T. A. and P. S. White. 1985. The ecology of natural disturbance and patch dynamics. Academic Press, New York.
Plumwood, V. 1998. Wilderness skepticism and wilderness dualism. Pages 652–690 in Callicott and Nelson, editors. The great new wilderness debate. University of Georgia Press, Athens.
Public Law 88-577. 1964. The wilderness act.
Pyne, S. J. 1982. Fire in America: A cultural history of wildland and rural fire. Princeton University Press, Princeton, N.J.
Pyne, S. J. 1997. Burning bush: A fire history of Australia. University of Washington Press, Seattle.
Robinson, S. K., F. R. Thompson III, T. M. Donovan, D. R. Whitehead, and J. Faaborg. 1995. Regional forest fragmentation and the nesting success of migratory birds. Science 267: 1987–1990.
Rolston III, H. 1998. The wilderness idea reaffirmed. Pages 367–386 in Callicott and Nelson, editors. The great new wilderness debate. University of Georgia Press, Athens.
Scott, J. M. 1999. A representative biological reserve system for the United States. Society for Conservation Biology Newsletter 6 (2): 1, 9.
Shelford, V. 1920. Preserves of natural conditions. Transactions of the Illinois State Academy of Science 13: 37–58.
Shelford, V. 1933. The preservation of natural biotic communities. Ecology 14: 240–245.
Snyder, G. 1996. Nature as seen from Kitkitdizze is no "social construction." Wild Earth 6 (4): 8–9.
Soulé, M. E. 1985. What is conservation biology? Bioscience 35: 727–734.
Standing Bear, L. 1998. Indian wisdom. Pages 201–206 in Callicott and Nelson, editors. The great new wilderness debate. University of Georgia Press, Athens.
Sumner, F. 1920. The need for a more serious effort to rescue a few fragments of vanishing nature. Scientific Monthly 10: 236–248.
Sumner, F. 1921. The responsibility of the biologist in the matter of preserving natural conditions. Science 54: 39–43.

Thomas, E. M. 1990. Reflections: The old way. New Yorker, October 15: 68–80.
Turnbull, C. M. 1972. The mountain people. Simon and Shuster, New York.
Whitelock, D. 1985. Conquest to conservation. Wakefield Press, Melbourne.
Wright, G., B. Thompson, and J. Dixon. 1933. Fauna of the national parks of the United States: A preliminary survey of faunal relations in national parks. Government Printing Office, Washington, D.C.

Dave Foreman

The Real Wilderness Idea (2000)

I COME NOT TO PRAISE the Received Wilderness Idea, but to bury it. The very name, "the Received Wilderness Idea," conjures up a mystical origin. If the Wilderness Idea that Baird Callicott, Bill Cronon and other postmodern deconstructionist scholars so eagerly banish with Milton's Lucifer has been received, I think it has been received as they hold hands in a darkened room around a séance table, trying to hear voices from the misty shades of Jonathan Edwards and Henry David Thoreau.

But, first, why should you lend me your ears on the idea of wilderness? Well, it's because I'm an expert on the Real Wilderness Idea—the one that created the National Wilderness Preservation System. I've been a wilderness back packer for 40 years, a wilderness river runner for more than 30. During the several thousand days and nights I've spent in wilderness for fun and for conservation, I've had a few hundred companions (not all at once!). I've heard their thoughts about wilderness while plodding up dusty switchbacks, floating past canyon walls aglow in sunset flame, and passing Scotch around the campfire. On many of these trips, my friends and I were checking out the wilderness qualities of unprotected areas and putting together boundary proposals to send to Congress for designation. In the 1970s, I wrote a widely-used guide, "How to Do a Wilderness Study." From all this, I got a very clear idea of wilderness, one that is widely shared with other conservationists doing the same thing.

In 1971, as I dove into wilderness issues in New Mexico, I found a complete set of the Wilderness Society's magazine, *The Living Wilderness,* in the basement of the University of New Mexico library. I read every issue all the way back to the first ones in the 1930s. During the early 1960s, *The Living Wilderness* covered the campaign for the Wilderness Act in great detail, including the arguments for and against wilderness protection. Since then I have read uncounted magazines, newsletters, and action alerts from many wilderness protection groups. I have read dozens upon dozens of brochures and maps about wilderness areas from government agencies.

My mentors in the conservation movement were people who had led the campaign for the Wilderness Act and later efforts to protect mandate areas (Forest Service Primitive Areas and National Park and Wildlife Refuge roadless areas) and Forest Service roadless areas. I was trained as a grass-roots organizer by Clif Merritt, who organized Westerners to support the Wilderness Act, Ernie Dickerman, who wrote the Eastern Wilderness Areas Act, and Harry Crandell, who wrote the wilderness provision for the BLM organic act. Dave Brower, Ed and Peggy Wayburn, Stewart Brandborg, and Celia Hunter taught me about wilderness battles stretching back to the 1930s. I talked at length with old-timers in Silver City, New Mexico, who had led the successful citizen fight against the Forest Service's proposed dismembering of the Gila Wilderness in 1952 (to allow logging). I have been privileged to know Bob Marshall's brothers, Aldo Leopold's daughter, Mardie Murie (Olaus Murie's widow), and Sig Olson. I applied their experience and wisdom when I became a national leader in the wilderness campaigns on RARE II, the BLM wilderness review, and the Alaska Lands Act.

I have sat through dozens of public hearings—agency and congressional, field and DC—about wilderness area designation. I believe I have known people involved in every wilderness designation bill passed by Congress. For 30 years, I have been involved in strategy meetings and public presentations about wilderness areas in nearly every state. During the past 15 years, I have given more than 200 lectures about wilderness at colleges in 35 states and Canadian provinces and afterwards discussed wilderness with small groups of students at local bars. I have stood with Earth First!ers, risking arrest and physical injury in nonviolent civil disobedience, to protect wilderness from bulldozers and chain saws. I have attended a dozen professional meetings on wilderness organized by federal and state agency wilderness managers, and I know key wilderness people in the agencies.

In my personal archives are three shelf-feet of congressional hearing records and committee reports on wilderness area designation; every Forest Service primitive area, Park Service and national wildlife refuge wilderness area recommendation document; every RARE II state document; every BLM wilderness study document for each of the Western states; the responses by conservation groups to all of these; and 23 file drawers of wilderness area issues dating back to the 1960s (this does not count a similar number of file drawers on other conservation issues). Believe it or not, I have read all of this stuff.

During 20 years as an editor, executive editor, or publisher of the *Earth First! Journal* (1980 to 1988) and *Wild Earth* (1990 to the present), I have read, rejected, accepted, and edited more wilderness articles than I want to remember from all over North America and the world. I spent eight years researching my book (with Howie Wolke) on lower-48 roadless areas, *The Big Outside*.

During the past 15 years, I have been closely involved with the key conservation biologists working on protected area design and protection strategy. My wilderness work and close colleagues now reach into Mexico, Costa Rica, Canada, Chile, Argentina, and southern Africa.

I have been personally involved in defending unprotected wilderness from dam building, water diversion, logging, road building, hard rock mining, oil and gas exploration and development, uranium mining, off-road vehicle abuse, poaching of reintroduced wolves, overgrazing, juniper chaining, observatory construction, and introduction of exotic species. I have helped defend designated wilderness areas from dam building, overgrazing, grazing developments, administrative vehicle use, non-commercial logging, government predator killing, sabotage of endangered species recovery (Gila trout), mountain bike and snowmobile invasion. We conservationists have not always been successful in this defense, and I know wild rivers now drowned behind dams, grand forests clear-cut, stunning badlands strip-mined.

In short, I know something about the only wilderness idea that matters on the ground—the one that has led thousands of people to devote their time, money, and sometimes their freedom and even lives to protect wilderness from exploitation. This is the Wilderness Idea that has created the National Wilderness Preservation System of the United States of America.

This Real Wilderness Idea is very different from the Received Wilderness Idea invented and then lambasted by Baird Callicott, Bill Cronon and other deconstructionist social scientists. The literary and philosophical writings they draw from have had little influence in the wilderness protection movement; in fact, intellectual and academic discussions about wilderness have pretty much been ignored by wilderness defenders. Since 1920, wilderness conservationists have been motivated primarily by two things: One, they like a particular wilderness; two, they see a need to protect it from development and exploitation.

As Samuel Hays (1996), the great historian of resource conservation, Nature conservation, and environmentalism, writes, "Cronon's wilderness is a world of abstracted ideas . . . but divorced from the values and ideas inherent in wilderness action."

This Received Wilderness Idea is a straw dog; it does not exist on the ground. It is not the idea of wilderness that led to the Wilderness Act and the National Wilderness Preservation System and spurred thousands of citizen conservationists from Alabama to Alaska. When one fights a phantom, it is easy to claim you have mortally wounded the monster.

Twenty-five hundred years ago, Socrates told Phaedrus, "I'm a man of learning and trees and open space teach me nothing, while men in towns do." More recently, Nobel Laureate Linus Pauling (1995) wrote,

> I remember reading a book on philosophy in which the author went on, page after page, on the question: If there is a leaf on a tree and you see that it is green in the springtime and red in fall, is that the same leaf or is it a different leaf? Is the essence of leafness still in it? Words, words, words, but 'chlorophyll' and 'xanthophyll'—which are sensible in this connection of what has happened to that leaf—just don't appear at all.

This so-called Received Wilderness Idea comes from Socrates and his buddies in town, not from the wilderness of trees and open country. And among all the words about the Received Wilderness Idea—words about living landscapes and the political reality that threatens them—don't appear.

I have spent my life fighting the lies, blather, and myths of extractive industry about wilderness. I have concluded that their pitiful arguments against wilderness are actually more legitimate, rational, and grounded in reality than those of the postmodern deconstructionists.

I am not going to respond point by point to the academic left's complaints about wilderness. I've done it before, most recently in the Callicott/Nelson anthology, *The Great New Wilderness Debate,* and I have not noticed anyone rebutting my specific points (Foreman 1998). (I will, however, respond in detail in my book-in-progress, *The War on Nature.*) What I would like to do is present not the Received Wilderness Idea, but the Real Wilderness Idea of the citizen conservation movement and how it is still robust after all these years, blending both experiential and ecological values and purposes.

SELF-WILLED LAND

In our slacker era, when rigor in thought and ethics is too much to ask for, we often get into a snarl because of poorly defined words. Bud Man on his motorized tricycle, academic grandees, and just about everybody in between use the word *wilderness* in sloppy ways, muddying the wrangle about conservation.

In a 1983 talk at the Third World Wilderness Conference in Scotland, philosopher Jay Hansford Vest sought the meaning of wilderness in Old English and further back in Old Gothonic languages. He showed that wilderness means "'self-willed land' . . . with an emphasis on its own intrinsic volition." He interpreted *der* as *of the.* "Hence, in wil-der ness, there is a 'will-of-the-land'; and in wildeor, there is 'will of the animal.' A wild animal is a 'self-willed animal'—an undomesticated animal; similarly, wildland is 'self- willed land.'" Vest shows that this willfulness is opposed to the "controlled and ordered environment which is characteristic of the notion of civilization." The early northern Europeans were not driven to lord over Nature; thus, wilderness "demonstrates a recognition of land in and for itself" (Vest 1985). Thanks to Vest, we are able to understand that this word, wilderness, is not a coinage of modern civilization; it is a word brewed by pagan barbarians of the Bronze and Iron Ages.

This self-willed land meaning of wilderness overshadows all others. Wilderness means land beyond human control. Land beyond human control is a slap in the face to the arrogance of humanism—elitist or common man, capitalist or socialist, first worlder or third; for them, it is also something to be feared.

I've called wilderness areas the arena of evolution. However, Aldo Leopold, as usual, was way ahead of me. Fifty years ago, he saw wilderness as the "theater" for the "pageant of evolution" (Leopold 1987). Evolution is self-willed. The land where evolution can occur is self-willed land, especially for large species.

THE WILDERNESS ACT

The civilized world's greatest embrace of self-willed land came in the form of the 1964 Wilderness Act in the United States. This legislation was the product of eight years of discussion and revision in Congress and in public hearings across the nation. It was pushed by hikers, horse packers, canoeists, hunters and fishers. It contains at least four definitions of wilderness. I believe that all four of these definitions are thoroughly in keeping with self-willed land. The first definition of wilderness is found in the statement of purpose for the Wilderness Act in Section 2(a):

> In order to assure that an increasing population, accompanied by expanding settlement and growing mechanization, does not occupy and modify all areas within the United States and its possessions, leaving no lands designated for preservation and protection in their natural condition, it is hereby declared to be the policy of the Congress to secure for

> the American people of present and future generations the benefits of an enduring resource of wilderness.

Was Congress, prodded by American citizens, setting up a National Wilderness Preservation System to preserve a mythical past wrapped up in literary romanticism, Manifest Destiny bravado, and Calvinist dualism? Well . . . no. It was much simpler. Wilderness areas needed to be protected because all of the remaining backcountry of the United States was threatened with development and industrial exploitation driven by population growth, mechanization, and expanding settlement. Here and throughout the wilderness conservation movement, the motive force has been to protect land from development. Hays (1996) writes, "[W]ilderness proposals are usually thought of not in terms of perpetuating some 'original' or 'pristine' condition but as efforts to 'save' wilderness areas from development." Wilderness areas, then, are lands protected from industrial civilization's conquest. Isn't that easy to understand?

The second definition is the ideal:

> A wilderness, in contrast with those areas where man and his works dominate the landscape, is hereby recognized as an area where the earth and its community of life are untrammeled by man, where man himself is a visitor who does not remain. Section 2(c).

Written by Howard Zahniser of the Wilderness Society, who, as a professional editor and writer, understood the importance of word selection, this definition agrees with self-willed land. First, wilderness is not where the works of man dominate the landscape. It is not under human will. Second, Zahniser chose the obscure word "untrammeled" carefully, and not just because it rolls off the tongue pleasantly. A *trammel* is a fish net and also a hobble for a horse, thus a thing that hinders free action. As a verb, *trammel* means to hinder the action of something. *Untrammeled,* then, means that the will of something is not hobbled; it is self-willed. Untrammeled land is the arena of evolution. Third, humans are only visitors in wilderness; there are no permanent human settlements. Many kinds of wilderness foes especially bristle at this barring of human habitation. However, I believe this lack of long-lasting settlement is key to wil-der-ness. Where humans dwell long, we trammel or hinder the willfulness of the land around our living sites and outward. How far? This hinges on the population size and technological sophistication of the group.

The third definition of wilderness immediately follows the second. It is the specific, practical definition of wilderness areas protected by the Wilderness Act and sets out the entry criteria for candidate areas:

> An area of wilderness is further defined to mean in this Act an area of undeveloped Federal land retaining its primeval character and influence, without permanent improvements or human habitation, which is protected and managed so as to preserve its natural conditions and which (1) generally appears to have been affected primarily by the forces of nature, with the imprint of man's work substantially unnoticeable; (2) has outstanding opportunities for solitude or a primitive and unconfined type of recreation; (3) has at least five thousand acres of land or is of sufficient size as to make practicable its preservation and use in an unimpaired condition; and (4) may also contain ecological, geological, or other features of scientific, educational, scenic, or historical value. Section 2(c).

Although in keeping with self-willed land ("undeveloped," "primeval character and influence," "without permanent improvements or human habitation," "natural conditions"), this is a practical definition that acknowledges that even mostly self-willed land may not be pristine ("generally appears," "affected primarily," "substantially unnoticeable"). Indeed, the word *pristine* does not appear in the Wilderness Act.

This down-to-earth view of wilderness answers the often silly question, "What is natural?" It understands that *natural* is not a single point opposed to the single point of *unnatural.* Rather, I think it sees that land falls on a continuum from wholly yoked by human will to altogether self-willed. At some point, land quits being mostly dominated by humans; at some other point, land begins to be controlled primarily by the forces of Nature. There is a wide gray area in between, where human and natural forces both have some sway. After natural forces become dominant, the land is self-willed. Because we humans have limited and differing understandings of ecology and depths of wisdom, we may find the changeover to self-willed land in different places on this unnatural-natural line. But this does not mean we cannot say, "This place is primarily natural." And let us not fall into the woolly-headed trap of thinking that naturalness is merely a human idea. Naturalness exists out there. A falling tree in a forest does not need a human ear to be.

Ecological wounds suffered by the land come from humans trying to impose their will. The severity of these wounds and their full impact settle whether the land is mostly self-willed (affected primarily by the forces of Nature) or not. Some kinds of wilderness foes falsely believe that conservationists see wilderness as pristine (an absolute word). Other anti-conservationists, in order to limit protection, argue that places must be pristine in order to qualify as wilderness areas. Neither gospel is true.

If we read Section 2(c) of the law closely, we see that there are really two definitions of wilderness twined about each other. One is a definition of the human experience in wilderness areas ("appears," "unnoticeable," "solitude," "a primitive and unconfined type of recreation," "educational," "historic," "scenic"). The other is an ecological definition ("undeveloped," "primeval character and influences," "forces of nature," "ecological," "scientific"). Understanding that these descriptions of ecological conditions and values are prominent in the Wilderness Act belies the persistent rap that the act and the National Wilderness Preservation System created by it are only about scenery and recreation. Even some conservationists and scientists have criticized the Wilderness Act for an overwhelming recreational bias. It's important to understand that this is not the aim of the act, although federal agencies have often managed wilderness areas as if it were.

The two lessons we need to draw from Section 2(c) are that wilderness areas are not expected to be pristine and that the ecological values of wilderness areas are strongly recognized along with experiential values.

The fourth definition of wilderness comes with rules for managing land after it comes under the protection of the Wilderness Act:

> Except as specifically provided for in this Act, and subject to existing private rights, there shall be no commercial enterprise and no permanent road within any wilderness area designated by this Act and except as necessary to meet minimum requirements for the administration of the area for the purposes of this Act (including measures required in emergencies involving the health and safety of persons within the area), there shall be

> no temporary road, no use of motor vehicles, motorized equipment or motorboats, no landing of aircraft, no other form of mechanical transport, and no structure or installation within any such area. Section 4(c).

(Elsewhere, the Wilderness Act provides for certain exceptions to the above prohibitions, such as firefighting, rescue, livestock grazing and prospecting for minerals until 1984, all of which were political compromises that supporters of the Wilderness Act had to make before Western members of Congress would allow passage. Thus, the Wilderness Act is somewhat flawed and sometimes at odds with itself.)

The use prohibitions try to keep the land untrammeled (self-willed). They are more strict than the entry criteria in Section 2(c). For example, there is no requirement that candidate wilderness areas have to be roadless or unlogged, but Section 4(c) holds that they must be managed as roadless after they are placed in the National Wilderness Preservation System. In other words, existing roads must be closed and no further commercial logging allowed after designation of an area as wilderness. There are many cases of once-roaded or earlier-logged areas in the National Wilderness Preservation System—including some of the classic big wilderness areas in the West.

If what wilderness means and what the Wilderness Act says are clearly worded, many misunderstandings about wilderness should melt away. However, as we too often find, muddying the meaning of wilderness is not always due to simple ignorance, but is a witting tactic by anti-conservationists.

The brawl over conservation is at heart about whether we can abide self-willed land.

THE RIVER WILD

In "Rewilding and Biodiversity: Complementary Goals for Continental Conservation," Michael Soulé and Reed Noss (1998) clearly show that science-based Nature-reserve design does not come to bury traditional wilderness area designation, but to marry it. To see how this is so, we need both a lookout that takes in the whole conservation movement and a metaphor that can limn it.

The metaphor I use for the conservation movement is that of a river's watershed, with streams dropping from high saddles and cirques and flowing down to mix as currents in the river. A good perspective is that of an eagle, which allows us to see the watershed spread out before us.

The headwater streams that flow together to make the River Wild are wildlife protection, stewardship, beauty protection, and forest protection. Downriver, the streams of wilderness protection, ecosystem representation, carnivore protection, connectivity, and rewilding flow in. Nearby, but apart, are watersheds for the rivers of resourcism and environmentalism. I see environmentalism (pollution fighting), conservation (wildlife and wildlands protection), and resourcism (efficient exploitation of resources) as separate movements, with different views about humans and Nature. Some of the headwaters of the Resourcism River come off the same ridges and peaks as those that feed the River Wild, but they flow in a different direction. The Environmentalism River does not spring from the same divides as the River Wild, but its course later runs parallel to the River Wild, with only a low ridge between the two.

All the streams feeding into the conservation movement spring from protecting land and wildlife from threats of development and exploitation.

From the farthest mountain pass flows the sturdy stream of Wildlife Protection. Contrary to the common wisdom, American conservation began with wildlife protection, not with forest protection. English aristocrat William Henry Herbert came to America in 1831 and brought with him the "code of the sportsman." In his woodsy role as "Frank Forester," Herbert fought the era's rapacious market hunting and spurred sportsmen to band together to fight game hogs. National hunting magazines began in the 1870s, and they joined the battle against commercial exploitation of game and fish and for habitat protection. Sport hunters and their magazines raised a din against the senseless slaughter of the buffalo. The first national conservation group was not the Sierra Club (founded in 1892), but the Boone and Crockett Club, founded in 1887 by Theodore Roosevelt and his fellow hunters. The role of Boone and Crockett in creating the first national parks, wildlife refuges, and forest reserves has generally been overlooked by historians as well as by today's conservationists (Reiger 1990).

The second headwater stream is that of Stewardship. One of the most remarkable Americans of the 19th century was Vermont's George Perkins Marsh. As Lincoln's ambassador to Turkey and later Italy, Marsh took in the sights of the Mediterranean, where among the ruins of classical civilizations he found ruins of the land. The rocky, treeless hills of Greece were as much a testament to a fallen civilization as the crumbling Acropolis. His 1864 book, *Man and Nature; or, Physical Geography as Modified by Human Action,* is one of the benchmarks of both history and science. He wrote, "But man is everywhere a disturbing agent. Wherever he plants his foot, the harmonies of nature are turned to discord." Former *New York Times* foreign correspondent and later environmental reporter Phillip Shabecoff (1993) writes, "Marsh was the first to demonstrate that the cumulative impact of human activity was not negligible and, far from benign, could wreak widespread, permanent destruction on the face of the earth." However, I also see a spring called Malthus contributing to the flow in the Stewardship Creek. Stewardship is needed to combat soil erosion and other careless land management; more recently, it has tried to deal with the threats of human population growth and depletion of resources.

The third headwater stream is Beauty—protection of national parks and other places to safeguard their spectacular, inspiring scenery. Yosemite Valley in the Sierra Nevada of California was not discovered by white settlers until 1851, and the mighty sequoias near it were not described until 1852. Within a few years, both were attracting visitors who wanted to see their splendor. In 1859, Horace Greeley, editor of the *New York Tribune,* visited the Yosemite Valley and wrote to his readers that it was "the most unique and majestic of nature's marvels" (Runte 1987). Five years later, on June 30, 1864, taking time from the burden of the Civil War, President Abraham Lincoln signed a bill transferring beautiful Yosemite Valley and the Mariposa Grove of sequoias to the state of California as a public park.

American citizens supported setting aside Yellowstone, Yosemite and the other early national parks primarily because of beauty, although other factors, including the support of railroads, helped lead to the political decisions. Conservationists feared that all of America's natural wonders were threatened by tawdry tourist development and industrial exploitation because of what had happened to Niagara Falls from 1830 on. Alfred Runte (1987) writes, "In the fate of Niagara Falls, Americans found a compelling reason to give preservation more than a passing thought. . . . A continuous parade of European visitors and commentators embarrassed the nation by condemning the commercialization of Niagara." This all holds true for the closely related national parks movement in Canada.

The fourth and final headwater stream is Forest Protection. It falls out of a cirque-held tarn, but cascades only briefly before a great sharp ridge splits the stream. One side pours off into the Resourcism River with Gifford Pinchot and the other falls into the River Wild with John Muir. In the 1880s, business interests in New York City called for protecting the Adirondacks to ensure a good water supply from the headwaters of the Hudson River. In the West, irrigators and towns worried about watershed destruction by overgrazing and logging in the high country and asked for protection. Forest lovers, led by John Muir, feared that all natural forests would soon be scalped by logging companies. New York protected state lands in the Adirondacks, and Congress authorized the President to withdraw forested lands in the West.

The 1891 Forest Reserve Act "merely established reserves; it did not provide for their management," explains Samuel Hays (1979). Conservationists ranging from Muir to the sportsmen of the Boone and Crockett Club hoped to keep the forest reserves off-limits to commercial logging, grazing and other uses. They wanted the reserves protected for their watershed, recreational and scenic values, as well as for wildlife habitat. Gifford Pinchot, however, demanded "management" that would include logging, grazing, and dam building. The 1897 Organic Act, which Pinchot pushed, opened the reserves for commercial exploitation. However, for both Muir and Pinchot, forest protection was a response to the threat of uncontrolled and wasteful logging.

Down the River Wild another stream—Wilderness—comes in. The specific movement to preserve wilderness areas came first from Forest Service rangers, such as Art Carhart and Aldo Leopold. Leopold, who railed against "Ford dust" in the backcountry, feared that growing automobile access to the national forests would destroy and replace the pioneer skills of early foresters. He wanted to protect the experience he enjoyed when he came to Arizona's Apache National Forest in 1909. "Wilderness areas are first of all a series of sanctuaries for the primitive arts of wilderness travel, especially canoeing and packing," said Leopold (1987). In 1921, he defined wilderness as "a continuous stretch of country preserved in its natural state, open to lawful hunting and fishing, big enough to absorb a two weeks' pack trip, and kept devoid of roads, artificial trails, cottages, or other works of man" (Leopold 1921). The backcountry was threatened by automobiles and roads. It needed protection. In the 1930s, conservationists like Bob Marshall called for wilderness protection in the national parks because the parks were threatened by proposals for scenic highways from the National Park Service and the tourist industry.

On the other side of the River Wild, just below the confluence with the Wilderness stream, the Ecological Representation stream joins in. As early as 1926, the *Naturalist's Guide to the Americas,* edited by prominent biologist Victor Shelford, called for protecting ecologically representative natural areas. Both the National Audubon Society and the Nature Conservancy have tried to buy and protect ecosystems not represented in federal and state protected areas. The National Park Service and conservationists have tried to establish national parks for all major ecosystems, admittedly without total success. The 1975 Eastern Wilderness Areas Act, which established wilderness areas on national forests east of the Rockies, was explicitly about ecosystem representation. During RARE II, the Forest Service, with conservationist support, sought to establish new wilderness areas that would protect hitherto unprotected ecosystems. The push here came because of development threats. Ecosystem representation, however, has not gotten the heed it needs. In a special report for the Department of the Interior, Reed Noss and his co-authors (1995) have detailed our poor record in protecting representative ecosystems.

Soon after, the Predator Protection stream splashes down as a stunning waterfall. In "A Nature Sanctuary Plan" unanimously adopted by the Ecological Society of America on December 28, 1932, Victor Shelford wrote, "Biologists are beginning to realize that it is dangerous to tamper with nature by introducing plants or animals, or by destroying predatory animals or by pampering herbivores. . . ." The Ecological Society said we needed to protect whole assemblages of native species, including large carnivores, and the natural fluctuations in numbers of species (Shelford 1933). At that time, protecting wolves and mountain lions was—well, bold, hence my seeing it as a waterfall. Large carnivores were clearly threatened with extirpation from the United States, including from the national parks.

Another conservation stream began in the 1960s with work by E. O. Wilson and Robert MacArthur on island biogeography. Closely tied to island biogeography is the *species-area relationship.* Michael Soulé (1995) writes, "One of the principles of modern ecology is that the number of species that an area can support is directly proportional to its size. A corollary is that if area is reduced, the number of species shrinks." The species-area relationship has been shown with birds, mammals, reptiles and other kinds of animals on the Greater Sunda Islands (the Indonesian archipelago), Caribbean islands and elsewhere. An ecological rule of thumb is that if a habitat is reduced 90 percent, it will lose 50 percent of its species.

In 1985, University of Michigan ecologist William Newmark looked at a map of the western United States and Canada and realized that our national parks were islands. As the sea of settlement and logging swept over North America, national parks became islands of ecological integrity surrounded by human-dominated lands. Did island biogeography apply?

Newmark found that the smaller the national park and the more isolated it was from other wildlands, the more species it had lost. The first species to go had been the large, wide-ranging critters—such as lynx and wolverine. Loss of species (*relaxation* in ecological lingo) had occurred *and was still occurring.* Newmark (1987) predicted that all national parks would continue to lose species (as Soulé had previously predicted for East African reserves). "Without active intervention by park managers, it is quite likely that a loss of mammalian species will continue as western North American parks become increasingly insularized." Even Yellowstone National Park isn't big enough to maintain viable populations of all the large wide-ranging mammals. Only the total area of the connected complex of national parks in the Canadian Rockies is substantial enough to ensure their survival.

Bruce Wilcox and Dennis Murphy (1985) wrote that "habitat fragmentation is the most serious threat to biological diversity and is the primary cause of the present extinction crisis." Reed Noss, then at the University of Florida, acted on their warning by designing a conceptual Nature reserve system for Florida consisting of core reserves surrounded by buffer zones and linked by habitat corridors. In a paper presented to the 1986 Natural Areas Conference, Noss (1987) said, "The problems of habitat isolation that arise from fragmentation can be mitigated by connecting natural areas by corridors or zones of suitable habitat."

This connectivity stream came into being because of fragmentation threats by dams, highways, clearcutting, and other development.

Those of us who float rivers know that it can take a long time before the water from an incoming stream mixes fully with the main current. We see this when a creek full of glacial milk dumps into the gin-clear waters of a river in the Yukon. A similar scene occurs in the Southwest when a clear mountain stream plunges into a red river full of silt. For miles, there may be two currents shown by their distinct tints.

So it has been with our river. The wildlife protection, stewardship, beauty, forest protection, and wilderness streams mixed fairly well, but the currents of ecosystem representation, predator protection, and connectivity did not mix as well.

Now a new stream—Rewilding—has entered. Unlike the other currents, this rewilding stream mixes all the other currents together into a deep, wide, powerful river.

Soulé and Noss (1998) "recognize three independent features that characterize contemporary rewilding:

- Large, strictly protected core reserves (the wild)
- Connectivity
- Keystone species."

In shorthand, these are "the three C's: Cores, Corridors, and Carnivores."

This rewilding approach is built on recent scholarship showing that ecosystem integrity often depends on the functional presence of large carnivores. Michael Soulé and his graduate students (1988) have shown that native songbirds survive in large suburban San Diego canyons where there are coyotes; they disappear faster when coyotes disappear. Coyotes eat foxes and prowling house cats. Foxes and cats eat quail, cactus wrens, thrashers and their nestlings.

In the East, David Wilcove, staff ecologist for the Environmental Defense Fund, has found that songbirds are victims of the extirpation of wolves and cougars. As we have seen, the population decline of songbirds as a result of forest fragmentation is well documented, but Wilcove (1986) has shown that songbird declines are partly due to the absence of large carnivores in the East. Cougars and gray wolves don't eat warblers or their eggs, but raccoons, foxes, skunks and possums do, and the cougars and wolves eat these midsize predators. When the big guys were hunted out, the populations of the middling guys exploded—with dire results for the birds. Soulé calls this phenomenon—mid-sized predators multiplying in the absence of large predators—*mesopredator release*.

John Terborgh of Duke University (in my mind the dean of tropical ecology) is currently studying the ecological effects of eliminating jaguars, pumas, and harpy eagles from tropical forests. He tells us that large carnivores are major regulators of prey species numbers—the opposite of once-upon-a-time ecological orthodoxy. He has also found that the removal or population decline of large carnivores can alter plant species composition, particularly the balance between large- and small-seeded plants, due to increased seed and seedling predation by superabundant herbivores that are normally regulated by large carnivores. This is called *top-down regulation* (Soulé and Noss 1998). There is compelling evidence for such top-down regulation in forests outside the tropics as well.

Rewilding is "the scientific argument for restoring big wilderness based on the regulatory roles of large predators," according to Soulé and Noss.

> Three major scientific arguments constitute the rewilding argument and justify the emphasis on large predators. First, the structure, resilience, and diversity of ecosystems is often maintained by 'top-down' ecological (trophic) interactions that are initiated by top predators (Terborgh 1988, Terborgh et al. 1999). Second, wide-ranging predators usually require large cores of protected landscape for foraging, seasonal movements, and other needs; they justify bigness. Third, connectivity is also required because core reserves are typically not large enough in most regions; they must be linked to insure long-term viability of wide-ranging species. . . . In short, the rewilding argument posits that large predators are often instrumental in maintaining the integrity of ecosystems. In turn, the large predators require extensive space and connectivity (Soulé and Noss 1998).

If large native carnivores have been extirpated from a region, their reintroduction and recovery is central to a conservation strategy. Wolves, grizzlies, cougars, lynx, wolverines, black bears, jaguars and other top carnivores need to be restored throughout North America in their natural ranges.

Although Soulé and Noss (1998) state, "Our principal premise is that rewilding is a critical step in restoring self-regulating land communities," they claim two nonscientific justifications: (1) "the ethical issue of human responsibility," and (2) "the subjective, emotional essence of 'the wild' or wilderness. Wilderness is hardly 'wild' where top carnivores, such as cougars, jaguars, wolves, wolverines, grizzlies, or black bears have been extirpated. Without these components, nature seems somehow incomplete, truncated, overly tame. Human opportunities to attain humility are reduced."

What Soulé and Noss have done here is of landmark importance for the wilderness conservation movement as well as for those primarily concerned with protecting biological diversity. They have developed the *scientific basis* for the need for big wilderness area complexes. Here, science buttresses the wants and values of wilderness recreationists. Big wilderness areas are not only necessary for inspiration and a true wilderness experience, but they are absolutely necessary for the protection and restoration of ecological integrity, native species diversity and evolution. Elsewhere, Soulé calls wilderness areas self-regulated, another way of saying self-willed or untrammeled.

Metaphors are never perfect, but this view of conservation as the watershed of the River Wild, with different side streams adding power, diversity, and nutrients, is pretty darn good. It allows us to see that new streams did not replace old streams. It recognizes that the headwater streams that initially formed the River Wild did not disappear when new streams flowed in. It shows the compatibility of the "scientific" streams with the aesthetic and recreational streams. And it proves that the threat of destruction drove all of these conservation currents.

Wilderness and biodiversity conservation are not airy-fairy flights of romantic fantasy to recapture a mythical past of purity and goodness, but real-world efforts to protect self-willed land from damage by increasing population, expanding settlement, and growing mechanization.

REFERENCES

Foreman, Dave. 1998. Wilderness areas for real. In: Callicott, J. Baird; Nelson, Michael P., eds. *The great new wilderness debate.* Athens, GA: The University of Georgia Press: 395–407.

Hays, Samuel P. 1979. *Conservation and the gospel of efficiency: the progressive conservation movement 1890–1920.* New York: Atheneum. 297 p.

Hays, Samuel P. 1996. The trouble with Bill Cronon's wilderness. *Environmental History.* 1(1): 29–32.

Leopold, Aldo. 1921. The wilderness and its place in forest recreational policy. *The Journal of Forestry.* 19(7): 718–721.

Leopold, Aldo. 1987. *A sand county almanac.* Oxford, UK: Oxford University Press. 228 p.

Newmark, William D. 1987. A land-bridge island perspective on mammalian extinctions in western North American parks. *Nature.* 325: 430–432.

Noss, Reed F. 1987. Protecting natural areas in fragmented landscapes. *Natural Areas Journal.* 7(1): 2–13.

Noss, Reed F.; LaRoe, Edward T., III; Scott, J. Michael. 1995. Endangered ecosystems of the United States: a preliminary assessment of loss and degradation. Washington, DC: USDI National Biological Service. Biological Report 28. 58 p.

Pauling, Linus. 1995. *Science.* 270: 1236.

Reiger, John F. 1990. The sportsman factor in early conservation. In: Nash, Roderick Frazier, ed. *American environmentalism: readings in conservation history.* New York: McGraw-Hill Publishing Company: 52–58.

Runte, Alfred. 1987. *National parks: the American experience second edition revised.* Lincoln, NE: University of Nebraska Press. 335 p.
Shabecoff, Philip. 1993. *A fierce green fire: the American environmental movement.* New York: Hill and Wang: 55–59.
Shelford, Victor E. 1933. The preservation of natural biotic communities. *Ecology.* 14(2): 240–245.
Soulé, Michael E.; Boulger, D. T.; Alberts, A. C.; Sauvajot, R.; Wright, J.; Sorice, M.; Hill, S. 1988. Reconstructed dynamics of rapid extinctions of chaparral-requiring birds in urban habitat islands. *Conservation Biology.* 2(1): 75–92.
Soulé, Michael E. 1995. An unflinching vision: networks of people defending networks of land. In: Saunders, D. A.; Craig, J. L.; Mattiske, E. M., eds. *Nature conservation 4: the role of networks.* Surrey: Beatty & Sons: 1–8.
Soulé, Michael E.; Noss, Reed F. 1998. Rewilding and biodiversity: complementary goals for continental conservation. *Wild Earth.* 8(3): 18–26.
Terborgh, John. 1988. The big things that run the world—a sequel to E. O. Wilson. *Conservation Biology.* 2(4): 402–403.
Terborgh, John; Estes, J. A.; Paquet, P.; Rails, K.; Boyd-Heger, D.; Miller, B. J.; Noss, R. F. 1999. The role of top carnivores in regulating terrestrial ecosystems. In: Soulé, Michael E.; Terborgh, John, eds., *Continental conservation design and management principles for long-terra, regional conservation networks.* Washington, DC: Island Press: 39–64.
Vest, Jay Hansford C. 1985. Will of the land. *Environmental Review.* 9(4): 321–329.
Wilcove, David S.; McLellan, C. H.; Dobson, A. P. 1986. Habitat fragmentation in the temperate zone. In: Soulé, Michael E., ed. *Conservation biology: the science of scarcity and diversity.* Sunderland, MA: Sinauer: 237–256.
Wilcox, Bruce A.; Murphy, Dennis D. 1985. Conservation strategy: the effects of fragmentation on extinction. *American Naturalist.* 125: 879–887.

Jill M. Belsky

Changing Human Relationships with Nature (2000, 2008)

Making and Remaking Wilderness Science

IN THIS PAPER I DISCUSS two themes that I, as an environmental sociologist, view as pivotal with regard to changing ideas of nature and wilderness (social) science and their implications for the practice of conserving and managing large ecosystems.

The first theme centers around to what extent wilderness studies has been limited by an ontological tension between those who approach the relationship between humans and nature on the basis of material factors and constraints, or through concepts and ideas. There is a dialectical tension that manifested itself in tension and polarization in the talks of Baird Callicott, Dave Foreman, and most strongly Gary Snyder. Callicott emphasized how our ideas of nature and wilderness have changed over time (i.e., a social constructionist or "idealist" approach), whereas Foreman spoke about the physical threats to wild nature and wilderness conservation (i.e., the materialist or "realist" approach to nature). Rather than pitting these ontological approaches and their related social science orientations against each other, I argue that each taps into an important dimension of wilderness studies. A dialogue between how nature has been culturally constructed and physically disturbed and/or preserved is critically needed.

A subtheme of this first point is that the study of nature and wilderness is deeply political. Wilderness scientists, like scientists everywhere, have downplayed the politics behind how "nature" and what is considered "natural" are defined and deployed on behalf of particular human interests.

My second major theme is that while important concepts and strategies for protecting ideals of "wilderness" have changed, there has been a tendency to substitute old sets of "received wisdoms" or "discourses" with new ones. I discuss how wilderness science has shifted between two ideal concepts and management strategies, that is, as a "pristine," delicately balanced ecosystem, devoid of people and managed for solitude, recreation and re-creation, to wilderness as "humanized" landscapes, manipulated ecosystems, especially by Native and rural peoples marginalized by development and coerced by violent protected-area management policies and practices. In the latter view, wilderness protection brings people in, especially via community-based approaches to conservation. I argue that neither position is inherently true or preferable. Whether a protectionist or community-based approach is desirable and workable is an empirical question that must be examined in the context of particular places, peoples, issues, and ecosystems.

In this paper I hope to illuminate the above themes and suggest instances where I see glimmers of hope that efforts are under way across the globe that utilize multiple approaches and adaptive management strategies tailored to particular social contexts and histories. In the conclusion I provide a brief mention of such efforts.

POSITIONING MYSELF IN THE DEBATE

Like everyone else, I have specific filters through which I make meaning of these topics. These include my formal education as an environmental sociologist to honor both materialist and idealist orientations. I have also become sensitive to cross-cultural and transnational perspectives, having spent most of my professional career studying social and environmental interactions in remote tropical places. My research has also been highly applied and geared toward seeking practical solutions and policies for bridging conflicts between development and conservation, park protection and resident peoples' cultural and economic survival—no easy task.

I am also learning how hard it is to achieve the often-mentioned goal of becoming interdisciplinary. Whether teaching, researching, or collaborating on a project, I am usually working side by side with physical scientists and officials, often from different nations and cultures. I am constantly explaining and defending why attention to social forces and social organization is relevant to ecological change and park management. I am still learning how to effectively communicate and get along with people who are vastly different from me in terms of language, disciplinary methods, technical terms, perceptions, and objectives—among others. And like everyone in this room, a personal connection to nature underlies why and how I do my job. I am an avid backpacker, biker, sea kayaker, and "nature" lover.

All of us are comprised of multiple, overlapping identities and interests that affect how we understand human-nature interactions. I hope the ones I've shared with you confuse and complicate your ability to pin a theoretical or ideological label on me or my thinking.

MATERIALIST AND IDEALIST APPROACHES TO NATURE AND WILDERNESS STUDIES

Throughout the conference, the idea of wilderness, and why we should discredit or support this idea, has been reemerging and making a lot of people squirm in their seats. It keeps popping up because social scientists in the 1980s and 1990s have been rekindling attention to ideas, culture, moral values, and social experience in their studies of society, and, not surprisingly, they are applying this approach to their examinations of environmental change. Established and accepted terms of discourse are being critically examined and their ideological origins and purposes exposed. While we've heard the terms "social construction of nature" and "discourse" bandied about at this conference, I don't think anyone has defined them for us or provided a more balanced sense of their applications, advantages, and disadvantages.

"Social construction" refers to the idea that how people "see" or understand nature or landscapes is very important and depends in large part on our own social context and perspective on social life (Greider and Garkovich 1994). This often occurs unconsciously and unwittingly when we think we are being completely objective. As our perspective changes across time and place, history and culture, the meanings we confer on nature change along with it. Social constructionists would say this is a universal human condition. Both laypeople and scientists "see" the world through socially influenced filters. As with where and how we grew up and the values taught to us and the stories told to us, our academic disciplines provide a filter to how we see and understand the world. Indeed, the very mission of science is to explicitly teach us how to see and represent the world appropriate to the assumptions and methods of our respective disciplines. Thus, our view of nature and what we see as natural is partially a product of our culture and its influence on the "construction" of what nature is perceived to be.

The social constructionist approach, according to Michael Bell, author of *An Invitation to Environmental Sociology* (1998), alerts us to the highly political and partial way we conceive of nature. This is because our understanding of nature depends on *social selection* and *social reflection.* We all tend to select particular features of nature to focus on, ignoring those that do not suit our interests and worldviews. Over forty years ago in *The Structure of Scientific Revolutions* (1962), Thomas Kuhn described how scientific theories, methods, and research topics are closely linked to the paradigms (as well as funding biases) of the existing scientific establishment, which changes reluctantly at a turtle's pace and only when contradictions and new questions expose the limitations of existing paradigms. Because of social selection and social reflection, "nature" and science are inescapably social and political phenomena.

This view suggests that all of our ideas about nature and environmental change are partial. That is, any one of us only sees part of the phenomenon, and that meaning is only complete when understood within the context and agenda of a community of like-minded thinkers. Despite assertions of objectivity, scientists obscure some portion of reality when they narrate the history and results of their studies. The narrative succeeds to the extent that it can hide the discontinuities and contradictory experiences that would undermine the intended meaning of the "story." Science is political because inevitably some aspects of what scientists see, hear, and record are sanctioned while others are denigrated or silenced.

For example, we are all aware of how attention has been redirected in the forestry sector over the past decade to how different publics make meaning of forests: as a source of living, connection to spiritual heritage, place for recreation and hunting, or aesthetic appreciation. Though not without extreme controversy, even Congress has made these variable meanings a legitimate consideration of forest policy on public lands. While we may argue over the sense of holding each view equal or as relative "truths," the point is that we all have forest images in our minds, and these images affect how we think forests should be managed. We see conflict, therefore, not only over the prioritization of what values the forest should be managed for but also over what the forest is and how it should be understood.

Let me suggest a more subtle example, one with far-reaching implications for how we understand nature and ecological processes (Bell 1998). It has often been told that Karl Marx wanted to dedicate his famous work on capitalism to Darwin. He wanted to do this to recognize Darwin's observation of competition in nature and how it influenced Marx's view of class conflict and struggle. For decades this anecdote symbolized the debt social scientists feel to ecologists. We often use biological metaphors. For example, an early and highly influential approach in sociology is "human ecology," and there is cultural ecology, social ecology, and, most recently, political ecology. For many years an intellectual dependence on the biological sciences also denoted an acceptance of the superiority of the physical over social or interpretive sciences.

But times change, and so does our narration of them. The influence of Darwin on Marx is being reframed to emphasize instead how social forces and contexts influenced Darwin himself. A review of Darwin's biography and personal letters describes how he hit upon the theory of natural selection. In 1838 he "happened to read for amusement Malthus on Population, shortly after returning from his voyage around

South America on board the HMS Beagle" (Darwin 1858:42–43, cited in Hubbard 1979:24). In his letters Darwin acknowledged an intellectual debt to as well as the phrase "survival of the fittest" from the writings of Herbert Spencer, a mid-nineteenth-century social theorist (Hubbard 1982).

When Karl Marx and his longtime friend and collaborator, Friedrich Engels, read Darwin's book on natural selection, *On the Origin of Species* (1859), their correspondence about it noted its close resemblance to the economic theories of free-market capitalism that were so fundamentally altering the character of English society and, increasingly, world society at the time. Marx noted to Engels in a private letter in 1862, "It is remarkable how Darwin recognizes among beasts and plants his English society with its division of labor, competition, opening of new markets, inventions, and the Malthusian struggle for existence" (Meek 1971:195, cited in Bell 1998:222).

The latter refers to Thomas Malthus's theory that the population grows faster than our ability to produce food. I might also add that Darwin's Malthusian image persists today with the tendency—especially among ecologists—to view population dynamics deterministically and monolithically as *the* cause of ecological change. Population growth is highlighted even when evidence suggests that other processes such as consumer demand, the treadmill of capitalist production, and the maldistribution of resources also set the wheels of environmental change and degradation in motion (e.g., Ehrlich 1968).

The point here is that the two scientists who first hit upon the theory of natural selection—Darwin and his lesser-known contemporary, Alfred Russel Wallace—were living in the midst of the world's first truly capitalist industrial society: 1840s and 1850s England. How might they have "seen" nature and ecological processes if they were living in a more communitarian, cooperative, and socially homogeneous society? It is clear that their most influential work, their view of what nature is and how natural systems operate, reflects not only scientific observation but also the social and political milieu in which those observations and subsequent theoretical explanations were made.

Moreover, Marx and Engels were bothered by the way Darwin's work enabled science to be used as a source of political legitimization. Their concern was with a process that some refer to as "naturalization"—the claim that if something is natural, it can be no other way, it is inevitable. If capitalism resembled so closely the laws of nature, the argument could and was being made that it also is inevitable. Bell (1998) points out that we routinely talk about the economic "forces" of capitalism, such as innovation and competition, as if they were pseudonatural processes, implying that any other arrangement would be somehow unnatural. We also talk about the marketplace as a "jungle" in which we have to "struggle to survive."

Others have gone on to prove their concern that "naturalization" arguments could and would be used and misused. Many so-called laws of human nature by self-labeled social Darwinists and others (including the Nazis) have been justified on the basis of "human nature." Arguments attempting to prove inherent differences in the capabilities of different human "races" have been used to justify social programs, brutal racism, and the annihilation of people, which is defended as "ethnic cleansing." At different times "survival of the fittest" has been used as a rationale to defend the transfer of wealth from one group of people to another, often under conditions where the social structure of opportunity is highly unequal. Naturalization arguments disguise underlying political and economic interests, conflicts, and competitions.

Furthermore, an emphasis on seeing certain human actions and nature as "natural" and hence innate, essential, eternal, nonnegotiable, and off-limits to critical questioning and scrutiny also flows from the appeal to nature as a stable external source of nonhuman values against which human actions can be judged without ambiguity. This is very compelling. However, this becomes far more problematic when you consider that scholarship across many fields has demonstrated that our views of nature—human and in the natural world—are far more dynamic, malleable, and enmeshed with human history than popular beliefs about some "balance of nature" have assumed (Botkin 1990). Many studies call into question the validity of appealing to nonhuman nature as an objective measure of ourselves and our relationships with nature.

The stance of viewing human nature and various other aspects of our world as "natural" is, in fact, a centuries-long dispute entailing "realists" versus "constructionists." The tension was in full evidence in the papers written by Callicott and Foreman (in this volume) and in the reading by Gary Snyder (a version of which is also in this volume). Realists, characterized by Foreman, focus their attention on material processes and factors such as consumption, economy, technology, development, population, and especially how biophysical processes shape our environmental situation. They stress that environmental problems cannot be understood apart from "real" material processes and believe that scientists can ill afford to ignore the material "truth" of environmental problems and the material processes that underlie them. Realists tend to view nature and what is natural as a self-evident truth that we should open ourselves to see and appreciate. Constructionists do not necessarily disagree, but they emphasize the influence of social forces and ideas in how we conceptualize those "threats" or the lack of those "threats." Constructionist approaches, illustrated in the talk by Callicott, emphasize the ideological origins of environmental problems, including what becomes defined and accepted as problems (or as nonproblems). Though strongly criticizing constructionists, Gary Snyder nonetheless illustrated how social construction in the form of the human imagination and poetry serves major roles in our relationships with nature. He reminded us that a map is not a territory or a menu the meal; instead, these are symbolic representations of real, material phenomena.

BEYOND DUALISM

Each of the above approaches defines and seeks to understand a dimension of nature, wilderness, and the threats to wild places and processes. Therefore, each position taps a partial reality; each has certain strengths and certain weaknesses. A major benefit I see of the materialist position is its grounding in particular people and places and on particular ecological processes and consequences. In contrast, a benefit of the social constructionist approach is its recognition that what we understand as nature, natural, or even problems is also based on a long and complicated human cultural and political economic history. I think it is an important insight to recognize that while nature, indeed, has a physical reality, how we apprehend that reality never occurs outside a social context. The meanings and measures people assign to nature cannot help but reflect that context.

But what are the limitations of each approach? When social constructionists do not seriously and dynamically draw material processes and ecological consequences into their analysis, I think they are flawed. The result is an untenable relativist position. For example, while a clear-cut may appear innocuous and even beautiful to a resident of Forks, Washington (whose interest is served by seeing it as a temporal if not "natural" part of his or her landscape), it does have physical effects on the ground: on soils, vegetation, wildlife, etc. These material consequences need to be incorporated into management decisions. But when materialists do not consider how social selection, reflection, and self-interest affect their visions and that their vision is one of many others, their position is also flawed and limited. I think there is much to be learned by examining the charge that wilderness advocates created a movement based on a partial view of nature and set of meanings that have become what Callicott referred to as the "received wisdom" of wilderness. Attention to this critique can open up and already has opened up space for broadening areas of concerns and the types of places and people involved in wilderness studies. For example, in Foreman's talk he explicitly included values besides recreation as a goal of wilderness management, particularly ecological function, and he specified that wilderness lands can and should include nonpristine places across the matrix (i.e., outside core areas). Lastly, he deliberately included photos of females in wilderness (though they were just female versions of "macho" rafters/recreationists).

Again, it behooves us to define our terms. What exactly does the wilderness "received wisdom" or "discourse" entail? I prefer to talk about "discourse" because it has become part of the lexicon and methodology ("discourse analysis") in critical analysis of the making and unmaking of "the" idea of wilderness. In everyday speech, discourse is used as a "mode of talking." Yet as Maarten Hajer notes in *The Politics of Environmental Discourse* (1995), in the social sciences discourse analysis aims to understand why a particular understanding at some point gains dominance and is seen as authoritative while other understandings are discredited. Discourse analysis is concerned with analyzing the ways in which certain problems are understood and represented to others, how conflicting views are dealt with, and how coalitions on specific meanings somehow emerge. Most importantly, a discourse expresses ideas, images, and words that are handed down to us as self-evident truths, as natural—it just couldn't be otherwise. But of course it can. Baird Callicott provides a rich discussion of the major substance of the wilderness discourse and the charges against it; so does Daniel Botkin. I do not need to repeat them here. According to them, the dominant wilderness discourse has been based on wilderness as balanced ecosystems, beautiful, inspirational places that are devoid of people, though others would say that wilderness is based on naturalness, remoteness, and solitude.

The dimension of wilderness that I have worked most closely with is the role of people within wilderness, especially people whose livelihoods are tied to natural resources. In the wilderness discourse human action is often pitted against the well-being of the natural environment. Wilderness, by law and practice, is a place where people can visit, recreate, but not remain, and they surely cannot work there. "Work" versus "play" is another one of those binary juxtapositions that has historically been associated with wilderness debates and has served to widen rather than bridge understanding and advocacy of livelihoods that are compatible with ecological processes. Richard White in "Are You an Environmentalist or Do You Work for a Living?: Work and Nature" (1996) takes on the fallacies of this duality directly. By failing to examine and claim work within nature, environmentalists have been seen as insensitive to the needs of labor, he says, especially to those working-class people whose livelihoods have been tied in the past to extractive enterprises. The failure to bring work—or labor and class issues in general—into the environmental conversation has ceded valuable cultural capital to the so-called wise-use movement. But as White points out, proponents of the wise-use movement are not significantly concerned with work and the issues of the working class. Instead, they turn issues of real work into those of invented property rights; they pervert the legitimate concerns of rural people—maintaining ways of life and getting decent returns on their labor—into the special "right" of large property holders and corporations to hold the natural world and the public good hostage to their economic gain. Acknowledging a place for people and work in nature is about identifying and supporting practices that tie livelihoods to maintenance of ecological function. Work that does not support and sustain the integrity of large ecosystems is not fostered. Gary Snyder's charge that environmental historian Bill Cronon represents the intellectual "high end" of the wise-use movement falls into this simplistic, dualistic, and ultimately unproductive gulf. If Cronon can conceive of work in nature, surely he is one with the wise-use movement and its earth-devouring, corporatist, invented property right arguments. But Cronon's works never make this point. Quite the contrary, his essays on nature and wilderness speak to the social and political factors that lead different peoples and corporations to conceive of and use natural resources as they do, often in highly environmentally degrading ways.

I would like to note that not only environmental philosophers and environmental historians acknowledge and critically examine the dominant wilderness discourse. Botanists Gomez-Pompa and Kaus (1992) identify and discuss a "wilderness myth" and, furthermore, the need to "tame it." Never once using the phrase "wilderness discourse," they nevertheless squarely capture its meaning when they suggest that "through time and generations, certain patterns of thought and behavior have been accepted and developed into what can be termed a Western tradition of environmental thought and conservation" (Gomez-Pompa and Kaus 1992:271). These biologists ask, Whose "ideal" or "idea" is this, and who benefits or loses from it? Baird Callicott's analysis of the wilderness myth amply demonstrates that women, Native peoples, and an array of different values and traditions of living with nature have been denigrated, usurped, or ignored because of the logic of the dominant wilderness discourse and its incorporation into international park-planning models. These injustices have been particularly true in tropical developing countries, where park planning has been based on Western protected-area models that, until recently, did not incorporate the meaningful participation and vested interest of resident peoples (West and Brechin 1991).

But in the critique and refashioning of our ideas of wilderness and protected-area management, have we replaced one set of partial images and self-selected dogma with another? I turn now to my second theme.

REMAKING WILDERNESS AND NATURE PROTECTION: NEW POSSIBILITIES, NEW RISKS

As academics and planners rethink ideas of wilderness and the practice of wilderness management, attention is being redirected to how peoples and communities with interests in these areas (or living within or next to "buffer zones") can be involved in their comanagement (West and Brechin 1991; Western and Wright 1994). Attention is also being directed to how to include people, communities, and natural processes on "matrix" lands—places that connect "core" protected areas and move beyond islands of biodiversity to protecting, restoring, and managing landscape-level ecosystems. While we can discuss the degrees to which some of these places are "self-willed" or pristine, we have increasingly recognized that there is no place on the planet not subject to some degree of human action. As discussed above, active ecosystem manipulation may be hidden behind a screen of naturalization arguments. Many of us social scientists are happy to accept the view that all places are manipulated by human action because it fits with places we have studied and, more fundamentally, because it provides a revisionist view of nature that squares with our political and social justice goals. These include contesting coercive forms of conservation and helping to reclaim resident and working people's history, land rights, and livelihoods.

However, biologists such as Vale (1998) warn that whether a landscape's fundamental ecological processes have been altered by human actions, significantly or not, needs to be empirically examined and not determined because of our commitment or lack thereof to a social ideology. He sees debates regarding the prevalence of "humanized" landscapes as crystallizing into two polar opposite positions: either one sees nature as self-willed, as largely untouched by human action and reserved for recreation, *or* one sees nature as guided by human hands, as personal, subjective, and a landscape of everyday living and work. But isn't it unrealistic to expect that only two categories of human-ecological interaction—nature as pristine or humanized, nature as stranger or home—are sufficient to capture the complexity and messiness of the real world? We need to be self-critical and honest about how our science is affected by our political goals and ever cautious of the seduction of binary categories.

But this new debate raises questions of how resident and working peoples and rural communities have been constructed in the "old" received wisdom and how they are being reconstructed in what may be understood as a "new" humanized wilderness discourse. I apply the term *rural* to the people who reside within or near wilderness areas and/or their buffer zones. In the introduction to their 1996 book, *Creating the Countryside: The Politics of Rural and Environmental Discourse,* DuPuis and Vandergeest warn that rural peoples and communities—just like landscapes—are often portrayed in simplistic and binary terms. Specific words are chosen and deployed to communicate these dualistic meanings and to give privilege to one set of meanings over others. Rural peoples are represented as either destroyers of nature, "slash-and-burn" farmers, "addicts" to extractive industries, uneducated, irrational, backward, traditional, and in need of outside "progressive" assistance or as living closer to nature, holders of "indigenous knowledge," sacred, located in the past and the periphery, and able to sustainably manage their local environment through local customs and social institutions (the classic "ecological noble savage" image). In both cases, the tendency is to view rural people and places generically and as having some essential characteristic rather than to understand them within their particular historical and social contexts. In addition, I think there is a pattern for rural peoples and communities to be viewed as destroyers of nature in the United States, given their reliance on extractive industries such as mining, logging, grazing, and commercial, petrochemical-based farming; and they have provided political action in support of these industries. Given this history, it is not surprising that there has been a reluctance on the part of conservationists to envision how rural peoples and rural livelihoods could have played any significant role in the formation of wildlands or in any potential role they could play in the restoration and protection of large wildlands in the future. In the United States policy emphasizes *ecosystems* and *ecosystem management.* But while I understand this logic, I think it underestimates the importance of rural places, peoples, and livelihoods in the management of large wildlands. I'll return to this point in the conclusion.

In contrast, in the tropics the tendency is to highlight the role of rural peoples, livelihoods, and communities in altering landscapes and to place the lion's share of hope for tropical conservation in them. This has led to an emphasis on *agroecosystems* and *agroecosystem management.* In the 1980s attention to the critiques of coercive conservation based on the wilderness discourse led to a reframing of environmental protection as compatible with economic development. Operating under the rubric of "sustainable development," projects have been funded around the world to "integrate" local livelihoods with environmental management (Wells and Brandon 1992). The idea of sustainable development legitimates "green" production, capitalist expansion, and accumulation that tread lightly on the earth. We can have our cake and eat it too. The positioning of development and environmental protection in the 1980s as compatible rather than as in conflict (as was the case during the 1960s and early 1970s) is one of the most important and shrewd shifts in human-nature thinking during my time. Many suggest it remains a contradiction in terms (Redclift 1987).

A modification of integrating economic development with environmental protection, especially to meet goals of "local participation," is focusing attention on "community" as the social management unit for implementing sustainable development. "Community-based conservation" or "community-based natural resource management" has become a shining light of conservation efforts in the tropics (Getz et al. 1999; Western and Wright 1994). Community- and citizen-led conservation efforts are also sprouting up across this country. The Sonoran Institute, for example, emphasizes "community stewardship" as its approach to integrating environmental protection and community economic development.

Support for emphasizing community in conservation stems from a variety of factors, including recognition of the role of rural communities—largely in the tropics—in developing sophisticated common-managed property and resource management customs that, until the intrusion of the modern state, market, and demographic pressures, were able to sustain both livelihoods and fundamental ecological processes. Advocates of community-based conservation argue that resident or rural peoples have a greater vested interest in the long-term condition of local environments than absentee corporate managers, have intimate local knowledge that can be applied, and are less bureaucratic and hence more efficient at implementing conservation and development efforts. In any event, they point out that it is worth paying attention to the man or woman with the shovel. They, not the erudite social theorist or biologist sitting in an office, will ultimately decide the fate of the forest, as the saying goes. I find it very interesting that many of the people utilizing a variation of this approach in the United States (e.g., the Greater Yellowstone Coalition, Y2Y, and the Sonoran Institute) have considerable prior experience working in the tropics, many in Latin America.

Sociologist Arun Agrawal (1997) also suggests that our enchantment with community in conservation across the globe builds on our current dissatisfaction with theories of progress and centrally planned development and conservation. As Callicott and others have described, the designation of parks and the implementation of protected-area management policies have often displaced resident peoples, "coerced" conservation, and sanctioned violence, especially in ex-colonies and places where indigenous peoples do not have economic or political voice (West and Brechin 1991; Peluso 1993). Community conservation has rekindled hope around the world that concerns for place, devolution of power, and revival if not initiation of new democratic institutions based on civic activism can and will take a place in environmental management.

While strongly supporting the rationale for community-based conservation and the value of local, place-based conservation efforts, I offer the suggestion that we need to be careful not to replace one monolithic understanding of rural peoples, communities, and dynamics of ecological change and development with another. More specifically, I think we cannot presume the existence of "ecological noble communities" or universally position them as the cornerstone of every conservation effort, whether in the tropics or elsewhere. Let me give you three reasons why I think so.

First, not all marine or forest-dwelling communities have the local governing bodies, educational skills, technologies, social customs, or conflict resolution skills (or the social capacity or social capital) to sustainably manage their environments and natural resources. Some have. Some haven't. Some had at one time. Some never had. In some instances there may be other local institutions or governing bodies (i.e., besides "the community") that should be considered in the local or co-management of natural resources.

Second, the celebration of community in conservation has taken the limelight off of more powerful actors and global trends, such as the actions of transnational corporations, international monetary lending institutions, multilateral trading treaties (such as NAFTA and GATT), and organizations (such as the WTO) that exert tremendous influence on the way "nature" is converted, commodified, and compromised. IMF debt-restructuring policies are creating environmental and socioeconomic structures that compel if not determine choices and actions in the rural hinterlands.

A fatal implication of the social constructionists' ascendancy is lack of attention to how political and economic institutions and relations operating at the global or "non-place-based level" affect social and ecological interactions at multiple scales. Even where community-based efforts may be able to mitigate local impacts of global threats to sustainable living, they merely treat symptoms and do not necessarily resolve underlying causal mechanisms (or contradictions) operating at broader levels. The products of such contradictions are merely transported or felt elsewhere.

Third, those in control of conservation policy and practice do not often have an accurate understanding of communities and ecological processes or of the supracommunity political and financial constraints under which they operate. I do not think all images are equal. Thus, even in the good name of community (or the discourse of sustainable development, I should add) many social and ecological disasters have been produced. For example, in the Amazon the insistence on "seeing" the tropical rain forest as exuberantly fertile negates the reality of infertile tropical soils and the disasters of large-scale grazing and colonization schemes. Slater (1996) suggests that we are fascinated with rain forests and rain forest peoples because they represent an "Edenic Narrative" or new Garden of Eden stories, complete with tales of natives living in complete harmony with nature and divine creatures, dramatic falls from grace, and subsequent nostalgia for paradise. But based on her research, she contests these images as skewed and static. Furthermore, modern construction of these images is increasingly controlled and manipulated by corporate interests, such as travel agencies, fertilizer companies, media networks, etc.

Having lived the past fifteen years off and on in rural communities in Southeast Asian tropical environments and more recently in a remote rain forest community in Central Sulawesi, I am acutely aware of how careful we need to be about imposing static categories and strategies on people in the name of conservation and development or thinking that nature is merely an abstract idea. In this village residents respond in diverse ways to the political and economic changes occurring in Indonesia. While some cling tenaciously to strategies to maximize food production and security, others are rapidly transforming traditional agroforestry systems to sun-grown cacao monocrops, a commodity trading high on the global market and a cultivation method, like sun-grown coffee, aimed at maximizing quick returns (Belsky and Siebert 2003). Some are embracing political opportunities to be citizens in "New" Indonesia, while others resist political reform as just more of the same.

This is also a community with few traditional forest management customs and social institutions. How to build on community values and practices while working to maintain rain forest ecological processes is a dilemma facing myself and my physical scientist colleague/husband in our collaboration with the Nature Conservancy and Indonesian Park and Forestry officials to develop strategies that integrate conservation and development. There was no presumption or image to uphold, however, for my ten-year-old son, who directly lived the Edenic experience. Not expecting the people to act one way or another toward nature, he was very disturbed during our stay when village kids shot colorful songbirds with their slingshots or tied their legs to sticks as toys. He disliked immensely using the river for human waste disposal, bathing, and drinking. But he was most alarmed at his dad's near-fatal illness, caused by a virulent strain of new biodiversity: chloraquine-resistant cerebral malaria. We were indeed living closer to nature, but not the kind he could romanticize.

So what is my point? The tendency to see rain forest peoples as either in the state of original innocence and harmony with nature or after the fall misses the messy reality of the diversity of peoples, desires, experiences, and (changing) relationships with (changing) nature. Our view needs to remain wide enough to contemplate the broader political and corporate forces affecting local peoples and local environments. Generic understandings do not capture the dynamic, often chaotic, and complex nature of social forces and their interactions with nature (or how people interact with changed or "second" nature). And generalizations do not alert us to the disasters created when policies are based on imaginary communities and imaginary people-nature relationships.

The case of Gales Point, a rural ecotourism project in Belize, illustrates the complexities involved in creating "community"-based conservation (Belsky 1999). Conceived by a group of very well intentioned wildlife biologists and others, the project was informed by generalizations of an essentialized, traditional creole community and how "links" could be formed between ecotourism and community support for hunting regulations and conservation. The planners paid little attention to local history, politics, social change, or the ties between this local community and the broader political economy. Lacking the social institutions and material resources to support the mental picture the planners carried in their head (and at least for a while successfully communicated to funders), the project was able to help a few households reach the goal of replacing income from hunting with that from ecotourism activities and fostering attention to environmental conservation. But it also exacerbated intracommunity rivalries and incited a backlash by households left out of the new ecotourism economy to the very conservation values it had hoped to foster. And despite what was occurring in this and other ecotourist communities, dominant groups in Belize continued to exercise control not only over land, labor, and other productive resources, but also over the production of meaning concerning the so-called ecotourism revolution in their country. Sometimes these dominant groups are classes and states; in other contexts they include environmental organizations, scientists, or well-meaning social and ecological activists, perhaps like ourselves. We often impose our modern (or postmodern) understandings of people and the landscapes and draw strict if not inaccurate boundaries between multiple, fluid categories of people and space; we then justify these partial and self-interested actions by claiming that they are "natural." When backed by power and capital, dominant groups are able to control the meanings that bolster policy and practice, even when a larger, less powerful majority thinks otherwise. What, then, are the policy implications for the ways we think and rethink humans' relationships with nature? I consider this question as we turn to my last theme of wilderness policy.

IMPLICATIONS FOR WILDERNESS MANAGEMENT AND POLICY

I have emphasized the interplay between material factors and ideas in the development of wilderness studies, science, and policy and why neither a focus solely on ideas, ideologies, and cultural constructions nor a focus solely on material processes and physical threats to environmental protection is sufficient. Attention to their interaction is critical, as is how such interaction occurs at multiple scales (i.e., across space and over time). Serious discussion and dialogue, not just casting aspersions on opposing positions and their advocates, is necessary. However, doing so, as Foreman and others point out, is a political act in itself and undermines the authority of some standpoints.

My second point applied critical attention to the opportunities and dangers in replacing the "received wisdom" on wilderness and strategies to protect wilderness and ecological processes with a new set of assumptions and policy prescriptions. I suggested there are both potentials and pitfalls with uncritical acceptance of thinking of all landscapes as either "self-willed" or altered by human action. Similarly, we cannot know without examination of a particular social context if local participation can be accomplished through community institutions or some other local institutions or how viable is a particular approach to integrating conservation and local development (e.g., developing rural ecotourism, nontimber forest products, or value-added enterprises). The emergent discourse on the benefits of collaboration over confrontational politics and litigation is another "received wisdom" that may also depend on context and the particular dispute. In all of these cases I suggest that analyses need to embrace the interplay between materialist and idealist approaches.

From the social constructionists I applied the insight that we all operate out of partial understandings based on our own processes of social selection, social reflection, and self-interest as well as the suggestion that the labels we use and the stories we tell about nature and social relationships to it are more than just mental constructs or images. They form the institutional basis for conservation missions, policies, and interventions. We need to pay attention to them. For these reasons, while not sufficient to make a movement, it does matter what you call the movement. The idea and term *wilderness,* regardless of its biases and problems in practice, has mobilized a global movement. I think it will continue to motivate people to seriously consider the movement, more so, I suspect, than if we replace the term *wilderness* with *biodiversity reserves,* as suggested by Callicott. Despite the fact that reserves were created as a response to privatized hunting reserves, there is still something disturbing to me about "reserving nature." The term begs the difficult question, Reserved for what and for whom? The term is also limiting because it suggests that ecological and other values should guide action only in designated "reserves" rather than across the landscape. I like Foreman's imagery of rivers and blended currents. It conjures the ecological dialogue and integration of approaches I also support. However, a colleague reminded me that rivers are also full of turbulence and the possibility of getting swamped.

Our discussion of ideas and words is not just an academic enterprise. When particular viewpoints are backed by political power and funding, they move out of our heads and into the realm of action. They have important consequences for people and habitats (Zerner 1996). I'd like to be very specific about what I see as policy consequences resulting from the different discourses I've raised in this paper. I should also emphasize that the discourses I'm talking about are also imagined models. They are not static. They have been influenced by these debates. Below I summarize some of the ways our thinking about human relations to nature have shifted and their policy implications. To the extent I am aware of particular efforts that incorporate these insights I briefly acknowledge them.

1. *The concept of "wilderness" has multiple meanings. We need to make visible or less "mystified" how human actions and social processes affect both the concept as well as the actual places we label as "wilderness."* I think this point creates much confusion, anger, and backlash. It is also the most subtle but perhaps the most powerful. For the many reasons discussed above we need to be cautious in seeing certain human actions and places as "natural," inevitable, inherent, and hence off-limits to critical questioning and scrutiny. A failure to examine and reveal the history of particular peoples and places, including the history of our ideas of them, enables naturalization arguments to exist and persist. We also need to acknowledge that different understandings serve different interests and hence that wilderness science involves political choices.

2. *Because of past conceptions of wilderness, places with people have received considerably less attention in wilderness science. However, as conceptions of wilderness expand, including their role in protecting and restoring ecological processes across broad landscapes and ecosystems, places within and beyond "core" areas are being incorporated into wilderness science.* We see this shift in many projects that try to pay attention to people and places outside of "core" protected areas and reserves and to envision how "living" and "working" landscapes can contribute to conservation in these areas. I think this is a welcome and necessary trend but also one that poses many challenges. For example, it must include lowlands that provide critical habitats and biological corridors between core areas but that also are most populated. It will be difficult for new conservation economies to be fostered, especially ones that are inclusive. Not every ex-forestry or ex-mining town can become an ecotourism haven, nor does such an economy always provide the equity, quality of life, or conservation ethics and practices associated with historic natural-resource-based livelihoods. Residents need to play a leading role in crafting their own futures, though they will require assistance to make this transition. For this reason, efforts such as those provided by the Sonoran Institute to work with community leaders and others to put their ideas into practice and in so doing help to build community stewardship and sustainable

livelihoods are critically important. I have been personally involved in a project aimed at a transnational and transcommunity approach to protected-area management in the Maya Forests across Belize, Guatemala, and Mexico. But this project has been quite limited in space and scope in large part because it has failed to work with grassroots efforts and build conservation on historic land uses and livelihoods. Another example, the Northern Rockies Ecosystem Protection Act (NREPA), which sought to develop legislation and venues to implement conservation across broad landscapes, also failed because it did not significantly link ecological and economic policies across its targeted area, nor did it work closely with landowners and managers. Notably, its plans included measures to include private working ranches in biological corridors. However, NREPA did not emphasize direct involvement of private landowners who reside in the various proposed corridors, nor has it developed its policies around the proprietary and other concerns of these private landowners (Wilson and Belsky 1999).

3. *The new wisdom is critical of the view that casts working rural people and development as enemies of environmental conservation.* We need to maintain a healthy skepticism about what livelihoods and which economic practices are compatible with (particular) ecological processes. Long-term social and ecological monitoring is critical. How to build collaboration between rural peoples and scientists as well as with corporate private landowners remains a fertile area for experimentation and adaptive management. Mandating collaboration between historical adversaries is not the answer.

4. *Much sensitivity has been developed over proposing universal wilderness protection following a "hands-off" policy. Such a policy will be unsustainable under particular demographic, economic, and customary property rights.* Until recently, the largest conservation organization in the world, the International Union for the Conservation of Nature (IUCN), provided the conceptualization and blueprint for protected-area management. IUCN's schema divides space in terms of a set of categories and prescribed behaviors: core areas and buffer, production, and use zones. Some now may include biological corridors. These models are still universally applied without specific understanding of particular

rural peoples' colonial, ethnic, and customary property rights, local knowledge, and involvement in the global political economy or without sufficient rigorous ecological assessment.

5. *As the concerns of wilderness science expands, the toolbox of techniques for studying, managing, and protecting large wildlands and ecological processes must also be broadened.* Merely mapping, zoning, and restricting human use are not sufficient for managing wilderness and large ecosystems. Even where designations are made, any one place may not be able to honor every wilderness value. Nor can any management tool or strategy be assumed to be inherently appropriate. In some instances management may best entail individual (landowner) strategies such as placing conservation easements on particular properties, or they may entail community-based solutions built on viable community institutions, such as employing planning boards to develop zoning schemes. In other cases the lawyers may have to be brought back in. We need to be careful not to pick a favorite strategy and become the kid with the new hammer—everything we see needs to be hammered!

We need to recognize and move beyond simplistic and narrowly paradigmatic (or singularly disciplinary) ways of conceptualizing problems and imagining solutions. In particular we need to transcend thinking in binary, opposing categories and be wary of the seduction of universalist solutions and models. These are not easy tasks. Discussions on the relationship between humans and nature favor extreme positions, sound bites, and avoidance of self-criticism. It is hard for most of us to know how to analyze complex linkages and multiscaled phenomena such as environmental change whose causal mechanisms are not place or discipline bound. Most of us can gain only a partial understanding of these phenomena. Rather than become humble in the face of such an awesome undertaking, we take sides. We make enemies of other viewpoints or positions. We encounter opposing perspectives not to understand them but to discredit them. To avoid controversy we learn instead to be cautious and to mute critical inquiry that stirs up challenging or difficult ways of framing discussions or that reveals our own limitations. We don't permit self-criticism for fear that we will threaten our cause. And we create the impression that others are either for the environment or against it. But there are many dangers when we refuse to critically assess our own assumptions and methods and recognize our own dogmas. As Cronon (1996:22) warns, "At a time when threats to the environment have never been greater, it may be tempting to believe that people need to be mounting the barricades rather than asking abstract questions about the human place in nature. Yet without confronting such questions, it will be hard to know which barricades to mount, and harder still to persuade large numbers of people to mount them with us." I hope this essay has raised some of those difficult and abstract questions and, more importantly, suggested some ways of beginning to shape responses to them.

REFERENCES

Agrawal, A. 1997. Community in Conservation: Beyond Enchantment and Disenchantment. Conservation and Development Forum Discussion paper, Gainesville, FL.

Bell, M. M. 1998. *An Invitation to Environmental Sociology.* Thousand Oaks, CA: Pine Forge Press.

Belsky, J. M. 1999. Misrepresenting Communities: The Politics of Community-Based Rural Ecotourism in Gales Point Manatee, Belize. *Rural Sociology* 64(4): 641–66.

Belsky, J. M., and S. F. Siebert. 2003. Cultivating Cacao: Implications of Sun-Grown Cacao on Local Food Security and Environmental Sustainability. *Agriculture and Human Values* 20(3): 277–85.

Botkin, D. 1990. *Discordant Harmonies: A New Ecology for the Twenty-First Century.* New York: Oxford University Press.

Cronon, W., ed. 1996. *Uncommon Ground: Rethinking the Human Place in Nature.* New York: W. W. Norton.

DuPuis, E. M., and P. Vandergeest. 1996. *Creating the Countryside: The Politics of Rural and Environmental Discourse.* Philadelphia: Temple University Press.

Ehrlich, P. R. 1968. *The Population Bomb.* New York: Ballantine Books.

Getz, W. M., L. Fortmann, D. Cumming, J. duToit, J. Hilty, R. Martin, M. Murphree, N. Owen-Smith, A. M. Starfield, and M. I. Westphal. 1999. Sustaining Natural and Human Capital: Villagers and Scientists. *Science* 283:1855–56.

Gomez-Pompa, A., and A. Kaus. 1992. Taming the Wilderness Myth: Environmental Policy and Education Are Currently Based on Western Beliefs about Nature Rather Than on Reality. *BioScience* 42(4): 271–79.

Greider, T., and L. Garkovich. 1994. Landscapes: The Social Construction of Nature and the Environment. *Rural Sociology* 59(1): 1–24.
Guha, R. 1989. Radical Environmentalism and Wilderness Preservation: A Third World Critique. *Environmental Ethics* 11:71–83.
Hajer, M. A. 1995. *The Politics of Environmental Discourse.* New York: Oxford University Press.
Haydon, R. 1997. A Look at How We Look at What Is "Natural." *Natural Areas News* 2(1): 2–5.
Hubbard, R. 1979. Have Only Men Evolved? In R. Hubbard, M. S. Henifin, and B. Fried, eds. *Women Look at Biology Looking at Women: A Collection of Feminist Critiques.* Cambridge, MA: Schenkman. 7–35.
Kuhn, T. 1962. *The Structure of Scientific Revolutions.* Chicago: University of Chicago Press.
Meek, R. L. 1971. *Marx and Engels on the Population Bomb.* Berkeley: Ramparts.
Peluso, N. 1993. Coercing Conservation: The Politics of State Resource Control. *Global Environmental Change* 3(2): 199–217.
Redclift, M. 1987. *Sustainable Development: Exploring the Contradictions.* London: Methuen Press.
Slater, C. 1996. Amazonia as Edenic Narrative. In W. Cronon, ed. *Uncommon Ground: Rethinking the Human Place in Nature.* New York: W. W. Norton. 114–31.
Vale, T. R. 1998. The Myth of the Humanized Landscape: An Example from Yosemite National Park. *Natural Areas Journal* 18(3): 231–36.
Wells, M., and K. Brandon. 1992. *People and Parks: Linking Protected Areas with Local Communities.* Washington, D.C.: World Bank.
West, P., and S. R. Brechin, eds. 1991. *Resident Peoples and National Parks.* Tucson: University of Arizona Press.
Western, D., and R. M. Wright. 1994. *Natural Connections: Perspectives in Community-Based Conservation.* Washington, D.C.: Island Press.
White, R. 1996. Are You an Environmentalist or Do You Work for a Living?: Work and Nature. In W. Cronon, ed. *Uncommon Ground: Rethinking the Human Place in Nature.* New York: W. W. Norton. 171–85.
Wilson, S., and J. M. Belsky. 1999. Voices from a Working Landscape: A Critical Assessment of an Approach to Biological Corridors on Private Lands in Montana. University of Montana, Missoula.
Zerner, C. 1996. Telling Stories about Biological Diversity. In S. Brush and D. Stabinksky, eds. *Valuing Local Knowledge: Indigenous People and Intellectual Property Rights.* Washington, D.C.: Island Press. 68–101.

David W. Orr

The Not-So-Great Wilderness Debate . . . Continued (1999)

> *"Something will have gone out of us as a people if we ever let the remaining wilderness be destroyed; if we permit the last virgin forests to be turned into comic books and plastic cigarette cases; if we drive the few remaining members of wild species into zoos or to extinction; if we pollute the last clear air and dirty the last clean streams and push our paved roads through the last of the silence, so that never again . . . can we have the chance to see ourselves single, separate, vertical, and individual in the world, part of the environment of trees and rocks and soil, brother to the animals, part of the natural world and competent to belong to it."*
> Wallace Stegner

IT IS ODD THAT ATTACKS on the idea of wilderness have multiplied as the thing itself has all but vanished. Even alert sadists will at some point stop beating a dead horse. In the lower 48 states, federally designated Wilderness accounts for only 1.8% of the total land area. Including Alaskan Wilderness the total is only 4.6%. This is less land than we've paved over for highways and parking lots. For perspective, at 27,000 acres, Disney World is far larger than many of our Wilderness Areas, roughly one-third of which are less than 10,000 acres in size (Turner, 619). Outside the United States there is little or no protection for the 11% of the Earth that remains wild. It is to be expected that attacks on the last remaining wild areas would come from those with one predatory interest or another, but it is disconcerting that in the final minutes of the 11th hour they also come from those who count themselves as environmentalists. Each of these critics claims to be for wilderness, but against the idea of wilderness. This fault line deserves careful scrutiny.

In a recent article, for example, novelist Marilynne Robinson concludes that "we must surrender the idea of wilderness, accept the fact that the consequences of human presence in the world are universal and ineluctable, and invest our care and hope in civilization" (Robinson, 64 [included in part 4 of this volume]). She arrives at this position, not with joy, but with resignation. She describes her love of her native state of Idaho as an "unnamable yearning." But wilderness, however loved, "is where things can be hidden . . . things can be done that would be intolerable in a populous landscape." Has Robinson not been to New York, Los Angeles, Mexico City, or Calcutta, where intolerable things are the norm? But she continues: "The very idea of wilderness permits . . . those who have isolation at their disposal [to] do as they will." Presumably there would be no nuclear waste sites and no weapons laboratories without wilderness in which to hide them. She ignores the fact that the decisions to desecrate rural areas are mostly made by urban people and support one urban interest or another. Robinson then comes to the recognition that history is not an uninterrupted triumphal march. There have been, she notes, a few dips along the way. The end of slavery in the United States produced a subsequent condition "very much resembling bondage." Now "those who are concerned about the world environment are . . . the abolitionists of this era" whose "successes quite exactly resemble failure." So with a few successes under their belt, unnamed conservationists propose to establish a global "environmental policing system" and serve in the role of "missionary or schoolmaster" to the rest of the world. But we cannot legitimately serve in that role because we, in the developed countries, "have . . . ransacked the world for these ornaments and privileges and we all know it." Accordingly, Robinson concludes that wilderness has "for a long time figured as an escape from civilization," so "we must surrender the idea of wilderness."

I have omitted some details, but her argument is clear enough. Robinson is against the idea of wilderness but she does not tell us whether she is for or against preserving, say, the Bob Marshall or Gates of the Arctic, or whether she would give them away to AMAX or Mitsubishi. She is against the idea of wilderness because it seems to her that it has diverted our attention from the fact that "every environmental problem is a human problem" and we ought to solve human problems first. Whether environmental problems and human problems might be related, she does not say.

The environmental movement certainly has its shortcomings. There are, in fact, good reasons to be suspicious of movements of any kind. But there is more at issue in Robinson's argument. The recognition that governments sometimes use less-populated areas for military purposes hardly constitutes a reason to fill up what's left of Idaho with shopping malls and freeways. Her assertion that abolition and environmentalism have produced ironic results is worth noting. But does she mean to say that we ought to ignore slavery, human rights abuses, toxic waste dumps, biotic impoverishment, or human actions that are changing the climate because we might otherwise incur unexpected and ironic consequences? Yes, rich countries have "ransacked the world," but virtually the only voices of protest have been those of conservationists aware of the limits of the Earth.

And what could she possibly mean by saying that "we are desperately in need of a new, chastened, self-distrusting vision of the world, an austere vision that can postpone the outdoor pleasures of cherishing exotica . . . and the debilitating pleasures of imagining that our own impulses are reliably good"? Are we to take no joy in the Creation or find no solace and refuge in a few wild places? Who among us imagines their impulses to be reliably good? Would she confine us to shopping malls and a kind of indoor air-conditioned introspection? Finally, Robinson seems not to have noticed that the same civilization in need of "rehabilitation" has done a poor job of protecting its land and natural endowment. Is it possible that human problems and environmental problems are reverse sides of the same coin of indifference and that we do not have the option of presuming to solve one without dealing with the other?

Marilynne Robinson's broadside is only the latest salvo in a battle that began years earlier with articles by Ramachandra Guha (1998), Baird Callicott (1991), and William Cronon (1995). The issues they raised were, to some extent, predictable. Professor Guha, for example, believes that the designation of wilderness in many parts of the world has led to "the displacement and harsh treatment of the human communities who dwelt in these forests" (273). His sensible conclusion is simply that "the export and expansion [of wilderness] must be done with caution, care, and above all, with humility" (277).

Philosopher Baird Callicott's views and their subsequent restatement raise more complex and arcane issues. Callicott [in Callicott and Nelson 1998] begins, as do most wilderness critics, by asserting that he is "as ardent an advocate" of wilderness as anyone and believes bird-watching to be "morally superior to dirtbiking." The idea of wilderness may be wrong-headed, he thinks, "but there's nothing whatever wrong with the *places* that we call wilderness" (587). He is discomforted by what he terms "the received concept of wilderness" inherited from our forebears who were all white males like Ralph Waldo Emerson, Henry David Thoreau, John Muir, Theodore Roosevelt, and Aldo Leopold. Callicott is unhappy with "what passes for civilization and its mechanical motif" that can conserve Nature only by protecting a few fragments. He proposes, instead, to rescue civilization by "shift[ing] the burden of conservation from wilderness preservation to sustainable development" (340). He proposes to "integrate wildlife sanctuaries into a broader philosophy of conservation that generalizes Leopold's vision of a mutually beneficial and mutually enhancing integration of the human economy with the economy of nature" (346). This does not mean, however, "that we open the remaining wild remnants to development" (346).

The heart of Callicott's argument, however, has to do with three deeper problems he finds in the idea of wilderness. It perpetuates, he thinks, the division between humankind and Nature. It is ethnocentric and causes us to overlook the effects tribal peoples had on the land. And, third, the very attempt to preserve wilderness is misplaced given the continual change that is characteristic of dynamic ecosystems. Callicott's critics, including philosopher Holmes Rolston, have responded by saying "tain't so." Humans are not natural in the way Callicott supposes. There are, in Rolston's words, "radical discontinuities between culture and nature" (370). The effects of eight million or so tribal people living without horses, wheels, and metal axes had a relatively limited effect on the ecology of North America. After the initial colonization ten thousand or more years ago, the effects they did have, such as burning particular landscapes, did not differ much from natural disturbances such as fires ignited by lightning. As for the charge that conservationists are trying to preserve some idealized and unchanging landscape, Rolston asserts that "Callicott writes as if wilderness advocates had studied ecology and never heard of evolution . . . wilderness advocates do not seek to prevent natural change" (375). To his critics, Callicott's dichotomy between wilderness preservation and sustainable development, as if these are either/or, makes little sense.

The dispute over wilderness went public in 1995 with the publication of an excerpt from William Cronon's essay "The Trouble with Wilderness, or, Getting Back to the Wrong Nature" in the *New York Times Magazine*. Cronon did not add much that had not already been said, but he did give the debate a post-modern spin and the kind of visibility that lent considerable aid and comfort to the "wise use" movement and right-wing opponents of wilderness. Remove the scholarly embellishments, and Cronon's piece is a long admonition to the effect that:

> We can[not] flee into a mythical wilderness to escape history and the obligation to take responsibility for our own actions that history inescapably entails. Most of all, it means practicing remembrance and gratitude, for thanksgiving is the simplest and most basic of ways for us to recollect the nature, the culture, and the history that have come together to make the world as we know it. (90)

Like Callicott, Cronon hopes that his readers understand that his criticism is "not directed at wild nature *per se* . . . but rather at the specific habits of thinking that flow from this complex cultural construction called wilderness" (81). In other words, it is not "the things we label as wilderness that are the problem—for nonhuman nature and large tracts of the natural world *do* deserve protection—but rather what we ourselves mean when we use that label." That caveat notwithstanding, he proceeds to argue that "the trouble with wilderness is that it . . . reproduces the very values its devotees seek to reject." It represents a "flight from history" and "the false hope of an escape from responsibility." Wilderness is "very much the fantasy of people who have never themselves had to work the land to make a living" (80). It "can offer no solution to the environmental and other problems that confront us." Instead, by "imagining that our true home is in the wilderness, we forgive ourselves the homes we actually inhabit" which poses a "serious threat to responsible environmentalism." The attention given to wilderness, according to Cronon, comes at the expense of environmental justice. Further, advocacy of wilderness "devalues productive labor and the very concrete knowledge that comes from working the land with one's own hands" (85). But Cronon's "principal objection" is "that it may teach us to be dismissive or even contemptuous of . . . humble places and experiences," including our own homes.

Cronon concludes the essay by describing why the "cultural traditions of wilderness remain so important." He asserts that "wilderness gets us into trouble only if we imagine that this experience of wonder and otherness is limited to the remote corners of the planet, or that it somehow depends on pristine landscapes we ourselves do not inhabit" (88). He admonishes us to pay attention to the wildness inherent in our own gardens, backyards, and local landscapes.

"The Trouble with Wilderness" later appeared as the lead chapter in *Uncommon Ground: Toward Reinventing Nature* (Cronon 1995). The authors' collective intention was to describe the many ways the concept of Nature is socially constructed and to ask: "Can our concern for the environment survive our realization that its authority flows as much from human values as from anything in nature that might ground those values?" The book is a slightly irritating collage of the obvious, the fanciful, the occulted,[1] and disconnected postmodernism contrived as part of a University of California–Irvine conference on "Reinventing Nature." The contributors were asked to summarize their thoughts in an addendum at the end of the volume titled "Toward a Conclusion," suggesting that they had not reached one.

In an insightful retrospective, landscape architect Anne Whiston Spirn, author of the best chapter in the book, lamented the fact that the discussions were "so abstracted from the 'nature' in which we were living . . . the talk seemed so disembodied." She wondered "how different our conversations might have been if they had not taken place under fluorescent lights, in a windowless room, against the whistling whoosh of the building's ventilation system" (448). Indeed, the entire exercise of "reinventing Nature" had the aroma of an indoor, academic, résumé-building exercise. And the key assumption of the exercise—that Nature can be reinvented—works only if one first conceives it as an ephemeral social construction. If Nature is so unhitched from its moorings in hard physical realities, it can be recast as anything one fancies.

Not surprisingly, wilderness critics have received a great deal of criticism (Foreman 1994, 1996, 1998, Rolston 1991, Sessions 1996/7, Soulé and Lease 1995, Snyder 1995, 1996, and Willers 1996/7). After the dust has settled a bit, what can be said of "The Great New Wilderness Debate"? First, on the positive side, I think it can be said that, under provocation from Callicott, Cronon, and others, a stronger and more useful case for wilderness protection emerged (Foreman 1995, Grumbine 1996/7, Noss 1998, Waller 1998). The conjunction of older ideas about wilderness providing spiritual renewal and primitive recreation with newer ones concerning ecological restoration and the preservation of biodiversity offers a better and more scientifically grounded basis to protect and expand remaining Wilderness Areas in the 21st century. It is clear that we will need to fit the concept and the reality of wilderness into a larger concept of land use that includes wildlife corridors, sustainable development, and mixed-use zones surrounding designated Wilderness or ecological reserves. But the origin of these ideas owes as much to Aldo Leopold as to any contemporary wilderness proponent. And, yes, environmentalists and academics alike need to make these ideas work for indigenous peoples, farmers, ranchers, and loggers. The development of conservation biology, low-impact forestry methods, and sustainable agriculture suggests that this is beginning to happen. For these advances, wilderness advocates can be grateful for their critics.

On a less positive note, the debate over wilderness resembles the internecine, hair-splitting squabbles of European socialists between 1850 and 1914. Often the differences between the various positions of that time were neither great nor consequential. Nonetheless positions hardened, factions and parties formed around minutiae, and contentiousness and conspiracy became the norm on the political left. As a result, by 1914 the left had coalesced into ideologically based factions, firmly and irrevocably committed to one impractical doctrine or another. It was a great tragedy that in the early decades of the 20th century, when the world needed far better ideas about the organization of property, government, and capital, it had few from the left. Instead, socialists of whatever stripe gave the strong impression to mainstream society that they had nothing coherent or reasonable to offer. Their language was obscure, their proposed solutions often entailed violence, their public manners were uncivil, and their tone was absolutist. It was in this environment that Lenin and his Bolsheviks concocted the odd brew of socialism, intolerance, brutality, messianic pretensions, and ancient czarist autocracy that became known as Marxism-Leninism. And the rest of the story, as they say, is history.

The world now more than ever needs better ideas about how to meld society, economy, and ecology into a coherent, fair, and sustainable whole. The question is whether environmentalists can offer practical, workable, and sensible ideas—not abstractions, arcane ideology, spurious dissent, and ideological hair-splitting reminiscent of 19th century socialists. In this regard, the most striking thing about the ongoing "great wilderness debate" is the similarity that exists between positions that have been cast as either/or. There is no necessary divide, for example, between protecting wilderness and sustainable development. To the contrary, these are complementary ideas. And there are some issues, such as the old and unresolvable question about whether and to what degree humans are part of or separate from Nature, that are hardly worth arguing about over and over again. Nor do we need to hear truisms that wilderness must be adapted to the circumstances, culture, and needs of particular places. These are obvious things that deserve to be treated as such. Finally, since all participants profess support for the place called Wilderness, as distinct from the idea of it, we are entitled to ask: what is the point of the great wilderness debate? If we intend to influence our age in the little time we have, we must focus more clearly and effectively on the large battles that we dare not lose. The time and energy invested in our "great debates" should be judged against the sure knowledge that while we argue among ourselves, others are busy bulldozing, clearcutting, mining, building roads, and, above all, lobbying the powers that be to ensure that these destructive activities continue.

Third, the effort to find common ground by "reinventing Nature" along postmodernist lines seems to me to have the same foundational perspicacity as, say, the effort to extract sunbeams from cucumbers for subsequent use in inclement summers—a project of the great academy of Lagado descried by Jonathan Swift. Most surely we see Nature through the lens of culture, class, and circumstance. Even so, it is remarkable how similarly Nature is in fact "constructed" across different classes, cultures, times, and circumstances. This is so because gravity, sunlight, geology, soils, animals, and the biogeochemical cycles of the Earth are the hard physical realities in which we live, move, and have our being. We are free to describe them in different symbols and wrap them in different cultural frameworks, but we do not thereby diminish their reality.

The idea that we are free to reinvent Nature is, I think, an indulgence made possible because we have temporarily created an artificial world based on the extravagant use of fossil fuels. But that idea will not be particularly useful for helping us create a sustainable and sustaining civilization, however useful it may be as a reason to organize conferences in exotic places and for keeping postmodernists employed at high-paying indoor jobs. "Reckless deconstructionism," in the words of Peter Coates, "cuts the ground from under the argument for the preservation of endangered species" (185). More broadly, it prevents us from taking any constructive action whatsoever. The postmodern contribution to environmentalism has privileged (in their word) an arcane, indoor, and ivory tower kind of environmentalism with more than a passing similarity to views otherwise found only on the extreme political right. Separated as it is from both physical and political realities—as well as the folks down at the truck stop—postmodernism provides no realistic foundation for a workable or intellectually robust environmentalism.

Looking ahead to the 21st century, the debate over wilderness has illuminated the fact that we will need larger—not smaller—ideas about land, Nature, and ourselves. We will need more, not less, ecological imagination. We certainly need to be mindful of the "otherness" in our backyards, as Bill Cronon reminds us, but that reminder is a small idea that comes at a time when we must cope with global problems of species extinction, climatic change, emerging diseases, and the breakdown of entire ecosystems. We need a larger view of land and landscape than is possible where "It's mine and I'll do with it as I damn well please" is the prevailing philosophy. As Aldo Leopold pointed out decades ago, we need well-kept farms and home places, well-managed forests, *and* large Wilderness Areas. None of these needs to compete with any other. Of the four, wilderness protection is by far the hardest to achieve. It is a societal choice that requires an ecologically literate public, political leadership, economic interests with a long-term view, and above all, the humility necessary to place limits on what we do. Until we have created a more far-sighted culture, the conjunction of these forces will always be rare, fragile, and temporary.

The battle over wilderness will grow in coming decades as the pressures of population growth and alleged economic necessity mount. There will be, someday soon, urgent calls to undo the Wilderness Act of 1964 and release much of the land it now protects to mining, economic expansion, and recreation facilities. At the same time it is entirely possible that much of our affection for wilderness, rural areas, and wildness will decline if we continue to become a tamer and more indoor people. In *Brave New World* (1932), Aldous Huxley described the effort to "condition the masses to hate the country" while conditioning them "to love all country sports." This process is already well under way and we are the less for it. As D. H. Lawrence put it:

> Oh, what a catastrophe for man when he cut himself off from the rhythm of the year, from his unison with the sun and the earth. Oh, what a catastrophe, what a maiming of love when it was made a personal, merely personal feeling, taken away from the rising and setting of the sun, and cut off from the magical connection of the solstice and equinox. This is what is wrong with us. We are bleeding at the roots. (quoted in Bass, 1996, 21)

In the century ahead, the battle over wilderness will become a part of a much larger struggle. We have entered a new wilderness of sorts, one of our own making, consisting of technology that will offer us a "virtual reality" (an oxymoron if there ever was one), fun, excitement, and convenience. Caught between the ugliness that accompanies ecological decline and the siren call of a phony "reality" cut off from soils, forests, wildlife, and each other, we will be hard pressed to maintain our sanity and the best parts of our humanity. The struggle for wilderness and wildness in all of its forms is no less than a struggle over what we are to make of ourselves. For my part, I believe we need more wilderness and wildness, not less. We need more wildlands, wildlife, wildlife corridors, mixed-use zones, wild and scenic rivers, and even urban wilderness. But above all, we need people who know in their bones that these things are important because they are the substrate of our humanity and an anchor for our sanity.

NOTES

The title of this essay is borrowed loosely from the book *The Great New Wilderness Debate,* edited by J. Baird Callicott and Michael P. Nelson (University of Georgia Press, 1998).

1. The word is one used by Gary Snyder describing the "Reinventing Nature" conference, "an odd exercise" he thought. See Gary Snyder, *A Place in Space* (Washington, DC: Counterpoint, 1995), p. 250.

LITERATURE CITED

Bass, Rick. 1996. *The Book of Yaak.* Boston: Houghton-Mifflin.

Callicott, J. Baird. 1991. "The Wilderness Idea Revisited: The Sustainable Development Alternative." *The Environmental Professional* 13:235–247.

Callicott, J. Baird. 1991. "That Good Old-Time Wilderness Religion." *The Environmental Professional* 13:378–379.

Callicott, J. Baird, and Nelson, Michael P., eds. 1998. *The Great New Wilderness Debate.* Athens: University of Georgia Press.

Coates, Peter. 1998. *Nature.* Berkeley: University of California Press.

Cronon, William. 1995. "The Trouble with Wilderness, or, Getting Back to the Wrong Nature," in William Cronon, ed. *Uncommon Ground: Toward Reinventing Nature.* New York: W. W. Norton.

Foreman, Dave. 1994. "Wilderness Areas Are Vital." *Wild Earth* 4(4):64–68.

Foreman, Dave. 1995. "Wilderness: From Scenery to Nature." *Wild Earth* 5(4):8–16.

Foreman, Dave. 1996. "All Kinds of Wilderness Foes." *Wild Earth* 6(4):1–4.
Foreman, Dave. 1998. "Wilderness Areas for Real," in Callicott and Nelson, eds. 395–407.
Grumbine, Edward. 1996/7. "Using Biodiversity as a Justification for Nature Protection." *Wild Earth* 6(4):71–80.
Guha, Ramachandra. 1998. "Radical American Environmentalism and Wilderness Preservation: A Third World Critique," in Callicott and Nelson, eds. 231–245.
Guha, Ramachandra. 1998. "Deep Ecology Revisited," in Callicott and Nelson, eds. 271–279.
Noss, Reed. 1998. "Sustainability and Wilderness," in Callicott and Nelson, eds. 408–413.
Noss, Reed. 1998. "Wilderness Recovery: Thinking Big in Restoration Ecology," in Callicott and Nelson, eds. 521–539.
Robinson, Marilynne. 1998. "Surrendering Wilderness." *Wilson Quarterly* 22(4):60–64.
Rolston, Holmes. 1991. "The Wilderness Idea Reaffirmed." *The Environmental Professional* 13:370–377.
Sessions, George. 1996/7. "Reinventing Nature? The End of Wilderness?" *Wild Earth* 6(4):46–52.
Snyder, Gary. 1995. *A Place in Space.* Washington, DC: Counterpoint.
Snyder, Gary. 1996. "Nature as Seen from Kitkitdizze Is No 'Social Construction.'" *Wild Earth* 6(4):8–9.
Soulé, Michael, and Lease, Gary. 1995. *Reinventing Nature? Responses to Postmodern Deconstruction.* Washington, DC: Island Press.
Spirn, Anne W. 1995. "Toward a Conclusion," in Cronon, ed. *Uncommon Ground.*
Stegner, Wallace. 1960. "Wilderness Letter," in Page Stegner, ed. *Marking the Sparrow's Fall: Wallace Stegner's American West.* New York: Henry Holt, 1998.
Turner, Jack. 1998. "In Wildness Is the Preservation of the World," in Callicott and Nelson, eds. 617–627.
Waller, Donald M. 1998. "Getting Back to the Right Nature: A Reply to Cronon's 'The Trouble with Wilderness,'" in Callicott and Nelson, eds. 540–567.
Willers, Bill. 1996/7. "The Trouble with Cronon." *Wild Earth* 6(4):59–61.

Wayne Ouderkirk

On Wilderness and People (2003)

A View from Mt. Marcy

RECENTLY, PHILOSOPHERS AND others have had a lot to say about the relationship between people and wilderness. The dominant theme in much of that continuing discussion is that, contrary to what used to be the prevailing view, humans and wild nature are not essentially separate. In philosophical terms, this new theme claims that humans and wild nature are not ontologically distinct. Though I agree with that view, I have a caution to add: The recently perceived and defended connection between humans and wilderness does not imply, and ought not be interpreted as saying, that modern human civilization and wilderness are ecologically compatible. In other words, we modern humans should leave what little wilderness is left on our planet to its own devices. Keeping our distance in that way is not a tacit return to or acceptance of the traditional dualism between humans and nature—although, as we shall see, dualism has its defenders among environmental philosophers. Rather, keeping our distance is an acknowledgment that our current human ways and numbers threaten to overwhelm our ecological and evolutionary home.[1]

A brief review of formerly prevailing opinions about people and nature will help put the recent debates about wilderness in context. However, the ideas of some of the "unifiers"—for example, William Cronon, J. Baird Callicott—seem to indicate that, once we abandon the traditional dualism, wilderness, whether as concept or as policy or both, is no longer needed. I think that interpretation is shortsighted.

THE TRADITIONAL VIEWS

Ideas die hard, and ideas that are parts of humanity's understanding of itself die hardest. So before explaining the traditional views about wilderness, I need to note that although they originated in the past they are all still very much with us in one way or another. The new understanding of wilderness (discussed below and which I partially support) is in direct conflict with some or all of these older ideas; and there is no final account of wilderness. Rather, that concept has entered the arena of philosophically disputed issues.

At 5344 feet, Mt. Marcy is the highest peak in New York State. Comparatively, its elevation is not particularly notable; more than forty other peaks in the Eastern United States are taller.[2] However, what it lacks in elevation, it gains in stature, for it is the symbolic centerpiece of the unique Adirondack Park, with its large wilderness areas that are explicitly protected in the state's constitution. (By the way, for perspective, the Adirondack Park is larger than Yellowstone, Grand Canyon, and Yosemite National Parks combined.)[3] We can use the view from Mt. Marcy's summit, along with its recent history, to illustrate four accounts of wilderness.[4]

Looking in several directions from that summit (especially south and west), one sees only wilderness: peaks, valleys, lakes, ponds, streams, bogs, and forest for as far as the eye can see. If some anthropologists and philosophers are correct, the original human concept of wild nature understood such a vista as part of our sacred home, the "Magna Mater," the "Great Mother."[5] I know of no Native American worldviews that specifically regard what we call wilderness as Mother, but they certainly regard it as both sacred and home. So although as far as we now can tell, no Native Americans made the heart of the Adirondacks their permanent residence (probably because of the harsh winter climate),[6] we have no reason to think that, prior to the European invasion, the indigenous peoples—Iroquois, Algonquin, and others—did not regard the region as part of their homes. Thus, the indigenous description or interpretation of the vista from Marcy might well say (and now I admit that I am guessing) that here are places where, at least for half of the year, one can live, find food, water, and fuel, and where one can communicate with the spirit world, just as one can do everywhere else. To see it as a source of sustenance is not, in this understanding of wilderness, to see it as mere resource for exploitation, but as the source of all things, including ourselves.

Some recent thinkers advocate that we return to such a view of the wild,[7] but that is unrealistic, for at least two reasons. First, although many contemporary people do indeed view nature as sacred, to advocate the wild-as-divine as our unifying understanding of wilderness would ignore the beliefs of huge numbers of people, both non-religious and religious, and including some wilderness advocates, for whom the natural world is not identified with the divine. Second, this view is allegedly part of the self-understanding of hunter-gatherer peoples. A few such peoples still exist and should always retain their right to continue their traditional cultures in what we now refer to as wild areas. Nevertheless, even if we restored to wilderness a sizable amount of farmland and added it to presently designated "wild-zones," those areas still could not possibly feed the current human population, magically transformed into hunter-gatherers, for the foreseeable future. A return to the Magna Mater is impossible for us.

To abbreviate history a bit (and hence to distort it!), the second influential concept of the wild was the opposite of the first. Through fundamental sweeping changes in human culture, centered around the development of agriculture and the correlated increase in human population, we came to regard wilderness not as home but as our counterpoint—something against which we have to struggle. With the eventual ascendancy of Christianity in Europe, that struggle received religious interpretation and sanction as wilderness became defined as the home of the devil and of evil.[8] Given this understanding, it is easy to see that opposition to the wild would be understood as good, and that eking out a human life would be a way to resist, perhaps to conquer, the devil. Though there are other attitudes toward nature in Christianity, variants of this idea certainly influenced early European colonists on this continent, many of whom saw it as their duty to conquer what they perceived not as sacred home but as Satan's home.[9]

Returning to Mt. Marcy and looking to the north and east, although the vista remains impressive, we can see human encroachments. Lake Placid Village and its surrounding area are partially visible, and the great cone-shaped Whiteface Mt. shows the alpine ski trails carved into its side to serve human recreation. More to the east, a small part of the Lake Champlain Valley is visible, including farms, orchards, and buildings. Nothing could seem more wonderful to this second religious understanding of wilderness than these signs that people are pushing against nature—subduing it. As for the rest of the view, well, it is simply a challenge that remains for us. What some might regard as beautiful is merely the wild, chaotic realm of unfettered nature, awaiting the "saving hand" of the righteous human. If we look more closely, we might notice an old fire tower or two—on Mt. Adams or Hurricane Mt., for example. In this understanding of wilderness, the towers are not, as the New York State Department of Environmental Conservation (NYSDEC) calls them, "non-conforming structures" to be removed in order to restore the wilderness to its allegedly pristine state. Rather, they are footholds, affirming the possibility of human control.

One cannot see all of Marcy's lower slopes from its summit, but several stream drainages are evident, including that of Feldspar Brook with its source, Lake Tear of the Clouds, the highest source of the Hudson River. Feldspar flows into a branch of the stunning Opalescent River, which in turn feeds the northernmost Hudson. Exploring those drainages and consulting Adirondack history, one will find some signs and symbols of yet another traditional concept of the wild. About three miles west of Marcy is "Flowed Land," a small, human-made lake. Just west of that is what looks like a dry creek bed which is actually a channel dug in 1854 by the McIntyre Iron and Steel Co. In combination, the lake and channel redirected all the water from the Opalescent River into another stream and thence to the company's iron furnaces at Tahawus, five miles away.[10]

These human creations illustrate the third concept of the wild, called by many the Modernist concept. Because the term "wild" seems to connote something alive, its use might be inappropriate in this context, for the central strand of early Modernist thought to which we are linking says that nature is not alive; it is a dead machine, devoid of any moral or other meaning. It is thus something that we may exploit in any way we like. Whatever their positive legacies, thinkers like Descartes, Bacon, and Hobbes also bequeathed us this mechanistic attitude toward nature.[11]

The effort to reroute the Opalescent, and the nineteenth and twentieth century clear-cutting of the Adirondack forest are striking illustrations of the mechanistic attitude toward nature. Human desire and intention are paramount in this tradition, and nature is a mere stock of literally raw material upon which we are to impose our designs and improve it. Notice that this concept of wilderness is compatible with the previous view; and when given religious blessing, it becomes a nearly irresistible force. Needless to say, this concept has gotten us into some extreme environmental trouble, and we are beginning to abandon it.

For a Modernist, the view from Mt. Marcy might represent many things. As in the previous concept of wilderness, the natural areas are challenges to human ingenuity. But the unregenerate Modernist would primarily see the area, as did the early miners and loggers, only as something with potential for development. Without the resourceful hand of human industry, the wilderness area is, then, a symbol of wasted opportunity. Any suggestion that there might be some non-use value here, or that here a set of vital processes, on infinitely varying scales, sets a moral limit to our actions, is foreign to the Modernist.

Again simplifying history, we come to what many of us think of as a better approach to wilderness. It is perhaps the concept that many of us in the early twenty-first century, including many defenders of wilderness, assume when we think about wilderness; and it is enshrined in national and state policy. The United States Wilderness Act of 1964 defines wilderness as "a place where the earth and its community of life are untrammeled by man where man is a visitor who does not remain."[12] New York State's management plan for the wilderness area encompassing Mt. Marcy basically assumes the same definition.[13] Here wilderness and humans are again defined as different, as belonging to separate realms. However, to that perceived dichotomy this approach adds a more positive evaluation of wilderness than does its purely Modernist predecessor. Rather than demonizing or mechanizing it, this fourth view sees nature's independent and continuing functioning as something good, something we should preserve and protect. Certainly this understanding improves on the previous two concepts because it accords some respect to nature, recognizing that modern humans need to control their actions, that some natural areas should remain as close to their natural states as possible. J. Baird Callicott calls this the "received" concept of wilderness; and those who defend it have offered various justifications for the call for moral consideration, some of them prudential, some based in human ethics, some based in ecocentric concerns.[14]

If we ascend Marcy yet again and use this more contemporary idea of wilderness, the vista takes on new meaning. The spectacular views are as they should be: The forest has re-grown, reclaiming the clear-cuts, obscuring the engineers' tinkering. This is not our home. If anything, Lake Placid, with all its modern amenities, is our home. But neither is this our enemy, or the devil's home, or a lifeless, inert resource. It is wilderness, an increasingly rare thing in the twenty-first century, something different from us which we are somehow morally obliged to respect.

Perhaps surprisingly, it is criticism of this, the "received" concept of wilderness, by some environmental thinkers that leads to my defense of wilderness. As I said earlier (and simplifying yet again!), I agree with the basic ontological conclusion that nature and humans are not ontologically separate, but not with the implications some thinkers have drawn from it.

THE CURRENT DEBATE

What could possibly be wrong, from an environmentalist's perspective, with the idea of wilderness that has enabled us to preserve millions of acres of nature? When we examine it carefully, the "received" concept turns out to be problematic after all. Though Callicott, a philosopher, and William Cronon, a historian, have different emphases and overall outlooks, their analyses do agree on some points.[15] Among other problems, they say, the "received" concept is ethnocentric. It ignores the fact that what Euro-Americans call pristine wilderness was in fact home to millions of indigenous North Americans. So, for example, when a US law says wilderness is a place where humans are only visitors, it continues the legacy of oppression against First Peoples, ignoring their claims to the land as well as their influence on it.[16]

This problem is not only of historical interest; it is built into the "received" concept of wilderness. Some Native Americans have been denied their hunting rights in National Parks; and now that we naively and arrogantly believe it appropriate to export our ideas, sometimes imposing them on the rest of the world, indigenous peoples have been excluded from newly created National Parks because the parklands are supposed to be wild in this "received" concept of wild, which includes in its meaning, "devoid of permanent human residents." Designation of areas as National Parks in India and Africa have resulted in expulsions or the prohibition of traditional indigenous uses of those areas.[17]

The underlying, crucial problem lies in a philosophical assumption embedded in the "received" concept of wilderness. Specifically, in another of their criticisms of that concept, Cronon and Callicott point out that it assumes an ontological dualism between humans and nature. This dualism is kin to René Descartes' separation of mind and body. It says that we are one sort of thing and wilderness another. In philosophical terminology, the difference is a metaphysical or ontological one. There is no more fundamental or complete kind of difference. Because that kind of difference is so basic, having to do with the very nature of the types of things that fall into each category, the differences must be recognized and maintained in all other contexts. Recall the words of the Wilderness Act: "where earth and its community of life are untrammeled by man." In this understanding, "earth and its community of life" clearly are distinct from humans. Thus, if we want wilderness, we must have no people living in it! To make matters worse, it was the fundamental dualism of the Christian and the mechanistic views of wilderness that legitimated environmental destruction. With us on one side of the alleged ontological divide and nature on the other, and with the assumption of human superiority, wilderness was simply fair game. And now, in the "received" view, we have a concept that purports to be more benign yet still includes that same destructive dualism!

But this alleged problem might not be a problem at all. If dualism is the metaphysical theory that best captures the structure of reality, then we will have either to construct a benign dualism or to live with dualism, trying to avoid the kinds of environmental destruction it may have caused or contributed to in the past. Others have extensively analyzed and critiqued the earlier forms of dualism, represented here by the nature-as-evil and nature-as-machine views of wilderness.[18] Instead, I want to concentrate now on the dualism of the "received" view, which has its defenders among environmental philosophers. For example, Holmes Rolston, III explicitly defends an ontological separation between humans and non-human nature.[19]

In several of his publications, Rolston does not deny that humans, like any other species, are natural in the sense that we evolved through the same processes as did all other species on earth. However, he asserts that we have also evolved "*out of* nature" (Rolston's emphasis).[20] He argues that we can see the fundamental difference, a difference in being, between us and nature in many ways: for example, the way that information is transmitted in human culture and in biological nature. Biologically, information is transmitted "intergenerationally on genes." That transmission can take considerable time, is limited in its reach to the offspring of the organism, and does not include acquired learning. But in culture, information "travels neurally," can be extremely rapid (even before the computer revolution), includes acquired learning not only of the parent organisms but also of huge numbers of other humans, both living and dead. Humans "form cumulative transmissible cultures," and no other creature does so. In addition, humans have choices in their lifestyles not available to other species, and that deliberative capacity "is without real precedent in nature," as are our highly developed conceptual abilities and moral capacity.[21] Among other things, Rolston explicitly applies his dualism to wilderness, which he sees as distinct from human culture.[22]

Here, two crucial questions arise: First, assuming for the moment that his arguments are cogent, is Rolston's brand of dualism environmentally benign or problematic? Second, are Rolston's arguments for dualism indeed cogent? Given that Rolston is a central figure in environmental philosophy, defending the intrinsic value of nature and explaining the nature and force of our moral obligations toward the environment, how, specifically, could his version of dualism be problematic? One possible response is to claim that *any* form of dualism is environmentally problematic. That is Val Plumwood's response. She argues that all dualisms introduce a hierarchy of their two poles, with one pole seen as more valuable than the other; and she attempts to show that the hierarchy is necessitated by the very logic of dualism.[23] If she is correct, then even Rolston's seemingly benign dualism conceals the same destructive hierarchy as did Descartes' mind/body dualism and we must evaluate it accordingly. However, although Plumwood may be right, a thorough evaluation of her analysis of dualism and a careful application of it to Rolston's views would take us too far afield. There is a shorter way to see how problems can arise.

To do so, we only need to re-examine what dualism is. In a human/nature dualism there is, necessarily and obviously, a separation between us and the world. That separation, equally obviously, puts us on one side of the dichotomy and nature on the other. Once that separation is accepted (and remember, *we* are the beings who accept it) and is adorned with an honorific description like "true" or "valid," we are alienated from nature. "Alienated" might seem too strong a word in the context of Rolston's environmental ethic, but it seems accurate. It is he who says that we evolve "*out of* nature" (Rolston's emphasis). Once out of nature, our relationship with it is necessarily a relationship to something other than ourselves; and while we might have an available theory of intrinsic value in nature, that value can never equal or surpass our own. That imbalance will enter all our decision-making. Nature is at a permanent disadvantage, and that disadvantage is easily exploitable. Thus, even a "benign" dualism is problematic in coming to terms with wilderness.[24]

Rolston worries that thinkers who argue against dualism miss important differences between humans and wild nature: "Environmental philosophy needs to see the difference in being human, and only after we get clear about that, do we also want to see the senses in which, although evolved out of it, culture has to remain in relative harmony with nature."[25] As I shall shortly argue, that worry is legitimate when considering Cronon's and Callicott's solutions to the problem; but there is a way we can acknowledge those differences without positing a metaphysical dualism.[26] But before offering my criticisms of Cronon and Callicott and presenting a sketch of that alternative account, I have to answer the second question, namely, are Rolston's arguments for dualism cogent? This is particularly important because the possibility remains that, although dualism, even Rolston's version of it, might be environmentally problematic, it might nevertheless be true.

A pertinent question to ask about Rolston's arguments is whether all the human characteristics he notes are constituted of physical entities and processes. For example, is information transfer within human cultures a physical process? Rolston uses such transfer as evidence for a metaphysical difference between us and nature. However, without the introduction of non-physical factors and the requisite, explicit argumentation for such introduction, we can legitimately understand human information transfer as a physical process and as such a part of nature.[27] That is, unless Rolston provides further argumentation to show that human information transfer involves something that is non-physical or out of nature, he has only pointed out an important phenomenon. He has not shown that that phenomenon forces us to accept dualism.

Rolston's worry noted above indicates that he thinks about the question in a very constricted way: Either there is dualism or all human differences are erased. But if we deny dualism we are not obliged to minimize such human abilities as information transfer. Nor are we forced to say that such transfers are "nothing but" the biochemical interactions of molecules. In order to claim that human information transfer, despite its complexity, remains a physical process and as such is not "out of" nature in Rolston's metaphysical sense, we needn't endorse a narrowly reductionist account of human communication.

In supporting or defending any distinction, it is, of course, a legitimate argumentative strategy to note characteristics of one of the distinguished phenomena not possessed by the other. That is Rolston's strategy in describing the various cultural phenomena which he thinks metaphysically different from wild nature. However, there are other plausible accounts of those phenomena, accounts which do not require dualism. For example, besides information transfer, he notes human intentionality, deliberative choice, conceptual ability, transmissible culture, and moral capacity.[28] Although many thinkers have considered intentionality as demonstrating a metaphysical difference between the mind and "mere" physical processes, others include intentionality in physical theories of mind. Rolston himself, when speaking of human deliberation, admits that "animals may have some precursor options in what they shall do." Other thinkers have noted the cognitive abilities of other animal species and see them as more continuous with our own abilities. It may even be true that some other animals possess and transmit their own culture. Finally, our supposedly unique moral capacities may not be so unique after all.[29] It is not my intention to exhibit even the outlines of those other, non-dualistic accounts of these important phenomena. The point is that such alternative accounts exist and therefore Rolston's argument for dualism requires much more evidence to establish the metaphysical distinction he asserts. Without such evidence, we are entitled to seek an alternative to dualism.

Moreover, if we take evolution seriously, it is difficult to see how dualism is even possible. Recall that dualism posits the existence of two kinds of entity that are different in being—fundamentally, essentially different. But if we evolved through the same processes as any other species, there seems no way to explain how a product of evolution (humans) could become something fundamentally different, *different in being,* from its other products. This is not to deny that humans differ biologically and perhaps in other ways (in our social relations) from other species. That would be silly. It is just to say that we are another manifestation of complex physical organization. Neither is it to commit, as Rolston charges, the genetic fallacy, "which confuses what a thing now essentially is with what its historical origins once were."[30] Again, to argue that humans are continuous with, not ontologically separate from, nature is not to deny that we possess differences. Humans are indeed different; for example, we differ from other primates, though we undoubtedly evolved from primates.

It is beyond the scope of this essay to pursue these arguments and counter arguments in further detail. My purpose in describing and raising problems for them is to emphasize that we need not interpret the differences Rolston sees between human culture and nature as constituting a metaphysical difference. In other words, the arguments for dualism are not cogent. Plus, when we recall the destruction that past dualisms justified and note the same potential lurking in "benign" dualism, it becomes even more dubious a candidate for a central role in an environmental philosophy.

In very broad strokes, the counter view is that humans are not separate from nature but are a part of it. There is growing support for that conclusion. Callicott bases his arguments for it on evolution and ecology, the former demonstrating our biological origins in and connections to the rest of life on earth, the latter showing that we remain members of complex, integrated ecological communities or systems. Moreover, as previously noted, there are other contemporary philosophical supports for the idea that we are part of nature, in philosophy of mind, in epistemology, and in other domains of philosophy.[31] I note these other philosophical theories not to indicate that they defend a concept of wilderness. Few mainstream philosophers explicitly discuss the environment in their published writings. Rather, those theories represent views about related philosophical questions that harmonize with the claim in environmental philosophy that humans are essentially tied to nature, not just in a practical sense, but in our very being.

But there's the rub: If we're part of nature, if there is no separation between humans and nature, then wilderness seems to be at best a human projection on the world. That is Cronon's conclusion. Certainly his social history of the concept of wilderness demonstrates clearly that the idea is not a neutral one, that it has had all sorts of value assumptions packed into it, not all of them desirable from our contemporary standpoint (sexist, classist, racist, reactionary). His conclusion is that wilderness is merely a social construct and that all of us, especially environmentalists, need to get beyond it so we can attend to the important environmental and social issues closer to hand. Paradoxically, Cronon professes no antipathy for lands currently preserved as wilderness areas; in fact, he thinks we should continue to protect them. Thus, when Cronon looks out from Mt. Marcy, he certainly sees something he wants to keep; but it is at the same time not something different from us with value independent of us. It is something we have created or invented, an artifact, an extension of us. Indeed, Cronon's critique of wilderness is the lead essay in a book he edited, *Uncommon Ground,* which, in its first printing, had the revealing subtitle, *Toward Reinventing Nature.*[32]

It is fair to say that Cronon's view assimilates wilderness into culture. Callicott reverses the assimilation, arguing instead that culture is nature—as natural, he says, as the works of termites and beavers, other species that radically alter their environments. As I mentioned earlier, he bases this conclusion on evolutionary and ecological theory. Interestingly, and again paradoxically, Callicott, like Cronon, advocates keeping our currently preserved wilderness areas. He proposes to change their designation from wilderness to "biodiversity reserves," a name he thinks will automatically forestall any question whether the areas can be exploited for human commercial or industrial purposes. Along with that proposal, Callicott envisions a somewhat utopian—ecotopian—"sustainable development alternative" to wilderness preservation. In this alternative vision, human settlements, with all their various activities—but with environmentally benign and appropriately scaled commercial, industrial, and agricultural activities—co-exist within and around areas that we would currently refer to as wilderness.[33] Thus, given his naturalistic arguments, Callicott would have to say that the view from Mt. Marcy reveals only various parts of nature. The spectacular mountain vistas, the Village of Lake Placid, the Olympic ski jumps, the fire towers, the farmlands—all of it is equally "natural" to him.

While I share Callicott's hope that we might someday live in an environmentally enlightened and benign society, I still object to his and Cronon's views. I find incoherent the notion, present in both their arguments, that there is something wrong with the idea of wilderness, but not with the areas preserved as such. A mere change of name ("biodiversity reserves") cannot eliminate the incoherence. Here is that incoherence as clearly as I can state it: These two thinkers object to the dualism they find in the "received" concept of wilderness, a dualism that identifies wilderness as different from human-altered environments. Instead, they argue that we are not separate from nature. At the same time, they wish to keep, and urge respect for, wilderness areas (by whatever label). But those areas are identifiable only by their differences from the human-altered environment, *differences* that these two thinkers deny.

I think this incoherence derives from the same source in both thinkers. Although they object to dualism, their thinking is determined by it. Wishing to avoid dualism, they each create monistic-seeming theories, but those theories carry the "ghost" of dualism. Faced with the human/wilderness dualism of the "received" concept, Cronon denies the wilderness pole, assimilating nature to culture. Callicott makes a parallel move, but identifies culture with nature. I think both identifications are problematic because they start with dualism as a model, and negate one of its poles, thereby denying important differences.[34] The motivation to avoid dualism is a good one, but dualism determines both of these theories and the results are unacceptable.

So I seem to be at an impasse. On the one hand, I agree with the ontological claim that dualism is both troublesome and false. The world is one kind of thing and humans are part of it. On the other hand, I find problematic the proposals that we should jettison the idea of wilderness as something separate from us. Do I contradict myself, then, and advocate a return to a dualistic concept of wilderness? Unlike Walt Whitman, I cannot rest content with an apparent contradiction. What to do then?

AN ALTERNATIVE PROPOSAL

When in doubt, take a hike! I return to Mt. Marcy. I have already described the spectacular view. Let me stress that besides the vast wilderness areas visible from the summit, there is also the unmistakable, unambiguous view of a major resort town and other human alterations. Here we already have part of Callicott's vision, wilderness and culture side by side, each continuing according to its own processes. Indeed, in the Adirondacks, the two grade off into one another in accordance with the various patterns created in the land use plans and zoning regulations. The Adirondack Park is six million acres of land, a mix of private and public lands, with the public lands designated forever wild by the state constitution. It is a grand preservationist experiment, not clearly and entirely envisioned as such by those who began it in the nineteenth century; and although its status is still contested, it holds great promise as a model for a corrected human/wilderness relationship.[35]

However, that is not to endorse Callicott's "sustainable development" proposal, to which I object on two grounds. First, on the practical level, a change of name will not provide any added protection for wilderness areas. Until respect for the non-human becomes much more widespread and is incorporated into our culture's everyday decisions, neither designation as wilderness, nor as "wildlife refuge," nor as "biodiversity reserve" will protect wild areas.

Second, on the conceptual level, and as already argued, Callicott is wrong simply to identify culture with nature. The view from Mt. Marcy shows that clearly. If Callicott's proposals prevail and we abandon the recognition of difference within the overall unity with nature, the human habitation represented by Lake Placid Village would not be only in the far off distance from the summit. Like the mining and logging operations of the nineteenth century, it would reach right into Mt. Marcy's wilderness area, which would no longer be what it is. In other words, if we only emphasize the connections between humans and wild nature, we begin to undermine the most basic rationale for preserving any natural area. After all, if we are, in Callicott's words, "completely natural," what could be objectionable about any of our uses of nature, which, to follow his logic, must also be completely natural? There seems then to be little reason to restrain ourselves. We are just as "natural" as wilderness so our presence there, including our habitations and industry, makes no difference. The protection that Callicott wants for "biodiversity preserves," or whatever we call wilderness areas, vanishes.

We get a similar result if we follow Cronon. If wilderness is a purely human construct, it carries with it no value of its own, only what we confer on it. We could easily change our minds, as many propose we should do in the case of the Arctic National Wildlife Refuge, and as many continue to propose in the Adirondacks. So accounts that acknowledge, from whatever theoretical vantage point, only the connections or identity between humans and the non-human world represent a real and substantial threat to the wild, even those, like Cronon's and Callicott's that purport to be environmentally enlightened.

I propose instead that we continue to respect the wild because it is simultaneously something of which we are a part and which lives on in its own way, continuing the processes that spun us and uncounted other species out of its incredible complexity. It is also home to those other species, beings who have their own ways and need wilderness—a place without human domination—for their own basic needs. On the level of terminology, I can think of no better name than "wilderness" for such areas of our planet. It denotes clearly that those areas are places we humans choose not to inhabit, in which we do not use our evolved capacities for exploitation, where we consciously recognize that we have many companion species that need their own habitats to continue their own natural histories. That is, in wilderness we do not alter or destroy ecological processes to our own ends but recognize that there are many, many ends besides our own.[36]

The term "wilderness" also carries the romantic connotations of a place where there are mysteries, adventure, excitement, challenge, danger; and I think that is also a good thing for us to keep.[37] A world made only of our own artifacts and totally under our control would be, I think, immeasurably impoverished. Where there is wilderness there is richness in the form of the unknown, the uncontrolled, the chance for novelty, the unexpected, the possibility of discovery. Eric Katz writes, "[V]alue exists in nature to the extent that it avoids the domination of human technological practice. Technology can satisfy human wants by creating the artifactual products we desire, but it cannot supply, replace, or restore the 'wild.'"[38] Moreover, we can jettison, and I believe are already in the process of jettisoning the ethnocentric connotations of the term. For example, there are other more recently created parks containing designated wilderness areas whose creation did not include expulsion of humans already living there. Abruzzi National Park in Italy, named the Adirondack Park's "sister" park, is apparently managed along similar lines, with the human and the wild coexisting. Gros Morne National Park in Newfoundland, Canada, is a recently created relative as well, as is Bolivia's Madidi National Park, begun at least with the promise of indigenous cultures' continuing in their ways within the park even while ecotourism intrudes.[39] Whether any of these experiments survive as initially designed, only time will tell; but they are promising models for the future, partly illustrating, I think, that dualism and ethnocentrism are not necessarily a part of the concept of wilderness.

None of my support for the concept of wilderness retracts my rejection of dualism. Rather, it points out that even given our unity with nature, we must recognize that there are simultaneously important differences within that unity between the non-human and us. If we ignore those differences, we invite the homogenization of the world, a continuation of the mechanistic conquest of the non-human.

No doubt some readers are wondering by now where my view fits in the range of ontological theories, other than in the broad and unenlightening category of non-dualism (or even crypto-dualism!). Is it a form of materialistic monism? Or neutral monism? Or perhaps a pluralism? Without wanting to be evasive, I can only say that I don't think that there is an exactly matching ontological category in Western philosophy. Materialistic monism claims that ultimately everything is the same kind of thing; but the view I am proposing, though seeing and emphasizing connections and unity, resists the reduction of everything to one. Pluralism seems to me to have the opposite defect, namely, that it emphasizes the separations among kinds; that is, it sees differences without connections. With its rejection of dualism and its attempt to avoid both materialistic monism and idealism, neutral monism has some positive features; but its neutral "stuff" seems either mysterious or ad hoc.[40] I am attempting to follow Plumwood's ecofeminist analysis, which stresses that an adequate view must acknowledge both connections and differences within an overall unitary nature.[41]

As a very tentative attempt to satisfy the request to situate such a view within Western philosophical traditions, I suggest a partial *parallel* between it and a portion of Aristotle's metaphysics, specifically his discussion of psychology. There, Aristotle points out that living things all have souls—the Greek word is *psyche*. Plants possess a nutritive soul, animals both a nutritive and sensitive soul, and humans a rational soul that includes the other two types.[42] So Aristotle recognizes that there are important differences among the kinds of living things he knew about; but he also sees a continuity, a commonality, a *psyche,* present in all of them.

Let me hasten to add that I am not advocating any form of panpsychism. I am only noting that, at least in this one place, Aristotle recognizes both continuity and differences, as does the view I am proposing. The kind of continuity and the kinds of difference contrast because there are tremendous divergences between an evolutionarily and ecologically informed theory and that of Aristotle, which, with its fixed essences, did not allow for anything like evolution as we understand it and which was unabashedly hierarchical and anthropocentric. Indeed, because he believed that nature is made for man, he might well think absurd the idea that humans should restrain their power and set aside wilderness areas for any other than human interests. And though as an astute biologist he would no doubt appreciate the concept and processes of an ecosystem, his ethical theory leaves no room for direct moral duties toward non-human entities like species and ecosystems, duties that are a cornerstone of contemporary environmental ethics.[43]

But Aristotle's perception of continuity-with-difference, limited as it is, sets a precedent, I think, for the kind of theory Plumwood and I are advancing. Speaking for myself, I have never thought of Aristotle as either a materialist or a dualist. He clearly abandoned Plato's radical dualism, but he never accepted anything like the monistic, materialistic worldviews of many pre-Socratics. I stress that I offer this comparison only tentatively; it may ultimately fail; there may be better precedents or analogues, which I am overlooking. And even if there proves to be no specific precedent in our traditions, I would still maintain that any acceptable ontological account must acknowledge the reality of both connection and difference without reducing one to the other.[44]

CONCLUSION

I am consciously treading a fine line here, and some readers may think that my arguments with Rolston and Callicott amount to hair-splitting. I greatly admire both of their work. However, I remain convinced that both of their theories contain serious problems. Rolston's dualism is ultimately environmentally problematic; and his arguments for it, in my estimate, are not cogent. Callicott's ultra-naturalism (and Cronon's ultra-culturalism) sacrifices too much to achieve unity. In opposition to my view, others may believe that, despite the undeniable connections between us and the rest of nature, once we note differences, we have acknowledged dualism. But that is too quick a reaction—and too easy, given our culture's long history of accepting a human-dominated dualism. I again refer to evolution. At various moments in the past, new characteristics emerged in the organic world: photosynthesis, locomotion, vision, social behavior. Those emergent characteristics do not each introduce another layer of being, another metaphysical category, even though they differ from their precedents. Yet those and other evolved capacities created levels of complexity that went beyond—in some cases, well beyond—those precedents. Arguing that humans' evolved characteristics represent something entirely new, so unique as to cause, and to require explicit recognition of, a new category of being, is to be dazzled by ourselves and our accomplishments. Instead of that dualist reaction, with all its inherent problems, we should recognize our continuity with the world. With the dualists, I do not want to reduce our differences to some "nothing but," simply and destructively equating them with their evolutionary precedents. Rather, I want to account for them as emergent within nature, requiring different and perhaps more complex kinds of understanding and explanation, but not metaphysically different from nature.

Am I trying to have my cake and eat it too? If that means retaining wilderness without dualism, then my response is yes. However, that is not, as the aphorism implies, to embrace a contradiction. It is, rather, to recognize the human place *within* nature and to recognize and celebrate the fact that there are many interrelated "places" within nature, many of which exist, continue, and ought to continue, autonomously, without our interference and control. In that kind of account, wilderness is still wild; but it is not so much opposed to us as needed, and needed in many, many ways. However, for us modern humans, wilderness excursions must end, and we have to descend from our imaginative climb of Mt. Marcy. So we will not have the opportunity here to review the many reasons offered for wilderness preservation.[45] As we descend, we should rejoice in the fact that the Adirondack wilderness and other similar areas exist,[46] that they are what they are, and that they form an integral, essential part of our world. Some readers will agree with me that we need still more such areas, not only for ourselves but also for myriad other species, and for yet other reasons. Perhaps other readers will disagree on the need for more wilderness and with all the arguments for it. Either response acknowledges the significance of wilderness in our thinking about humans' relationship with the rest of the world.

As discussed above, there are several competing ideas of wilderness. My own view is that we need a wilderness concept that acknowledges both the connections and the differences between wild nature and us. Emphasizing only the differences has given us a dualism alienating us from non-human nature and resulting in environmental destruction. Emphasizing only the connections ignores important differences and undermines the valid reasons for limiting human interference with non-human nature. Only by acknowledging both the differences and connections will we develop an accurate, workable ontology and maintain a livable planet.

NOTES

This essay is a substantially revised version of a lecture presented to the faculty of Empire State College, State University of New York, in 2001, in fulfillment of the author's receipt of the college's Susan H. Turben Award for Excellence in Scholarship. The author thanks his ESC colleagues as well as participants in the North American Interdisciplinary Conference on Environment and Community at Weber State University, Ogden, Utah, March, 2002, for suggestions for improvement. He also thanks an anonymous reviewer for *Philosophy and Geography* for helpful comments and criticisms.

1. This acknowledgment is not an unconscious dualist assertion on my part. Other species are capable of overpopulating and destroying their environments, though our capacities here are perhaps unprecedented. For example, it is well known that white tail deer, in the absence of natural predators, will overgraze and destroy their ranges. Analogous things can happen on the microscopic level as well. For an example, see Ralph Buchsbaum and Mildred Buchsbaum, *Basic Ecology* (Pittsburgh: Boxwood Press, 1957), 103.

2. There are at least forty-one 6000-foot peaks in the Eastern US. For a listing, see www.americasroof.com/6000.shmtl. In addition, there are four peaks in New Hampshire higher than Marcy but less than 6000 feet. Mt. Washington

in New Hampshire is over 6000 feet and is counted in the first total. Given the many 6000-foot peaks in the Southern Appalachians, I would guess that there are also many 5000-foot peaks in that region that are higher than Marcy, though I have been unable to find a listing of them.

3. Paul Schneider, *The Adirondacks: A History of America's First Wilderness* (New York: Henry Holt & Company, 1997), xii.

4. The following historical taxonomy of theories of wilderness is based on Max Oelschlaeger's account in *The Idea of Wilderness: From Prehistory to the Age of Ecology* (New Haven: Yale University Press, 1991). The presentation is my own.

5. See Oelschlaeger, *The Idea of Wilderness,* especially Chapter 1.

6. Philip Terrie, *Contested Terrain: A New History of Nature and People in the Adirondacks* (Syracuse, NY: The Adirondack Museum and Syracuse University Press, 1997), 1.

7. Oelschlaeger urges something like this when he calls upon us to become "posthistoric primitives" and to adopt a "new-old way of being." See *The Idea of Wilderness.* Dave Foreman comes close to doing so in *Confessions of an Eco-Warrior* (New York: Harmony Books, 1991).

8. Oelschlaeger, *The Idea of Wilderness,* Chapter 2.

9. See Chapter 1, "Old World Roots of Opinion," in Roderick Nash, *Wilderness and the American Mind,* 3rd ed. (New Haven: Yale University Press, 1982).

10. Barbara MeMartin, *Discover the Adirondack High Peaks,* 2nd ed. (Canada Lake, NY: Lake View Press, 1998), 100.

11. For a similar assessment see Carolyn Merchant, *The Death of Nature: Women, Ecology and the Scientific Revolution* (San Francisco: Harper, 1990). Frequently, Descartes is singled out as the main offender; but for a discussion of Hobbes's contribution, see Keekok Lee, *The Natural and the Artefactual: The Implications of Deep Science and Deep Technology for Environmental Philosophy* (Lanham, MD: Lexington, 1999), 23–31.

12. Public Law 88-557, 88th Congress, September 3, 1964, Section 2(c). Reprinted in *The Great New Wilderness Debate,* ed. J. Baird Callicott and Michael P. Nelson (Athens: University of Georgia Press, 1998), 120–30. The definition of wilderness is on page 121.

13. New York State Department of Environmental Conservation, *High Peaks Wilderness Complex Management Plan: Wilderness Management for the High Peaks of the Adirondack Park* (Albany, NY, 1999), 1.

14. J. Baird Callicott, "The Wilderness Idea Revisited: The Sustainable Development Alternative," in *The Environmental Professional,* 13 (1991): 235–47. Reprinted in Callicott and Nelson, 337–66; this anthology contains writings representing just about every perspective on wilderness.

15. Callicott, "The Wilderness Idea Revisited"; William Cronon, "The

Trouble with Wilderness; or, Getting Back to the Wrong Nature," in *Uncommon Ground: Toward Reinventing Nature,* ed. William Cronon (New York: Norton, 1995), 69–90. After its first printing, the subtitle of Cronon's anthology was altered to *Rethinking the Human Place in Nature.*

16. For an account of the extent of Native American influence on the land, see William Cronon, *Changes in the Land: Indians, Colonists and the Ecology of New England* (New York: Hill and Wang, 1983). See also William M. Denevan, "The Pristine Myth: The Landscape of the Americas in 1492," in Callicott and Nelson, 414–42; and Steven R. Simms, "Wilderness as a Human Landscape," in *Wilderness Tapestry: An Eclectic Approach to Preservation,* ed. Samuel I. Zeveloff, L. Mikel Vause, and William H. McVaugh (Reno: University of Nevada Press, 1992), 183–202. A small example is that after Yosemite National Park was created, its white managers had to discover that the Native Americans who had lived in Yosemite Valley had actively managed the vegetation there. See www.modbee.com/yosemite/.

17. For one case of Native Americans excluded from traditional hunting grounds, see Cronon, "The Trouble with Wilderness," 79. Regarding India, see Deane Curtin, *Chinnagounder's Challenge: The Question of Ecological Citizenship* (Bloomington: Indiana University Press, 1999). For Africa, see Aaron Sachs, *Eco-Justice: Linking Human Rights and the Environment,* Worldwatch Paper 127 (Washington, DC: Worldwatch Institute, 1995). But others have argued that, though it might at times have been unjustly or mistakenly used, the concept of wilderness is not confined to Euro-Americans. See, for example, Dave Foreman, "Wilderness Areas Are Vital: A Response to Callicott," *Wild Earth,* 4 (1994): 67. The essays in *The World and the Wild: Expanding Wilderness Conservation Beyond Its American Roots,* ed. David Rothenberg and Marta Ulvaeus (Tucson: University of Arizona Press, 2001) are additional indications that the concept is not exclusively Euro-American. Regarding the ethnocentricity charge, it is also interesting to note that there is a triennial World Wilderness Conference, with delegates from all over the world. Of course, that does not refute the general ethnocentricity charge, but it at least calls it into question. The expulsion of indigenous people from their lands to "create a wilderness" would remain an oppressive, ethnocentric interpretation of wilderness, however.

18. One can read much of twentieth and twenty-first century philosophy of mind as a series of attempts to dispel Cartesian dualism. For its original formulation, see René Descartes, *Meditations on Philosophy,* especially Meditation II. For a lucid explanation of Cartesian dualism and a useful survey of contemporary views, including other forms of dualism, see John Heil, *Philosophy of Mind: A Contemporary Introduction* (London & New York: Routledge, 1998). The environmental philosophy literature is replete with varying critiques of dualism. Besides Callicott's writings, a few examples are: Carolyn Merchant,

The Death of Nature: Women, Ecology and the Scientific Revolution (San Francisco: Harper & Row, 1980); Val Plumwood, *Feminism and the Mastery of Nature* (London: Routledge, 1993); Pete Y. A. Gunter, "The Disembodied Parasite and Other Tragedies; or: Modern Western Philosophy and How to Get Out of It," in *The Wilderness Condition,* ed. Max Oelschlaeger (Washington, DC: Island Press, 1992), 205–19. Oelschlaeger's *The Idea of Wilderness* is thoroughly anti-dualist.

19. Holmes Rolston, "Naturalizing Callicott," in *Land, Value, Community: Callicott and Environmental Philosophy,* ed. Wayne Ouderkirk and Jim Hill (Albany, NY: State University of New York Press, 2002), 107–22. See also Rolston, *Conserving Natural Value* (New York: Columbia University Press, 1994), especially Chapter 1. Eric Katz is another dualist environmental philosopher. See his collection of essays, *Nature as Subject: Human Obligation and Natural Community* (Lanham, MD: Rowman & Littlefield, 1997), especially "The Big Lie: Human Restoration of Nature," 93–108. In my essay, "Katz's Problematic Dualism and Its 'Seismic' Effects on His Theory," I critiqued Katz's dualism. That essay, Ned Hettinger's similar critique, and Katz's response to both of us are in *Ethics and the Environment,* 7, no. 1 (2002): 102–46.

20. Rolston, "Naturalizing Callicott," 110; and *Conserving Natural Value,* 5.

21. Rolston, "Naturalizing Callicott," 110–11.

22. In his "The Wilderness Idea Reaffirmed," *The Environmental Professional,* 13 (1991): 370–77. Reprinted in Callicott and Nelson, 367–86.

23. Plumwood, *Feminism and the Mastery of Nature.* See especially Chapter 2, "Dualism: The Logic of Colonisation." For Plumwood, the problem with dualism is not limited to its implications for the environment but extends to gender and other social and political relations, into all of which dualism introduces an oppressive value hierarchy.

24. This argument echoes, in very abbreviated form, Plumwood's account.

25. Rolston, "Naturalizing Callicott," 110.

26. So I do not think that Rolston has to be a dualist; but that is a topic for another essay.

27. Here I am using "nature" to mean something like "that which occurs without supernatural influence." The word and its cognates are notoriously ambiguous. For discussions, see Neil Evernden, *The Social Creation of Nature* (Baltimore: Johns Hopkins University Press, 1992) and Kate Soper, *What Is Nature? Culture, Politics and the Non-Human* (Oxford: Blackwell, 1995).

28. Rolston, "Naturalizing Callicott," 111.

29. On intentionality in a physical theory of mind, see John Searle, *The Rediscovery of Mind* (Cambridge, MA: MIT Press, 1992); and Plumwood, *Feminism and the Mastery of Nature,* 136–40. On animal belief and cognition, see Tom Regan, *The Case for Animal Rights* (Berkeley: University of California Press, 1983), Chapters 1 and 2, and 243 ff. Regarding animal morality, see Frans

de Waal, *Good Natured: The Origins of Right and Wrong in Humans and Animals* (Cambridge, MA: Harvard University Press, 1997), and S. F. Sapontzis, *Morals, Reason and Animals* (Philadelphia: Temple University Press, 1987), 27 ff. De Waal also claims that chimpanzees have and transmit a culture, in his *The Ape and the Sushi Master: Cultural Reflections of a Primatologist* (New York: Basic Books, 2001).

30. Rolston, "Naturalizing Callicott," 110. I am not convinced that this actually is an instance of the genetic fallacy, which is usually defined as rejecting a viewpoint because of its source or history.

31. In this connection, to the works cited in note 30, we can add Donald Davidson, "Psychology as Philosophy," in *Philosophy of Psychology,* ed. S. Brown (New York: Macmillan, 1974); Richard Rorty, *Objectivity, Relativism, and Truth* (Cambridge, England: Cambridge University Press, 1991); Daniel Dennett, *Elbow Room: The Varieties of Free Will Worth Wanting* (Cambridge, MA: MIT Press, 1984).

32. Cronon, "The Trouble with Wilderness." Cronon criticizes the concept of wilderness throughout this essay. One reference to retaining wilderness preserves is on page 81.

33. The following essays by Callicott contain discussions of these ideas: "The Wilderness Idea Revisited" (see note 15); "Should Wilderness Areas Become Biodiversity Reserves?" in Callicott and Nelson, 585–94; and "A Critique of and an Alternative to the Wilderness Idea," *Wild Earth,* 4 (1994): 54–59.

34. Neil Evernden in *The Social Creation of Nature* (see note 28) criticizes materialist theories of human nature and what he calls "idealism" (by which I think he means social constructionism), on the grounds that they recapitulate dualism in the same manner I am accusing Cronon and Callicott.

35. For the history of the Adirondack Park's creation, see Terrie, *Contested Terrain* and Schneider, *The Adirondacks.* See also Alfred S. Forsyth and Norman J. Van Valkenburgh, *The Forest and the Law II* (Schenectady, NY: The Association for the Protection of the Adirondacks, 1996). For the Adirondacks as a modal of living with wild nature (and for other models), see Bill McKibben, *Hope, Human and Wild* (St. Paul, MN: Hungry Mind Press, 1995).

36. Ned Hettinger and Bill Throop argue that the wildness of natural systems is a central value for an ecocentric ethic, in "Refocusing Ecocentrism: De-Emphasizing Stability and Defending Wildness," *Environmental Ethics,* 21 (1999): 3–21.

37. A colleague, Elana Michelson, noted the sexist connotations of some of these features of wilderness I applaud here. She referred to Bacon's exhortation to "penetrate" nature to force "her" to reveal some of her "mysteries." I hope it is clear from the context that my rationale for retaining the term wilderness is to recognize and respect its mysteries as real limits to our subjugation and control of nature. I believe that, just as the term "wilderness" is losing its

ethnocentric and dualistic overtones, so too will it lose its sexist overtones. See the rest of my discussion and Plumwood, *Feminism and the Mastery of Nature,* 161–64.

38. Eric Katz, "The Call of the Wild: The Struggle against Domination and the Technological Fix of Nature," *Nature as Subject,* 110. I agree with Katz's admiration for and valuing of the wild, but not with his full evaluation of wilderness restoration. See my essay cited in note 20.

39. For some introductory information on Abruzzi, see www.pna.it/index.htm; for Gros Morne, see www.grosmorne.pch.gc.ca/default.htm; for Madidi, see Steve Kemper, "Madidi National Park," *National Geographic,* 197 (March 2000). For international perspectives on the relationship between people and wilderness, see Rothenberg and Ulvaeus, *The World and the Wild.*

40. Neutral monism is historically associated with Bertrand Russell and William James. See Frederick Copleston, *History of Philosophy,* Vol. 8, *Modern Philosophy: Bentham to Russell* (Garden City, NY: Image, 1967), 209–13. In contemporary philosophy of mind, John Heil accepts the label, "neutral monism" for his theory of mind. See his *Philosophy of Mind* (cited in note 19), especially the final chapter. How much his theory has in common with Russell's and James's and how appropriate the label for his theory are interesting questions, which I cannot address here.

41. Val Plumwood, *Feminism and the Mastery of Nature* (London: Routledge, 1993).

42. Aristotle, *De Anima,* especially Book II.

43. The environmental ethics literature is full of assertions and justifications of such duties. As one example, see Holmes Rolston, III, *Environmental Ethics: Duties to and Values in the Natural World* (Philadelphia: Temple University Press, 1988). Aristotle's is, of course, a virtue ethic, and there have been several proposals for a virtue-based environmental ethic. However, such proposals seem to me to focus too much on human excellence (virtue) rather than features of the non-human that might cause us to include it in our thinking about morality. For one example of a different view from mine, see Louke van Wensveen, "Ecosystem Sustainability as a Criterion for Genuine Virtue," in *Environmental Ethics,* 23 (2001): 227–41.

44. Since the original version of this essay was published, I have found another precedent for the type of view I am advocating here, namely, theories of emergence in the philosophy of mind. Although emergence is sometimes interpreted as dualistic (e.g., by J. Kim), I mention it here because most versions of it emphasize the kind of tight connection I see between culture and nature (within emergence theory, that is expressed as mind and body) while still allowing for differences. For thorough discussions, see the essays in *Emergence or Reduction? Essays on the Prospects of Nonreductive Physicalism,* ed. Ansgar Beckerman, Hans Flohr, and Jaegwon Kim (Walter de Gruyter, 1992); *The Emergence*

of Consciousness, ed. Anthony Freeman (Imprint Academic, 2001); a special issue of the *Journal of Consciousness Studies* 8 (2001); and Timothy O'Connor and Hong Yu Wong, "Emergent Properties," in *The Stanford Encyclopedia of Philosophy* (Summer 2005 edition), ed. Edward N. Zalta, http://plato.stanford .edu/archives/sum2005/entries/properties-emergent/.

45. For a compendium of such arguments see Michael P. Nelson, "An Amalgamation of Wilderness Preservation Arguments," in Callicott and Nelson, 154–98. I do not necessarily endorse Nelson's evaluations of all of the arguments.

46. Many people associate American wilderness areas with the western United States and Alaska. However, there are large tracts of wilderness in the east, including the areas so designated in the Adirondack Park and elsewhere, as well as some large de facto wilderness areas in the Northeast. See Christopher McGrory Klyza, "Public Lands and Wild Lands in the Northeast," and other essays in Klyza's edited volume, *Wilderness Comes Home: Rewilding the Northeast* (Hanover, NH: Middlebury College Press, 2001).

Jonathan Maskit

Something Wild? (1998, 2008)

Deleuze and Guattari, Wilderness, and Purity

In Jane Austen's *Mansfield Park* a group of wealthy young people, suitably chaperoned of course, sets off in search of amusement in a tour of Sotherton, a local estate. After looking over the house itself the party takes in the grounds. Having taken several turns around the terrace, they arrive again at "the door . . . which opened to the wilderness." Finding the door unlocked, "they were all agreed in turning joyfully through it, and leaving the unmitigated glare of day behind. A considerable flight of steps landed them in the wilderness, which was a planted wood of about two acres, and though chiefly of larch and laurel, and beech cut down, and though laid out with too much regularity, was darkness and shade, and natural beauty, compared with the bowling-green and the terrace."[1]

The wilderness, as trope of the English garden, is ground that is planted and then left to take its own course. This particular wilderness comes in for criticism from Austen: the trees are wrong, perhaps not impressive enough; it is "laid out with too much regularity." Nevertheless, she praises its "natural beauty," although not without the caveat that this is only in comparison with the lawns. But we are liable to offer other criticisms of this picture. We might object that, having been planned and planted, this is no wilderness at all. This is an illusion, part of a planned garden offering itself as the image of that which is unplanned and, in this case, failing to pull off the trick. But having recognized this description as being of something other than wilderness, does this mean that we have an articulable concept here? Perhaps, but perhaps not.

Wilderness, it sometimes seems, is like right action or pornography: we may have difficulty in defining it, but we think we know it when we see it. We know, upon reading Austen's description, that while this planted wood may have qualified as wilderness for her (and even that is unclear), it will not do for us. For a place to qualify as wilderness it must have a certain purity to it—the word "pristine" is often used. It often must have a certain remoteness to it. It should show no (or few) visible signs of civilization—roads, houses, power lines, and so on are all things that make a place less wild.

While the discussion below will be couched in terms of space, an abstraction, concern for wilderness begins as concern not for the abstract but for the concrete. We do not encounter and come to care about wilderness in the abstract; we come to care about particular places. Something about them strikes us, or some of us at least, as important and worth being preserved. One of the implications of this essay is that the very possibility of a place being wild will be called into question, for to be a place is to be located (or at least locatable), named, known. The characteristics to which I will point as those that are of concern in discussions of wilderness are those that, were they all present, would make of any wilderness no place at all. In other words, to the degree to which we can term a wilderness a place, it fails to be wild; and insofar as it is wild, it is no-place. Our failure to find places that live up to these standards will mean that (1) we will need to reconsider what we mean by wilderness and (2) anything we call wilderness (despite its prior cultural shapings) will, because of those cultural shapings and despite its wildness, turn out to be a place.

In what follows I argue that our reaction to the traditional view of wilderness is not so easy as it seems. What I would like to do is to pose *the* metaphysical question, What is wilderness? The investigation that follows we might term, after Heidegger, ontogeography or, if one prefers, geo-ontology. I begin by discussing the links between *wilderness* and *wildness* through an etymological investigation. I then turn to some views on wilderness as found in the ecophilosophical literature and suggest that we find a set of positions there beset with metaphysical problems. I then introduce several of the notions from Gilles Deleuze and Félix Guattari's *A Thousand Plateaus,* primarily those of smooth and striated space, and use them to discuss the distinction between wilderness and wildness. I then suggest that "wilderness" is an example of striated rather than smooth space and that our politics should focus on the broader idea of wildness rather than narrowly on wilderness. Finally, I suggest a new way of thinking about wilderness that seeks to overcome many of the problems found in other conceptions.

"WHERE THE WILD THINGS ARE"

The importance of wilderness as both a philosophical and a cultural concept is not at all new.[2] The Jews wandered for forty years in the wilderness before being allowed to enter Palestine. Kant links wilderness, under the name *rohe Natur* (raw nature), with the sublime and thus with morality (albeit through a circuitous path).[3] Roderick Nash, although a bit optimistic in his conclusions, did groundbreaking work tracing the development of the concept of wilderness in American thought.[4] My discussion of wilderness is not rooted in the historical manifestations of this term in thought. It rests instead in a tension in both contemporary and historical usage of the word. Wilderness, it seems clear, carries with it connotations of wildness. It suggests a nature untrammeled by humanity: pristine, apart, uncultured, and uncivilized. Crops do not grow in the wilderness; production for human purposes forms no part of what wilderness is all about. Wilderness, in this sense, functions as the other of culture.[5] It is a place that we can visit, but we do not belong there. It is dangerous and unpredictable. With only a touch of irony, one might remark that, unlike in our cities, bastions of security, one could get hurt in the wilderness.

The word "wilderness" is not at all new in English, and it can be found in recognizable forms in multiple precursors of our modern tongue, for example, the Middle Low German and Middle Dutch *wildernesse* or the (conjectured) Old English *wild(d)éornes.* The *Oxford English Dictionary* (*OED*) offers two possible etymologies. One of them traces "wilderness" to *wilder* or *wil(d)déor* (wild deer). The other, more probable one traces it to *wilddéoren* (wildern).[6] What is clear in both cases is that the word goes back to *wild,* which itself has a long history in many Teutonic languages.[7] In both Dutch and German *wild* means, among other things, "game," as in "wild game." While most of the roots of *wild* had essentially the same meaning as the modern word (the Old English, Old Frisian, and Middle Dutch *wilde,* the Old High German *wildi*), some of them are a bit different. The Old Norse *villr* meant "bewildered, astray." The modern Norwegian *vill* means "wild," as in "wild child"; the Swedish *vill,* "confused" or "giddy." But the *OED* is cautious with *wild,* noting that "the problem of the ulterior relations of this word is complicated by uncertainty as to its primary meaning."[8] We cannot know whether it is first and foremost an adjective or a noun. I hope we can agree that in the case of wilderness the essential meaning of *wild* is that of wild nature, where wild means untrammeled, uncultivated, uncivilized, raw, perhaps free. Wilderness is certainly not someplace bewildered or confused (although it can have that *effect*).

If this is what we mean by wilderness today, then we both appeal to and reinstantiate in our usage the dualist distinctions between culture and nature or, if one prefers, "man" (human) and nature.[9] We introduce a rift between a wild person and a wild place and insist that the place has a certain purity to it—as they used to say in advertising, "untouched by human hands." In order to see this more clearly, we need to leave the field of etymology behind and turn to some of the contemporary ecophilosophical literature on wilderness.

DIFFERING VIEWPOINTS

The literature on wilderness often does not bother with questions of etymology.[10] In fact, most of the literature on the topic does not even grapple with the most basic question, What is wilderness? We do not address the metaphysical question par excellence either because we are doing ethics or because we see no need for it.[11] I ask here after wilderness, and its connection with wildness, because I believe that wilderness presents a very serious problem for us today. The point, I take it, of writing about wilderness is often to offer resources for conservation. Wilderness is threatened and is in need of being protected. If we are to protect it, we must be able to offer reasons why. These reasons usually take one of two forms: wilderness is worthy of protection either because of something inherent to it (intrinsic value arguments) or because it offers a way to attain things that are themselves worthwhile and could not be attained without it (instrumental value arguments). While both sides have their proponents, my sense of the literature is that it is generally agreed upon that intrinsic value arguments would be stronger or more convincing. Such arguments, if successful, might necessitate, for example, the granting of rights (perhaps to self-determination or to life) to wilderness. But even those who believe that what is needed is an argument for intrinsic value will often be willing to offer instrumental value arguments with the acknowledgment that such arguments are likely to hold more sway in the realm of contemporary politics.[12]

Now these two positions vis-à-vis wilderness preservation tend to line up with two metaphysical positions concerning wilderness. One of these sticks closely by the etymology of the word (usually without invoking it) and insists that wilderness is the other of culture, a sphere apart. Wilderness, for these thinkers, has a certain ontological solidity: it was before we were, it will be after we are gone. Its essence is wildness, and wildness means self-subsistence and independence. To bring laws and rules and all the other trappings of civilization to wilderness is to tame it and, in effect, to destroy it. In part because of its ontological stature, wilderness, for these thinkers, carries with it an intrinsic value. Independent and self-directed, it should be treated as, to borrow Kant's phrase, an end in itself. These authors, amongst whom I include Thomas Birch, Eric Katz, Peter Reed, Holmes Rolston, III, Gary Snyder, David Orr, Alan Shields, and others, I call the "wilderness ontologists."[13]

The other camp—"wilderness constructivists"—follows Kant's epistemological vision more closely. Wilderness for writers such as Robert W. Loftin, Bill McKibben, Philip M. Smith and Richard A. Watson, David M. Graber, J. Baird Callicott, William Cronon, and Steve Vogel is not something independent and free; rather, it is a human construct. Wilderness is not the other of culture but a "product" of culture. This production of wilderness happens at two levels. The first is that of language: we name something "wilderness" because it is someplace other than where we live and work. Wilderness is a place that is not ours, to which we do not belong. We thus linguistically produce wilderness as that which seems not to have been produced. But at the same time wilderness is produced in a second way: geographically. We draw lines on maps, we designate some spaces as wilderness and not others. Of course, we do not plan wilderness as extensively as Central Park or the Tuileries. Nevertheless, it is a human production. For those authors who hold that wilderness is an artifact, the grounds to be offered for its preservation are often similar to those we might give for great works of art. At other times it is argued that wilderness is important as a "playground" or for the survival or leadership skills it can teach. But in all cases the arguments proffered treat wilderness as an instrumental good.[14]

The first camp, the wilderness ontologists, run into certain problems, some of a metaphysical nature, others practical. The metaphysical problem is that this position requires a pre-Kantian metaphysics. It is no wonder that the figure in the history of philosophy most often cited with praise by deep ecologists and other wilderness ontologists is Spinoza. But if one accepts the Kantian turn against substantialist metaphysics (as almost all of post-Kantian philosophy does), then a position that holds that wilderness is tenable as both an ontological and a value category is going to need argumentation. Unfortunately, such argument is more often than not lacking or unconvincing. The practical problem for the wilderness ontologists is that wilderness as they understand it is either rapidly disappearing or gone already.[15] If there is no such thing as wilderness so understood, then arguing for its value or preservation makes little sense. In order to address this difficulty, some wilderness ontologists appeal to the connection between wilderness and wildness and suggest that protection of wildness simply *is* protection of wilderness. But this solution is not without problems of its own. I will return to this in my final section.

The wilderness constructivists are also not without their problems, and again, these problems are both metaphysical and practical. The metaphysical aspect of the problem arises from the difficulty in saying what constitutes an artifact with value. Why is this area and not that one worthy of preservation? Most of the wilderness constructivists beg this question entirely either by beginning with the assumption that there is such a thing as wilderness or, more problematically, by making appeals to some kind of clear value in wilderness, a value that, by their own theories, could not be there. The practical problem of this view is straightforward. Humanity, as a series of cultures, has shown little skill in preserving that which was valued by previous cultures. Only if something could be picked up and transposed into a new symbolic order was it worthy of being saved. In Istanbul Haggia Sophia has seen life as both a church and a mosque and has, for this reason, been preserved for centuries. The temples of Rome have not fared so well. And the Parthenon's stint as a fifteenth-century munitions depot has scarred it to this day. If we have come to value these things today, this does not mean that they will continue to hold such a position. And if wilderness matters to some of us now, we have a difficult task ahead of us to pass on that sense of value to those who are to come.

In order to try to help move the discussion forward, I would like now to appeal to the work of two thinkers whose names do not often come up in discussions of ecophilosophy: Deleuze and Guattari.

DELEUZE AND GUATTARI: THE SMOOTH AND THE STRIATED

Strictly speaking, it is not quite right to say that Deleuze and Guattari never come up. Even if mentioned only sporadically in English-language discussions, their work is, nevertheless, often ecologically motivated. Guattari was, at the time of his death, attempting to bridge the ideological differences between France's two ecologically oriented political parties: the Right Greens and the Left Greens. He is also the author of several important ecophilosophical texts.[16] However, the notion of wilderness does not appear in their work. What does appear and what plays a central role are the notions of *space* and *territory.* In their last coauthored book, *What Is Philosophy?* they describe the process of philosophy itself as a transformation of earth (*terre*) into territory (*terri-*

toire). Such a process, described already in *A Thousand Plateaus,* goes by the name *territorialization.* But, like Heidegger's play of revealing and concealing, there can be no pure territorialization; rather, any territorialization is always a reterritorialization. Conversely, there is always also deterritorialization. The two occur together: "There are two components, the territory and the earth, with two zones of indiscernability, the deterritorialization (from territory to earth) and the reterritorialization (from earth to territory). It is impossible to say which is primary."[17] We can use this structure of de/reterritorialization to grasp the ways in which the human understanding of wilderness has come into being. But in order to do so it will be helpful to introduce another set of terms that Deleuze and Guattari use in the description of space: smooth and striated. To give a full explication of what either of these terms means would require a detailed exegesis of some very dense texts. They are linked with an entire complex of concepts that serve to open up not merely possibilities for thinking but new ways of living as well.[18] I cannot undertake such an exegesis here; I seek instead to treat Deleuze and Guattari's thought the way Foucault, a great admirer of Deleuze, wished his own work to be treated: as a toolbox. I borrow here several screwdrivers and files; and, as anyone who has worked with their hands knows, sometimes a tool can be put to uses other than those for which it was designed.

In brief, smooth space is space that allows movement in a multiplicity of directions without impediments. The image Deleuze and Guattari offer of smooth space is usually the desert (the biblical wilderness) or the sea. However, it will soon become clear that neither of these spaces is as smooth as they may at first appear. Striated space is that which has been codified, gridded, subjected to concepts. The possibilities for movement within striated space are limited, sometimes severely. If one wants to travel by car, one must follow the roads. Trains only travel on tracks.[19] And airplanes, moving through the apparently smooth space of the ether, follow strictly regulated flight paths.

Like material correlates (although this understanding twists Deleuze and Guattari's text somewhat) of de/reterritorialization, smooth space and striated space do not exist in pure form. Space is always in a state of becoming. Old striations are replaced with new, what was severely striated becomes smoother, what was smooth becomes striated. If we wished, we could line up smooth space with the earth and striated space with territory, but always only with the understanding that there is no pure earth, no territory that is not also earth, and no space that is not smooth, striated, and in transition. This is a system not of being but of becoming, a Nietzschean or Heraclitean world.

Before returning to the question of wilderness, which I shall do by linking the concept of smooth space as developed by Deleuze and Guattari with the idea of wildness we find in the ecophilosophical literature, I should like to say a bit more about these notions of smoothness and striation—do a little philosophical geography.

What is at issue in Deleuze and Guattari's work is the task of thinking in a way that is as nonmetaphysical as possible. This means that the basic oppositions that we accept all too easily are questioned from the outset. The distinctions between thought and action, space and time, culture and nature, mind and body, and so on are all put in play here. So in the course of a discussion that concerns itself explicitly with the notion of *space,* we find discussions not only of "physical" space (the seas, the deserts) and of "theoretical" space (geographical space, geometrical space) but also of musical space (Pierre Boulez), technological space (different varieties of cloth making), and aesthetic space (different types of vision). I cannot treat all these aspects here, and, at the risk of subjecting Deleuze and Guattari's text to an interpretation that forces it back within the sphere of metaphysical dichotomies but also with the awareness that what I am doing here is an appropriation for specific purposes, I will focus primarily on their discussions of what we might term physical space.

In their discussion of the difference between smooth and striated spaces, Deleuze and Guattari offer several different ways of contrasting them. The first and most thorough I take the liberty of quoting at length:

> Smooth space is filled by events or haecceities, far more than by formed and perceived things. It is a space of affects, more than one of properties. It is *haptic* rather than optical perception. Whereas in the striated forms organize a matter, in the smooth materials signal forces and serve as symptoms for them. It is an intensive rather than extensive space, one

> of distances, not of measures and properties. . . . A Body without Organs instead of an organism and organization. Perception in it is based on symptoms and evaluations rather than measures and properties. That is why smooth space is occupied by intensities, wind and noise, forces, and sonorous and tactile qualities, as in the desert, steppe, or ice. The creaking of ice and the song of the sands. Striated space, on the contrary, is canopied by the sky as measure and by the measurable visual qualities deriving from it.[20]

The striated: capturable, describable, reducible to language, measurable, constraining movement. The smooth: explosive, aesthetic (in the Greek sense of *aesthesis*), expressive itself (but perhaps not capturable in other expressions), a space of happenings and becomings rather than of Being. Striated space is the space of the State. It is the space of laws and principles, of maps and roads. Smooth space, for Deleuze and Guattari, is the space of nomads—those who live outside the state. It is the space in which movement is not channeled and directed or subjected to the strictures of instrumental reason. Which brings us to geometry:

> The smooth and the striated are distinguished first of all by an inverse relation between the point and the line (in the case of the striated, the line is between two points, while in the smooth, the point is between two lines); and second, by the nature of the line (smooth—directional, open intervals; striated—dimensional, closed intervals). Finally, there is a third difference, concerning the surface or space. In striated space, one closes off a surface and "allocates" it according to determinate intervals, assigned breaks; in the smooth, one "distributes" oneself in an open space, according to frequencies and in the course of one's crossings. (*MP*, 600/480–81)

Here the difference becomes clearer. Striated space is *Newtonian* or *Cartesian,* an infinite array of points awaiting the plotting of lines. The point always preexists the line, so to move in such space is to travel from one preexistent point to another. Striated space's array of points and linkage of directions of movements define the surface of the space "in advance" of any movement. The movement itself is only "necessary" insofar as it serves to bridge the space between the two points. This is why speed is so important in travel: the faster you go, the "shorter" the line.[21] In smooth space the speed of travel is not so important, and maybe where you are going isn't even that important either. You head this way or that with no particular destination whatsoever, because if one could find oneself in a pure smooth space, there would be no destinations preexistent to one's arrival. The points are "determined" by the lines. Movement in this direction and then that creates a point where the direction of motion changes, a linkage of rays. The partitions of smooth space follow upon motion and then erase themselves. The water does not show the path of a boat for too many minutes; the wind covers one's tracks in the desert sands.

Although they are quite insistent that there is no such thing as purely smooth or striated space (these are concepts), Deleuze and Guattari refer to several examples repeatedly: the desert, the steppes, the ice (of the Eskimos), and the seas.[22] It is the last of these that comes in for the most extended discussion. At first thought, the open sea presents us with a wonderful example of smooth space: motion is possible in any direction, points of reference are nonexistent; wherever one chooses to go one leaves no tracks; directionality is all-important, for there are no points. As Deleuze and Guattari put it, "the sea is a smooth space par excellence, and yet," they continue, "[it] was the first to encounter the demands of increasingly strict striation" (*MP,* 598/479). The sea, while apparently a smooth space, is actually gridded and plotted, charted and divided. Because of the absence of natural "landmarks," the sea requires the imposition of a grid from outside. There is no pretense here that the striations of the sea follow natural patterns or express what was already there. This transformation from smooth to striated is necessitated because of the sea's very smoothness: the problems of navigation in open water motivated the development of better maps and means of navigation.[23]

Now this striation of the sea, like any striation, is never complete and always gives rise to new smooth spaces. We can imagine the sort of "primordial" smoothness of the sea arising once more for those lost in lifeboats with neither maps nor navigational aids. Tossed on the waves, they can move in whatever direction they like, but there is almost no point in moving at all if they don't know the best way to go. But striated space can give rise to new smooth spaces in other ways as well. The channeling of shipping into routes and lanes opens up avenues for pirates, who will cross those lanes and sail where they will. But it also allows the possibility for groups like Greenpeace and the Sea Shepherd Society to play against those striations (as well as whole other sets of striations) and produce a form of movement that does not fit in with the patterns the striations should allow. "Nothing is ever done with: smooth space allows itself to be striated, and striated space re-imparts a smooth space, with potentially very different values, scope, and signs" (*MP,* 607/486).

What I would like to suggest here is that this notion of smooth space seems to have a certain element of *wildness* about it. It is a space of openness and freedom, unconstrained by laws and regulations. And part of why I think that this idea of smooth space can be helpful here is that, if it is not clear already, Deleuze and Guattari much prefer smooth space to striated. Part of the goal of their philosophical enterprise is to find ways to smooth out spaces that are now striated. It remains to be seen where wilderness fits into this picture.

THE IMPOSSIBILITY OF WILDERNESS

One thing is clear from Deleuze and Guattari's text: one never finds a pure space. All space "in the world" is an admixture of smooth and striated, and the mixture is always unstable and changing. Space that is more smooth than striated can change: maps are drawn, roads are built, trees are cut. But the transformation can work in the other direction as well: demonstrations clog the streets, buildings crumble, new trees sprout, new possibilities arise within (and against) the striations. It is this alloyed character of these spaces that blocks this from being just another metaphysical dualism. The smooth and the striated are not nature and culture. The concept of smooth space is intended to capture simultaneously the understanding of wildness that is traditional to the West as well as the understanding of space that is traditional to nomadic cultures.

The analysis of space presented by Deleuze and Guattari allows us to understand the ways in which, even within striation, smoothness reerupts. Thus, where there have been planned spaces, wild ones can reappear. In other words, no smoothing is ever final. City dwellers and suburbanites know this all too well. The manicured lawn is "rudely" interrupted by dandelions; the order and cleanliness of the home are shattered by roaches and mildew. Perhaps one finds rabbits and deer eating in the garden. But we must be careful here, for wildness is not the same as wilderness; there never was an originary smooth space, and *re*smoothed space certainly doesn't qualify. The etymological link between wilderness and wildness often allows us not to consider the difference between originary smooth space and resmoothed space. But etymology is not destiny, and the etymological confusion leads to a notion of *purity* at the core of the entire discussion of wilderness in ecophilosophy. Now purity is one thing if it concerns concepts, but it is quite another when what is at issue is space. When we strive to protect wilderness, we do so not only because it is wild or, in Deleuze and Guattari's terms, smooth. Not at all. What we strive to protect is *both* the wildness and the purity. We resist notions of multiple use, we resist road building, we may even resist trail construction. We seek to have a wilderness whose wildness is of a character other than human. But how do we go about this?

We pass laws. We fight for stricter and stricter controls on what can and cannot be done in the wilderness. We fight to keep out the RVs and the mountain bikes, the pack animals and the snowmobiles. We cultivate a wilderness ethic and are quick to chastise those who do not abide by its rules of conduct. Of course, we must do these things to preserve the wilderness, for the wilderness is the space of the other than human and, as such, must be protected from the ravages we are likely to inflict on it. But in this process of drawing boundaries and passing laws, of controlling conduct and making maps, of cutting trails and providing rescues, we change something important. Let us assume that the space in question is, before we begin, more smooth than striated. In the process of making it into a wilderness—and let us be clear that this is what is going on here—we are striating this space. We are making it less wild. If what we mean by wilderness is wild space, and if, in turn, that means smooth space, then the one way we are not going to achieve this is through the processes we have followed for wilderness protection.[24]

Wilderness cannot be brought about by processes of striation. Fair enough. So what is needed are processes of smoothing. But processes of smoothing do not produce wilderness. They produce wildness. Like a dialectic with neither guiding principles nor telos, the interplay of smoothing and striating can never give rise to spaces with the purity we would like to see in our wilderness.

In what I consider to be the best piece yet in the literature on wilderness, Thomas Birch argues for the importance of protecting wilderness through processes of striation because what is needed is the preservation of possibilities of wildness that can take hold against the imperium or the state. But the wildness that is at stake here is not of a piece with wilderness as pure space of becoming outside the control of humanity. The wildness is, on the one hand, that of anarchist politics and, on the other, the space of a wild becoming in which nature reerupts where it had been suppressed. But is the reeruption of nature along abandoned railway lines wilderness? Are the grasses insistently poking through the sidewalks and shattering their coherence wilderness? Birch seems to think so, and Eric Katz cites him approvingly on just this point: "Birch thus recommends that we view wilderness, wherever it can be found, as a 'sacred space' acting as 'an implacable counterforce to the momentum of totalizing power.' Wilderness appears anywhere: 'old roadbeds, wild plots in suburban yards, flower boxes in urban windows, cracks in the pavement. . . .' And it appears, in my life, in the presence of the white-tailed deer of Fire Island."[25] This entire passage depends on a failure to distinguish between wilderness and wildness. Just four pages before this passage Katz, this time citing Eugene Hargrove, praises the notions of authenticity and continuity with its past that distinguishes wild nature from restored nature. But if wild nature—wilderness—depends on a continuity and authenticity, and this continuity cannot contain acts by humanity, then we are not apt to find wilderness in old roadbeds or backyards. Wildness, yes; wilderness, never. These are striated spaces in the process of becoming smooth, but they are not the smooth spaces that were there before striation. And they certainly don't have any sort of purity or authenticity about them.

None of this means that we cannot have spaces that are more or less wild. But even for those spaces where no one ever goes there exist maps and, of course, satellite location devices. Let me repeat: there can be for us no truly smooth space.[26] Now, I agree wholeheartedly that it is a worthwhile project to try to smooth out some of the spaces that we do have. And I agree as well that we should resist further striations in spaces that are now relatively smooth. But I am also in agreement with Birch here in my skepticism that the state—*the* greatest force of striation we have—is the place to look for smoothing.

If we are interested in smoothing out space, then we have to create a politics, a politics of smooth space, what Deleuze and Guattari might term *nomad politics*. Such a nomad politics not only plays off the smoothnesses opened up in striated space (Tiananmen Square in 1989, Greenpeace, 1968, Earth First!), it creates (or may create) new smooth spaces as it goes. But these smooth spaces may also be restriated, and, of course, they will have neither purity nor authenticity to them. Pure wilderness, I am afraid, is not a possibility.

THE SMOOTH AND THE WILD

In this final section I would like to suggest a way of thinking about wilderness that might move us beyond some of the problems discussed above. This requires a mediation of the two camps (wilderness ontology and constructivism) currently engaged in the debate. This mediation depends on a reformulation of the notion of wilderness in the wake of the Deleuzo-Guattarian analysis offered here. I also add a further element to the mixture in order to give a normative ground that some may find lacking in Deleuze and Guattari's distinction.

First, a nonmetaphysical notion of wilderness.[27] We might call it smooth wilderness, but I prefer historically smooth space.[28] Not all smooth spaces qualify as historically smooth space. Birch's old roadbeds and cracked pavement may well be smooth spaces, but I would hesitate to call them historically smooth space, just as others would hesitate to call them wilderness. But in order to be able to make this distinction on more solid ground than mere intuition, we need the introduction of a notion of history or temporality. Deleuze and Guattari's distinction between smooth and striated is useful because it shows the fruitlessness of many discussions about wilderness, but without the addition of this second axis of analysis it does not yield normative grounds for preferring one smooth space over another. The notion of historically smooth space is designed to do just that. Deleuze and Guattari maintain that there are no pure spaces, neither smooth nor striated. But even if there are no purely smooth spaces (wilderness in the metaphysical sense), there are surely different degrees of striation. If, having grasped that the distinction is drawn along a spatial continuum, we then fail to see that the *processes* of smoothing and striation happen along a temporal continuum, we then risk lapsing back into the sort of thinking that we were trying to avoid. We can't get back to some preconceptual Eden, but, we might ask ourselves, aren't there some places more Edenic than others? Aren't there some places that have historically been less striated? Surely the answer to this question must be yes. While we might be interested in smooth spaces in general, the notion of historically smooth space seeks to distinguish those spaces that have been severely striated and then resmoothed from those that have seen no such striation.[29] For something to be historically smooth space, it must have some sort of historical continuity to it. For example, a clear-cut and replanted "forest" cannot qualify as historically smooth space; a selectively logged and "naturally" regenerated forest might. Because there can be no pure notion of historically smooth space, discussions as to what constitutes such space is problematic. However, even if we (and who this "we" is remains to be decided) were to decide that a particular "restored" place did not constitute historically smooth space, it might still be valued as a resmoothed space.

In addition, we need to take account of the fact that not all human actions are interchangeable or equivalent. If we accept Deleuze and Guattari's distinction between nomadic (organized but nonstatist) and state action and politics, we might be able to see many places as historically smooth space even though they have been lived in and manipulated by human beings over the course of time. We gain a further advantage from this distinction as well: we can see these places as lived in and changed while still maintaining the character of historically smooth space *without* lapsing into a naturalization of the "noble savage" so common in the eighteenth century.

What should be clear by now is that this notion of historically smooth space, which makes no appeals to purity, is a far more unwieldy concept than other notions of wilderness; it cannot allow us to draw the clear and sharp lines we might like. But if such a concept is less clear-cut than we might like, it is not without its usefulness. For starters, such a concept accords better with the world we live in. Recent evidence is quite clear that many of the places taken as archetypal wildernesses show signs of human alteration.[30] These signs may not be dramatic. The events that caused them may have happened a long time ago. But if we insist that wilderness is a space free of human action, then such spaces must be disqualified.[31] But if what we mean by wilderness is historically smooth space, then we are not necessarily constrained in this way. The problem remains of *how* and *where* we draw the lines between that which constitutes historically smooth space and that which does not. If an inability to clarify where these lines are to be drawn at a theoretical level appears as a shortcoming, it is worth remembering that no previously offered "definition" of wilderness has shown itself to be acceptable, and thus offering a more fluid way of looking at these issues may not be such a bad option after all. As a preliminary formulation, let me suggest that no space that has undergone severe striation in the recent past qualifies, and only those that have been resmoothed a long time ago—for example, the Mayan jungles *before* the restoration of the ruins qua ruins—could. It is not the presence, either currently or historically, of humans that "taints" a space; it is the processes of striation that matter. And since no striation is permanent, even those places that have been severely striated can be smoothed again. The concept of historically smooth space serves only to delimit those spaces that either have never undergone severe striation or did so such a long time ago that processes of resmoothing have had sufficient time to rework the space. The relevant question is whether a space is smooth enough (and the specification of this "enough" I leave open).

Second, the notion of historically smooth space allows us better to take account of restoration ecology than other notions of wilderness might. Seen from a Deleuzo-Guattarian perspective, restoration ecology is not about putting things back the way they were. There is no recoverable "way they were." Instead, the best to be hoped for is that we can find spaces that have not been striated too much and can endeavor to resmooth them. The space "restored" in this way is thus not the same as the space that was "there" precedent to the striation. But this brings me to the greatest problem of all and returns us to the link between wilderness and wildness with which this essay began.

What is needed at this point is thoughtful reflection on what forms of wildness or smoothness are of value. I have suggested that the link between wilderness and wildness is problematic at best and that a better way to think of wilderness issues is using Deleuze and Guattari's distinction between smooth and striated space. However, if we accept this distinction as merely a spatial one, without the addition of an element of temporality or historicity, then we risk losing all grounds for distinguishing one kind of smooth space from another. In order to overcome this difficulty I have suggested historicizing the notion of smooth space into historically smooth space. This addition can help us to take account of degrees and shadings of wildness or smoothness. However, I would like to conclude this essay with two problems that are in need of further work. The first is more theoretical, the second more practical.

The first hinges on the traditional distinction between nature and culture. If the notion of smooth space seemed to make this distinction no longer valid in these discussions, the notion of historically smooth space seems to bring it back with a vengeance. This worries me. Traditionally from the standpoint of the West, the activities of aboriginal peoples (read non-Westerners) were often perceived as "natural." The peoples of Africa, the Americas, Australia, and the South Pacific in particular were taken as "not there," that is, part of the landscape. For discussions about wilderness, the valid questions remain *who* (or *what* in the case of institutions) was doing *what* to *which* landscapes for *how long* (and *how long ago*) and *why*. Coming up with answers to these questions is a tall order indeed. What seems clear at this point are two things: (1) the world of nature is not at all a static one, and (2) wherever people have been they have changed that place. The first point means that there is no stable nature we can preserve or leave alone. The second means that we have to draw finer distinctions between sorts of actions. Perhaps, and this is only a suggestion, the distinction between smooth and striated might again be helpful here in assessing actions. Some actions are smoothing (or resmoothing), others are striating. An area that has seen primarily smoothing actions historically could qualify as historically smooth space, one that had seen severe striation followed by resmoothing might only be newly smooth space (and not historically smooth space), and so on.

Ancillary to this and still in need of argumentation is the preference for smooth over striated space (or historically smooth space over other sorts of space). If my preference is clear enough to me, that does not mean that it is clearly articulated enough to convince others that they ought to share it. And even if others are convinced of the value of smooth space, there still remains the question of the distinction between historically smooth spaces and other smooth spaces and why one ought to prefer one rather than the other. A humanist, anarchist, nomad politics could well opt for smooth spaces in urban areas and care not a whit for historically smooth spaces. If, however, the value of smooth space remains to be demonstrated, the converse also remains the case: striated space is also in need of justification by its apologists (the state, industry, etc.). If rather than justification we have had, up until now, the unstoppable march of "progress" (i.e., striation) backed up by force, that is not, in and of itself, satisfactory justification for the status quo.

The second problem is the more vexing of the two. If at the level of discussions about wilderness, ecology, and environmental issues in general there has been much progress (or at least much said), at the level of practice we are often lacking in concrete ideas. The analysis of smooth and striated space offered here only complicates matters. If earlier voices (e.g., John Muir) could suggest state action as the appropriate means for wilderness protection, it is no longer so clear that this avenue is open to proponents of either smooth space *or* historically smooth space. One of the central points in Deleuze and Guattari's analysis, and one with which I tend to agree, is that the state is primarily a force of striation. As such, proponents of the state as a force for smoothing find themselves in a paradoxical position. Birch has suggested this already, but the problem is even worse than he acknowledges. Wilderness areas are not simply prisons in which wildness is incarcerated. Like good modern prisons, which do not simply throw walls (whether physical or otherwise) around a region and then allow unrestricted movement within it, wilderness areas are striated within. Some species are kept out, others controlled. Human activities and motions are channeled. One need only look at what has become of Yosemite Valley, Yellowstone, and the Grand Canyon for evidence of state striation in the name of protection. Given this, the question is, Can the state be a force for smoothing? If the answer to this is yes, some account needs to be given of how this can be the case. If, however, the conclusion we reach is that the state cannot be a force for smoothing, then a rethinking of political practice is in order. Although Deleuze and Guattari hint broadly in the direction of a nomad politics, much work still needs to be done in this area in particular in order to see what such a politics would look like and how it would work.

NOTES

This is a slightly modified version of an article that originally appeared in *Philosophy & Geography* 3 (1998): 265–83. I have slightly revised the literature review section as well as that on etymology, removed one unfortunate phrasing, added several new endnotes, and altered the essay's subtitle. The argument remains unchanged. The paper was originally written under the financial auspices of the Belgian American Educational Foundation. An earlier version was read at the Society for Philosophy and Geography meetings in New York City in December 1995. The paper has been greatly improved thanks to helpful suggestions from Noam Cook, Barbara Fultner, Katrina Korfmacher, Andrew Light, Ulrich Melle, Dan Smith, Rudi Visker, Steve Vogel, Michael Zimmerman, and two anonymous reviewers for *Philosophy & Geography.*

1. Jane Austen, *Mansfield Park* (London: Oxford University Press, 1960), 91.

2. The title of this section is borrowed, of course, from Maurice Sendak's wonderful children's book, which I loved so much as a child.

3. Eliane Escoubas discusses *rohe Natur* as the showing of nature without concepts in her "Kant or the Simplicity of the Sublime," in *Of the Sublime: Presence in Question,* ed. Jean-François Courtine et al., trans. Jeffrey S. Librett (Albany: SUNY Press, 1993), 55–70.

4. Roderick Nash, *Wilderness and the American Mind* (New Haven: Yale University Press, 1967). This book is now in its fourth edition (Yale University Press, 2001).

5. Nash makes this point, invoking the "fact that civilization created wilderness" (ibid., 4th ed., xi).

6. *Oxford English Dictionary,* 2nd ed., s.v. "wilderness."

7. Jay Hansford C. Vest argues that *wilderness* is rooted in an early notion of "will-of-the-land" and already contains within it a notion of agency. See his "Will-of-the-Land," *Environmental Review* 9 (1985): 323–29 as well as his

"The Philosophical Significance of Wilderness Solitude," *Environmental Ethics* 9 (1987): 303–30.

8. *Oxford English Dictionary,* 2nd ed., s.v. "wild."

9. I acknowledge but will not address here the ecofeminist criticisms of the conflation between man and human.

10. There are, in addition to Vest, three further notable exceptions to this claim. The first is Roderick Nash, who spends almost three and a half pages discussing the etymology of wilderness at the very beginning of *Wilderness and the American Mind.* Nash's conclusions are, however, largely the same as my own: "The definition of wilderness is complex and partly contradictory" (4). The second exception is Gary Snyder, who largely, as I have, follows the *OED* for his English etymology. He then, however, gives a fascinating etymology of the Chinese and Japanese words for nature and wilderness ("The Etiquette of Freedom," in *The Practice of the Wild* [San Francisco: North Point Press, 1990], 3–24). Finally, Dave Foreman, who is both less thorough than Nash and Snyder yet more sure of his conclusions in "Wilderness Areas for Real," simply asserts: "The root for 'wilderness' in Old English is *wil-deor-ness:* self-willed land" (in *The Great New Wilderness Debate,* ed. J. Baird Callicott and Michael P. Nelson [Athens: University of Georgia Press, 1998], 405). This seems to be too quick and too convenient, putting wilderness in the status not only of other culture but capable of autonomous action and thus, on Kantian grounds (although Foreman doesn't mention Kant), deserving of respect.

11. There are also those who offer what we might term a phenomenology of wilderness and thus, following the dictates of pre-Heideggerean phenomenology, do not see the need to ask ontological questions. For example, Philip M. Smith and Richard A. Watson assert that "wilderness is not a simple geographic concept" and depends upon experience ("New Wilderness Boundaries," *Environmental Ethics* 1 [1979]: 61–64, 61).

12. The clearest case of this is William Godfrey-Smith, "The Value of Wilderness," *Environmental Ethics* 1 (1979): 309–19. For a succinct overview of the various types of arguments generally given for wilderness protection see Michael P. Nelson, "An Amalgamation of Wilderness Preservation Arguments," in Callicott and Nelson, *The Great New Wilderness Debate,* 154–98.

13. See Thomas H. Birch, "The Incarceration of Wildness: Wilderness Areas as Prisons," *Environmental Ethics* 12 (1990): 3–26; Eric Katz, "The Call of the Wild: The Struggle against Domination and the Technological Fix of Nature," *Environmental Ethics* 14 (1992): 265–74; Peter Reed, "Man Apart: An Alternative to the Self-Realization Approach," *Environmental Ethics* 11 (1989): 53–69; Holmes Rolston, III, "Values Gone Wild," *Inquiry* 26 (1983): 181–207 and "Valuing Wildlands," *Environmental Ethics* 7 (1985): 23–48; Allan Shields, "Wilderness, Its Meaning and Value," *Southern Journal of Philosophy* 11 (1973):

240–53; Gary Snyder, "The Etiquette of Freedom," and "The Rediscovery of Turtle Island," in *A Place in Space: Ethics, Aesthetics, and Watersheds* (Washington, D.C.: Counterpoint, 1995), 236–51. Since this paper was originally written (in the fall of 1995), the debate between constructivists and ontologists has become, to say the least, quite heated. Sadly, I think some of what has passed for debate has been either talking past each other or failing to grasp the other side's position. I leave for a future work a more detailed engagement with this issue. See Holmes Rolston's response to J. Baird Callicott, "The Wilderness Idea Revisited: The Sustainable Development Alternative," in Callicott and Nelson, *The Great New Wilderness Debate,* 337–66 entitled "The Wilderness Idea Reaffirmed," and Callicott's response to Rolston's response, "That Good Old-Time Wilderness Religion," both in ibid., 367–86, 387–94; Gary Snyder, "Is Nature Real?" in *The Gary Snyder Reader: Prose, Poetry, and Translation, 1952–98* (Washington, D.C.: Counterpoint, 1999), 387–89; Michael E. Soulé and Gary Lease, eds., *Reinventing Nature? Responses to Postmodern Deconstruction* (Washington, D.C.: Island Press, 1995); Eileen Crist, "Against the Social Construction of Nature and Wilderness," *Environmental Ethics* 26 (2004): 5–24.

14. See Robert W. Loftin, "Psychical Distance and the Aesthetic Appreciation of Wilderness," *International Journal of Applied Philosophy* 3 (1986): 15–19; Bill McKibben, *The End of Nature* (New York: Random House, 1989); Philip M. Smith and Richard A. Watson, "New Wilderness Boundaries," *Environmental Ethics* 1 (1979): 61–64; David M. Graber, "Resolute Biocentrism: The Dilemma of Wilderness in National Parks," in Soulé and Lease, *Reinventing Nature,* 123–36; Callicott, "The Wilderness Idea Revisited"; William Cronon, "The Trouble with Wilderness, or, Getting Back to the Wrong Nature," in Callicott and Nelson, *The Great New Wilderness Debate,* 471–99; Steve Vogel, *Against Nature* (Albany: SUNY Press, 1996) and "Habermas and the Ethics of Nature," in *The Ecological Community,* ed. Roger Gottlieb (New York: Routledge), 175–92.

15. McKibben and Vogel both argue this point as concerns *nature.* But if there is no nature, then, ipso facto, there can be no wilderness (as a category of that nature).

16. Félix Guattari, *Les trois écologies* (Paris: Éditions Galilée, 1989) (*The Three Ecologies,* trans. Ian Pindar and Paul Sutton [New Brunswick, N.J.: Athlone Press, 2000]) and *Chaosmose* (Paris: Éditions Galilée, 1992) (*Chaosmosis: An Ethico-aesthetic Paradigm,* trans. Paul Bains and Julian Pefanis [Bloomington: Indiana University Press, 1995]).

17. Gilles Deleuze and Félix Guattari, *Qu'est-ce que la philosophie?* (Paris: Éditions de Minuit, 1991), 82.

18. There is already a large body of literature on Deleuze and Guattari. See, for example, Ronald Bogue, *Deleuze and Guattari* (London: Routledge,

1989); Constantin V. Boundas and Dorothea Olkowski, eds., *Gilles Deleuze and the Theater of Philosophy* (New York: Routledge, 1994); Michael Hardt, *Gilles Deleuze: An Apprenticeship in Philosophy* (Minneapolis: University of Minnesota Press, 1993); Brian Massumi, *A User's Guide to Capitalism and Schizophrenia: Deviations from Deleuze and Guattari* (Cambridge, Mass.: MIT Press, 1992). *Gilles Deleuze and the Theater of Philosophy* contains an extensive bibliography of the literature both on Deleuze and on Deleuze and Guattari.

19. I write these words in a hotel room in Strasbourg. The space of France, striated in the extreme, is at this moment disrupted by a general strike. Train travel is utterly impossible in all of France. The strike disrupts the possibilities for movement allowed by the striations and thus brings motion to a halt. At the same time, new possibilities are opened up in the space where trains might ordinarily be.

20. Gilles Deleuze and Félix Guattari, *Mille plateaux: Capitalisme et schizophrénie 2* (Paris: Éditions de Minuit, 1980), 598 (*A Thousand Plateaus: Capitalism and Schizophrenia,* trans. Brian Massumi [Minneapolis: University of Minnesota Press, 1987], 479). All further references are given in the text as *MP* with the paginations to the French first and then the English after a solidus.

21. Paul Virilio has written extensively on the idea of speed. See, for example, his *Vitesse et politique* (Paris: Éditions Galilée, 1977) (*Speed and Politics: An Essay on Dromology,* trans. Mark Polizzotti [New York: Semiotext(e), 1986]) and *Défense populaire et luttes écologiques* (Paris: Éditions Galilée, 1978) (*Popular Defense and Ecological Struggles,* trans. Mark Polizzotti [New York: Semiotext(e), 1990]). This obliteration of space through speed finds it conclusion in that perennial love of the science fiction author: the transporter chamber.

22. Deleuze and Guattari's understanding of the links between space, concepts, and thinking is rather complex. Thus, while their notions of space are clearly conceptual, their notions of conceptuality are themselves spatial. In a section entitled "Geophilosophy" they write: "Absolute deterritorialization is not without reterritorialization. Philosophy reterritorializes itself through the concept. The concept is not an object, but a territory. It has no Object, but a territory" (*Qu'est-ce que la philosophie?* 97).

23. While Deleuze and Guattari offer the open ocean as a smooth space, it is no doubt true that there are certain regularities such as currents that make it more prone to some striations than others. Currents and prevailing wind patterns serve, just as mountains and rivers do, as either facilitators or inhibitors of motion in certain directions. Nevertheless, until the maps are drawn, establishing shipping lanes and bringing nautical hazards within the sphere of human knowledge, the seas remain, from the standpoint of politics and trade, a space that is without striations.

24. Thomas Birch's "The Incarceration of Wildness: Wilderness Areas as

Prisons" makes a similar argument, although in different terms. I agree with Birch's analysis of wilderness areas; it is his mildly optimistic conclusions I take issue with (see below).

25. Eric Katz, "The Call of the Wild: The Struggle against Domination and the Technological Fix of Nature," *Environmental Ethics* 14 (1992): 273.

26. A completely smooth space would be one that had never been mapped, traversed, named, transformed, or brought within the sphere of state power in any way. There may someday be smooth spaces again. But such a radical resmoothing would (1) require a break in the continuity of civilization (and perhaps the disappearance of humanity) and (2) still fail to reinstantiate the smooth spaces we insist on thinking wildernesses should be.

27. This suggestion was made by Noam Cook in an e-mail to the author of February 2, 1996.

28. John O'Neill argues persuasively for taking time and history seriously in environmental debates in "Time, Narrative, and Environmental Politics," in Gottlieb, *The Ecological Community,* 22–38.

29. Such resmoothing could happen in a number of ways. Whenever striation stops, spaces "naturally" resmooth themselves. Jan E. Dizard discusses a reservoir as, in my terms, a striated space that has resmoothed itself in some ways but not others (see "Going Wild: The Contested Terrain of Nature," in *In the Nature of Things: Language, Politics, and the Environment,* ed. Jane Bennett and William Chaloupka [Minneapolis: University of Minnesota Press, 1993], 111–35). But there are also, as discussed above, cultural processes of resmoothing. Perhaps the strongest forms of these are connected with warfare. War severely restriates space in some ways but also smoothes it as well (e.g., destruction of lines of communication and transportation). In its most extreme form thermonuclear war is an almost purely smoothing process. However, I would hesitate to equate space smoothed in this way, wild as it might be, with wilderness.

30. See Gary Paul Nabhan, "Cultural Parallax in Viewing North American Habitats," in Soulé and Lease, *Reinventing Nature,* 87–102.

31. Part of the problem here is that for Deleuze and Guattari not all human actions are striating. Actions by the state are almost always striating; actions by others may be smoothing. For a recent discussion of ecological restoration, an issue not unrelated to wilderness preservation, which justifies restoration by an appeal to the value of the outcome of natural processes *without* appealing to those processes themselves, see William Throop, "The Rationale for Environmental Restoration" in Gottlieb, *The Ecological Community,* 39–55.

Irene J. Klaver

Wild (2008)

Rhythm of the Appearing and Disappearing

Wilderness: you don't go there to find something, you go there to disappear. Wallace Stegner

The antelope are a strange people. . . . They appear and disappear; they are like shadows on the plains.
Pretty Shield

WILD IS THE SMOKE RISING, turning in slow movements, tracing out invisible currents of air with its gray white elegance before disappearing. Wild are the tracks of the mountain lion and the deer in the snow and the scratches on my arms and legs, painful traces of following the animal tracks through the thickets of their trajectories. The deer standing out against the tree line vanished as suddenly as it came to the fore, leaving indeterminate whether the deer leaped into the forest or the trees absorbed it. Wild is this very play between appearance and disappearance, the slipping in and out of the limits of presence. To be wild is to stand out *and* to disappear, a rhythm of foreground and background, of light and shadow. Untamed and not named, the wild escapes the frames of our knowledge—miles of fungal filaments in a handful of soil, the feeling of frost in the air, cold touching one's skin, the rich fragrance after a summer rain, seagulls screaming at the coast.

In this essay I explore the relation between this notion of wild and the concept of wilderness. I argue that wilderness functions as a reification of the wild. Safely locked up, far away from our everyday, the wild is supposed to be by itself in designated wilderness areas. Hence, our relation with the wild has been reduced to an organized expedition and is in that sense domesticated or tamed. Given the pervasive presence of modern culture, the creation of wilderness areas has been important to safeguard habitat for other natures than human nature and even other cultures than human culture. But, insofar as designated wilderness areas are instantiated to preserve a possibility for us to experience the wild, their creation has had the paradoxical effect of further alienating us from the wild. I don't advocate decommissioning these areas; on the contrary, I want to show how their existence detracts from an understanding of the wild, nor are they sufficient to protect the wild. I hold a plea for an expansion of the experience of the wild in our everyday existence. This can be thought of as an active everyday engagement with the world around us by affording space for the presence of otherness. It entails fostering a mentality that takes otherness, including nature, seriously and affords it a place to coexist with and in human culture. To accomplish this we need a better understanding of the various meanings of wild and wilderness and their complex relation.

Many of our cultural structures are impervious: they don't allow for the water to percolate into the ground. We humans have forgotten how to stand out *and* simultaneously disappear. We cannot leave it to wilderness areas for water drops to embark on a wild journey into the soil on their way to an aquifer; we need to find the interstices of wildness in the grass of our backyard, even in the semipermeable asphalt of our parking lots and interstates.

Besides being an area of land untrammeled by humans, wilderness is a story—a relatively new story, historically idiosyncratic to the nineteenth-century West, when modern civilization got a firm grip on our lives.

It is a story sliding and stumbling over a range of meanings as wide as the mountains of Montana. Despised and feared as the darkest and most vile, adored and desired as the most pure and pristine, wilderness evokes strong passions of love and hate. Ultimately, as any passion, it deals with the place of the other and, thus, with the place of ourselves and how we afford place to otherness.

Initially, wilderness was synonymous with threat, a hostility to be mastered. As a by-product of Western culture, especially agri-culture, wilderness stood for what was not-(yet)-cultivated, the dark and dangerous, the unlimited, lurking at the limits, on the verge of overgrowing the fragile new structures of culture, be they crops or laws, with their respective outcasts of weeds, witches, and (were)wolves: gleaming green eyes staring holes into the walls of our cities and souls. But by the time European forests were clear-cut, the last frontier was won, mysterious and tropical jungles were divided into profitable colonies, and the various outlaws and savages were assigned their proper places in mental institutions, prisons, reservations, zoos, and anthropology books, the Big Bad Wolf reappeared, disguised as the desire of Little Red Riding Hood. Caged, trapped, burned, a devil set on fire, the wolf kept coming back to us; and we keep coming back to the wolf, the hungry shadow of our imagination.

With the woods disappearing and its wild denizens gone or domesticated, the stories that spoke of a life *with* these others edged toward extinction: tales told by lisping old women living at invisible borders between villages and fields, with no one listening except some eager assistants of enterprising philologists, the Brothers Grimm. The wild drifted into fairy tales, collected and preserved in the Big Book, as dried flowers in the herbarium, pinned specimens of rare butterflies behind glass. Neatly separated, ordered, and systematized, they became objects for study and research. In this second "life" they themselves no longer move but are moved—either by the categories of sciences, such as linguistics, botany, and biology, or by the symbols of a cultural imagination. With the hide of the wolf fading above the fireplace, Father tells the children, before they go to sleep, tales of the Wolf and the Seven Kids. They take the wolf with them to one of the last places left to meet him—their dreams. Freud's Wolf-man met him there when he was a little boy. It was in the middle of a cold Russian night when the window of his bedroom flew open, and in the leafless tree, dark against the white snow, seven wolves stared at the boy, their ears attentive, listening for something already there and about to come. Screaming, the child woke up, responding to the wolves' ears that already knew, already heard, his fear.[1]

The disappearance of European woods and open lands provided the young American nation with a new identity. It found a sense of self in celebrating its pristine, "virginal" landscapes. They filled the felt lack of national and cultural history, countering a cultural inferiority vis-à-vis Europe. Congress officially acknowledged the renewed self-presentation of the nation by acquiring in 1847 Moran's majestic painting of the Grand Canyon for its Senate lobby. The Hudson River school (1825–75) consolidated, nationally and internationally, an American artistic identity through its sublime imagery of the wild American "landscape."

Ironically, the same period saw the most cruel persecution of the American wild: between 1850 and 1880 over 75 million bison were killed as much to cripple the Native culture that depended on them, lived with them, as for their hides and tongues. It took out the heart of the Plains Indians, who used to travel with the migrating buffalo herds. Like butterflies pinned into place, Native Americans were put into reservations. A culture was decimated by eliminating its lifeway. It made room for the cowboy with his cattle and for the farmer with his 160-acre lot.

Packs of wolves followed the American Fur Trading Company to scavenge the countless carcasses, the carnage that the buffalo hunters left behind. They formed an easy target for the same hunters. It was the livestock industry, however, that meant the end of wolves on the plains. Cattlemen conquering the West replaced the indigenous grazers of the prairies with their own tame animals and hired commercial wolfers to eliminate the varmints, who, having lost their original prey, turned to domestic stock. Strychnine-laced meat was spread all over the prairie, and the poisonous saliva of dying wolves, foxes, and coyotes dried in the grass and was stored for months, years, killing those that fed on the prairie such as mustangs, elk, and antelopes. If one includes the passenger pigeons that were used for target practice, it is conceivable that between 1850 and 1900 some 500 million animals were slaughtered on the plains. Barry Lopez speaks of an "American pogrom: . . . Perhaps one million wolves; two million. The numbers no longer have meaning."[2]

Where the cattle industry took the vast interior grasslands, mining companies took the mountains, and the timber industry took the trees. Together they left a legacy of exploitation and destruction on a scale never gauged fully but with a result that has become understood and legitimized as economic "necessity," manifest destiny. To temper this intemperate exploitation, wilderness areas were belatedly established.

The deepest mystery remains the wolf. His eradication was so far beyond what was necessary and the way in which it was done so rich in cruelty and perversion that one can speak of a holocaust in the sense of a totality of destruction—the Greek *holokauston,* "that which is completely burnt." The wolf turned into a sacrificial animal—a burnt offering to some unspoken, some unspeakable cause. By the end of the nineteenth century the wolf was virtually exterminated in the United States. Poisoned, trapped, maimed, tortured, fed ground-up glass, set on fire.

The myth of the Wild West speaks with a double tongue. The ideal of wilderness arose when Americans no longer knew anymore how to live with the wild. From its beginning wilderness has been the abstraction of wildness, sacrificing wildness on its iconic altar.

Where "wilderness" refers to an area with well-defined boundaries, "wildness" is "by definition . . . intractable to definition, is indefinite."[3] It signifies what is not determined and not easily grasped, much like mist lingering in the landscape. The deer takes up the forest as much as the forest takes up the deer.

The rise of the concept of wilderness coincides with the first serious thematization of the unconscious, the mental equivalent of wild land and of sexuality, the bodily equivalent—usually, and not accidentally, manifest as the female body. Woman, like wilderness, figures as an abstract ideal that channels profoundly ambivalent emotions: as the mysterious other, she is uncontrollable and therefore desirable, and at the same time she is the other in need of being conquered and domesticated. As with wilderness, she becomes an emblem of life instead of a partner and participant. Thus, for Nietzsche, the prototypical nineteenth-century philosopher, she is the abyss, the veiled one, the Truth. If she breaks out of this idealized and abstracted paradigm, all too often that meant that she became unintelligible and undesirable as a woman. Most women writers and scientists were unmarried. And of what does the high rate of hysteria or suicide among the gifted sisters of famous men speak? Similarly, when nature breaks out of the ideal imagery of wilderness, it loses its imputed magic and becomes reduced to a mere resource rather than a regenerative source. Woman and the wild, thus hypostasized in static concepts, reduce reason's fear of disappearing in its own longing for its opposite but leave at the same time the desire unfulfilled, an unfulfillment that makes the force of the other, the mystery of animality, wildness, and bodily desire, burn only more strongly.

Rethinking otherness requires an understanding of the place the other occupies. The very externalization of the other subtends the modern notion of subjectivity, grounded, as it is, in a sense of lack. The modern subject is subject to desire, the desire to become a subject vis-à-vis the postulation of an object. Thus, both identities are frozen, and so they become "fixed," as Luce Irigaray says, "not 'free as the wind,'" because this subject "already knows its object and controls its relations with the world and with others" instead of living *with* them.[4] However, "any *finalization* of identities is an anathema that destroys otherness. . . . This is why wildness, which contradicts any finalization in identification, is at the heart of otherness, as well of course at the heart of any *living* self or society."[5]

Thus, wildness is as vital for the self as it is for the other. It affords life, *vita,* in both self and other; it vitalizes both of them in one and the same moment. It prevents any closure of communities with its imminent danger of stigmatizing the other into Jew, woman, wolf, or woods. Wildness guarantees the openness of any being in common, the sharing of commonality by different entities in a common place or project. As soon as this commonality closes its borders, that is, eradicates its wildness, it is reified into a "common Being" of a more or less totalitarian cast.[6] Wildness, so easily itself stigmatized as "something out there in nature," may be crucial in a rethinking of contemporary forms of sociality. Preserving wildness implies a political and ethical commitment of living with otherness versus incorporating or domesticating the other. It requires an ethics of attentiveness, flexibility, adjustment, change.

The question is, How do we preserve this kind of wildness, or, more generally, how do we retrieve otherness from being turned into a symbol?

One mode of opening up the here and now happens through a heightened sensitivity to sense-experience. It opens everyday life into more and unpredictable—wild—layers of experience, especially when one pays close attention to the specificities of others, be they human, animal, placial, or elemental. This creates a sense of the unexpected precisely through a familiarity with the intricacies of the other. Exploring the particularities of the places we inhabit makes room for an awareness of an endless set of performative possibilities of the other. A richness of per-formative behavior manifests itself implicitly through complex patterns of adaptive and generative activities, as we see in wolves roaming *through* the land, *through* the seasons, and, yes, *through* our dreams. Rick Bass invokes this performativity when he points out that "the thing that defines a wolf more than anything—better than DNA, better than fur, teeth, green eyes, better than even the low mournful howl—is the way it *travels*."[7] Lingering on the traveling of wolves implies a porosity of conceptual borders that dissolve in the complexity of different modes of participation in the landscape. Only when there is an openness to the experience, that is, a participation in particular activities and places, can we keep the wildness of the other—even in wilderness areas. Wilderness, then, "is not only a place you go. Wilderness is what happens to you. Shivered, sweated."[8] In participation things happen to you: walking the strenuous gradient of a steep slope, the rhythm of the mountain pulses through your blood, and your body sweats the heat of the day till night comes in and cold covers the rocks, the trees, your bones. Wildness pervades us if we are open to it and participate in it. It is implicit in us and we are implicit in it.

The unnamed and untamed slip between borders into the "darkest woods, the thickest and most interminable swamp," or "impermeable, unfathomable bog," escaping human measurements of words, definitions, numbers, surveys, buildings, and roads.[9] The wild is "refreshing" or re-creating precisely because it is not fixed but tends to transcend the frames of our thoughts. The unframed is the place where our thoughts and our desires can roam freely, joining that which is already unfathomable: nobody knows where a mountain begins or where the sky ends. It is a tacit knowing that is to be determined along the way instead of caught in stable definitions.

Neither defining nor confining, the silent and wild leave open a space for wandering and wondering; it is where our clear and distinct grasp transcends into impermeability, where our knowledge and being appear in the mode of dis-appearing. As soon as one stakes out the wild, it is gone. As long as one can be a presence that leads to its own dis-appearance, one stays wild, disappearing not in the sense of simply vanishing from existence but as a beginning *not-to-appear,* a not standing out anymore.

Rain waiting in the trees, branches dark, wet and soaked through with water, clear crystals hanging down, carrying the whole world shining and upside down . . . All gone when you grasp it, leaving the branches dark and alone with their shapes. We have forgotten how to disappear, too much of what our words and hands grasp stays present, sticks to our future, covers our past. We stand out but have lost our capacity to disappear. And therefore so much around us and ultimately so much of ourselves has in fact disappeared or has been drained, clear-cut, paved over, eradicated, eroded.

Forgetting to disappear means not knowing how to stand out. Not knowing how to die means not knowing how to live. Not knowing how to be silent means not being able to listen or speak. Standing out, living, listening, speaking is a question of respecting the porosity of limits. In sheer standing out, identity fixates itself as well as "its" otherness. Standing out in dis-appearance, however, is making room for the other to appear: the foreground acknowledges the background by permanently shifting ground.

Elusive presences. Brown velvet wings, a butterfly lands on my knee, explores with its tongue my skin. Instinctively I whisper "hey," as welcoming a friend. Disturbed, it flaps away and lands quietly on a flower. I feel inadequate: I do not know how to greet a butterfly in the spring.

A warm summer evening. I spread my sleeping bag out on a soft patch of moss and sit down on an old tree stump, enjoying the quiet evening. Suddenly leaves rustle and a big raccoon wobbles out of the bushes. Agile with its little hands, it inspects my camping spot and pulls the pad from underneath my sleeping bag. I grin. The raccoon looks up; I try to make it feel comfortable again with some soft sounds that usually calm my cat and dog. In a flash the animal disappears in the woods, leaving me alone with my lack of subtlety, my inadequacy in knowing how to speak to a raccoon.

Only when I am silent do those animals take me up in their presence; only when I stand out *and* disappear at the same moment can they be with me. Only when I am just there, without being present too much, do they come back.

As with seasons, the being of the wild is change; it is a fluttering companion located between waking and dreaming, a colorful connection between knowing and not knowing; never caught in rigidity, it always moves, comes and goes.

Silence. The very absence of content or, for that matter, of *telos* opens up the possibility for other things to appear, to show themselves. Silence opens "the doors of the music to the sounds that happen to be in the environment. This openness exists in the fields of modern sculpture and architecture. The glass houses of Mies van der Rohe reflect their environment, presenting to the eye the images of clouds, trees, or grass, according to the situation."[10] The voice of silence opens to different voices, its pace to different rhythms; its character or definition is formed by the possibility of receiving different characterizations.

Silent is Donegal's dark soil. Without "definition" it carries neither coal nor gold and is as indeterminate as Heany's Irish peat:

> Melting and opening underfoot,
> Missing its last definition
> By millions of years.
> They'll never dig coal here.[11]

Silent is Santayana's landscape. He was one of the few philosophers in the previous century who thought that landscape was worthy of philosophical reflection; he considered its primary trait to be *indeterminacy:* "A landscape to be seen has to be composed." The ability to appreciate wilderness landscapes he called the "mastery of the formless."[12]

Silent is the stream's roaring path through the forest, not because there is no sound but because the water's speaking is immediate, unmediated, without representation. The river does not name. Wildness is not-naming, it is letting things appear without interpreting, translating, or casting them in static forms. Silence affords a place for many sounds. Indiscriminately the river takes up what comes along and has its say by washing away.

Silence, the "resting place," where the attention given to words, an attention that has stolen the secret of reminiscences, can evaporate: "This secret is only the inner presence, silence, unfathomable and naked."[13] Silence slips between representing something and being the something that is represented.

"The wild does not have words." Its "unwritten pages spread themselves out in all directions!"[14]

Sitting outside at the back door of the house. Jacob-dog lying at my feet. Wind hiding in the trees. All quiet. Monday-cat sleeping on my lap. My hands resting on her fur. "Fur," I whisper. Puff of air. That is all there is to it. Fur. Air. But words have to be found for the book.

To break the silence that speaks for itself, without voice, *unbestimmt,* undetermined.

Wilderness is not silence. Silence does not exist. In complete isolation John Cage hears his blood flowing and his nerve cells firing.[15] Silence is not wilderness. Wilderness does not exist. It is an abstraction. It is time to leave wilderness behind us. It has had its function. Now is the time for the wild, the specific, the here and the now. Wild is the scent of your sweat, of salt, of sea. Wild is the mountain without roads. Wild is: No Roads. Wild means we relearn to travel.

There are different modes of change. Things, including living things, have duration, they exist over time, they follow the heat and cold of the seasons, becoming warmer and colder themselves. Not so the cold itself; becoming warmer, it will *disappear* in the warmth and is as such more versatile, more fluid, than any of the things through which it manifests itself. Only the snowman is true to the cold: along with the vanishing winter he will lose his substantive presence in an indeterminate pool of water saturating the thawing earth. To him, in his quality of nonadherence, Wallace Stevens dedicates the poem "The Snow Man." For him who does not speak, who does not name but listens, Stevens let the winter wind blow with its sound of a bare place.

> For the listener, who listens in the snow,
> And, nothing himself, beholds
> Nothing that is not there and the nothing that is.[16]

The snowman is no-thing and hence beholds nothing. Silently he stands and listens until he hears the spring coming with soft winds and green grass shines through his disappearance.

In February winter settled. Even the salty sound froze, its waves stilled in silent ice, firmly held by the cold. Just by the passing of time, of season, a strip of wildness emerged along the northern coast of Long Island. Water turned solid—an elemental change—and we walked over it. Jacob, the dog, immediately rose to the occasion: agile as a polar hound, he inspected the newly formed arctic tundra, chasing screaming seagulls over white slates of snowy ice. They left him alone, their loud laughter resonating sarcastically in the thin air while their white bodies circled safely away in the blue sky, only to return with the geese, who carry warm winds under their wings, winds that blow the ice into floes and disperse the Nordic wild over the sea to be taken up by the wild of the water again.

Wild is what comes and goes; a flock of geese, Heraclitean flux. Mist in the early morning, creating a thin slice of wildness, land of layers, of mist and mystery, the intimacy of still time before the rush-hour traffic hits the road.

The silent and wild appear *and* disappear, ever evading closed definitions. As soon as they are defined into a special name, you have to be on your guard. For wild animals will just appear, like the butterfly, the raccoon; or we follow their disappearing tracks through snow, bushes, deserts, over the mountains. They don't come when we call them, as Jacob does or Monday the cat. When we call wild animals by a name they disappear—sometimes forever.

Nondomesticated animals like birds are just around, a general presence, a sound in the air. When we see them regularly around the house we begin to recognize particular ones, like the reckless Carolina wren who eats Monday's cat food from her bowl while she is snoozing next to it or the blue jay couple that nested in our garage. Being in our life, they are in our language: "Did you see that the blue jay's eggs are hatched?" We talk about them but rarely give them a personal name. They are birds. They are around: in the trees, in the air.

On their way to extinction they often become distinct, get a name, and become true individuals—in-divisible, no further division possible anymore, they are single specimens. They don't fly around anymore, they don't sing anymore, they are singled out. Caged and named, they stand before us, and, after attempting special breeding programs, we all too often have to witness their death, amounting to their extinction.

Martha was the last passenger pigeon; she died in 1913 in the Cincinnati Zoo. Lady Jane and Incus were the last two Carolina parakeets. This native parrot of the United States (one of a very few) was a "pest" for agriculture and thus doomed to be wiped out. The last two died in the zoo in 1918. Orange, the last dusky seaside sparrow, died old and infertile in the eternal youth of Disney World. He was driven from the air to make room for Cape Canaveral space explorations.

Named, they lost their time, their place. Their last days were spent out of place, in zoos and amusement parks. Martha, Incus and Lady Jane, and Cody, the buffalo in Kevin Costner's *Dances with Wolves.*

Intimacy is experiencing something together, a *shared* grammar, not a *declaration,* spoken by me to you; it is not just inside me, neither inside you, but in and between us. Most of the time intimacy is not even in words but is enacted—the way animals love or the land takes us in. Monday the cat, arching her body to find the hollow of my hand, caresses in one and the same move me and herself. The caress creates an "us," a moment in which we belong together, are part of each other, participate in each other. Me, you, the cat, the caress. Intimacy is not a participation in the sense of a relation between two separated entities; rather, it is an implicit connection in which individuality at once arises and dissolves.

Wild is what travels through our skin, through our borders. Clouds roaming the sky, worms slowly inching through heavy earth, way under any property lines. Crossing our borders, however, all too often means the end of the wild. Roadkills take much of America's wildlife. But borders go far beyond interstates, property lines, fences, the iron bars of the cages in the zoo. The most dangerous and insidious borders are the implicit ones, the vast invisible borders we radiate before we know that the cyanide we use to separate gold from the mountain slowly seeps into the groundwater, evaporates and condenses into acid rain, pouring down, hundreds of miles away, for years to come. Sweeping movements of unintended borders spread wildly and out of control through time and through places. The clear-cut mountain not only lost its trees but everything that cohabitated, participated with the trees, from spotted owls flying through the canopy, to mycorrhizal fungi and worms buried deeply beneath the roots. Eroded, there is even no soil left to leave one's tracks in. It's all gone, all disappeared, including the regenerative capacity of appearing again.

Again and again, it's the same old story. As soon as one of the few remaining, or shyly, slyly, returning, wolves eats something that "belongs" to one of us, such as a calf, lamb, or dog, the outrage is complete and a witch hunt starts again. That innumerable more dogs and cattle are killed on roads seems a futile detail. Rick Bass follows these flames of irrational hate in his gripping account of the Ninemile Wolves in Montana: "What was lost, in this whole story—two steers, and two lambs? May we all never be judged by anything so harshly or held to as strict a life or unremitting of borders as the ones we try to place on and around wolves."[17]

What is judged here is the wild, the other, culminating in this charismatic symbol, the wolf. Wolves do not need us. The wild does not need us. It does need us to be capable of disappearing. We need the wild, our body running, loving, the storm at sea, the geese screaming and flying on their migrating path, the gentle sound of spring rain on new leaves. We need the wild, the other, to enhance and intensify our everyday life, to experience that there is always more to the here and now and that this more is always already *here.* It is here in our body, in our thoughts, in our lives, in the geese, in the ground under our feet, in the boulders lying on the beach since glacial times, in the howl of the wolf heard by the Montana wind. The Rocky Mountain gray wolf, *Canis lupus irremotus,* "The Wolf Who Is Always Showing Up," appears to disappear, only to appear again.[18]

NOTES

1. Sigmund Freud, "The Occurrence in Dreams of Material from Fairy-Tales" (1913), in *Collected Papers,* vol. 4, trans. Joan Riviere (New York: Basic Books, 1959), 236–43.
2. See part 3, "The Beast of Waste and Desolation," especially chapter 9,

"An American Pogrom," in Barry Lopez, *Of Wolves and Men* (New York: Charles Scribner's Sons, 1978), 167–99.

3. Thomas Birch, "The Incarceration of Wildness: Wilderness Areas as Prisons," *Environmental Ethics* 12 (Spring 1990): 3–27, 8.

4. Luce Irigaray, *An Ethics of Sexual Difference,* trans. Carolyn Burke and Gillian C. Gill (Ithaca, N.Y.: Cornell University Press, 1993), 185.

5. Birch, "Incarceration of Wildness," 11.

6. See Jean-Luc Nancy, *The Inoperative Community,* ed. Peter Connor (Minneapolis: University of Minnesota Press, 1991), esp. chap. 1. See also Giorgio Agamben, *The Coming Community,* trans. Michael Hardt (Minneapolis: University of Minnesota Press, 1993).

7. Rick Bass, *The Ninemile Wolves* (New York: Ballantine Books, 1992), 6.

8. David Strong, *Crazy Mountains: Learning from Wilderness to Weigh Technology* (Albany: State University of New York Press, 1995), 110.

9. Thoreau, "Walking," in *Walden and Other Writings of Henry David Thoreau,* ed. Brooks Atkinson (New York: Modern Library, 1950), 615–17.

10. John Cage, *Silence* (Middletown: Wesleyan University Press, 1973), 7–8.

11. Seamus Heany, "Bogland," in *Door into the Dark* (New York: Oxford University Press, 1969), 55–56.

12. George Santayana, *The Sense of Beauty: Being the Outlines of Aesthetic Theory,* ed. W. Holzberger and H. Saatkamp, Jr. (Cambridge, Mass.: MIT Press, 1988 [1896]), 85.

13. George Bataille, *Inner Experience,* trans. Leslie Anne Boldt (Albany: State University of New York Press, 1988), 16.

14. Tomas Tranströmer, *Selected Poems 1954–1986,* ed. Robert Hass (New York: Ecco Press, 1987), 159.

15. John Cage, "Experimental Music" (1957), in *Silence,* 8. Cage adds: "Until I die there will be sounds. And they will continue following my death. One need not fear about the future of music."

16. Wallace Stevens, "The Snow Man," in *The Palm at the End of the Mind,* ed. Holly Stevens (New York: Vintage Books, 1972), 54.

17. Bass, *The Ninemile Wolves,* ix.

18. Ibid., 14.

Eileen Crist

Against the Social Construction of Nature and Wilderness (2004)

THE POSTMODERN constructivist perspective on nature holds that cultural, economic, political, linguistic, scientific, and other practices mold the meanings of *nature* and *wilderness*. For constructivists such practices inescapably underlie all perceptions and valuations of the natural world. They argue that there exist no unmediated representations of nature, for the latter are anchored in social contexts—contexts indelibly inscribed within the ways of knowing that generate such representations.[1]

Constructivism considers it to be axiomatic that the intrinsic meaning of natural phenomena is unavailable, and that human semiotic and material work bestows meaning to them. Since interpretive and practical work is quintessentially social, constructivists further maintain that the emergence and character of beliefs, including true beliefs, about nature can be accounted for by sociocultural factors—be they economic conditions, political circumstances, paradigms, interests, networks, discursive practices, and the like. Since all beliefs are accounted for on sociocultural grounds, the constructivist position implies some degree of epistemic relativism—beliefs are not immutable or universal, but relative to the locations and time of their production. In the words of Phil Macnaghten and John Urry, "there is no single 'nature,' only natures. And these natures are not inherent in the physical world but discursively constructed through economic, political and cultural processes."[2]

This paper is a critique of the postmodern constructivist view of nature. As Ian Hacking has noted, a host of things and ideas have been argued to be socially constructed—from "gender" and "literacy," to "quarks" and "reality." Constructivism comprises a large and heterogeneous body of literature. My aim, here, is not to take on postmodern constructivism *tout court,* but specifically to critique its application to "nature" and "wilderness." By "postmodern constructivism," I characterize literature that evinces the following themes: an emphasis on cultural ideas, narratives, power constellations, politics, and the like as primary driving forces behind the establishment of knowledge; the repudiation that there exist foundations to knowledge that transcend socio-historical contexts; an epistemic predilection for the relativization and pluralization of "knowledges"—stressing their contingency and diversity; and skepticism toward "canonical knowledge" and/or "master narrative."[3]

While at face value the idea that knowledge is socio-historically situated seems trivially true, probing into the assumptions and repercussions of the "social construction of nature" reveals it to be intellectually narrow and politically unpalatable. Despite a predilection for uncovering the sociocultural roots of representations, constructivists about "nature" and "wilderness" do not deconstruct their own rhetoric and underlying assumptions to consider what fuels the credibility social constructivism musters as a "knowledge/power configuration."[4] I argue that recent applications of social constructivism to environment-related issues reflect the recalcitrance of anthropocentrism and buttress the drive to humanize the Earth. As an intellectual looking glass of these trends constructivism functions as ideology—and it is, as conservation biologist Michael Soulé has pointed out, as dangerous to the goals of conservation, preservation, and restoration of natural systems as bulldozers and chainsaws.[5]

INTELLECTUAL GRIEVANCES WITH CONSTRUCTIVISM

In articulating how the natural world is represented, constructivists are partial to formulations that stack the deck in favor of social constructivist conceptions. Metaphors of human labor regarding the creation of knowledge abound—familiar examples are building, constructing, assembling, manufacturing, inventing, or producing knowledge. Such vocabulary trades heavily on received distinctions between nature/natural and culture/artifactual, and through its semantics pushes the constructivist envelope—viz., that knowledge is primarily man-made, not imparted by nature. Another loaded vocabulary used with respect to knowledge creation is that of claims-making, contesting, and negotiating—a semantics transferred from political and litigation affairs, and designed to construe knowledge as perennially provisional or, to cite the constructivist idiom again, "contingent." Finally, in articulating how nature is represented, constructivists tend to be partial to ascriptive formulas: they maintain that human beings assign, impute, or attribute meaning to the natural world.

In one formulation, the constructivist analysis of nature is described as "a concern with how people *assign* meaning *to* their world."[6] This sort of wording is so automatically associated with constructivism that it is also used when paraphrasing its perspective: "We cannot experience nature except through the lens of meanings *assigned to* it by particular cultures," writes environmental ethicist Anna Peterson.[7] The choice of the verb "assign" is implicitly presented as a neutral descriptor of the interface between representations and nature. But this semantic choice is neither neutral nor unproblematic. Not only is such wording loaded to favor constructivist conceptions; it also embeds the assumption that people operate on an existentially distinct plane vis-à-vis the natural world; and it blankets over a manifold of language-games describing how knowledge and the natural world relate. These points are elaborated in what follows.

Constructivist scholars sometimes admit that nature itself delimits how it is represented—maintaining, for example, that knowledge is "hybrid" or "co-produced" by cultural processes and natural constraints.

But two things subsequently cancel out this empty gesture of what David Demeritt calls "constrained constructivism"[8] toward the deciding power of the natural world. Firstly, in the analyses themselves the bulk of the focus and credit goes to economic, discursive, network, rhetorical, and other sociocultural factors through which (ever-"contingent") representations are said to be constructed, negotiated, contested, black-boxed, and the like. Secondly, in (meta)descriptions of the constructivist project semantics that surreptitiously support a human-centered viewpoint are employed—such as "assigning meaning" to nature: from the outset, ascriptive ways of framing the interface between representations and nature plainly *assert* that meaning-making is a one-way affair from human arenas to the natural world.

The idea of imputing meaning to the natural world presumes a standpoint separate from it. While constructivists aver that only from specific standpoints can representations be created—that a "view from nowhere" is chimerical[9]—on a more fundamental level, by systematically eliding the substantive role nature plays in how it is represented, constructivists existentially divorce the human perspective from the natural world and describe meaning-making as acts of delegation emerging out of alliances, competition, negotiations, networks, rhetoric, or techniques of human arenas. Openly or implicitly, the natural world is portrayed as mute, intrinsically meaningless, ontologically indeterminate, epistemologically unavailable, and aesthetically indistinct—white noise, which prior to representation exists either as the proverbial blooming buzzing confusion, or as an elusive trickster amenable to indefinite registrations. Nature becomes narrated, theorized, inventoried, and comprehended—birthed into signified existence—by human activity. Prior to this representational animation the natural world is epistemically, aesthetically, ethically, and in all ways without intrinsic or participatory voice.

In one of his last essays, Paul Shepard lambasted this perspective as asphyxiating and provincial.[10] One way to point to its prevarication is by means of a little word-play: the assumptions underlying the supposed neutral inquiry into "how people *assign* meaning *to* the world" may be pried open by countering its mirror-image formulation of inquiry into "how people *receive* meaning *from* the world." The former sounds more sonorous to the Western intellectual ear not because it is ultimately more cogent, but because it is rooted in a dominant humanist-Cartesian tradition of subject-object separation that grants human cognitive sovereignty over everything. But there exist potent contemporary and premodern traditions, which, in contrast to the anthropocentric gospel of Man-the-Meaning-Maker, have regarded meaning as already afforded *within* the world—and human beings, as well as other animals, are able to tune into, tap, decipher, or directly receive those meanings.[11]

Another way to make this anti-constructivist point is that the representational structures people work with are *derived from* the world within which the human species evolved. The composition of language co-evolved as, and with, the emanations and exigencies of the natural world—it is neither an alien installation nor a quantum leap beyond nature accomplished by the human brain.[12] It is not as if we have been beamed onto this planet from another dimension and must struggle to represent a nebulous world in "our" terminologies. Rather, such comprehensive ideas and universal preoccupations with truth, goodness, and beauty are integral with the natural universe within which they originated and within which their applications lean.

The difference between the typecast alternatives "assigning meaning" and "receiving meaning" is heuristically important in yet another way. Anyone can assign meaning to nature, arbitrarily or to serve whatever purposes or motives. Not everyone is in position to receive meaning from the natural world with equal alacrity or acumen. People receive meaning with divergent depth and accuracy according to whether they are equipped with pertinent knowledge, relevant training, prior experience, tuned awareness, passionate interest and attention, breadth of understanding, care, or sufficient self-cultivation.

When vivisectionists, for example, claimed that the movements and cries of cut-up animals were mechanical reflexes, they were indeed assigning to nature a self-serving registration—projecting a "virtual reality" that allowed them to go about their business without the inconveniences that a true registration would have entailed. But in discussing dogs' ability to love, Charles Darwin noted that "everyone has heard of the dog suffering under vivisection, who licked the hand of the operator; this man, unless he had a heart of stone, must have felt remorse to the last hour of his life."[13] Darwin—for whom feeling, reason, intelligence, curiosity, wonder, aesthetics, and morality were evidenced within the animal world[14]—was neither "contesting" the vivisection perspective nor "negotiating" an alternative "narrative." He did not even bother to address its deluded opinions, but simply noted pain where pain is; he then almost casually remarked on the awakened conscience which *naturally* would haunt the vivisectionist provided he were open to the tidings of nature.

The choice of anthropocentrically slanted vocabularies—that construe knowledge through metaphors of labor, political/legal deliberation, or meaning-imputation—systematically erases the diversity of language-games available to describe representational activities. Representations of nature can be, and are, said to distort, imaginatively project, misconstrue, misinterpret, embellish, provisionally understand, approximate, work for all intents and purposes, intuit, predict, accurately explain, or deeply discern. Representations are also variously describable as interesting, beautiful, suggestive, questionable, objectionable, persuasive, compelling, or obvious.[15] None of this variety of assessing how representations and nature intersect is heeded by postmodern constructivism, which, on the contrary, ousts the wealth of epistemic valuations that ordinary-language and practices work with (in science and other arenas) in favor of a narrow, skewed set of metaphors. A diversity of predications is stifled under the monolithic formula that knowledge is "socioculturally constructed," as though the latter somehow enlightens more than the range of epistemic differentiations that it smothers.

Natural language embodies an eclectic array of descriptions about how knowledge and belief interface with the natural world: from delusive, biased, and self-serving, to provisional, good-enough, or approximate, to profound, stable, accurate, and even (heaven forbid) universally true. The concepts "knowledge" and "belief," themselves, predicate the epistemic standing of phenomena with qualitatively distinct degrees of certainty:[16] but the divergence between "knowing" and "believing" is either openly disavowed in constructivist thought or whitewashed under the gloss of representations as "contingent." The erasure of diverse representational modalities—in favor of a one-dimensional human-hegemonic vocabulary of knowledge as sociocultural "construction" and/or "narrative"—is the ubiquitous linguistic move upon which the constructivist understanding of nature rests.

But the moment that the manifold language-games capturing the gamut of relations between knowledge and nature are readmitted—as a *bona fide* map rather than epiphenomenal—we are delivered from the suffocating picture of a lone, representation-constructing being projecting meaning either on a blank screen (strong constructivism) or an elusive nature differentially construable according to social position (standpoint epistemology). This view, as Shepard noted, is as oppressive as the positivism it has sought to discredit.[17] The two perspectives have more in common than either would care to acknowledge: they share what David Ehrenfeld has famously called "the arrogance of humanism," and Vicki Hearne aptly described as "humania."[18] In assessing the art of interfacing scientific knowledge and natural reality, positivism and constructivism both acclaim human representational and interventionist capacities as the centerpiece. Neither school of thought has ever counseled students about the significance of humility and respect toward the natural world. This is not coincidental: these perspectives are what historian Lynn White has called "post-Christian,"[19] in the sense that for both the primary locus of meaning is human categories-*cum*-techniques—in Biblical terms, naming-and-working.

The hidden ties of constructivism to the Judeo-Christian worldview reveal the "social construction of nature" as a post-Christian viewpoint. The first similarity involves the striking family-resemblance between the constructivist supposition that nature is intrinsically voiceless and the Biblical myth in which Adam is given the task to name the Creation. The second similarity involves the alleged special status of human beings: in Biblical terms Man was made in God's image, while in constructivism as symbol-possessing and technology-producing beings humans stand apart from all animals. The third similarity between the Judeo-Christian and constructivist views is that for both the natural world is devoid of native meaning, being, order, mystery, value, or feeling. Indeed, it was the Judeo-Christian worldview that evacuated immanent significance from the natural world, thereby desacralizing it and making it a place to be dominated and used, virtually unrestrainedly, by human beings.[20]

The exorcism of *anima* from nature—after two millennia of a dominant material and religious culture of European Judeo-Christianity—constitutes a (by now) undetectable pillar of postmodern constructivism: the silencing of wild nature through long-term colonization and through what sociologist Max Weber discerned as "the disenchantment of the world"[21] is deep inside the belly of an amnesiac paradigm that exalts human cultural "readings" and "practices" as font of all knowing.

The constructivist perspective has inherited, in secularized form, key elements of the religion that White called "the most anthropocentric of the world." A major difference between constructivist and Christian viewpoints is that the former acknowledges the diversity and flux of narratives, while the latter has often sought to impose a single doctrine. Nevertheless, the two partake of the same worldview: that the basis of the human relationship to nature has far more to do with meaning projection and instrumental intervention, than it does with *the cultivation of receptivity* — opening oneself, listening, watching, being within, letting be, or merging into. Secular and religious (respectively), their story is the same "old story of the tail wagging the dog," as deep ecologist George Sessions notes about postmodern anthropocentrism.[22]

The bottom line of the humanist mindscape—of which postmodern thought is the latest outgrowth—is that knowledge is a human franchise from which we naturally draw a sense of cognitive supremacy over the rest of creation and/or cognitive sovereignty over the world. According to constructivist Andrew Ross, for example, "there are no 'laws' in nature, only in society, because 'laws' are made only by us and can therefore only be changed by us. Nature, in short, does not always know best."[23] An ecocentric sensibility recoils from such supercilious parochialism: knowledge is a boon from nature not a human project about or projection onto it; and knowledge is evidenced throughout the animal world as naturalist, wilderness, and increasingly scientific writings attest.[24]

The constructivist assumption that the natural world is devoid of immanent meaning is neither self-evident nor uncontested. For the cultures, individuals, and ecological movement that have embraced an ecocentric understanding nature is suffused with feeling—with love, joy, grief, curiosity, pain, wonder; nature is suffused with intelligence—

awareness, attention, communication, reason, cunning; nature is suffused with energy perceived as aesthetic elation; nature is suffused with mystery experienced as transcendental feeling; and nature is suffused with spectacular order—complex, autopoietic, ever-changing, dynamically temporal, and emergent. The cavalier rejection of the natural world as intrinsically meaningful rests on the historical extirpation of peoples who have regarded and treated plants, animals, and the land as possessing native intelligence in dialogue with human beings; and it rests on its contemporary dismissal as New Age atavism.

When nature is understood as the emanating source of meaning and knowledge—rather than the object, playpen, or epistemic outcome of cultural endeavors—what common sense mostly intuits also follows logically: that there exist ways of representing the world that are *essentially* more profound, more true, more insightful, more enduring—not to say more respectful and more beautiful—than others for neither sociocultural nor "knowledge/power" related reasons, but because they align with nature in valid, perceptive ways. Western science has created such knowledge in spades, as have other and far older knowledge systems. Moreover, not only intersubjective knowledge traditions but also individuals through self-cultivation can transform themselves into mediums of "personal knowledge"—the human mind-heart-body, being itself a piece of the world, can become a transparent instrument for understanding and expressing nature.

POLITICAL GRIEVANCES WITH CONSTRUCTIVISM

The constructivist agenda has been described as the objective to understand "the social history of nature"; this agenda is the converse of, and quite inimical to, the objective to understand "the natural history of society/humanity."[25] My interest is not to defend a *naturalistic* account of human society over a *social* account of nature. Rather, I consider the political ramifications of focusing on sociocultural accounts of nature at this particular historical juncture.

Attending to the social history of nature, by default, skirts an ending of natural history that we are bearing witness to today: the quickening, worldwide ruination of natural systems such as wetlands, waterways, tropical, temperate, and boreal forests, grasslands, deserts and tundra, coastal and ocean habitats, and their native biodiversity. This ecological destruction—whether examined at the levels of habitat, ecosystem, species (as well as subspecies and varieties), organisms' recent natural ranges and migration routes, population numbers, genetic diversity, or evolutionary viability—is being documented and vociferously protested by life scientists from evolutionary biology, ecology, wildlife science, botany, and other disciplines. Indeed, a new "conservation biology"—defined as science in the service of conservation of life's native diversity—was created in the 1980s to oppose and mitigate the biodiversity crisis.[26]

At a time when unprecedented developments in the world and the life sciences call for a thoughtful openness toward the scientific enterprise, students in the humanities are taught to deconstruct and translate natural science discourses into the idioms of their own fields. The project is not to learn from science about the (state of the) natural world. Instead, it is to kindle skepticism toward taking scientific claims at face value in order to understand the genesis of those claims as products of political negotiation, network action, ideological or ethical motivation, technological determination, or other social variables depending on the specifics. On this view, the self-presentation of "scientific knowledge" is like the tip of the iceberg: what is not visible, but brought to light by constructivist analyses, is the submerged part that constitutes the sociocultural underpinnings which scientists disregard or screen out in formal presentations of facts, theories, or products.

In revealing the importance of social factors in science, and making scientists more aware of them, this project is intellectually and pragmatically valuable. It becomes incoherent, however, when built upon a stout allegiance to skepticism toward the realist status of scientific claims—for the apparent purposes of either disclosing the natural-scientific enterprise as a branch of the human sciences or defrocking scientific claims as having no special status, being "one set of stakeholder claims" among others. Questioning scientific and technological developments is desirable for eschewing blind faith in the scientific establishment and cultivating critical-mindedness; but constructivism goes beyond this welcome goal to place the scientific enterprise under the siege of skepticism.[27] But skepticism about the veridicality, and (where applicable) the universality, of scientific knowledge does not serve the art of critical thinking: rather, it collides head-on with the voice of reason which states that an enterprise dedicated to the pursuit of (universal) truth(s) about nature must, at least some of the time, hit bull's eye.[28]

The project of "the social history of nature" is not intrinsically at odds with what has been called the end of natural history, the end of nature, the extinction holocaust, the death of birth, biological meltdown, or biological Armageddon.[29] At the level of analysis, however, instead of attending to the degradation of natural systems, constructivism focuses exclusive attention on *human discourses* about it.[30] This approach to environmental issues obeys standard constructivist moves, which either bracket "nature itself" as extraneous to sociocultural exegeses about it, or regard "natural reality" as outcome rather than source of scientific representations.[31] But the epistemological construal of sociocultural input as sufficiently explanatory of, or constitutive force behind, "natural reality" grants power to human practices that reflects and reinforces our species' capacity for colossal arrogance; it generates the familiar logical and political problems associated with relativism;[32] and funnels all fascination about knowledge creation as a story about *people*—rather than revelation, conjecture, distortion, etc., regarding *nature*.

Taking a human-driven ending of natural history seriously presupposes admitting the independent reality of what is ending; and it requires trust in the scientific discourses charged with understanding the building-blocks and processes of natural history. Insurmountable roadblocks to these prerequisites seem built into constructivist reasoning—for both scientific inquiry and its submitted views about natural history are regarded as socioculturally negotiated, provisional configurations. But coming to terms with the predicament of complex life on Earth necessitates that the relevant biological knowledge be taken at face value—a very different stance from deconstructing and/or bracketing its status as realistic representation, or regarding its content as the outcome (rather than source) of inquiry. Taking science seriously means that instead of an exclusive meta-discursive focus on *how* scientific "claims" are made, there is receptivity to the validity of biological findings; and instead of focusing on how scientific assessments are "contested"—a favorite constructivist tack—what scientists are agreeing on is (also) attended to.

Crucially for the argument presented here, life scientists concur that we are in the midst of a human-driven biodiversity crisis.[33] The gravity of this diagnosis is not marred by the caveat that scientific estimates of extinction rates often diverge widely. The significant point is that biological science—conservation biology, especially—is the key source of knowledge about biodiversity losses, regardless of the obstacles in producing precise quantitative expressions.[34] The reality of this crisis is documented with urgency by a burgeoning biological literature; as E. O. Wilson puts it, "the evidence is persuasive: a real problem exists, and it is worthy of your serious attention."[35]

Yet constructivist analyses of "nature" favor remaining in the comfort zone of zestless agnosticism and noncommittal meta-discourse. As David Kidner suggests, this intellectual stance may function as a mechanism against facing the devastation of the biosphere—an undertaking long under way but gathering momentum with the imminent bottlenecking of a triumphant global consumerism and unprecedented population levels. Human-driven extinction—in the ballpark of Wilson's estimated 27,000 species per year—is so unthinkable a fact that choosing to ignore it may well be the psychologically risk-free option.

Nevertheless, this is the opportune historical moment for intellectuals in the humanities and social sciences to join forces with conservation scientists in order to help create the consciousness shift and policy changes to stop this irreversible destruction. Given this outlook, how students in the human sciences are trained to regard scientific knowledge, and what kind of messages percolate to the public from the academy about the nature of scientific findings, matter immensely. The "agnostic stance" of constructivism toward "scientific claims" about the environment—a stance supposedly mandatory for discerning how scientific knowledge is "socially assembled"[36]—is, to borrow a legendary one-liner, striving to interpret the world at an hour that is pressingly calling us to change it.

A key claim that constructivism trades on is the fluidity of scientific knowledge—as Mick Smith puts it, "science changes; its opinions are not permanent."[37] This view, along with the fact that there exist disagreements and clamorous (sometimes highly politicized) debates within science, are cited as conspicuous indications that the image of science as "impartial, consensual, and universally valid" is belied by empirical studies of scientific inquiry that reveal it to be shifting, polemical, political, value-relevant, theory-laden, technologically-mediated and -oriented, or paradigm-dependent. While the constructivist project thus broadens the understanding of science, and at first glance seems a tenable substitute for a previously idealized view, on closer examination it often conceals that stable scientific facts about the natural world are legion and amassing.[38] Constructivism tends to promote an image of science as ever-changing and disputatious, endeavoring to replace the idealization of consensus-driven linear progress with an equally fictitious picture of contentious contingency.[39] Indeed, "contesting," "contested," and "contentious" are prominent buzzwords of constructivism.

A germane (for this paper) example of stable knowledge about nature, which has enhanced the horizons of humanity immeasurably, has been the discovery of evolution. Even as debates about the mechanisms and speed of speciation have raged for a century and a half, it is equally the case that in 1859 Charles Darwin opened a floodgate through which evidence confirming common descent has not ceased flowing. One can foresee theories about evolution gaining and losing ground, but one would be hard pressed to imagine *the gigantic fact* of common descent by modification one day chucked into the bin of obsolete beliefs. After 150 years of supporting evidence from every province of biological science, all odds favor that the evolutionary kinship of Earth's life forms is here to stay as universal fact.[40] To put it unambiguously, common descent by modification as "universal fact about life" means that it holds true for those who lived before, and who presently ignore or oppose, Darwin's discovery; it may even hold true for life in the universe at large, for without a mechanism of transmutation to enable adaptation, even if life emerged on a planet it would be unlikely to survive the titanic forces of environmental change, in the long run.

A case about stable scientific knowledge can also be made regarding the understanding of ecosystems. It is well known that views about the stability versus flux of ecosystems, and the relationship between biological diversity and ecological resilience, have markedly shifted; they are likely to shift again.[41] But the general insight into—along with innumerable concrete facts about—what Darwin called "the entangled bank" of organisms interlocked in food pyramids, relationships of symbiosis, tolerance, and competition, conversion of nutrients, waste assimilation and decomposition, and element cycling is so solid as to have become nearly prosaic: it constitutes the *ground* from which debates about the relative stability versus dynamism of ecosystems are launched. To focus on how perspectives within ecology have shifted may be intellectually stimulating, but to obscure the background of accruing ecological knowledge in relation to which scientific analysis has changed is to elide a huge portion of the spectrum that composes "scientific knowledge."

Connected to established knowledge about evolutionary and ecological processes is a wealth of recent conservation biology studies regarding: consequences of habitat destruction and fragmentation for ecosystems and their biodiversity; area-species requirements, especially for viable populations of predators and other keystone species; impact of invasive species; connection between genetic variability and evolutionary viability; the assessment of overall declining biological integrity of ecosystems; estimates of population thresholds beneath which species and subspecies enter the red zone of potential annihilation; exacerbating effects of climate change on the biodiversity crisis.[42] These scientific findings, among innumerable others, educate about the state of the biosphere: they reveal that without requisite changes in human affairs, cornerstone dimensions of natural history—namely, evolutionary processes, ecological integrity, robust populations of nonhumans, and biodiversity—will continue to be dismantled.

Epistemological focus on the "social history *of* nature," at a time when the catastrophic impact of "social history *on* nature" is swelling, may reasonably be charged as a diversion of intellectual and political energies away from the main event.

A more severe censure of the constructivist approach to nature is that not only does it distract attention from the environmental predicament, but it also supports that predicament. Constructivists diagnose radical ecological views as "an artefact of current social circumstances"[43]—a charge to which radical ecologists plead guilty since they aim to redress these circumstances. But social constructivism is also "an artefact of current social circumstances"—albeit a far cry from protest: the most troubling facet of the constructivist paradigm is that, as an approach to understanding nature, it is boosted by (and in that sense cashes in on) the social destruction of nature.

In her tempered critique of constructivism, Peterson observes that nature can be regarded as "socially constructed" in two ways, ideational and material: ideas about nature are shaped through culturally diverse lenses; and natural landscapes are physically altered by human technologies and activities.[44] Peterson sees these as distinct facets of "the social construction of nature." What others have added to this analysis is that the two resonate with one another especially at this historical juncture.[45]

The notion that the Earth's natural systems are only graspable in "mediated" terms strikes a cheerless chord with the global undertaking to convert the planet into a *Homo sapiens* outpost: if nature is sufficiently pliable to be molded by human work, then it can be deemed passive enough to be fully constituted through cultural discourses; and as nature is increasingly simplified by human incursions, it not only seems but *becomes* more susceptible to conceptual subordination. These are the tacit harmonies between the social destruction and social construction of the planet's natural systems. And thus an order of things indictable as corrupt is, instead, implicitly tapped by constructivism to bolster its epistemology.

With the human impact on the planet escalating, the autonomous self-organization of the natural world is correspondingly obliterated, and alongside this obliteration, the idea that there exists no "essential nature" beyond cultural mediations entrenches itself as robustly realistic. As the biosphere is colonized—settled, paved, mined, burnt, dammed, drained, over-fished, poached, and roundly used—diversified conceptions of how "nature" and "society" (should) relate are more facilely bulldozed by a monolithic image of "nature-society" hybridization. The idea that "we have moved from thinking of nature and society as distinct realms or regions to thinking of them as interlaced or entangled"[46] is typically redeemed through icons of a domesticated, impoverished, or technologically remade world.

For example, Steve Hinchliffe provides a pictorial illustration—skeptically captioned "Natural parks?"—showing a denuded aspect of Snowdonia National Park, with pastureland and a fence in the foreground, informing us that "this scene is as social as it is natural." Along with a hypothetical example of cloning (in which we are similarly edified that biology and society would contribute to a cloned person's identity), he apparently hopes that "these examples may have convinced you that nature and society are indeed two sides of the same coin."[47] Indeed, they are two sides of the same coin as long as, in wilderness advocate Bob Marshall's words, "the tyrannical ambition of civilization to conquer every niche on the whole Earth"[48] is either left undisturbed or implicitly condoned as an acceptable historical course.

Leaning heavily on Latour's thesis in *We Have Never Been Modern,* Hinchliffe censures the separation of nature and society as "pure" categories.[49] But the hybrid (constructivist) model of entangled nature/society and the purified (objectivist) model of distinct nature/society share the *totalizing design* characteristic of all ideological and/or overtheorized formulations: we are invited to buy into them hook, line, and sinker. From an ecologically informed environmentalist perspective both models are deficient; both are "purifications"—wholesale academic kits with ready-made semantics and concepts that spare students the trouble of creating their own tooling.[50]

The alternative is to regard the received umbrella categories of "nature" and "society," such as they are, as referring to an array of empirical phenomena and conditions. The character of their relation is not to be decided *a priori* by grandiose theoretical schemes, but rather diversely defined and understood depending on what is at stake—on specific contexts of analysis, values, and action. It is under such auspices that wilderness advocates defend areas "where the earth [*sic*] and its community of life are untrammeled by man, where man himself is a visitor who does not remain."[51] Whatever the flaws of this definition may be, the intent of those who so crystallized the understanding of wilderness forty years ago is a *key site of resistance* against both the realization of a humanized, biologically degraded planet and its epistemological handmaiden of everything as a "hybrid," "cyborg," and "nature-society" hodgepodge.

THE ENDANGERED IDEA AND REALITY OF WILDERNESS

There is nothing intellectually or socially innocent about the timing of the disclosure that "wilderness" is a cultural concept: as wild nature sinks into the quicksand of all manner of development, the idea itself starts to feel like gossamer. What poses as a sophisticated argument—that wilderness is a construct since it has been a (non)idea amenable to historically diverse conceptions—in socio-historical context can be understood as an unsurprising ideological reverberation of the appropriation of wild nature.

In his work *Grizzly Years,* Doug Peacock observed that wilderness was becoming an endangered idea well before it became academic fashion to question its essence. "After Vietnam I saw the world changing with amazing rapidity, with a violent tempo I had not noticed before 1968. The pace I had heard as a slow drumbeat in the fifties was now a rapid staccato . . . Everywhere you looked, you saw a microcosm of the entire buzzing globe—even in the woods, in grizzly country. *The entire concept of wilderness as a place beyond the constraints of culture and human society was itself up for grabs.*"[52] As early as the late 1960s, Peacock sensed how the deflowering of wilderness was paving the way to its conceptual emasculation.

The tightening blockade on wild nature is a fitting existential background to the idea of wilderness as sociocultural construct. Because of this snug historical fit the constructivist view of wilderness functions as ideology—regardless of whether it is so intended. "Wilderness" qua construct conceptually erases the objective reality of the word's referent thereby fortifying its physical eradication by the very civilization that spawned constructivist thought. As Soulé puts it, the siege on nature has become two-fold: the overt physical siege and the "covert assault [which] serves to justify, where useful, the physical assault."[53] In a similar vein, Kidner argues that constructivism "provides a model of nature which fits seamlessly into the industrialist view of the world."[54]

The argumentative strategy of the social construction of wilderness proceeds in line with what Vandana Shiva has called "the politics of disappearance."[55] The main tactic is obscuring from view that the meaning of a concept is not composed only of its *sense* but also of its *reference*. What wilderness refers to is systematically left out of discussion as constructivist analyses remain at the level of people's (culturally and historically divergent) ideas, as though beliefs and sentiments about wilderness fully exhausted the meaning of the concept. To borrow a well-worn example from linguistics, it is as though analysts documented the divergent beliefs of two tribes about the "morning star" and the "evening star": finding that narratives about these "stars" differ profoundly, analysts concluded that either they cannot possibly refer to the same celestial object, or they do not refer to anything (really knowable) beyond the discourses about the "stars" themselves.[56]

In disregarding the reference dimension of wilderness, constructivist thinking renders its meaning completely in the abstract.[57] The meaning of wilderness is, of course, not solely its referent(s): but as encroachment into virtually all land and ocean habitats escalates, this ancient facet of the concept of wilderness—which has threaded through its diverse cultural senses—is being hacked just as surely as its physical counterpart. By treating "wilderness" as an abstract idea, constructivists are both reflecting and condoning the eclipse of its reality.

Another tactic in the politics of disappearance is that insofar as reference to wilderness as self-organizing, self-determining nonhuman habitats is at all admitted, it is denied any essential existential/ontological standing.[58] The negation of essentialism is promoted by presenting ecological knowledge as perennially controversial and tentative, and more generally, by undermining the credentials of biological science to speak with ultimate authority about natural systems. Constructivist literature is also replete with *en passant* assertions of the supposedly obvious—that there is no essential core to "wilderness" beyond the play of culturally diverse narratives or socially negotiated constructions. The anti-essentialism of postmodern constructivism is presented as the high-ground of the intellectually elite. An essentialist view of wilderness is deemed an anachronism held by naïve romantics—or by those uninitiated into the abstruse meditations of postmodern illuminati.

Wilderness as an essential reality independent of human presence, will, and control is also rejected as "one pole of a dualism," reflecting a reified separation between pristine nature and impure humanity.[59] Critics of the wilderness idea make a lot out of the historical roots, and ostensible chimera, of understanding wilderness as a pristine realm untouched by people. In fact, such analyses assess the human separation from wild nature as *the driving force* behind environmental destruction: it was from such a disconnected mindset that the conquest of the New World, for example, was launched. This argument is sound insofar as it is evoking the connotation of "separation" from wilderness *sensu* human attitudes and actions alienated from, superior over, and thereby entitled to indiscriminate use of wild nature.

However, if the colonizing *modus operandi* is looked at from a different angle, the problem is equally well-defined as a *deficient sense* of appropriate dimensions of human separation from wild habitats. Conquistadors have always striven to annex both wild nature and people through violating rightful boundaries—first annihilating and then assimilating the other, whether nonhuman or human. So, while much is made of the supposed problem of human separation from wilderness—or of "society" from "nature"—little attention is paid to the virtuous face of separation. In a world where all are honored, a respectful observance of separation is also honored as the complement of intimacy with nature not its negation. This sense of separation does not stem from an ideology of human-wilderness dualism, but from the cultivation of an ecological ethic as Aldo Leopold understood it: a self-imposed limitation on our actions flowing from love, respect, and admiration of the land.[60]

It is in this spirit that radical ecologists advocate wilderness as an essential reality largely independent of human presence and control: wilderness areas of the Earth are the homelands of nonhumans—in scientific terminology, they are biodiversity reserves where native life can continue to flourish and evolve.[61] Without the range of conditions that wilderness avails, we are faced with the dismal possibility of a human-inaugurated biogeological era of an indigent natural history of wild native animals, plants, and ecosystems. Life will continue of course, but the *flame of life*—fanned by the bellows of evolutionary surging, immeasurable ecological complexity, prodigal numbers of living beings, and a diversity of life forms still unknown to the nearest order of magnitude—is in very real peril of being snuffed out.

CONCLUSION

What Max Weber called modern civilization's "disenchantment of the world"—which critical theorists Max Horkheimer and Theodore Adorno bitterly interpreted as that "arid wisdom that holds that there is nothing new under the sun"[62]—is materializing into a mundane, homogenized reality which everywhere bears (or, as affairs proceed, will bear) the human stamp.

In procession with this emerging new reality order, the memory (or future possibility) of a time when the natural world emanated an essence that was thickly fragrant, unbelievably fresh, profligate, seemingly indomitable, diverse, significantly unknown, enchanted, and wild is swiftly dimming in the human psyche. Toward the late nineteenth century, British poet Gerard Manley Hopkins saw through the human transmogrification of the world with piercing words:

> Generations have trod, have trod, have trod;
> And all is seared with trade; bleared, smeared with toil;
> And wears man's smudge and shares man's smell: the soil
> Is bare now, nor can foot feel, being shod.

Over a century after these lines were penned, it is becoming increasingly unlikely that we may long be comforted by the presentiment, which the poet expressed later in his sonnet, that "for all this . . . There lives the dearest freshness deep down things." Indeed, longing for such freshness is increasingly reckoned an embarrassment—labeled as romantic, atavistic, and unrealistic. The dismissive power of such labels mirrors the brawn of the prevalent socioeconomic system in which, as Herbert Marcuse incisively discerned, "not only radical protest, but even the attempt to formulate, to articulate, to give word to protest assume a childlike, ridiculous immaturity."[63]

If resistance against the endpoint of a colonized planet has hope of succeeding, we should be exceptionally wary of the postmodern call to put aside childish concepts like "purity," "essence," and "the romantic idea of wilderness."

NOTES

1. See Steve Hinchliffe and Kath Woodward, *The Natural and the Social: Uncertainty, Risk, Change* (London: Routledge, 2000); Arturo Escobar, "After Nature: Steps to an Anti-Essentialist Political Ecology," in *Current Anthropology* 40/1 (1999): 1–6; Phil Macnaghten and John Urry, *Contested Natures* (London: Sage, 1998); Jozef Keulartz, *The Struggle for Nature: A Critique of Radical Ecology* (London: Routledge, 1998); Sheila Jasanoff and Brian Wynne, "Science and Decisionmaking," in Steve Rayner and Elizabeth Malone, eds., *Human Choice and Climate Change* (Columbus, OH: Battelle, 1998); Philippe Descola and Gisli Palsson, *Nature and Society: Anthropological Perspectives* (London: Routledge, 1996); John Hannigan, *Environmental Sociology: A Social Constructionist Perspective* (London: Routledge, 1995); William Cronon, ed., *Uncommon Ground. Toward Reinventing Nature* (New York: W. W. Norton, 1995); Andrew Ross, *The Chicago Gangster Theory of Life: Nature's Debt to Society* (London: Verso, 1994).

2. Macnaghten and Urry, *Contested Natures,* p. 95.

3. Jean-François Lyotard, *The Postmodern Condition: A Report on Knowledge* (Minneapolis: University of Minnesota Press, 1989); Ian Hacking, *The Social Construction of What?* (Cambridge, MA: Harvard University Press, 1999); Michael Lynch, "Towards a Constructivist Genealogy of Social Constructivism," in Irving Velody and Robin Williams, eds., *The Politics of Constructionism* (London: Sage Publications, 1998); André Koukla, *Social Constructivism and the Philosophy of Science* (London: Routledge, 2000); David Demeritt, "What Is the 'Social Construction of Nature'? A Typology and Sympathetic Critique," in *Progress in Human Geography* 26/6 (2002): 767–790.

4. See George Sessions, "Postmodernism and Environmental Justice," in *The Trumpeter* 12/3 (1995): 150–154; David Kidner, "Fabricating Nature: A Critique of the Social Construction of Nature," in *Environmental Ethics* 22 (2000): 339–357.

5. Michael Soulé and Gary Lease, *Reinventing Nature? Responses to Postmodern Deconstruction* (Washington, D.C.: Island Press, 1995), p. xvi. Gary Snyder, "Is Nature Real?" in Tom Butler, ed., *Wild Earth: Wild Ideas for a World out of Balance* (Minneapolis, MN: Milkweed Editions, 2002).

6. Hannigan, *Environmental Sociology,* p. 33.

7. Anna Peterson, "Environmental Ethics and the Social Construction of Ethics," in *Environmental Ethics* 21 (1999): 339–357, p. 341.

8. Demeritt, "What Is the 'Social Construction of Nature'?"

9. Donna Haraway, "Situated Knowledges: The Science Question in Feminism and the Privilege of Partial Perspective," in Mario Biagoli, ed., *The Science Studies Reader* (New York: Routledge, 1999).

10. Paul Shepard, "Virtually Hunting Reality in the Forests of Simulacra," in Michael Soulé and Gary Lease, eds., pp. 17–29.

11. Wilderness advocates, deep ecologists, naturalists, poets, farmers who live with the land, scientists, and phenomenologists, in differing ways, have expressed opposition to the worldview of a passive natural world rendered meaningful by the human *cogito*.

12. Cf. Gary Snyder, *The Practice of the Wild* (New York: North Point Press, 1990).

13. Charles Darwin, *The Descent of Man and Selection in Relation to Sex* (Princeton: Princeton University Press, 1981 [1871]), p. 40.

14. Charles Darwin, *The Descent of Man and Selection in Relation to Sex; On the Expression of Emotions in Man and Animals* (Chicago: The University of Chicago Press, 1964 [1872]); *On the Formation of Vegetable Mould by Worms with Observations on their Habits* (Chicago: The University of Chicago Press 1985 [1881]).

15. E. O. Wilson, *Consilience: The Unity of Knowledge* (New York: Vintage Books, 1999), p. 64.

16. Jeff Coulter, *Mind in Action* (Oxford: Polity Press, 1989).

17. Shepard, "Virtually Hunting Reality," p. 20.

18. David Ehrenfeld, *The Arrogance of Humanism* (New York: Oxford University Press, 1978); Vicki Hearne, *Adam's Task: Calling Animals by Name* (New York: Vintage Books, 1987 [1982]).

19. Lynn White, Jr., "The Historical Roots of Our Ecologic Crisis," in *Science* 155 (1967): 1203–1207.

20. White, "The Historical Roots of Our Ecologic Crisis"; Roderick Nash, *Wilderness and the American Mind* (New Haven: Yale University Press, 1967); Carolyn Merchant, *The Death of Nature* (San Francisco: Harper: SanFrancisco, 1983). For more ecocentric interpretations of Christianity, see Holmes Rolston III, "Wildlife and Wildlands: A Christian Perspective," in Dieter Hessel, ed., *After Nature's Revolt: Eco-Justice and Theology* (Minneapolis: Fortress Press, 1992). White himself ended his classic paper by proposing Francis of Assisi as patron saint of ecologists.

21. Max Weber, "Science as a Vocation," in H. H. Gerth and C. Wright Mills, eds., *From Max Weber: Essays in Sociology* (New York: Oxford University Press, 1946 [1919]), p. 155.

22. Sessions, "Postmodernism and Environmental Justice," p. 153.

23. Ross, *The Chicago Gangster Theory of Life,* p. 15.

24. Donald Griffin, *Animal Minds* (Chicago: The University of Chicago Press, 2001).

25. Mick Smith, "To Speak of Trees: Social Constructivism, Environmental Values, and the Future of Deep Ecology," in *Environmental Ethics* 21 (1999): 359–376.

26. Michael Soulé, "What Is Conservation Biology?" in *BioScience* 35 (1985): 727–734; Reed Noss, "Is There a Special Conservation Biology?" in *Ecography* 22 (1999): 113–122.

27. Skepticism has been injected into constructivism by the appeal to two philosophical theses as (ironically) sweepingly true: the "underdetermination thesis" (all theories are underdetermined by evidence), and the "theory-ladenness of observation" (data are always mediated by interpretation, techniques, paradigms, etc.). Hacking, *The Social Construction of What?* p. 73.

28. For philosophical expositions of the incoherence of skepticism, see the late Wittgenstein-influenced analyses of Coulter, *Mind in Action;* Stanley Cavell, *Disowning Knowledge: In Six Plays of Shakespeare* (Cambridge: Cambridge University Press, 1987); and Hearne, *Adam's Task.*

29. It reflects the severity of the biodiversity crisis that recent scientific literature often characterizes the human-driven annihilation of plants, animals, and ecosystems in such value-oriented terms. For constructivists, expressions like "holocaust" or "Armageddon" would be construed as a "rhetoric of calamity" (Hannigan, *Environmental Sociology,* p. 36), or as environmentalist "morality play" (Ross, *The Chicago Gangster Theory of Life,* p. 31). Such a constructivist standpoint must remain blindly focused on the words, rather than looking at the realities compelling scientists and others to use them.

30. Constructivists even express skepticism about the diagnosis that we are in the midst of "an environmental crisis." For example, Hannigan, *Environmental Sociology,* p. 30.

31. Bruno Latour and Steve Woolgar, *Laboratory Life: The Construction of Scientific Facts* (Princeton: Princeton University Press, 1986 [1979]).

32. See James Proctor, "The Social Construction of Nature: Relativist Accusations, Pragmatist and Critical Realist Responses," in *Annals of the Association of American Geographers* 8 (1998): 352–376.

33. E. O. Wilson, *The Future of Life* (New York: Alfred Knopf, 2002); John Terborgh, *Requiem for Nature* (Washington, D.C.: Island Press, 1999); Paul Ehrlich, "Extinction: What Is Happening Now and What Needs to Be Done," in David K. Elliott, ed., *Dynamics of Extinction* (New York: John Wiley, 1986), pp. 157–164; Peter Raven, "Disappearing Species: A Global Tragedy," in *The Futurist* (October 1985): 9–14; "What Have We Lost, What Are We Losing?" in Michael J. Novacek, ed., *The Biodiversity Crisis: Losing What Counts* (An American Museum of Natural History Book. New York: The New Press,

2001); Stuart Pimm, "Can We Defy Nature's End?" in *Science* 293 (2001): 2207–2208; Wilson, *The Future of Life*.

34. W. Wayt Gibbs, "On the Termination of Species," in *Scientific American* (November 2001): 40–49; Eileen Crist, "Quantifying the Biodiversity Crisis," in *Wild Earth* (Spring 2002): 16–19.

35. E. O. Wilson, "Introduction" to Susan Middleton and David Liittschwager, *Witness: Endangered Species of America* (San Francisco: Chronicle Books, 1994), p. 17.

36. Hannigan, *Environmental Sociology,* p. 31.

37. Smith, "To Speak of Trees," p. 370.

38. Holmes Rolston, III, "Nature for Real: Is Nature a Social Construct?" in T. D. J. Chappell, ed., *The Philosophy of the Environment* (Edinburgh: University of Edinburgh Press, 1997), pp. 38–64; Hacking, *The Social Construction of What?*

39. Criticizing radical ecologists for drawing on scientific ecology, constructivist Keulartz maintains that "as an empirical and experimental science . . . [i]ts results are by definition controversial and tentative, so that ecology as such is fallibilist rather than fundamentalist in character." Keulartz's bizarre view of ecological science as "ever-uncertain" is inspired by the postmodern perspectives of Latour, Haraway, Derrida, and others (Keulartz, *The Struggle for Nature,* pp. 155, 2, 158).

40. Ernst Mayr, *The Growth of Biological Thought: Diversity, Evolution, and Inheritance* (Cambridge, MA: The Belknap Press, 1982); *One Long Argument: Charles Darwin and the Genesis of Modern Evolutionary Thought* (Cambridge, MA: Harvard University Press, 1991).

41. Daniel Goodman, "The Theory of Diversity-Stability Relationships in Ecology," in *The Quarterly Review of Biology* 50 (1975): 237–266; Daniel Botkin, *Discordant Harmonies: A New Ecology for the Twenty-first Century* (New York: Oxford University Press, 1990); Donald Worster, "The Shaky Ground of Sustainability," in George Sessions, ed., *Deep Ecology for the 21st Century* (Boston, MA: Shambhala Press, 1995).

42. Stuart Pimm and Peter Raven, "Extinction by Numbers," in *Nature* 403 (2000): 843–845; John Terborgh, "The Big Things that Run the World—A Sequel to E. O. Wilson," in *Conservation Biology* 2 (1988): 402–403; Reed Noss et al., "Conservation Biology and Carnivore Conservation," in *Conservation Biology* 10 (1996): 949–963; Greta Nilsson, *The Endangered Species Handbook* (Washington, D.C.: Animal Welfare Institute, 1983); Gary Meffe and Ronald Carroll, "Genetics: Conservation of Diversity within Species," in Gary Meffe and Ronald Carroll, eds., *Principles of Conservation Biology* (Sunderland, MA: Sinauer Associates, Second Edition, 1997), pp. 161–201; David Pimentel, Laura Westra, and Reed Noss, eds., *Ecological Integrity: Integrating Environment, Conservation, and Health* (Washington, D.C.: Island Press, 2000); Michael

Soulé, ed., *Viable Populations for Conservation* (Cambridge: Cambridge University Press, 1987); Robert Peters and Thomas Lovejoy, eds., *Global Warming and Biological Diversity* (New Haven, CT: Yale University Press, 1992); Stephen Schneider and Terry Root, "Impacts of Climate Changes on Biological Resources in Status and Trends of the Nation's Biological Resources," in Michael J. Mac et al., eds., *U.S. Department of the Interior, U.S. Geological Survey* (Reston, VA, Vol. 1, 1998), pp. 89–116; Reed Noss, "Beyond Kyoto: Forest Management in a Time of Rapid Climate Change," in *Conservation Biology* 15 (2001): 578–590.

43. Smith, "To Speak of Trees," p. 365.

44. Peterson, "Environmental Ethics and the Social Construction of Ethics"; Demeritt, "What Is the 'Social Construction of Nature'?" pp. 778–779.

45. Soulé, "The Social Siege of Nature," in Soulé and Lease, eds., pp. 137–170; Kidner, "Fabricating Nature."

46. Hinchliffe and Woodward, *The Natural and the Social,* p. 155.

47. Ibid., p. 3.

48. Robert Marshall, "The Problem of the Wilderness," in J. Baird Callicott and Michael P. Nelson, eds., *The Great New Wilderness Debate* (Athens, GA: The University of Georgia Press, 1998 [1930]), pp. 55–96.

49. Bruno Latour, *We Have Never Been Modern* (Cambridge, MA: Harvard University Press, 1994 [1991]).

50. In her "Cyborg Manifesto" (Simon During, ed., *The Cultural Studies Reader* [London and New York: Routledge, 1999], pp. 271–291), Haraway takes issue with "totalizing tendencies of Western theories of identity" (p. 279)—and then proceeds to propound exactly such a suffocating theory in the guise of the "ontology of the cyborg" (p. 272).

51. "Wilderness Act of 1964," in Callicott and Nelson, eds., p. 121.

52. Doug Peacock, *Grizzly Years: In Search of the American Wilderness* (New York: Henry Holt, 1990), p. 65.

53. Soulé, "The Social Siege of Nature," p. 137.

54. Kidner, "Fabricating Nature," p. 352.

55. Vandana Shiva, "Monocultures of the Mind," in *The Trumpeter* 10 (1993): 132–135.

56. Rolston makes a cognate point in noting that constructivist analyses of wilderness conflate epistemological and ontological dimensions of the concept. "Nature for Real," p. 54.

57. See Jack Turner, *The Abstract Wild* (Tucson: The University of Arizona Press, 1996), chapter 2.

58. In examining "the production of spaces in nature," Macnaghten and Urry include "the wilderness" with "zoos, Disney Worlds, nuclear plants, shopping centers, and military zones" (*Contested Natures,* p. 173).

59. J. Baird Callicott, "The Wilderness Idea Revisited: The Sustainable De-

velopment Alternative," in Callicott and Nelson, eds., pp. 337–366. See also Cronon, "The Trouble with Wilderness."

60. Aldo Leopold, "The Land Ethic," in *The Sand County Almanac* (Oxford: Oxford University Press, 1968 [1949]), pp. 201–226.

61. See Tom Butler, ed., *Wild Earth: Wild Ideas for a World out of Balance* (Minneapolis, MN: Milkweed Editions, 2002); Michael Soulé and Reed Noss, "Rewilding and Biodiversity: Complementary Goals for Continental Conservation," in *Wild Earth* 8/3 (1998): 18–28; Dave Foreman, "Wilderness Areas for Real," in Callicott and Nelson, eds., pp. 395–407; Holmes Rolston, III, "The Wilderness Idea Reaffirmed," in Callicott and Nelson, eds., pp. 367–386.

62. Max Horkheimer and Theodore Adorno, *Dialectic of Enlightenment* (New York: Continuum, 1969 [1944]).

63. Herbert Marcuse, *Eros and Civilization* (Boston: Beacon Press, 1966 [1955]), p. xxi.

John O'Neill

Wilderness, Cultivation and Appropriation (2002)

ENVIRONMENTALISM, ESPECIALLY as it has developed in the new world contexts of the USA and Australia, often centres on a defense of "wilderness" and "nature" understood as that which is untouched by human interference. Our environmental crisis is presented as one of "the end of nature" understood as the disappearance of a world that is not effected by human intention. "An idea, a relationship, can go extinct just like an animal or a plant. The idea in this case is 'nature', the separate and wild province, the world apart from man to which he has adapted, under whose rules he was born and died."[1] For the deep green movement, "nature" and "wilderness" are the central normative categories of the environmental movement.

These attitudes are not confined to explicitly deep green radicals. Something of them pervades the nature conservation movement. Natural landscapes, pristine and untouched by the marks of human beings, form the primary objects of appreciation. This is sometimes expressed in quasi-scientific terms, for example, in references to landscapes that exhibit "ecological integrity" characterized by states that show only minimal human intervention. It is also expressed in aesthetic and moral terms by reference to the experience of wilderness. Correspondingly "cultural landscapes," particularly those transformed by human labour, are characterized as second-best landscapes that have some environmental value to the extent to which they approximate to "the real thing." Thus one finds talk of "semi-natural" landscapes as landscapes of particular value. One does not find reference to "semi-cultural" landscapes. Where cultural landscapes are introduced they are often taken to have value for some more humanistic reasons that do not properly fall under an environmental or nature conservation ethic, say to do with reasons of cultural identity.

This attitude is often accompanied among European conservationists by something of an inferiority complex—there is very little if any "wilderness" in Europe. It is often accompanied by the visiting deep ecologist by incredulity at the ways in which landscapes in Europe are managed. Here, for example, is a comment from Richard Sylvan on his visit to the Three Peaks area of the Yorkshire Dales in the UK.

> [The] Three Peaks district is now prized for its recreational values, it is prized for its *comparative remoteness and wilderness,* its fewness of people and absence of industry, for the walks and wild meadows it offers. But it is a landscape far removed from its *pre-agricultural original.* It has been almost totally *stripped of its native vegetation,* and most habitats and much of *its ecology destroyed, the remainder substantially modified,* in the former quest . . . for agricultural advantage and optimal, or often excessive, grazing usage. The district remains starkly treeless . . . But woods, formerly with different wildlife, there formerly were, as a tiny protected strip at Colt Park pleasantly testifies. Most of the district still remains overrun by, and severely eroded by, sheep, which none but subsidized and distorted market system would support. Remarkably, however, there appears little pressure for economic adjustment and ecological restoration, for removal of some sheep and return of more woods. Many recreationalists appear to prefer impoverished grasslands, treelessness. Even environmental organisations like English Nature own sheep and lease out lands for sheep grazing.[2]

What is notable here, as the passages I have emphasized here indicate, is that the ideal by which landscapes are to be judged is the unmodified "pre-agricultural original." It is an ideal of nature independent of human intervention that forms the standard from which others are judged.

Two lines of criticism are often offered in response to this line of argument. The first is that from a variety of constructivists who deny there is something called "nature" to be defended. To take a few examples:

> Nature per se does not exist . . . Nature is only the name given to a certain contemporary state of science.[3]
>
> It is fair to say that before the word was invented, there was no nature . . .[4]
>
> We have no basis for distinguishing between Nature and our own changing historically-produced representations of nature. . . . Nature is a cultural product.[5]
>
> There is nothing outside the text.[6]

That constructivist line has been often used against appeals to nature and wilderness by environmentalists.[7] The second line of criticism is that developed in the environmental justice movement and concerns the role of appeals to "nature" and "wilderness" in the appropriation of land of often socially marginal populations, in particular their control and exclusion for the creation of "nature parks."[8]

These two lines of argument, from constructivism and from justice, are often found together. They are, however, logically independent. My own view is that the second line of argument is broadly right and my main purpose in this paper will be to contribute to it by placing recent appeals to "wilderness" in the context of a longer history of use of wilderness to justify the appropriation of land. However, the first line of argument has I think more problems. While there may be good reasons for believing we should start with cultural rather than natural landscapes in environmental valuation, I do not think that strong forms of constructivism of the kind expressed in the passages above are defensible. In developing those doubts I examine what survives of modern environmentalism after proper criticism of the wilderness ideal. I finish by putting these comments in the context of more general problems about the relationship between environmental philosophy and the particular cultural perspectives that emerge from disciplines like anthropology. There is a *prima facie* conflict between the aspirations of many philosophers for thin and cosmopolitan moral language that transcends local culture, and the aspirations of disciplines like anthropology to uncover a thick moral vocabulary that is local to particular cultures. I will argue that while there are real dangers in the philosophical project of presenting as universal what is a culturally specific and local set of values, a danger that is illustrated in the appeals to wilderness in recent environmental ethics, there is no incompatibility between a universal and objective ethical reflection of the kind many philosophers aim for and the project of uncovering interpretative depth in ethical life of the kind anthropology offers. Rather the general critical project of philosophy is enriched by engagement with the anthropological project.

WILDERNESS PROTECTION: POLITICAL PROBLEMS

The appeal to wilderness to appropriate land is not new to political argument. It has a long history in social and political thought. Indeed that history points to part of the problem with the wilderness ideal in deep green environmentalism. Two initial points need to be made here. First, the image of much of the "new world" as wilderness relies upon a colonial perception of European colonial settlers of an unspoilt pristine terrain dramatically different from the domesticated environments of Europe. Second, this image was associated with claims that were made for the justifiable appropriation of that land from its native population. Both points are illustrated well in the work of Locke and the characterization of America as wilderness.

Consider Locke's comparisons of the "wild woods and uncultivated waste of America left to Nature without any improvement, tillage or husbandry"[9] with the improved and cultivated lands of Britain and his corresponding account of the original appropriation of land:

> Whatsoever he tilled and reaped, laid up and made use of, before it spoiled, that was his particular Right; the Cattle, and Product was also his. But if either the Grass of his Inclosure rotted on the Ground, or the Fruit of his planting perished without gathering, and laying up, this part of the Earth, notwithstanding his Inclosure, was still to be looked on as Waste, and might be the Possession of any other.[10]

Locke's references to the "wild woods and uncultivated waste of America left to Nature without any improvement, tillage or husbandry" needs to be read in its historical context. It formed part of the denial of rights in land to the Aboriginal population.

Locke, through the patronage of the Earl of Shaftesbury, was the secretary to the Lord Proprietors of Carolina and the Council of Trade and Plantations. Locke's theory of property was shaped by the arguments of colonialists for an ethical justification to appropriate the land of the native American population, particularly justifications that appealed to the rational use of land that had remained wild and unimproved. Subsequently, appeal was made to Locke's arguments to justify further appropriation of native land.[11] Locke's justification of private property in land given the Christian premise of original common ownership is voiced in terms of appropriation through cultivation and enclosure.[12]

> God, who hath given the world to men in common, hath also given them reason to make use of it to the best advantage of Life, and convenience. . . . And though all the Fruits it naturally produces, and Beasts it feeds, belong to Mankind in common . . . yet being given for the use of Men, there must of necessity be a means *to appropriate* them some way or other before they can be of any use, or at all beneficial, to any particular Man. The Fruit or Venison which nourishes the wild *Indian,* who knows no Inclosure, and is still a Tenant in common, must be his . . .[13]

To claim that the "wild *Indian* who knows no Inclosure" is still a tenant of the common is to deny his claims to the land. Indigenous populations had rights only to that they had appropriated through their labour, and given the European image of them as hunters and gatherers which Locke reiterates here, appropriation extended only to that they caught and collected. The land being "a wilderness," uncultivated and unenclosed, the "vacant places of *America*" could be rightfully settled by Europeans without the consent of previous inhabitants or with their having "reasons to complain."[14]

> God gave the World to Men in Common, but . . . it cannot be supposed He meant it should always remain common and uncultivated. He gave it to the use of the Industrious and Rational (and *Labour* was to be *his Title* to it) . . .[15]

The right to the original acquisition of land is subject in Locke's theory to two well-known limiting provisos, that produce not spoil and that there be good enough left for others. Both provisos are overcome through the introduction of money which can be hoarded without spoiling and which, through commerce, serves to foster the improvement of agriculture and hence "to increase the common stock of mankind."[16] The appropriation of America is justified by its being brought into the world of commerce and hence cultivation. It is the absence of money and commerce that explains the lack of productive appropriation of the land in America:

> What would a Man value Ten Thousand, or an Hundred Thousand Acres of excellent *Land* . . . in the middle of the in-land Parts of *America,* where he had no hopes of Commerce with other Parts of the World, to draw *Money* to him by the Sale of the Product? It would not be worth the Inclosing, and we should see him give up again to the wild Common of Nature, whatever was more than would supply the Conveniences of Life, to be had there for him and his Family. . . . Thus, in the beginning, all the World was *America,* and more so than that is now; for no such thing as Money was anywhere known.[17]

By being brought into the world of commerce the wild common of America, uncultivated by the indigenous population is turned into a productive resource cultivated by the industrious and rational.

The Lockean account of the "vast wilderness" of America as land uncultivated and unshaped by the pastoral activities of the indigenous population formed part of the justification of the appropriation of native land. It is also false. The land had been shaped by its native populations. However, it is a myth that has survived and has led to conservation management policies that ignore the impact of indigenous pastoral and agricultural activities. The problems are evident in the well-discussed problems in the history of the management of one of the great symbols of American wilderness, Yosemite National Park. In their influential report "Wildlife Management in the National Parks," the Leopold Committee recommended that the goal of management be to maintain or recreate the biotic associations "as nearly as possible in the condition that prevailed when the area was first visited by the white man. A national park should represent a vignette of primitive

America."[18] What was that "primitive" condition? Here is their report of the first white visitors to the area:

> When the forty niners poured over the Sierra Nevada into California, those who kept diaries spoke almost to a man of the wide-space columns of mature trees that grew on the lower western slope in gigantic magnificence. The ground was a grass parkland, in springtime carpeted with wildflowers. Deer and bears were abundant.[19]

However, the "grass parkland" was in part the result of the pastoral practices of the indigenous people, who had used fire to promote pastures for game and black oak for acorn. After the Ahwahneechee Indians were driven from their lands in 1851, "Indian style" burning techniques were discontinued and fire suppression controls introduced. The consequence was the decline in meadowlands under increasing areas of bush. When Totuya, the granddaughter of chief Tenaya and survivor of the Ahwahneechee Indians who had been evicted from the valley, returned in 1929, she remarked on the landscape she found "Too dirty; too much bushy." Moreover, in the Giant Sequoia groves the growth of litter on the forest floor and competitive vegetation inhibited the growth of new Sequoia and threatened more destructive fires. Following the Leopold report, both cutting and burning were used to "restore" Yosemite back to its "primitive" state.

The shift from the land management policies embodied in the Leopold report and those found in the work of Locke reflects a change in attitudes to wilderness. The wilderness ideal is historically a local one. It is uncultivated land that the environmentalist now attempts to protect, not the cultivated landscapes that Locke praises. The dominant perceptions of land and landscapes have shifted. Recent environmental thought has echoed Mill's Romantic-influenced observation about the limits of agricultural expansion.

> It is not good for man to be kept perforce at all times in the presence of his species. A world from which solitude is extirpated, is a very poor ideal. Solitude, in the sense of being often alone, is essential to any depth of meditation or of character; and solitude in the presence of natural beauty and grandeur, is the cradle of thoughts and aspirations which are not only good for the individual, but which society could ill do without. Nor

> is there much satisfaction in contemplating the world with nothing left to the spontaneous activity of nature; with every rood of land brought into cultivation, which is capable of growing food for human beings; every flowery waste or natural pasture ploughed up, all quadrupeds or birds which are not domesticated for man's use exterminated as his rivals for food, every hedgerow or superfluous tree rooted out, and scarcely a place left where a wild shrub or flower could grow without being eradicated as a weed in the name of improved agriculture.[20]

It is something like Mill's post-Romantic vision that informs the modern environmental movement.

The problem with this wilderness ideal is that it is not just a historically local perception that was associated with the appropriation of land. It is also socially and geographically local as well and it retains its link with appropriation. However, the appropriation is now made in the name of wilderness rather than cultivation. It is invoked in the creation of "nature parks" for the new eco-tourism, at home and abroad, which is premised on the assumption that nature requires at least the absence of human activity and at best the absence of people. It has led to policies of exclusion and control of the indigenous human populations on the grounds that they do cultivate and shape the land. Thus the development of conservation parks in the third world through the eviction of the indigenous populations that had previously lived there. Consider the fate of some of the Masai who have been excluded from national parks across Kenya and Tanzania.[21] Attempts to evict indigenous populations from the Kalahari reveal the influence of the same wilderness model: "Under Botswana land use plans, all national parks have to be free of human and domestic animals."[22] Nor is the policy of enforced eviction confined to Africa. Similar stories are to be found in Asia where the same alliance of local elites and international conservation bodies has led to similar pressures to evict indigenous populations from their traditional lands. In India, the development of wildlife parks has led to a series of conflicts with indigenous populations. Thus, in the Nagarhole National Park, there are moves from the Karnataka Forest Department to remove 6,000 tribal people from their forests on the grounds that they compete with tigers for game. The move is supported by international conservation bodies. Hence the remark of one of the experts for the Wildlife Conservation Society: "relocating tribal or traditional people who live in these protected areas is the single most important step towards conservation."[23]

The control and exclusion of populations is a theme that runs through much anthropological work on the environment. Consider the comment from a person in the Makala-Barun National Park and Conservation Area in Nepal reported by Ben Campbell: "This park is no good. They don't let you cut wood, they don't allow you to make spaces for paddy seed-beds, they don't permit doing *khoriya* [a form of slash and burn agriculture]."[24] The wilderness model fails to acknowledge the ways parks are not wilderness but a home for its native inhabitants, the degree to which the landscapes and ecology of the "wilderness" were themselves the result of human pastoral and agricultural activity, and the cultural significance of particular landscapes, flora, and fauna to the local populations. Insofar as the indigenous populations are recognized they are often themselves treated as a kind of exotic fauna, who are a part of nature, rather than fellow humans who also transformed their landscapes.

The conflicts between the attempts to create nature parks that embody the wilderness, and the local often marginalized populations who have lived and worked in that wilderness are not confined to the third world. Consider, for example, the following comments of a local living by the natural park of Sierra Nevada and Alpujurra, granted biosphere status by UNESCO and Natural Park status by the government of Andalusia:

> [Miguel] pointed out the stonework he had done on the floor and lower parts of the wall which were all made from flat stones found in the Sierra. I asked him if he had done this all by himself and he said "Yes, and look, this is nature" ("Si, y mira, esto es la naturaleza"), and he pointed firmly at the stone carved wall, and he repeated this action by pointing first in the direction of the Sierra [national park] before pointing at the wall again. Then, stressed his point by saying: "This is not nature, it is artificial (the Sierra) this (the wall) is nature" ("Eso no es la naturaleza, es artificial (the Sierra) esto (the wall) es la naturaleza").[25]

Finally in the UK consider the Yorkshire Dales about which Richard Sylvan expressed his criticisms. There is a real conflict in the perceptions of the landscapes between farmers and conservationists.[26] Farmers on the one hand and landscape planners and conservationists on the other have different perceptions of what constitutes a good environment of which some are self conscious. Farmers' perceptions are often husbandry based. Hence the comment of a farmer in the Yorkshire Dales: "A farmer will look at someone else's farm and could tell whether it was well farmed or not. They wouldn't look at the view and think 'What a good view!'"[27] Given that perspective the wildness loved of the conservationist can be seen as a defect. "If a piece of land's conserved, it tends to get overgrown, it gets brown. I suppose people from off will tend to look at that and admire its tones, in autumn sort of thing. Or golden spring, or whatever. But a farmer will look at it and think—it's overgrown."[28] The farmer will sometimes look upon the land in a way that is different from both the nature conservationist and the visitor admiring the landscape. Attempts to fence off nature and allow it to grow wild are met with disapproval: it represents a "mess."[29] Hence, the resistance felt by some farmers to the authorities who represent conservationists and landscapers—outsiders who aim to mould the environment in ways that are alien to their own husbandry-based conceptions. Correspondingly there is the articulated threat to a community founded upon farming being transformed into a museum exhibit to conform to some idealized Romantic image of how the countryside should look: "National Parks, English Nature, they'll finish up with all the farmers running about in smocks, like museum curators. That's not a community. We have a community which is a working community . . ."[30] The worry here is that a particular conception of the way nature and landscape ought to be is being imposed from the outside on those who live and work in an environment, for whom nature is not primarily an object of scientific interest or aesthetic contemplation, but something with which one has a working relationship.

WHAT'S LEFT OF THE WILDERNESS?

Thus goes the case for the prosecution against the wilderness ideal. It is a historically and socially local vision, historically implicated in the colonial appropriation of land, and currently implicated in the exclusion or control of often poor and powerless groups that live in lands too marginal to sustain intensive agricultural activity. The case against the wilderness model in this context is I think powerful and one with which I broadly concur. It also suggests, as I noted earlier, that perhaps we should reverse the order of primacy between natural and cultural landscapes. For it might be argued, that the natural landscape itself is just a particular cultural landscape, one that has a particular social and cultural history. Landscapes themselves, like that in the Yorkshire Dales or the Yosemite, are managed to mould them to expected patterns. And even where landscapes are not directly managed, the perception of the landscape as "natural" or "wilderness" is itself a culturally specific achievement. The conflicts outlined in the last section are conflicts between different cultural landscapes. These can be direct material conflicts on how landscapes themselves should be transformed by human activity. However, they can also be conflicts in ways of seeing landscapes—consider the comments of the farmers above. Conflicts about appropriation can likewise be conflicts on who has legitimate powers to determine the material future of landscapes, with who has property rights and economic and political power to shape a landscape.[31] However, conflicts about appropriation can also have a cultural and symbolic dimension, as to which perceptions and understandings of environments predominate.[32] The conflicts between the farmers and the conservationists in the Yorkshire Dales, or between Miguel and the park authorities in Andalusia, or between the peasants and park authorities in Nepal have both dimensions. They are in part about who has rights to direct and control the land, but they are also about how the land is to be described and perceived. On this view then environmental conflicts are conflicts between different cultural landscapes. Natural landscapes are cultural landscapes that dare not speak their name. Pushing the line of argument further, it is sometimes argued that we should drop the notion of nature altogether from environmental discussion.[33]

There is I think much in that line of argument. However, the final conclusion, that nature disappear altogether from discussion is I think mistaken. Neither would I want to reject the environmentalist's view as simply internally incoherent. Indeed I think there is much to the environmentalist's position that can be rescued from the wilderness. An initial point to be made here is that even appeal to "wilderness" itself has a more ambivalent role in the politics of nature than my discussion this far might suggest. Wilderness, in particular in the romantic celebration of it, was sometimes appealed to in order to justify public access to what is common against the privatization of land. Consider, for example, Mill's comments on access to uncultivated land:

> [T]he exclusive right to the land for purposes of cultivation does not imply an exclusive right to it for purposes of access; and no such right ought to be recognised, except to the extent necessary to protect the produce against damage, and the owner's privacy against invasion. The pretension of two Dukes to shut up a part of the Highlands, and exclude the rest of mankind from many square miles of mountain scenery to prevent disturbance to wild animals, is an abuse; it exceeds the legitimate bounds of the right of landed property. When land is not intended to be cultivated, no good reason can in general be given for its being private property at all; and if any one is permitted to call it his, he ought to know that he holds it by sufferance of the community, and on an implied condition that his ownership, since it cannot possibly do them any good, at least shall not deprive them of any, which could have derived from the land if it had been unappropriated.[34]

The appeal to wilderness in this context forms part of an assertion of rights to common access for common enjoyment.[35] Such appeals were central for example to the struggles in the UK for access to mountains and moorland by the urban working class in the nineteenth and twentieth century that culminated in the mass trespass movement. It still animates parts of current nature conservation which appeals to the need to maintain boundaries around land that protects it from privatization and commercialization. Hence one of the worries of defenders of wilderness is that its critics remove constraints on the commercial development of currently protected areas. For the reasons outlined earlier, I do not believe that the concept of wilderness in the sense of places relatively untouched by human intervention is the appropriate concept to use in this context. There is little if any wilderness in this sense. Neither am I convinced that employing the concept in its older sense of "uncultivated land" fares particularly better. However, the defense of particular places and the maintenance of boundaries against commerce are entirely proper—and there are good reasons to hold that this should include places in which the albeit culturally specific experience of wildness is possible.

Moreover, much of the ethical vocabulary which environmentalists call upon to criticize features of some of our contemporary relations to the non-human world survives the rejection of the wilderness model: for example, reference to the cruelty inflicted on fellow creatures, of the failure of care involved in the wanton destruction of places rich in wildlife and beauty, of the pride and hubris exhibited in the belief that the world can be mastered and humanized, of our lack of a sense of humility in the midst of a natural world that came before and will continue beyond us. Nor do I think there are reasons to deny that both arts and sciences have developed the human senses in ways that allow humans to respond to the qualities that objects possess in a disinterested fashion to objects and that in doing so they have developed a human excellence.[36] What is I think true is that they have a particular local cultural origin.[37]

Does the fact that they have a particular cultural origin matter? There are certainly occasions when it appears to matter. Consider the following incident. Returning from a winter climbing trip in Glencoe in Scotland, two friends and I were passing Loch Lomond. It was a day of bright sunshine and without wind. There was not a ripple on the Loch and the mountains were reflected without flaw in the water. Two of us made the kind of comments full of expletives you would expect on such occasions. The third who had a training in the history of art then began an account of the development during the eighteenth and nineteenth centuries of the aesthetic responses to the landscape which we had just exhibited. And it completely ruined the moment. Should his comments have undermined our appreciation? Does knowledge of the cultural origins of our responses to the natural world destroy those responses? Clearly such histories can have that effect. It is the source of the power of genealogical criticisms of social practices and attitudes: the history of the use of the concept of wilderness outlined earlier is perhaps an example. Or to take another, once one has read the story of the highland clearances in Scotland, it is difficult not to see a depopulated landscape rather than a pristine wilderness. However, the cultural self-understanding of our attitudes, understanding and perceptions of non-human nature need not undermine it. They can be seen as cultural achievements. Gellner comments that anthropologists sometimes tend to be liberals at home and conservatives abroad.[38] The comment is not facetious. It raises important methodological issues for anthropology—most notably whether the suspension of criticism of those with whom one is engaged is a condition of understanding, and if not what the principles of interpretation ought to be.[39] I leave these methodological issues aside here, however. I want to add a variant to the point that sometimes perhaps they may also have a tendency to be celebratory abroad and deflationary at home. Even if the responses to landscapes have historically and socially local origins, they are ones that can have their own virtues, that make a contribution to a wider conversation about values.

There are two points I think need to be made here. First, the cultural sources of our responses need to be distinguished from the objects of our responses. The point is one that needs to be stressed against certain strong forms of constructivism. For the strong constructivist, once we are made aware of the cultural origins of our responses we realize that there is no "nature" there, that we are surrounded by a world of cultural objects. That strong constructivism is mistaken. There is a clear distinction to be drawn between the sources of our attitudes, which are economic, political and cultural, and the objects of our attitudes which can still remain non-cultural. That our capacity to appreciate and respond to the non-human natural world in a certain way is a cultural achievement, the outcome of social and cultural processes, does not entail that the object of our attitudes is a cultural object. This is not to deny that many landscapes that are presented as "natural" are cultural in a real material sense—they are the result of human activity. Hence, the proper redescriptions of wilderness outlined earlier. However, it would be simply false to hold that at the level of geology, for example, that all is a human product: it has a history before us. More generally, the world in which we live is the result of an interplay of human and non-human history. And at the level of processes rather than objects or end-states, we live in a world of unintentional non-human natural processes that proceed regardless of human intentions and indeed which often thwart them. This is a source of both human sorrow, for example, of life and land lost in flood, but also of human delight, for example, at the plant or bird that arrives uninvited in an industrial wasteland. It is possible for culture to foster appreciation of non-human objects that themselves are not cultural—to maintain a sense of the otherness of non-human nature and our place within it. The picture of a world in which humans can see nothing but the reflections of themselves is itself a peculiar modern human conceit that our constructivist times tends to encourage. There is a core of environmentalism that is properly critical of that conceit. We live in a larger world of which human life is just a part.[40]

Second, the particular local cultural origins of responses and vocabulary with which particular groups approach the natural world does not entail that they cannot belong to a wider conversation. All knowledge and value assertions have a local origin—they could have no other. That is consistent with some making claims on a wider audience. However, the possibility and nature of that wider conversation raises some important tensions between the universalizing tendencies of philosophical arguments and the more culturally local concerns that are represented most notably in anthropological perspectives.

ENVIRONMENTAL VALUES THROUGH THICK AND THIN

Is there an incompatibility between philosophical and anthropological reflection on environmental values? Philosophy often makes claims to wider aspirations to be part of general reflection on human values and predicaments that transcend particular times and places. There are a variety of reasons for those aspirations, but at least one source lies in a set of enlightenment emancipatory values which is taken to involve taking a critical standpoint that transcends what is local. Local cultural practices are sometimes oppressive to particular individuals and groups, for example, to women and subordinate castes and classes. To launch criticism of those practices one needs a standpoint and set of normative concepts that transcends the local, that allows the possibility of standing outside particular social practices to formulate sceptical questions about them. There is a strong strand in recent philosophical argument that holds that such a standpoint requires a minimal moral language—an ethical discourse written in terms that have "thin" or minimal meanings, that employ general abstract terms of rights and goods that are taken to be universal, transcending the specific ethical understanding of local culture. These contrast with thick concepts whose understanding requires immersion in particular local practices. There are a number of different expressions of that moral minimalism. One expression is something like the open question argument—that a thin vocabulary of the right and the good is a condition of being able to raise a question about specific moral concepts, for example, of being able to ask with Nietzsche of the virtue of humility: "is humility good?" Another is the kind of consideration that in part moves Rawls's distinction between a thick and thin theory of the good in order to offer an account of a justice which is not tied to any specific comprehensive theory of the good.

Mainstream environmental ethical reflection in philosophy for similar reasons often eschews thick, more specific ethical concepts in an attempt to create a universal environmental ethic that is not relative to time and place. The tendency is reflected in particular in the use of Kantian and utilitarian approaches to both animal welfare and the environment. For those concerned with environmental and animal welfare issues, the need to launch criticism of practices from a standpoint that transcends the local is often taken to be of particular significance in that part of the enterprise is to show that there are non-human individuals and groups who have an ethical standing that is not recognized in most existing social practices.[41] The enterprise is a critical endeavour in defense of those who are rendered ethically invisible in the existing social worlds. The need for cosmopolitan language of universal thin concepts is also taken to be more acute in the environmental sphere in virtue of the fact that environmental problems are global and hence require an ethical language that crosses cultures.

The work of anthropologists might appear to run largely in the opposite direction, to recovering the thicker local vocabularies and understandings that individuals and groups bring to the environments they inhabit. That focus might indeed be seen to be part of the anthropological enterprise. Clifford Geertz, for example, in *The Interpretation of Cultures* characterizes anthropological analysis in these terms: "What defines [anthropological analysis] is the kind of intellectual effort it is: an elaborate venture in, to borrow a notion from Gilbert Ryle, 'thick description.'"[42] On the line that Geertz develops, the exercise in uncovering layers of interpretative depth defines the anthropological enterprise. To the extent that recent environmental anthropology follows that project it appears to be in opposition to the more universalist claims of the environmental movement. In doing so it has or can have a political and ethical agenda of its own. While anthropology has its own complex relation with colonialism, recent critical anthropology often characterizes itself, at least implicitly, as representing the voice of local and often marginalized groups against the global discourse that is employed in the alliance that is sometimes forged between nature conservation bodies and international centres of power. In the international environmental policy and treaties the use of the language of science and economics that often serves the purpose of offering a language that transcends the local—hence the references in international policy directives to "biodiversity," "nature," and "sustainability" understood as the maintenance of "natural capital." In this context, global discourse, far from offering a standpoint to criticize illegitimate power and injustice embodied in local social practices, becomes rather a way in which global economic and political power is itself expressed.

It might appear then that there is a deep opposition at both theoretical and political levels between on the one hand the universalism of philosophy, and in particular environmental philosophy, and on the other the appeal to specific and local understandings of the anthropologist. However, that view would I think be a mistake. First, anthropological work which has a political position in defense of local cultural understandings often implicitly calls upon the same emancipatory values of the enlightenment as its opponents—and it is none the worse for that. The arguments about the ways in which particular voices and perceptions are silenced has critical power only to the extent it calls upon shared beliefs about the value of equality in standing, voice, and power. The problem lies not I think in the appeal to such values, but rather in the assumption that runs through much of debate that it is general, abstract, "thin" evaluative concepts to which such appeals must be made.

The problem here can be approached through an ambiguity in the way the terms "thick" and "thin" are themselves used in recent philosophical and anthropological literature. The terms are used in a number of distinct senses which are often elided:

1. *Interpretative depth.* Geertz's own use of the term has origins in Ryle's work in the philosophy of mind. Ryle used the terms thick and thin descriptions in part to capture differences in the interpretative depth of action descriptions. The thinnest description of an action is a description of physical behaviour:

   "Two boys fairly swiftly contract the eyelids of their right eyes. In the first boy this is only an involuntary twitch; but the other is winking conspiratorially to an accomplice. At the lowest or thinnest level of description the two contractions of the eyelids may be exactly alike. From a cinematograph-film of the two faces there might be no telling which contraction, if either, was a wink, or which, if either, was a mere twitch. Yet there remains an immense but unphotographable difference between a twitch and a wink."[43]

   Descriptions are thickened as interpretative depth is added. Hence Ryle's examples of successively thicker descriptions: a boy contracts his eyelid; a boy is winking; a boy is parodying another's attempt at winking; a boy is practising a parody of another's attempt at winking. Thickness in this sense refers then to the interpretative depth of descriptions of behaviour.

2. *Theories of the good:* Rawls uses the terms to draw a distinction between theories of the good, between a thin theory which refers to the good "restricted to the essentials" and determines the class of primary goods that rational individuals will necessarily require to pursue whatever ends they might have, and a thick theory of the good which specifies particular ends.[44]

3. *Specific versus general ethical concepts:* Another usage of the terminology, due largely to Williams, is to draw a distinction between ethical concepts like brave, cowardly, kind, pitiless, which are specific reason-giving concepts and world-guided in the sense that their application is "determined by what the world is like," and ethical concepts, like good, bad, right and wrong

which are general, abstract and not world-guided. The former are thick, the latter are thin.[45]

4. *Cultural specificity:* While Walzer explicitly picks up on the use of the term by Geertz, he uses the contrast primarily to refer to the cultural specificity of moral terms, where the thick refers to what is specific to a local place and historical context, and thin to what is claimed to be universal.[46] The assumption appears to be made that the cultural specificity and interpretative depth come together, that the anthropological project of uncovering layers of depth in understanding will take one at the same time to that which is particular to place and time. Walzer in introducing the notion of thick descriptions refers to Geertz's usage, that of interpretative depth, but then shifts immediately to the notion of cultural locality.[47]

In the context of the relation of philosophy and anthropology, it is the first and last uses that are of particular significance: to mark differences in interpretative depth; and to mark degrees of cultural specificity. Given the assumption that interpretative depth and cultural specificity come together, the conclusion appears to follow that the anthropological project moves in opposite direction to the more universalizing aspirations of philosophy. However, that assumption needs to be questioned. It is often, although not always, the other way around, that it is as one moves to descriptions with greater interpretative depth that values and practices that are shared emerge. Thus in the examples discussed earlier, as one moves from thinner to thicker descriptions—from "burning a forest" to "clearing the land" to "maintaining the agricultural land of a family"—that what is common begins to emerge. It is as specific meanings are uncovered, for example, the very concrete and particular ways land embodies the life of a community that what is more universal comes to the fore. The passages quoted earlier from peasants and farmers speak not just to a local audience, but a wider potential audience. It is in virtue of this fact that such passages should be an occasion for critical reflection on nature conservation and deep green attitudes to nature. Thus one effect of conversations with peasants and farmers of the kind quoted earlier is that through their expression of shared understandings that our attitudes and perceptions of what is natural are changed. We look on a herb rich meadow and see not only the flowering of biodiversity, but also the decline of an agricultural community; or we look upon a nature park and see not the protection of nature, but the disruption of a community's lived relationship with its environment through a state artifice. That interpretative depth often takes us to what is shared is not an accident. Shared practices and values are a condition of common action and communication. This is not to deny the possibility of disagreement or of social and moral practices that are radically different. However, disagreement and dialogue are possible only against a shared background of understandings.[48]

There is, then, no incompatibility between a universal ethical reflection and the project of uncovering interpretative depth. Where an incompatibility does exist, it is with a particular style of philosophical reflection on values of which Kantian and utilitarian ethical systems are typical expressions, as are the forms of ethical proceduralism that divorce the norms of engagement in argument from substantive ethical claims. This style of reflection abstracts from the thick vocabulary in which most particular judgements are expressed for a thinner language of "the right" and "the good." That language is, as I noted above, taken to be a condition of the possibility of standing outside our own ethical practices and formulating sceptical questions about them. We must be able to ask not only the question "but is x good?" for example, "but [also] is humility before nature good?" The claims are mistaken. It is not the case that the use of thick concepts rules out theoretical reflection and general principle in ethics. To ask "Is humility before nature good?" is not to ask whether it has some property of goodness or to ask for the expression of some preference, but to raise questions about the relationship of such humility to other particular evaluative claims we might make—for example, about its compatibility with other admirable human accomplishments. It is to place an ethical concept within the critical company of other evaluative thick concepts. It is particular thick concepts that have reason-giving power, not more abstract thinner concepts.

The programme of reducing thick to thin ethical concepts cannot I think be sustained as a general philosophical project.[49] It also fails in the particular case of the environment. Our environmental reflections need to begin with the rich thick vocabulary that informs particular claims and judgements and perceptions of places. While some of the thick concepts of environmentalism may not survive reflection on their own origins, in particular that of wilderness, there are others that do so. As I noted earlier there is a defensible core of environmentalism that lies in specific claims—concerning the cruelty inflicted on fellow creatures, of the failure of care, of the hubris exhibited in the belief that the world can be mastered and humanized and so on. Such claims employ a rich thick normative vocabulary. That they do so is a virtue that is compatible with the aim of more general reflection on ourselves and our practices. Given that style of philosophical reflection, the general critical project of philosophy is not only compatible with but is enriched by engagement with anthropology.[50]

NOTES

1. B. McKibben, *The End of Nature* (Harmondsworth: Penguin, 1990), 43–44.

2. R. Sylvan, "Dominant British Ideology," unpublished manuscript. Richard Sylvan wrote the paper after a seminar on his work which was held near Colt Park in the Yorkshire Dales. My criticism of his comments here continues a discussion he provoked on that occasion. It was a discussion that was sadly cut short by his death.

3. C. Larrere, "Ethics, Politics, Science, and the Environment: Concerning the Natural Contract," in *Earth Summit Ethics: Toward a Reconstructive Post-modern Philosophy of Environmental Education,* ed. J. Baird Callicott and F. de Rocha (Albany: SUNY Press, 1996), 122.

4. N. Evernden, *The Social Creation of Nature* (Baltimore: Johns Hopkins University Press, 1992), 89.

5. D. Cupitt, "Nature and Culture," in *Humanity, Environment and God,* ed. N. Spurway (Oxford: Blackwell, 1993), 35.

6. J. Derrida, *Of Grammatology* (Baltimore: Johns Hopkins University Press, 1976), 158.

7. See, for example, S. Vogel, *Against Nature* (New York: SUNY Press, 1996). For a view from the wilderness side see Holmes Rolston III, "Nature for Real: Is Nature a Social Construct?" in *Respecting Nature: Environmental Thinking in the Light of Philosophical Theory,* ed. T. D. J. Chappell (Edinburgh: University of Edinburgh Press, 1997).

8. For the debates about wilderness see J. Baird Callicott and M. P. Nelson,

eds., *The Great New Wilderness Debate* (Athens: University of Georgia Press, 1998) and W. Cronon, ed., *Uncommon Ground* (New York: Norton, 1995). On exclusions see R. Guha, "The Authoritarian Biologist and the Arrogance of Anti-Humanism," *The Ecologist,* 27, no. 1 (1997): 14–20.

9. J. Locke, *Two Treatises of Government,* ed. Peter Laslett (Cambridge: Cambridge University Press, 1988), 2.37.

10. Locke, *Two Treatises of Government,* 2.38.

11. For a detailed examination of the relation between Locke's theory of property and justification of colonial expansion see B. Arneil, "The Wild Indian's Venison: Locke's Theory of Property and English Colonialism in America," *Political Studies,* 44 (1996): 60–74; J. Tully, "Placing the 'Two Treatises,'" in *Political Discourse in Early Modern Britain,* ed. N. Phillipson and Q. Skinner (Cambridge: Cambridge University Press, 1993); J. Tully, *An Approach to Political Philosophy: Locke in Contexts* (Cambridge: Cambridge University Press, 1993), chap. 5. The influence of the Lockean view is evident not only in appeal to it in subsequent legal claims but also in economic theory. Consider for example the following from Smith:

> The whole of the savage nations which subsist by flocks have no notion of cultivating the ground. The only instance that has the appearance of an objection to this rule is the state of the North American Indians. They, tho they have no conception of flocks and herds, have nevertheless some notion of agriculture. Their women plant a few stalks of Indian corn at the back of their huts. But this can hardly be called agriculture. This corn does not make any considerable part of their food; it serves only as a seasoning or something to give a relish to their common food; the flesh of those animals they have caught in the chase . . . [In] North America, again, where the age of hunters subsists, theft is not much regarded. As there is no property amongst them, the only injury that can be done is the depriving them of their game. (A. Smith, *Lectures on Jurisprudence* [Liberty Press: Indianapolis, 1982], i.29, i.33, 15–16.)

12. The claims that cultivation gives a title to land is already in Aquinas who appeals in turn to Aristotle. "If a particular piece of land be considered absolutely, it contains no reason why it should belong to one man more than to another, but if it be considered in respect of its adaptability to cultivation, and the unmolested use of the land, it has a certain commensuration to be the property of one and not of another man, as the Philosopher shows (Polit. ii, 2)" (Aquinas, *Summa Theologica* II.II 57.3). In Locke it becomes *the* title.

13. Locke, *Two Treatises of Government,* 2.26.

14. Locke, *Two Treatises of Government,* 2.36.

15. Locke, *Two Treatises of Government,* 2.34.

16. Locke, *Two Treatises of Government,* 2.37.

17. Locke, *Two Treatises of Government,* 2.48–49.

18. A. S. Leopold, S. A. Cain, C. M. Cottam, I. N. Gabrielson and T. L. Kimball, *Wildlife Management in the National Parks,* US Department of the Interior, Advisory Board on Wildlife Management, Report to the Secretary, 4 March 1963, 4. Cited in A. Runte, *National Parks: The American Experience,* 2nd ed. (Lincoln: University of Nebraska Press, 1987), 198–99.

19. A. S. Leopold et al., *Wildlife Management,* 6. Cited in A. Runte, *National Parks,* 205.

20. J. S. Mill, *Principles of Political Economy,* bk. IV, chap. 6, sec. 2.

21. See G. Monbiot, *No Man's Land* (London: Macmillan, 1994), chaps. 4 and 5. Consider, for example the Masai suffering from malnutrition and disease on scrubland bordering the Mkomazi Game Reserve from which they were forcibly evicted from land in 1988 (*The Observer,* 6 April 1997, 12).

22. "Bushmen Fight to Stay on in Last Botswana Haven," *The Times,* 5 April 1996: 11.

23. Cited in R. Guha, "The Authoritarian Biologist and the Arrogance of Anti-Humanism," *The Ecologist,* 27, no. 1 (1997): 17.

24. B. Campbell, "Nature and Its Discontents in Nepal," *Contesting Nature: Anthropology and Environmental Protection,* workshop, Manchester University, 21 September 1998.

25. K. Lund, "What Would We Do Without Biodiversity?" *Contesting Nature: Anthropology and Environmental Protection,* workshop, Manchester University, 21 September 1998.

26. I draw here on M. Walsh, S. Shackley and R. Grove-White, *Fields Apart? What Farmers Think of Nature Conservation in the Yorkshire Dales,* a Report for English Nature and the Yorkshire Dales National Park Authority (Lancaster: Centre of the Study of Environmental Change, 1996). For discussions see J. O'Neill and M. Walsh, "Landscape Conflicts: Preferences, Identities and Rights," *Landscape Ecology,* 15 (2000): 281–89.

27. M. Walsh et al., *Fields Apart?* 22. Compare the remarks of farmers on conservation on Pevensey Levels in J. Burgess, J. Clark and C. Harrison, *Valuing Nature: What Lies Behind Responses to Contingent Valuation Surveys?* (London: UCL, 1995).

28. M. Walsh et al., *Fields Apart?* 22.

29. M. Walsh et al., *Fields Apart?* 23.

30. M. Walsh et al., *Fields Apart?* 44.

31. J. O'Neill, "Property, Care and Environment," *Environmental Planning C: Government and Policy,* 19 (2001): 695–711.

32. My thanks to Jacques Weber for putting this point to me.

33. This line is developed in S. Vogel, *Against Nature* (New York: State University of New York Press, 1996).

34. J. S. Mill, *Principles of Political Economy* (Oxford: Oxford University Press, 1994), bk. 2, chap. 2, sec. 6, the passages come in the context of Mill's rejection of certain landed rights:

> When the "sacredness of property" is talked of, it should always be remembered, that any such sacredness does not belong in the same degree to landed property. No man made the land. It is the original inheritance of the whole species. Its appropriation is wholly a question of general expediency. When private property in land is not expedient, it is unjust. It is no hardship to any one, to be excluded from what others have produced: they were not bound to produce it for his use, and he loses nothing by not sharing in what otherwise would not have existed at all. But it is some hardship to be born into the world and to find all nature's gifts previously engrossed, and no place left for the new-comer. To reconcile people to this, after they have once admitted into their minds the idea that any moral rights belong to them as human beings, it will always be necessary to convince them that the exclusive appropriation is good for mankind on the whole, themselves included. (J. S. Mill, *Principles of Political Economy* [Oxford: Oxford University Press, 1994], bk. 2, chap. 2, sec. 6.)

35. It is echoed for example in the failed attempts to introduce rights of access in the late nineteenth and early twentieth centuries: ". . . no owner or occupier of uncultivated mountain or moor lands in Scotland shall be entitled to exclude any person from walking on such lands for the purposes of recreation or scientific or artistic study, or to molest him in so walking" (Clause 2, Access to Mountains [Scotland] Bill [1884]).

36. J. O'Neill, "Science, Wonder and the Lust of Eyes," *Journal of Applied Philosophy,* 10 (1993): 139–46.

37. Compare Bernard Williams's comments: "a self-conscious concern for preserving nature is not itself a piece of nature: it is an expression of culture, indeed of a very local culture (though that of course does not mean it is not important)" (B. Williams, "Must a Concern for the Environment Be Centred on Human Beings," in *Making Sense of Humanity,* B. Williams [Cambridge: Cambridge University Press, 1995], 237).

38. E. Gellner, *Cause and Meaning in the Social Sciences* (London: Routledge and Kegan Paul, 1973), 29.

39. Gellner, *Cause and Meaning in the Social Sciences,* 38ff.

40. See R. Goodin, *Green Political Theory* (Cambridge: Polity Press, 1992), chap. 2.

41. Typical is Taylor: "The fact that most people actually have a higher respect for persons than they do for animals and plants only reveals that they have not freed themselves from the anthropocentric perspective that under-

lies their whole outlook on nature. This results, in my view, from a failure to achieve true objectivity of judgement" (P. Taylor, "Are Humans Superior to Animals and Plants?" *Environmental Ethics,* 6 [1984]: 149–60, 159).

42. C. Geertz, *The Interpretation of Cultures* (New York: Basic, 1973), 6.

43. G. Ryle, "The Thinking of Thoughts: What Is 'Le Penseur' Doing?" in *Collected Papers Volume II* (London: Hutchinson, 1971), 480–96, 480; cf. "Thinking and Reflecting," in the same volume, 465–79.

44. J. Rawls, *A Theory of Justice* (Oxford: Oxford University Press, 1972), sec. 60.

45. B. Williams, *Ethics and the Limits of Philosophy* (London: Fontana, 1985), 129–31 and 140ff. I think more adequate formulation of this distinction is between "thin" and "thick" uses of concepts: general thin concepts like "good" have specific world-guided uses when employed with a substantive—for example, where we talk of a "good farmer." Indeed, if it is the case that "good" is an attributive adjective, then thin uses of "good" are elliptical—one can always ask—"a good what?" (J. L. Austin, *Sense and Sensibilia* [Clarendon Press: Oxford, 1962], 69. Cf. P. Geach, "Good and Evil," in *Theories of Ethics,* ed. P. Foot [Oxford: Oxford University Press, 1967], 64–73.) The thick-thin distinction in this sense is tied mainly to various debates in meta-ethics about the defensibility of centralism—the view that thin concepts, general normative concepts, are conceptually prior to and independent of thick concepts (S. Hurley, *Natural Reasons* [Oxford: Oxford University Press, 1989], chap. 2.)—and the defensibility of non-cognitivism—the view that normative utterance express attitudes or preferences towards the world and are not assertions that are true or false of the world. (For discussions of both sides of the debate see S. Blackburn, "Through Thick and Thin," *Proceedings of the Aristotelian Society,* 66 [1992]: 258–99, and J. Dancy, "In Defence of Thick Concepts," *Midwest Studies in Philosophy,* 20 [1995]: 263–79.)

46. M. Walzer, *Thick and Thin* (Notre Dame: University of Notre Dame Press, 1994).

47. M. Walzer, *Thick and Thin,* xi.

48. Cf. M. Midgley, "Trying out One's New Sword," in *Heart and Mind* (New York: St. Martin's Press: 1981), 72–73.

49. S. Hurley, *Natural Reasons* (Oxford: Oxford University Press, 1989), chap. 2.

50. An earlier version of this paper was read to a workshop on *Humans in Nature: The Ethics and Aesthetics of Cultural Landscapes,* Norway, March 2000. It also owes a great deal to my involvement in the seminar workshop *Contesting Nature: Anthropology and Environmental Protection* held in Manchester, 21 September 1998. My thanks for the comments and conversations made on both occasions.

Michael McCloskey

Conservation Biologists Challenge Traditional Nature Protection Organizations (1996–1997)

SINCE THE MID-1980s the field of "conservation biology" has emerged with great impact. Its leaders have marshaled the findings of field research around a program representing their ideas of how to protect biodiversity. This research is valuable and its implications need to be heeded. The aim of protecting biodiversity is one that all those interested in protected areas will share.

However, their approach poses challenges to traditional organizations championing Nature and protected areas, such as national parks and wilderness areas. Many of the traditional approaches are criticized in the literature of conservation biology. In this emerging literature, the preservation of biodiversity is put forth as the *raison d'etre* for protected areas. Every other reason for having them is treated as secondary, if not trivial and old-fashioned. Little interest is shown in the rich culture of values accumulated over more than a century that explain why so many protected areas exist (over 70 such values are discussed in the taxonomy of wilderness values I have prepared).[1] The diversity of reasons for having protected areas has expanded the constituency for them.

Moreover, in conservation biology the systems of national parks and protected areas are not judged in terms of these reasons which explain why we have the areas that we do have and why they are located where they are and are of the size they are. Viewing the product of so much history through a new lens (representing a kind of presentism), these observers are quick to judge the existing system wanting in terms of achieving their new biodiversity goals. They assert that the parks are not in the right places; they are not large enough; they are often too far apart; they are not managed properly; they are not buffered from outside influences; and sometimes they represent damaged goods.[2] One commentator decries "... drawing lines around areas and trying in vain to hold them forever in the condition in which they were found."[3]

They observe that a large number of major habitat types are not represented at all in the US National Park system, pointing out that 33% of the potential natural vegetation types are not represented. They observe further that US Forest Service Wilderness Areas fail to represent 40% of the terrestrial ecosystems (as laid out in the Bailey-Kuchler scheme).[4] Having noted that 35% of designated Wilderness is still open to grazing, they then assert that only about 3% of the land area of the United States is really strictly protected (in contrast to 11–12% nominally protected).[5]

Their disdain for what has been achieved so far is evident. One of them characterizes the notion of reserves as "anachronistic."[6] He asserts that "Amenity preservation has resulted in parks as ecological islands, crown jewels without a crown."[7] He dismisses the achievements represented by existing systems as "... token environmental reform."[8] Those interested in esthetics and recreation are sometimes lumped together with those interested in profits and maximum yields.[9] The very idea of "set asides" is attacked because it could engender "... a feeling of free license elsewhere."[10]

This disdain for historic accomplishments has been encouraged by overblown credit given to the "worthless lands" theory propounded by Alfred Runte.[11] While there have always been those who wanted to limit reserves to lands devoid of economic value, it is a distortion of history to assert they always succeeded, and even Runte does not contend that has been the case. Certainly Yosemite contains valuable water power sites, as does the Grand Canyon. It is unfortunate that the report of the IUCN [International Union for Conservation of Nature and Natural Resources; more commonly, the World Conservation Union] on the Caracas conference gives credence to the notion that "In the history of protection in North America, parklands have largely been limited to what is considered barren and economically useless for other purposes."[12] If this thesis ever tended to be true, it was primarily in the early history of protected areas.

As the park and conservation movement came to play a more active role, reserves have included more of value. Hard fought contests in the 1930s over establishing Kings Canyon National Park and Olympic National Park certainly involved lands of value for dams and timber, as did earlier efforts which succeeded in expanding Sequoia National Park. While commercial interests often succeeded in getting some areas they coveted dropped from park proposals, this does not mean that conservationists got nothing, or that parks got only worthless lands.

Most of the contests over protected areas in the period after World War II involved spirited struggles over lands that were far from being commercially worthless. The battle over Redwood National Park stands as the ultimate example of valuable lands going into the park system; $1.3 billion worth of timber was set aside there. Most struggles today over wilderness designation involve conflicts over lands valuable for timber or minerals.

While we may easily acknowledge that the present system of protected areas is inadequate and reflects the historical interests that produced it, to heap scorn on systems that are, nonetheless, very valuable is counter-productive. Referring to the lands in the system as "worthless" suggests they are now worthless to the nation as protected areas. Additionally unhelpful is to have them pilloried as flawed, limited, mislocated and mis-managed.

It will be all too easy for the public to conclude that protected area systems should be dismembered if this is the message they get from leaders in the biological community. In fact, this is exactly what the National Park system's critics in Congress were advocating in the Hefley bill (HR 260). They were saying, in effect, "weed out the old system before anything more is added." Unwittingly, some voices in the biological community are playing into their hands. What these conservation biologists really want is a vast expansion of the system of protected areas, but their rhetoric may instead feed efforts to shrink the system's size.

In contrast to the 11–12% of the US currently in protected areas of some type, leading conservation biologists are calling for setting aside an average of 50% of every eco-region in protected areas (in wilderness or buffer zones).[13] Their ". . . calculations of the area necessary to represent all species and ecosystem types in a region can run as high as 99%, but are usually in the range of 25 to 75 percent."[14] They not only want to maintain viable populations of all native species, but to do so in their "natural patterns of abundance and distribution."[15]

Doing the latter requires devoting a majority of all habitat to its original uses, prior to European settlement. To restore the Florida Panther, these conservation biologists want to place 60–70% of the states of Florida, Georgia, Alabama and South Carolina in reserves.[16] To assure the Grizzly Bear's recovery in the lower 48 states, they want to allocate 60% of the Northern Rockies region to protected status.[17] They want to put 50% of the Oregon Coast bioregion into reserves.[18] And they feel that 25% of all rangeland belongs in reserves.[19] Perhaps the idea of protected areas is not so anachronistic after all.

And while they want to establish ". . . more or bigger parks, wilderness areas and other reserves," they also want to ". . . manage better the semi-natural matrix (multiple use public and private lands) that covers most of our country."[20] And they want to do all of this on a continental scale, with a planning horizon of 10,000 years.[21] They want a moratorium on all habitat degradation while these new plans are being put in place.[22]

Those putting forth these expansive ideas know they sound "utopian," but they warn of the consequences of thinking only ". . . in terms of what is politically 'reasonable [or] practical.'. . ."[23] They caution that ". . . we had better be very clear about the consequences of pragmatism for both species and ecosystems: They will soon disappear, along with Earth's habitability for *Homo sapiens.*"[24]

Thus, notice is served on society: follow their course or accept responsibility for the consequences. Biological imperatives are posited which demand what "must" be done, with some saying these demands are "non-negotiable."[25]

This posture represents a kind of neo-determinism: they are bearers of truth; society either conforms or pays the price. There is no room here for social choice or uncertainty.[26]

Yet, by their own admission, much is uncertain, and thinking on some matters has been reversed. How is one to know how much of their program may be undermined by new findings? How much is based on scientific knowledge and how much on personal preference?

The field's founder, Michael Soulé, has specified that conservation biology exists to do a job: it is mission oriented; yet *Conservation Biology* editor Reed Noss admits that some of what they advocate is "untested."[27] Noss says that "We do not yet (and may never) know what we are doing."[28] He states that "ecosystem conservation is problematic";[29] "we remain dangerously ignorant about natural ecosystems."[30] "Ecosystems are more difficult to classify than species"[31] and "no accepted classification of . . . ecosystems exists in the United States."[32] Moreover, "greater ecosystems are not self-evident."[33] These admissions leave one wondering about how to persuade the public that their case is sound.

For example, although the case for connections between preserved blocks is well established, the case for wildlife corridors as one way of providing connections is less well established. Corridors are suggested to facilitate genetic exchanges among populations in different locations; some think they ought to be wide enough to support their own resident populations.[34] Yet skeptics point out that little has been proven about their value, especially at a regional scale.[35] They point out that corridors may also spread disease, edge-loving species, and wildfires, and pose a higher mortality risk for wandering animals.[36] The skeptics observe that corridors may also be expensive to maintain.[37] Michael Soulé admits that there are "no answers yet" for these concerns.[38] Is this an idea ready for its debut in a broad-scale way (as contrasted to being tested and refined)? Notwithstanding these questions, many are ready to proceed.

All of these *idées fixes* come at a time when other long-held notions of ecology are being discarded. It was long believed that the more diverse ecosystems (in terms of species richness) were more stable and resilient, yet that correlation is now in doubt.[39] Diversity and stability may not go hand in hand. Even David Tilman's recent data defending a variation of the stability hypothesis are not reassuring. The variation in diverse ecosystems may leave individual species at risk.[40]

Among the earliest ideas of ecology was that habitats progressed through successional stages of vegetation to reach a climax which would represent an equilibrium state (except for disturbances that would start the process over). This notion of climax communities has now largely fallen out of favor among ecologists. Now equilibrium theories ". . . have been largely replaced by dynamic paradigms."[41] Nature is seen as ". . . full of uncertainty and episodic at different spatiotemporal scales."[42] ". . . Nature is a shifting mosaic . . . [and is essentially in flux. . . ."[43] "Species composition of vegetation varies continuously in time and space. . . ."[44] "Because chance [disturbance] factors and small climatic variation can apparently cause very substantial changes in vegetation, the biota and associated ecosystem processes for any given landscape will vary substantially over any significant time period—and no one variant is more 'natural' than the others."[45] Indeed, if the climates of regions are changing with the global buildup of greenhouse gases, all sorts of changes may be triggered in plant communities. Habitats on mountain tops and in low-lying coastal areas may no longer be suitable; plants in the US will need to migrate northward; some biologists have suggested that "preserves themselves may need to move."[46]

As implications of chaos theory in physics have seeped into ecology, one is left wondering whether reserves can be built around expectations that any given plant or animal community will be assured of a future there. Minor perturbations might displace them, and climate change may wreak havoc.

The collapse of the equilibrium model and the diversity-stability supposition, along with the cloud being cast by impending climate change, all raise fundamental questions about the context for tackling major new challenges to protect biodiversity. Instead of knowing more about what to do, it almost seems as if we know less about what to do and how to plan for the future.

Despite these uncertainties, the advocates of a major scale-up in preservation want to change some of the ways protected areas are managed. Most advocate "hands-on management," with some boasting that "we can engineer nature at nature's rate. . . ."[47] Cautions are issued, however, about avoiding "over-management."[48] Hands-on management is justified to block agents that would undermine biodiversity and to restore habitats.[49] Edward Grumbine advocates varying approaches according to local conditions, with no consistent standards.[50]

Michael Soulé advocates managing wildlife in protected areas through various means, including culling, artificial transfers, and immunization. He calls for eradication of exotic biota.[51]

Donald Waller urges re-thinking the prescription for managing wilderness so that it is managed from a biodiversity perspective rather than just for esthetics and recreation.[52] He also suggests that some additional human intrusions may be justified if more land can be preserved for biodiversity purposes, leaving the door open for limited snowmobiling and cutting firewood in designated Diversity Maintenance Areas (DMAs).[53]

Conservation biologists support active programs to eliminate exotic species in reserves and to re-introduce native species.[54] Restoration of damaged range habitats might also entail use of mechanical treatments or herbicides.

In discussing reserves on rangelands, Reed Noss and Allen Cooperrider suggest some restrictions on recreational activities, such as backpacking, which are perceived to have discernible impacts.[55] In buffer zones around reserves, they would allow light grazing, selection forestry, non-motorized recreation (including fishing and hunting), and small-scale subsistence agriculture.[56]

For forested areas in reserves, including new designations they contemplate, they would encourage natural fire regimes.[57] In small reserves, they would use prescribed burns;[58] in large reserves, they would take the "let burn" approach. Where necessary to create enough gaps for reproducing early successional habitats, they would fell trees to simulate treefalls,[59] though generally only at the edge of reserves or in buffers. Donald Waller would try to ". . . sustain disturbance regimes typical of the region without losing species. . . ."[60] This would entail efforts "to maintain patterns of disturbance and habitat patches similar to those that have occurred historically. . . ."

William Baker, however, warns against rushing into heavy burning programs while so little is known about the historical frequency and size of high intensity wild-fires. He says ". . . it is premature to undertake extensive manipulative restoration action using either prescribed disturbances or mechanical means, as these may only produce undesirable alteration."[61]

In general, these changes in management prescriptions involve more intrusiveness than is now authorized in Wilderness Areas and a de-emphasis on recreation. They also mean vesting more authority in managers to decide what is warranted in the name of biodiversity protection. The approach assumes that large amounts of data are available from monitoring to adjust approaches so that management can be adapted to apply new knowledge (i.e., adaptive management). Still unclear, however, is on what basis, or under what guidelines, we should trust managers with so much discretion, particularly in light of past mistakes and tendencies to cater to local commercial interests.

CONCLUSION

Leaders in both communities should foster a rapprochement between traditional Nature protection organizations and conservation biologists. The former need to learn more biology, and the latter need to learn more about how to get results. Both could benefit from listening to each other. Less hubris and presumption may also help. A symbiotic relationship might then develop.

NOTES

1. See Michael McCloskey, "Evolving Perspectives on Wilderness Values: Putting Wilderness Values in Order," in *Preparing to Manage Wilderness for the 21st Century* (Forest Service, Southeastern Forest Experimental Station, Athens, Georgia, General Technical Report SE-66, 1990).

2. Reed F. Noss and Allen Y. Cooperrider, *Saving Nature's Legacy: Protecting and Restoring Biodiversity* (Island Press, Washington, DC, 1992), p. 210.

3. Ibid., p. 92.

4. Ibid., p. 174.

5. Ibid., p. 172.

6. R. Edward Grumbine, *Ghost Bears: Exploring the Biodiversity Crisis* (Island Press, Washington, DC, 1992), p. 210.

7. Ibid., p. 183.
8. Ibid., p. 227.
9. Ibid., p. 29.
10. See J. M. Brussard, D. D. Murphy, and R. F. Noss, "Strategy and Tactics for Conserving Biological Diversity in the United States," *Conservation Biology* (Vol. 6, No. 2, 1992), pp. 157–159; cited in Donald M. Waller, William S. Alverson, and Walter Kuhlman, *Wild Forests: Conservation Biology and Public Policy* (Island Press, Washington, DC, 1994).
11. Alfred Runte, *National Parks: The American Experience* (Univ. of Nebraska Press, Lincoln, 1979), ch. 3.
12. J. A. McNeely, I. Harrison, and P. Dingwall, *Protecting Nature: Regional Reviews of Protected Areas* (IUCN, Gland, Switzerland, 1994), p. 282.
13. Noss, op. cit., p. 167.
14. Ibid., p. 168.
15. Ibid., p. 91.
16. Ibid., p. 164.
17. Ibid., p. 163.
18. Ibid., p. 128.
19. Ibid., p. 254.
20. Ibid., p. 130.
21. Ibid., p. 88.
22. Ibid., p. 187.
23. Ibid., p. 94.
24. Grumbine, op. cit., p. 187.
25. Ibid., pp. 158, 227, 247.
26. Duke University political science professor Lynn Maguire warns that proponents of this kind of thinking ". . . must be wary of confounding advocacy of [their] goals . . . with scientific analysis of various means of achieving them. Not only would this be an inappropriate use of scientific analysis . . . but it would be a perilous tactic, since much of the science of landscape ecology is more at the state of plausible hypothesis than of well established theory." See *Environmental Policy and Biodiversity,* ed. by R. Edward Grumbine (Island Press, Washington, DC, 1994), p. 270; but Edward Grumbine asks: "How can conservation biology survive the inevitable checks and balances of American politics?" id., p. 13.
27. Noss, op. cit., pp. 84, 94.
28. Ibid., p. 97.
29. Ibid., p. 90.
30. Ibid., p. 97.
31. Ibid., p. 107.
32. Ibid.
33. Ibid., pp. 84, 94.

34. Ibid., p. 154.

35. Waller et al., *Wild Forests,* op. cit., pp. 92, 112–113.

36. Ibid.

37. Ibid.

38. See Michael Soulé, "Conservation Biology in Context," in *Environmental Policy and Biodiversity,* ed. by Edward Grumbine, p. 99.

39. Noss, op. cit., p. 197.

40. See Anne Simon Moffat, "Biodiversity Is a Boon to Ecosystems, Not Species," *Science* (Vol. 271, 15 March 1996), p. 1497.

41. Ibid., p. 92; also see Donald Worster, "The Ecology of Order and Chaos," *Environmental History Review* (Vol. 14, No. 1–2, Spring/Summer 1990), pp. 1–18.

42. Grumbine, *Environmental Policy and Biodiversity,* op. cit.

43. Noss, op. cit., p. 166, quoting S. T. A. Pickett, V. T. Parker, and P. L. Fiedler, "The New Paradigm in Ecology: Implications for Conservation Biology Above the Species Level," in *Conservation Biology: The Theory and Practice of Nature Conservation, Preservation, and Management* (Chapman and Hall, New York, 1992), pp. 65–68.

44. Noss, op. cit., p. 247.

45. Grumbine, *Ghost Bears,* op. cit., p. 59.

46. Ibid., p. 62.

47. Advocated by Daniel Botkin (1990) in his book *Discordant Harmonies: A New Ecology for the Twenty-First Century;* discussed in Grumbine, *Environmental Policy and Biodiversity,* op. cit., p. 387.

48. Noss, op. cit., pp. 98, 207.

49. Ibid., pp. 92, 174.

50. Grumbine, *Ghost Bears,* op. cit., p. 216.

51. Michael Soulé, "What Is Conservation Biology," in Grumbine, *Environmental Policy and Biodiversity,* op. cit., p. 38.

52. Waller, op. cit., p. 236.

53. Ibid., p. 243.

54. Noss, op. cit., pp. 92, 206.

55. Ibid., p. 243.

56. Ibid., p. 254.

57. Ibid., p. 149.

58. Ibid., pp. 165, 209.

59. Ibid., p. 207.

60. Waller, op. cit., p. 44.

61. Grumbine, *Environmental Policy and Biodiversity,* op. cit., p. 92.

PART FOUR

Thinking through the Wilderness Idea

Marilynne Robinson

Wilderness (1998)

ENVIRONMENTALISM POSES stark issues of survival, for humankind and for all those other tribes of creatures over which we have exercised our onerous dominion. Even undiscovered species feel the effects of our stewardship. What a thing is man.

The oldest anecdotes from which we know ourselves as human, the stories of Genesis, make it clear that our defects are sufficient to bring the whole world down. An astonishing intuition, an astonishing fact.

One need not have an especially excitable or a particularly gloomy nature to be persuaded that we may be approaching the end of the day. For decades, environmentalists have concerned themselves with this spill and that encroachment, this depletion and that extinction, as if such phenomena were singular and exceptional. Our causes have even jostled for attention, each claiming a special urgency. This is, I think, like quarreling over which shadow brings evening. We are caught up in something much larger than its innumerable manifestations. Their variety and seriousness are proof of this.

I am an American of the kind whose family sought out wilderness generation after generation. My great-grandparents finally settled in

Idaho, much of which is wilderness now, in terms of its legal status, and is therefore, theoretically, protected. In the heart of this beloved, empty, magnificent state is the Idaho Nuclear Engineering Laboratory, among other things a vast repository for radioactive waste. Idaho, Utah, Nevada, New Mexico, beautiful names for vast and melancholy places. Europeans from time to time remark that Americans have no myth of landscape. In fact we have many such myths. People who cherish New England may find it difficult to imagine that Utah is cherished also. In fact, I started writing fiction at an eastern college, partly in hopes of making my friends there understand how rich and powerful a presence a place can be which, to their eyes, is forbidding and marginal, without population or history, without culture in any form recognizable to them. All love is in great part affliction. My bond with my native landscape was an unnamable yearning, to be at home in it, to be chastened and acceptable, to be present in it as if I were not present at all.

Moses himself would have approved the reverence with which I regarded my elders, who were silent and severe and at their ease with solitude and difficulty. I meant to be like them. Americans from the interior West know what I am describing. For them it is, or is like, religious feeling, being so powerful a reference for all other experience.

Idaho, Utah, Nevada, New Mexico. These names are all notorious among those who know anything at all about nuclear weapons. Wilderness is where things can be hidden, from foreign enemies, perhaps, but certainly from domestic critics. This effect is enhanced by the fact that wilderness dwellers everywhere are typically rather poor and scattered, not much in the public mind, not significant as voters. Wilderness is where things can be done that would be intolerable in a populous landscape. The relative absence of human populations obscures the nature and effect of programs which have no other object than to be capable of the most profound injury to human populations. Of course, even wilderness can only absorb such insult to the systems of life to a degree, for a while. Nature is very active—aquifers so vast, rivers so tireless, wind so pervading. I have omitted to mention the great Hanford Reservation in Washington State, with its ominous storage tanks, a whole vast landscape made an archaeological history of malign intent, and a great river nearby to spread the secret everywhere.

Russia is much more generously endowed with wilderness than America. Turn the globe, and there is an expanse that puts our little vastness in perspective. It is my impression that depredations of the kind we have been guilty of have been carried further in Russia and its former territories, at least in proportion to the permission apparently implied by empty spaces. But wilderness can be borrowed, as the coast and interior of Australia and, of course, Nevada have been by the British for their larger nuclear weapons projects, and wilderness can be relative, like the English Lake District and the northwest coast of France. And then there is the sea. We have all behaved as if there were a place where actions would not have consequences.

Wilderness is not a single region, but a condition of being of the natural world. If it is no longer to be found in one place, we assume it exists in other places. So the loss of wilderness always seems only relative, and this somewhat mitigates any specific instance of abuse. Civilization has crept a little farther; humankind has still to learn certain obvious lessons about living in the world. We regret and we repent and we blame, and we assume that things can be different elsewhere. Again, the very idea of wilderness permits us to evade in some degree a recognition of the real starkness of precisely the kind of abuse most liable to occur outside the reach of political and economic constraints, where those who have isolation at their disposal can do as they will.

Utah is holy land to a considerable number of American people. We all learn as schoolchildren how the Mormons, fleeing intolerance and seeking a place where they could live out their religion, walked into the wilderness, taking their possessions in handcarts, wearing a trail so marked that parts of it are visible to this day. We know they chose barren land by a salt lake, and flourished there.

It is a very pure replication of the national myth. So how did we make the mistake we made, and choose this place whose very emptiness and difficulty were a powerful proof to the Mormons of the tender providence of God—how could we make Utah the battleground in the most furious and terrified campaigns in our long dream of war? The choice kept casualties to a minimum, which means that if the bombs were dropped in populous places the harm would have been clearly intolerable. The small difference between our fantasies of war and war itself would be manifest. As it is, there are many real casualties, and no doubt there would be far more if all the varieties of injury were known and acknowledged.

This is a potent allegory. It has happened over and over again that promised land or holy land by one reckoning is wasteland by another, and we assert the sovereign privilege of destroying what we would go to any lengths to defend. The pattern repeats itself so insistently that I think it is embedded not merely in rational consciousness but also in human consciousness. Humankind has no enemy but itself, and it is broken and starved and poisoned and harried very nearly to death.

Look at England. They have put a plutonium factory and nuclear waste dump in the Lake District, a region so beautiful that it was set aside, spared most of the marring burdens of population. And what a misfortune that has been. Relatively small populations result in relatively small bases for interpreting public health effects, so emptiness ensures not safety so much as deniability. Wilderness and its analogues seem to invite denial in every form. In Utah and in Cumbria, it was the urgent business of years to produce weapons capable of inflicting every extreme of harm on enemy populations. Do they harm people who live where they are made and tested? If the answers "no," or "not significantly," or "it is too difficult to tell" were ever given in good faith, then clearly some mechanism of denial had come into play. The denial was participated in at a grand enough scale to make such answers sufficient for most of the public for a very long time, even though one effect was to permit methods of development and testing that assured widespread public contact with waste or fallout, and that will assure it into any imaginable future.

Denial is clearly a huge factor in history. It seems to me analogous to a fractal, or a virus, in the way it self-replicates, and in the way its varieties are the grand strategy of its persistence. It took, for instance, three decades of the most brilliant and persistent campaign of preachment and information to establish, in the land of liberty, the idea that slavery was intolerable. Strange enough. These antislavery agitators were understandably given to holding up Britain's ending of slavery in her colonies as the example of enlightened Christian behavior. But at the same time, British slave ships used the old slave routes to transport British convicts to Australia. Every enormity was intact, still suffered by women and children as well as men. Of course the color of the sufferer had changed, and it is always considered more respectable in a government to ravage its own population than others'. To this objection, I will reply that the arrival of the British was an unspeakable disaster to the native people of Australia and Tasmania. Slavery and genocide were only rechanneled, translated into other terms, but for the American abolitionists, and for the British abolitionists as well, this was nothing to pause over. It is understandable that Americans should wish to retain all the moral leverage that could be had from the admirable side of the British example. Still, this is another potent allegory, something to unriddle, or at least to be chastened by. After our terrible war, the people who struggled out of bondage, and were won out of bondage, found themselves returned to a condition very much resembling bondage, with the work all before them of awakening public awareness, in the land of the free, to the fact that their situation was intolerable.

Reform-minded Americans still depend on the idea that other countries are in advance of us, and scold and shame us all with scathing comparison. Of course they have no tolerance for information that makes such comparison problematic. The strategy, however generous in impulse, accounts in part for the perdurable indifference of Americans to actual conditions in countries they choose to admire, and often claim to love.

I have begun to consider Edgar Allan Poe the great interpreter of Genesis, or perhaps of Romans. The whole human disaster resides in the fact that, as individuals, families, cities, nations, as a tribe of ingratiating, brilliant, momentarily numerous animals, we are perverse, divided against ourselves, deceiving and defeating ourselves. How many countries in this world have bombed or poisoned their own terrain in the name of protecting it from its enemies? How many more would do so if they could find the means? Do we know that this phenomenon is really different in kind from the Civil War, or from the bloodbaths by which certain regimes have been able to legitimate their power? For a long time we have used dichotomies, good people/bad people, good institutions/bad institutions, capitalist/communist. But the universality of self-deceptive and self-destructive behavior is what must impress us finally.

Those who are concerned about the world environment are, in my view, the abolitionists of this era, struggling to make an unenlightened public aware that environmental depredation is an ax at the root of every culture, every freedom, every value. There is no group in history I admire more than the abolitionists, but from their example I conclude that there are two questions we must always ask ourselves—what do we choose not to know, and what do we fail to anticipate? The ultimate success of the abolitionists so very much resembled failure that it requires charity, even more than discernment, to discover the difference. We must do better. Much more is at stake.

I have heard well-meaning people advocate an environmental policing system, presided over by the member governments of the United Nations Security Council. I think we should pause to consider the environmental practices and histories of those same governments. Perhaps under the aegis of the United Nations they do ascend to a higher plane of selflessness and rationality and, in this instance, the cowl will make the monk. Then again, maybe it will not. Rich countries that dominate global media look very fine and civilized, but, after all, they have fairly ransacked the world for these ornaments and privileges and we all know it. This is not to say that they are worse than other nations, merely that they are more successful, for the moment, in sustaining wealth and prestige. This does not mean they are well suited for the role of missionary or schoolmaster. When we imagine they are, we put out of mind their own very grave problems—abandoning their populations, and the biosphere, in the very great degree it is damaged by them, to secure moral leverage against whomever they choose to designate an evildoer. I would myself be willing to give up the hope of minor local benefits in order to be spared the cant and hypocrisy, since I have no hope that the world will survive in any case if the countries represented on the Security Council do not reform their own governments and industries very rigorously, and very soon.

I think it is an indulgence to emphasize to the extent we do the environmental issues that photograph well. I think the peril of the whole world is very extreme, and that the dolphins and koalas are finally threatened by the same potentialities that threaten everything that creeps on the face of the earth. At this time, we are seeing, in many, many places, a decline in the wealth, morale, and ethos whose persistence was assumed when certain features of modern society were put in place, for example, nuclear reactors and chemical plants. If these things are not maintained, or if they are put to cynical uses by their operators or by terrorists, we can look forward to disaster after disaster. The collapse of national communities and economies very much enhances the likelihood that such things will happen.

We have, increasingly, the unsystematic use of medicine in the face of growing populations of those who are malnourished and unsheltered and grossly vulnerable to disease. Consider the spread of tuberculosis in New York City. Under less than ideal circumstances, modern medicine will have produced an array of intractable illnesses. In the absence of stability and wealth, not to mention a modicum of social justice, medicine is liable to prove a curse and an affliction. There are those who think it might be a good thing if we let ourselves slip into extinction and left the world to less destructive species. Into any imaginable future, there must be people to maintain what we have made, for example, nuclear waste storage sites, and there must be human civilizations rich and sophisticated enough to know how this is done and to have the means to do it. Every day this seems less likely.

And only consider how weapons and weapons materials have spread under cover of this new desperation, and how probable truly nihilistic warfare now appears. These are environmental problems, fully as much as any other kind.

Unless we can re-establish peace and order as values, and learn to see our own well-being in our neighbor's prosperity, we can do nothing at all for the rain forests and the koala bears. To pretend we can is only to turn our backs on more painful and more essential problems. It is deception and self-deception. It stirs a sad suspicion in me that we are of the Devil's party, without knowing it.

I think we are desperately in need of a new, chastened, self-distrusting vision of the world, an austere vision that can postpone the outdoor pleasures of cherishing exotica, and the first-world pleasures of assuming we exist to teach reasonableness to the less fortunate, and the debilitating pleasures of imagining that our own impulses are reliably good. I am bold enough to suggest this because, to this point, environmental successes quite exactly resemble failure. What have we done for the whale, if we lose the sea? If we lose the sea, how do we mend the atmosphere? What can we rescue out of this accelerating desperation to sell—forests and weapons, even children—and the profound deterioration of community all this indicates? Every environmental problem is a human problem. Civilization is the ecology being lost. We can do nothing that matters if we cannot encourage its rehabilitation. Wilderness has for a long time figured as an escape from civilization, and a judgment upon it. I think we must surrender the idea of wilderness, accept the fact that the consequences of human presence in the world are universal and ineluctable, and invest our care and hope in civilization, since to do otherwise risks repeating the terrible pattern of enmity against ourselves, which is truly the epitome and paradigm of all the living world's most grievous sorrows.

J. Baird Callicott

The Implication of the "Shifting Paradigm" in Ecology for Paradigm Shifts in the Philosophy of Conservation (2003)

INTRODUCTION

For nearly half a century now, ecology has been shifting away from a "balance-of-nature" to a "flux-of-nature" paradigm (McIntosh 1998, Pickett and Ostfeld 1995). By the mid-1970s the latter had begun to eclipse the former in ecology, but non-ecologists remained, for the most part, clueless that such a sea change was occurring. In the early 1990s, the new, fluxy way of understanding associations of organisms and ecological processes began to dawn on the laity (Botkin 1990, Worster 1990). Not surprisingly, fields of endeavor that have been informed by ecology will have to take account of the paradigm shift in ecology that is now virtually complete. Here I suggest how the philosophy of conservation might be affected. I begin with a review of the dominant schools of twentieth-century thought about conservation, go on to review the shift from the balance-of-nature to the flux-of-nature paradigm in ecology, and, lastly, suggest what the implications of that paradigm shift might be for an ecologically well-informed twenty-first-century philosophy of conservation.

Conservation *philosophy* has been primarily an American enterprise, precisely because the *practice* of conservation, traditional in many European and Asian societies and in pre-Columbian North American societies, was suspended after the conquest of the New World by the Old. The indigenous populations of the Western Hemisphere suffered a demographic disaster during the first century after European contact. Old World diseases, such as small pox and influenza, wiped out an estimated 90% of the microbially inexperienced human populations of the New World (Crosby 1976, Denevan 1992). With a proportionate reduction of cultural predation (i.e., human hunting pressure), horticulture, and cultural fire, wildlife populations soared and forests regenerated throughout the Nearctic (Denevan 1996). When European settlers began gradually to spread across the North American continent during the seventeenth, eighteenth, and nineteenth centuries, they encountered a nouvelle "wilderness condition" and super-abundant plant and animal resources. As long as more unharvested and unclaimed timber and game could be found over the next hill or across the next river, no one gave a thought to conserving forests or wildlife. With the completion of a transcontinental railroad, the conquest of the plains Indians, and the wholesale slaughter of the vast herds of bison—all during the last quarter of the nineteenth century—the mid-continental North American frontier palpably closed (Turner 1920). By the end of the nineteenth century, wildlife had actually become scarce and deforestation rampant in the United States. Conservation was necessary, but for nearly three centuries, no one had theorized it. The situation was ripe for the birth of a philosophy of conservation.

THREE PARADIGMS IN THE PHILOSOPHY OF CONSERVATION

From the mid-nineteenth century to the mid-twentieth, three distinct paradigms emerged in the philosophy of conservation: preservationism, resourcism, and harmonization.

PRESERVATIONISM

The first staunch protest against the unrestrained exploitation of American forests and wildlife was articulated by Henry David Thoreau. Thoreau was an associate of Ralph Waldo Emerson, a leading exponent of a religio-philosophical movement called Transcendentalism. In his essay, *Nature,* Emerson (1836) posited the existence of a noumenal reality that lies beyond the phenomena disclosed by the senses. One must leave "the streets of the city," all "society" and even one's private "chamber" to find a genuine solitude; and alone in "Nature," more particularly in "the woods" and in "the wilderness," a person might transcend the finitude of ordinary existence and become one with the "Universal Being" (Emerson 1836). There too one might encounter the purest and most prefect forms of beauty.

These ideas had been vaguely anticipated half a century earlier by the Puritan preacher, Jonathan Edwards, who found "the images or shadows of divine things" in nature (Edwards 1978). Edwards was also a warm advocate of the extreme Calvinist doctrine of original sin, according to which all human beings are fallen and depraved. Perry Miller (1964) argues that Transcendentalism actually evolved from Puritanism. The first generation of Puritans believed themselves to have been sent by God on "an errand into the wilderness," which they imagined to be the worldly stronghold of Satan, whose minions were the Indians. Taming and civilizing the wilderness was thus more than a utilitarian task for William Bradford, John Winthrop, Cotton Mather, and Michael Wigglesworth in the seventeenth century; it was a crusade (Nash 1967). By Edwards's time, a century after, the errand into the wilderness had been successfully run. In New England shining cities on hills had been built, farm fields cleared and planted to crops, the Indians and the large predatory animals exterminated or driven away. Sin and the abode of the devil had moved to town, in the eighteenth century Puritan imaginary. Nature in New England had become less an evil realm of violence, chaos, and undisciplined eros, than an unfallen realm of innocence, peace, beauty, divine order, and truth.

All these currents of thought gather in the writings of Thoreau advocating Nature preservation. In *Walden*—subtitled *Or Life in the*

Woods—Thoreau (1854) finds solitude, solace, perfect beauty, and higher laws and truths in Nature. Though not a Puritan in doctrine, his lifestyle was certainly puritanical: he was celibate, a vegetarian, a teetotaler, and a scold. Thoreau was also borderline misanthropic. His attitude toward his fellow "citizens" was a secular equivalent of Edwards' toward his fellow "sinners in the hands of an angry God." In "Walking," Thoreau richly and concretely retraces the movement from city streets to Nature as envisioned in Emerson's ruminations. In that essay, Thoreau wrote the enigmatic slogan which is so often quoted, "in Wildness is the preservation of the World" (and misquoted as "in Wilderness is the preservation of the World"). In the posthumously published *Huckleberries,* written in 1861, a year before his death, Thoreau (1970) actually proposes wilderness preservation, albeit a scaled-down version of it in comparison with twentieth-century ideas about an appropriate size for wilderness areas: "I think that each town should have a park, or rather a primitive forest, of five hundred or a thousand acres, either in one body or several—where a stick should never be cut for fuel—nor for the navy, nor to make wagons, but to stand and decay for higher uses—a common possession forever, for instruction and recreation."

In the late nineteenth and early twentieth centuries, John Muir popularized and politicized Thoreau's call for Nature preservation. Muir is best known for his rapturous celebration of the "glories" of Nature, especially in California's Sierra Nevada range (Fox 1981). But according to Donald Worster (1988), "there was a harshly negative side to Muir's vision, a disgust for human pretensions and pride that ran very close to misanthropy. Lord man, the self-proclaimed master of creation, was for him an ugly blot on the face of the Earth." Like Thoreau, Muir was also greatly influenced by the writings of Emerson. The intimate, albeit counter-intuitive, relationship between Transcendentalism and Puritanism is confirmed in Muir's biography. He was born in Calvinist Scotland, raised a Presbyterian, and force-fed the Bible by his father, Daniel, who could, without hyperbole, be called a religious psychopath.

Reduced to its essential elements the Preservationist philosophy of conservation comes to this: First, dualism—man and nature are separate. This dualism is highly value-charged. Man is fallen, depraved, and sinful; Nature is innocent, pristine, and virgin. For the most part, the very presence of fallen man, to say nothing of his works, rapes and desecrates pure and virgin Nature. For some select (or "elect") sensitive souls, however, Nature has "higher uses"—for aesthetic experience, and as the site for a monastic sojourn, whereupon the pilgrim gladly endures Nature's spiritually cleansing hardships, embraces its solitude, and retreats into it from a profane and fallen human society. The wilderness sojourner ideally becomes so enraptured and transcendent that he or she is barely present materially in Nature at all. In the words of Emerson (1836), "Standing on the bare ground,—my head bathed by the blithe air, and uplifted into empty space, all mean egotism vanishes. I become a transparent eyeball. I become nothing. I see all. The currents of Universal being circulate through me."

This Emersonian vanishing act is carried forward into the contemporary wilderness experience. As Val Plumwood (1998) notes,

> The presence and impact of the modern adventure tourist is somehow "written out" of focus in much of the land called wilderness. . . . The modern subject somehow manages to be both in and out of this virginal fantasy, appearing by wilderness convention as a disembodied observer (perhaps as the camera eye [or Emersonian "transparent eyeball"]) in a landscape whose virginity is forever magically renewed.

Maybe this explains the otherwise apparently hyperbolic and overwrought rule requiring transcendental tourists to pack out—rather than thoroughly and discreetly burn or bury—all material traces of their presence in designated wilderness areas, even their feces.

The principal policy supported by Preservationism is the cordoning off of national parks and designated wilderness areas, primarily for the "instruction and recreation" of the elite segment of the public capable of putting Nature to these "higher uses" identified and lionized by Emerson, Thoreau, and Muir. Such organizations as the Sierra Club (of which Muir was the founding president) and the Wilderness Society are the principal political manifestation of the Preservationist philosophy of conservation.

RESOURCISM

Resourcism is more democratic than Preservationism. The problem with the way hunters, loggers, farmers, and miners had used American natural resources was not that such uses were consumptive and vulgar—as Thoreau and Muir insinuated—but that they were inefficient and destructive, from the Resourcist point of view.

The first American thinker to make this case was George Perkins Marsh (1864). Two years before the word "ecology" had been coined by Ernst Haeckel and a quarter century before a distinct science known by that name had crystallized, Marsh's observations were essentially ecological. Marsh served as United States Ambassador to Turkey from 1849 to 1854 and to Italy from 1861 to 1882 (Lowenthal 1958, 2000). He attributed the eroded and desiccated landscapes he found in the Asian and European countries around the Mediterranean Sea to the deforestation of the region that occurred in the ancient Persian, Greek, and Roman empires. Marsh argued that ancient anthropogenic deforestation had caused soil erosion, flashy streams and rivers, and eventually regional climate change and partial desertification. The deforestation then in process in North America could cause his country eventually to look (and ecologically malfunction) like contemporary Turkey, Greece, and Italy. That in turn could have severe economic and geopolitical consequences. The decline of empires, he believed, had more to do with anthropogenic environmental change than with social decadence. In addition to his economic and more generally pragmatic brief for conservation, Marsh (1864) added a religio-ethical argument: "man," he wrote, "has too long forgotten that the earth was given to him for usufruct alone, not for consumption, still less for profligate waste."

In the concept of usufruct, Marsh gets at one of the core ideas of Resourcism. "Renewable" natural resources, such as trees and fish, can be consumed without depleting and eventually destroying them. But to do so successfully takes more than a concept. One must know the growth rates and the age at which their growth begins to slow of various kinds of useful trees, the reproductive rates of edible fishes and other wildlife, and thousands of other things about renewable natural resources to make them yield both plentifully and sustainably. Thus to realize Marsh's idea of only skimming the interest off of living natural capital requires the development, through research, of various applied sciences (such as forestry and fishery biology), the development of enforceable natural-resource public policy, and the creation of state and federal agencies to implement the findings of these sciences and enforce these public policies.

Further, public ownership of natural resources makes implementation of resource-related science and public policy a lot easier. Upon the assassination of his predecessor, William McKinley, Theodore Roosevelt—an outdoor enthusiast and nascent conservationist—became President of the United States in 1901. Among other signal accomplishments, he created the United States Forest Service in 1904 and used his executive powers to add immensely to the federally owned National Forests. He appointed Gifford Pinchot, who had been educated as a forester in Europe, as the first Chief of the Forest Service (Miller 2001).

It fell to Pinchot fully to articulate the elements of the Resourcist philosophy of conservation. He reversed the Preservationists' virtual apotheosis of Nature—which is always spelled with a capital "N" in the writings by Emerson, Thoreau, and Muir. Their "Nature" became Pinchot's "natural resources"—a name which slyly implies that the natural environment exists solely for human use. And in case one misses the point of that name change, Pinchot (1947) bluntly declares that "there are just two things on this material earth—people and natural resources."

The dualism evident in Preservationism is evidently intensified in Resourcism, but the values charging the dualism are different. As the meme-line of thought ancestral to Preservationism moves from seventeenth-century Calvinism to nineteenth-century Transcendentalism the Manichean value poles reverse in the way they charge the human-nature ontological dichotomy with good and evil. In Resourcism, it's not that people are good and natural resources are evil; rather, people are privileged and natural resources are other-ized and objectified. They lose their identities as individuals, even as species, and become but raw material for human transformation into humanly useful commodities (Plumwood 1998). Pinchot (1947) deliberately assimilates Resourcism to the broader utilitarian ethic then current in philosophy by echoing John Stuart Mill's summary moral maxim—"the greatest happiness for the greatest number"—with his own: "the greatest good of the greatest number for the longest time."

While Preservationism is politically expressed primarily through often oppositional non-governmental organizations, in the United States Resourcism is institutionalized in federal and state agencies (Fox 1981). The Forest Service (housed, tellingly, in the Department of Agriculture) was the first and the organizational and philosophical model for most of the others, including the US Fish and Wildlife Service, the Bureau of Land Management, and various state departments of natural resources. The National Park Service, more dedicated to Nature preservation and higher uses, may represent an anomalous governmental institutionalization of Preservationism in the United States, but, as an agency, it is the exception that proves the Resourcist rule in the public sector.

HARMONIZATION

Aldo Leopold single-handedly conceived a third philosophy of conservation. He began his life-long career in conservation firmly in the Resourcist camp. Leopold was educated at the Yale Forest School, which was founded by Gifford Pinchot, and joined the Forest Service immediately upon graduation in 1909, the penultimate year of Pinchot's tenure as Chief. Leopold's passion for hunting inclined him to be interested in "secondary" forest resources, especially game animals. His earliest sketches of the science of game management, for which he eventually wrote the first textbook, were modeled squarely on forestry—a game population census is analogous to timber reconnaissance, the game farm is analogous to the tree nursery, predator control is analogous to fire suppression, and so on and so forth (Leopold 1918).

Leopold's Resourcist identity is obscured by his interest in wilderness preservation, for which he began to campaign within the Service after about a decade in its employ. Leopold resigned from the Service in 1928 (Meine 1988). And in 1935 he helped to found the Wilderness Society and even flirted with the prospect of serving as its first president, an honor that went to Robert Sterling Yard instead, but only after Leopold had demurred (Meine 1988). Given this trajectory, we are tempted to see Leopold as gradually moving out of the Resourcist camp into the

Preservationist camp. But that would be mistaken. Rather, Leopold first rearticulated the Preservationist agenda in Resourcist terms, and then later in terms of his own novel philosophy of conservation.

While still with the Forest Service Leopold argued that wilderness recreation was the "highest use" of certain areas of the National Forests which were too rugged and remote to log, too poor to farm, and which had no mineral resources worth mining (Leopold 1925). And by "highest use," he did not mean anything like what Thoreau meant by "higher uses." In the Resourcist lexicon "highest use" means, rather, the use of a piece of land that yields the most utility. For example, if a patch of old-growth forest is clear-cut, should it be replanted with trees that will, after seventy-five years, yield more timber, should it be allowed to quickly grow back on its own to early fast-growing successional trees useful only for pulp, or should it be burned periodically to create a grassland useful for grazing domestic sheep and cattle? Which of these is its highest use? In this sense, Leopold (1925) argued that the highest use of some Forest Service lands was wilderness recreation, and by "wilderness recreation" he mainly had in mind big game hunting and primitive modes of travel (by pack train and canoe)—his own preferred uses—not Transcendental tourism.

Leopold's confidence in Resourcism was shaken by several experiences (Flader 1974, Meine 1988). The first was a disastrous consequence of his forestry-inspired Resourcist ideas about game management. In the mid-1920s on the Kaibab plateau and again in the late 1920s on the Gila national forest where predators had been exterminated—in the latter case at Leopold's own behest—deer populations irrupted and then, after overbrowsing and ruining their range, precipitously crashed. The second was a trip to Germany in the mid-1930s where Leopold personally observed the untoward end-point of intensive forestry. Generation after generation of even-aged spruce monocultures had sickened the German soil on which they stood and artificial maintenance of deer populations had impoverished the understory of German forests. The third was a hunting trip to the Sierra Madre Occidental of Mexico in the late 1930s where Leopold observed a robust but apparently stable population of deer in unmanaged forests coexisting with wolves and mountain lions.

That Marsh's proto-ecological insights lie at the fountainhead of Resourcism is ironic, because ecology was not among the sciences that Resourcism inspired and which, in turn, informed it. For lack of anything better, we might term the suite of Resourcist sciences "Newtonian Biology." Resource-management sciences focus on getting the maximum sustainable yield out of various species—ponderosa pine, rainbow trout, whitetail deer—largely in isolation from one another. Leopold (1939) noted the futility of this conception of conservation in his plenary address to the joint meeting of the Society of American Foresters and the Ecological Society of America. "Conservation," he said,

> introduced the idea that the more useful wild species could be managed as crops, . . . [but] utility attached to species rather than to any collective total of wild things. . . . The emergence of ecology has placed the economic biologist in a peculiar dilemma: with one hand he points out the accumulated findings of his search for utility, or lack of utility, in this or that species; with the other he lifts the veil from a biota so complex, so conditioned by interwoven cooperations and competitions, that no man can say where utility begins or ends.

By the time Leopold had begun to view the world through the combined lenses of ecology and evolutionary biology, he had moved to Wisconsin and joined the faculty of the state's flagship university (Meine 1988). A big part of his job—as Professor of Game Management in the Department of Agricultural Economics in the College of Agriculture at a land-grant institution—was extension services. Leopold worked with Wisconsin farmers to grow wild game "crops" ancillary to domesticated crops (Leopold 1999). In this middle landscape between the streets of the city and the wilderness, Leopold began to think of conservation as a harmony between people and land.

Both Preservationism and Resourcism were thoroughly infected with residues of the prescientific Judeo-Christian worldview. Preservationism is imbued with the Judeo-Christian idea of a fallen humanity and an Edenic Nature. Resourcism assumes the Judeo-Christian idea of human privilege coupled with fiduciary responsibility—"stewardship," in a word. But both, however differently value-charged, assert the radical man-nature dualism foundational to Judeo-Christian belief. Further, according to Brian Balogh (2002), Pinchot's zeal for maximizing the utility and efficiency of natural resource exploitation was largely motivated by the "social gospel . . . [that] salvation lay in collective good works," which pervaded Anglo-American Protestant thought in the late nineteenth century. From the point of view of evolutionary biology, however, people are a part of nature. Thus, in sharp contrast to both Preservationism and Resourcism, Leopold's harmonization philosophy of conservation was nondualistic, and, more generally, reflected a Darwinian scientific rather than Judeo-Christian religious worldview. Leopold conceived the human economy to be a subset of the economy of nature. His basic idea was to reform the human economy so that it complements and enhances the larger economy of nature in which it is embedded, instead of disrupting and degrading it. He himself worked to reform the small-hold agricultural economy of south-central Wisconsin as a starting point.

From the Preservationist point of view, the conservation norm, standard, or ideal is wilderness, where, as David Brower once quipped, "the hand of man had never set foot" (Fox 1981). Such a standard, of course, is not an achievable conservation goal in places, like most of Wisconsin, that are humanly inhabited and economically exploited—short of evicting the human population or exterminating it, as happened to the Indians all over the continent. Alternatively, Leopold posited an essentially functional ecological norm for conservation that he called "land health" (Leopold 1999). Conservation—that is, a harmony between people and land—could be achieved when human use of land did not negatively affect such natural ecological functions as soil building and retention; water retention, purification, and stream-flow modulation; nutrient retention and cycling; damping the amplitude of animal population cycles; excluding invasive exotics; retaining native biodiversity.

Wilderness, in Leopold's harmonization philosophy, was assigned two new roles. First, not all wild animals can coexist with human beings; and among these are the most charismatic—the big, fierce predators. Wilderness serves conservation as the vital habitat of "threatened species" (Leopold 1936). Second, it can provide "a base-datum of normality" for land health (Leopold 1941). How do we know what the normal ecological functions are of an ecosystem that is humanly inhabited and economically exploited? By studying the same functions in a similar area that is not so inhabited and exploited, Leopold answered. As a Resourcist he had once argued that every state should have a designated "wilderness playground," for the convenience of impecunious wilderness recreationists (Leopold 1925). As a Harmonist he argued that every sort of ecosystem should have a representative wilderness set aside. These wilderness ecosystems could serve as scientific controls in comparison with which ecologists could measure the departure from normal function—ostensively defined by reference to the control ecosystems—of similar ecosystems that were humanly inhabited and economically exploited. In reference to such controls, conservationists could also measure their own success in restoring those functions to normality (Leopold 1941). Leopold proposed the very place of his epiphany, the Sierra Madre hunting grounds he visited in Mexico, as a base datum of normality, against which to measure the ecological dysfunctionality of the ecologically similar but manhandled US Southwest after it had experienced three quarters of a century of logging, grazing, dam-building, and irrigated agriculture (Forbes and Haas 2000).

THE PARADIGM SHIFT IN ECOLOGY

Historically, ecology has been a contentious science, rich in competing paradigms—the superorganism, the biotic community, the ecosystem, salient among them. But a deeper-running shift from a conception of nature in a state of equilibrium, undisturbed by humans, to a conception of nature constantly changing and disturbed by many natural forces is now largely complete.

THE BALANCE OF NATURE

Early-twentieth-century ecology was almost as metaphysical as it was scientific. F. E. Clements (1905, 1916), the most influential ecologist of his era, believed that "plant associations" were actually organisms of the third kind. He suggested, in other words, that the earliest organisms to evolve were single-celled; and that after many generations of close, symbiotic association, some single-celled organisms of different species, evolved into multi-celled organisms; and finally that these two levels of biological organization, in turn, formed close symbiotic associations and eventually became components of a third level of biological organization—"superorganisms"—which were the proper objects of ecological study. As cells are to multi-celled organisms, so multi-celled organisms are to superorganisms; and as organs are to multi-celled organisms, so species populations are to superorganisms. Consequently, ecology was then often characterized as a branch of physiology (Cittadino 1980, McIntosh 1983). Whereas traditional physiologists study the function and coordination of the various organs in organisms, the new ecological physiologists study the function and coordination of the components (species populations) of superorganisms. Clements's own speciality was plant succession which he conceived to be the developmental biology—the ontogeny—of superorganisms.

Serious discussion of superorganisms in ecology was accompanied by serious discussion of another putative ecological entity, the biotic community. Species perform roles or even have "professions," according to Charles Elton (1927), in the "economy of nature." Just as in the human economy there are butchers, bakers, candlestick makers, doctors, lawyers, and college professors, so in the metaphorical economy of nature some organisms are producers (plants), others consumers (animals), and still others are decomposers (fungi and bacteria). Each of these great guilds is divided into myriads of specialists: various herbaceous, aquatic, and woody plants; herbivorous, omnivorous, and carnivorous animals coming in all sizes and shapes (or taxa). And just as human socioeconomic communities can be sorted into types (and subtypes)—hunting/gathering communities (adapted to tropical forests, temperate savannas, or arctic tundras), rural/agrarian communities (practicing horticulture, hydraulic agriculture, or aquiculture), industrial/manufacturing communities (making textiles, paper, or heavy machinery)—so biotic communities can be sorted into types: forests (tropical rainforests, mixed temperate hardwoods, boreal softwoods), grasslands (long-grass prairies, short-grass prairies, steppes), wetlands (swamps, marshes, bogs).

Exasperated by both the metaphysical and metaphorical tendencies of his discipline, Arthur Tansley (1935) suggested a nonmetaphorical organizing concept for ecology. He argued that the proper objects of ecological study should be entities he called "ecosystems." He was sharply critical of Clements's superorganism concept, especially as it was mannerized by John Phillips (1934, 1935a, 1935b), a South African ecologist enthralled with the holistic speculations of his statesman/philosopher countryman, Jan Smuts. However nonmetaphorical, Tansley's ecosystem concept was almost as metaphysical as Clements's superorganism concept. He denied that ecosystems were third-order organisms, but conceded that "mature well-integrated . . . plant associations had enough of the characters of organisms to be considered as quasi-organisms . . . [and that] a mature complex plant association is a very real thing" (Tansley 1935). These very real quasi-organismic ecosystems were "relatively stable," had "a more or less definite structure," and exhibited "dynamic equilibrium," according to Tansley (1935); in them could be found an "inter-relation of parts adjusted to exist in the given habitat and to coexist with one another."

The ecosystem concept was greatly developed by Raymond Lindeman (1942), who measured the amount of solar energy captured by the primary producers of an aquatic ecosystem and by the organisms at each succeeding trophic level (herbivorous, omnivorous, and carnivorous). Soon thereafter, the primary fields of ecosystem studies came to be how energy flows through and nutrients cycle within ecosystems. But even in these quantitative thermodynamical and chemical ecosystem studies, Clements's organismic idea was latently present. Lindeman's mentor, G. E. Hutchinson (1940), for example, characterized the study of energy flows through ecosystems as the study of their "metabolism."

The ecosystem concept was further greatly developed by Eugene Odum. Like Tansley, Odum (1969) accepted Clements's belief that ecological succession proceeds through an orderly and predictable series of steps to a "climax" community. For example, in well-watered temperate climates, opportunistic, invasive, herbaceous grasses and weeds are followed by woody bushes and shrubs, which are followed by sun-loving, fast-growing, short-lived trees, which are followed finally by shade-tolerant, long-lived trees that reproduce themselves in perpetuity. He then argued that as succession proceeds to climax, photosynthetic energy is increasingly allocated away from production of biomass to "maintenance" of the ecosystem—"namely, increase control of, or homeostasis with, the physical environment in the sense of achieving maximum protection from its perturbations" (Odum 1969). Odum in fact revived Clements's original superorganism concept, but in a much more scientifically sophisticated guise.

The balance-of-nature paradigm in ecology is characterized both by typology and teleology. Like individual members of a species, each superorganism or ecosystem is a member of a type—a juniper/ponderosa-pine forest, a tall-grass prairie, a tamarack/sphagnum-moss bog. Daniel Simberloff (1982) critically traces this mode of ecological thought through Linnaeus and Newton all the way back to Plato. As to teleology, an ecosystem will exhibit linear development through the same successional seres as other members of its type toward a single equilibrial end-point, the climax. Tansley (1935), Hutchinson (1959), and Odum (1969), even believed ecosystems to be sufficiently robust as biological entities to have evolved by natural selection—which favored ecosystems that could achieve homeostasis, resistance to perturbation, "symbiosis, nutrient conservation, stability, a decrease in entropy, and an increase in information" (Odum 1969). Hutchinson's evolutionary theory of ecosystems was intimately connected to the then prevailing belief that stability was proportional to diversity. Because biotic communities with the greatest diversity are the most stable, Hutchinson (1959) reasoned, they would outcompete less stable ones in the struggle for existence. He thought this explained the great diversity of life. In the balance of nature paradigm, moreover, ecosystems are conceived to be "closed"—open only to sunlight, rainfall, and air currents; otherwise they are supposed to be resistant to invasion by outside organisms, and to be internally self-regulating, through such thermostat-like negative feedback relationships as predator-prey population dynamics, and production-respiration equilibria. Further, in the balance-of-nature ecological paradigm, natural disturbances are regarded as exceptional events that are external to ecosystems. And finally, human beings are also regarded as disturbing agents external to ecosystems. Thus, the proper objects of ecological study are ecosystems that are substantially free of anthropogenic effects.

THE FLUX OF NATURE

Henry Gleason (1926) was an early critic of Clementsian ecology. He argued that while various examples of a putative type of plant community often looked alike, upon closer, quantitative inspection, the composition of each example differed so greatly from other examples of the same putative type as to confound organization by means of such a typology. Moreover, the spatial boundaries between putative types of communities were often so fuzzy that it was difficult to declare where one began and the other ended, Gleason (1926) noted, and the same was also true of the temporal boundaries between successional seres. Thus, Gleason (1926) asks rhetorically, "Are we not justified in coming to the general conclusion, far removed from the prevailing opinion, that an association is not an organism, scarcely even a vegetational unit, but merely a *coincidence?*" Gleason offered his own alternative explanation of vegetational associations. Each species is individually adapted to a suite of environmental parameters or gradients—soil pH and nutrients, temperature, moisture, light intensity and duration, herbivory, and so on. Plants that are frequently found together in the same association are those that are individually adapted to similar suites of environmental parameters. The unique—that is, type-defying—mix of species in a given association is attributable to several factors. First, the peculiar combination of environmental parameters at various generally similar sites will often differ. For example, while the temperature and moisture regimes of two generally similar sites may be nearly the same, the soils may differ. Second, plants living at the margins of their climatic, edaphic, or other environmental parameters may be present, but more vulnerable to competitive exclusion by those for whom the same conditions are optimal. Third, the plants one finds at a given site will be those whose "propagules" (seeds or spores) chanced to have found their way there.

Gleason was virtually ignored in his own day, but by mid-century his observations began to be confirmed and his theory of stochastic propagation and coincidental individualistic adaptation to environmental gradients began to be believed (Curtis and McIntosh 1951, Whittaker 1951). Gleason's individualistic model of vegetational groupings was reinforced by a growing consensus in evolutionary biology that rejected "group selection" (Williams 1966). Lacking inheritable DNA, typological biotic communities and ecosystems are not the sort of entities that can evolve by natural selection toward greater and greater internal "symbiosis, nutrient conservation, stability, a decrease in entropy, and an increase in information." Decisive support for the individualistic view came from palynology. As Tansley (1935) graphically put it, "If a continental ice sheet slowly and continuously advances or recedes over a considerable period of time all the zoned climaxes which are subjected to decreasing or increasing temperature will, according to Clements's conception, move across the continent 'as if they were strung on a string.'" But an analysis of pollens preserved in pond sediments indicates that the forest communities now existing assembled recently—no more than 5000 years ago—and that the trees that compose them migrated from different Pleistocene refugia, in different directions, and at different rates (Davis 1984).

A better appreciation of disturbance in ecology complemented the latter-day triumph of Gleason over Clements. Disturbance of biotic communities and ecosystems by such forces as fire, flood, wind, disease, and pestilence has come to be regarded as frequent and routine—so much so that ecologists began to recognize the existence of "disturbance regimes" (Pickett and White 1985). Disturbances at smaller temporal and spatial scales are often abnormal, external, and destructive, but at larger scales become "incorporated"—periodic, internal, and benign. Thus at a spatial scale of 250 acres and a temporal scale of 10 years, fire in a ponderosa-pine forest is abnormal, external, and destructive, but at a spatial scale of 25,000 acres and a temporal scale of 100 years periodic fire is itself as much a necessary and beneficial internalized ecological process as succession or nutrient cycling.

The ecological myth that diversity and stability are positively correlated was thoroughly debunked by Daniel Goodman (1975). Ecosystem processes are carried out by a relatively few "driver" species and the rest are more or less expendable "passengers," although ecology is not so exact a science that ecologists can know for sure which are which and passengers become drivers and vice versa under changed circumstances (Walker 1992). Moreover, actual populations of predators and their prey do not obey the mathematical laws that they were supposed to obey in the balance-of-nature representation, such that their populations fluctuate only slightly around a point of equilibrium (Botkin 1990). They might even exhibit mathematically chaotic behavior (May 1974, Worster 1990).

As open, ill-bounded, disturbance-ridden, directionlessly dynamic quasi-entities, ecosystems are hardly stable, that is, in any state of equilibrium whatsoever. Rather they may temporarily settle into a "domain of ecological attraction" and then suddenly "flip" to another domain in response to the vagaries of the disturbances and invasions to which they are continually subjected (Holling 1973). For example, in response to anthropogenic fire suppression and grazing by domestic livestock, the arid Southwest region of the United States flipped from a mosaic of grasslands at lower elevations and forests at higher to one in which the herbaceous grasses were replaced by woody brush and forests expanding downhill (Leopold 1924). And, for another example, in response to heavy commercial fishing and the creation of the Welland and Erie canals, which opened an invasion route from the Atlantic Ocean, the Laurentian Great Lakes flipped from biotic communities dominated by deep-water ciscoes and lake trout to ones dominated by alewife and sea lamprey invaders (Ashworth 1986).

Finally, the wilderness myth has been debunked (Callicott and Nelson 1998). *Homo sapiens* has been such a ubiquitous ecological presence for so many thousands of years that anthropogenic disturbances have been long incorporated into all the world's ecosystems (with the exception of those of Antarctica). For example, the great prairies covering the broad mid-longitudes of North America were historically maintained by anthropogenic fire (Pyne 1982); and the pre-Columbian ungulate populations of the intermountain West of North America were kept low by cultural predation (Kay 1994, 1995).

S. T. A. Pickett and R. S. Ostfeld (1995) summarize the new, consolidated flux-of-nature paradigm in ecology:

> Ecological systems are never closed, but rather experience inputs such as light, water, nutrients, pollution, migrating genotypes, and migrating species. . . . Stable equilibria are rare, although some systems of sufficient size and duration may exhibit stable frequency distributions of states. For example, a landscape may be a shifting mosaic of patches or community types, and in some cases, the number of young and old commu-

> nities can remain constant, even though specific spots change as a result of disturbance and succession. Successions are rarely deterministic, but are affected by specific histories, local seed sources, herbivores, predators, and diseases. Disturbance is a common component of ecological systems, even though some sorts of disturbance are not frequent on a scale of human lifetimes. . . . And finally, landscapes that have not experienced important human influences have been the exception for hundreds if not thousands of years.

IMPLICATIONS OF THE FLUX-OF-NATURE PARADIGM IN ECOLOGY FOR RECONSTRUCTING CONSERVATION PHILOSOPHY

Because the philosophy of conservation is informed by ecology, the shift from the equilibrium to the non-equilibrium paradigm in ecology—from the balance of nature to the flux of nature—requires a thorough review and reorganization of conservation philosophy.

CAVEATS AND QUALIFICATIONS

First of all, we must be careful not to throw out the ecological baby with the ecological bath water. Ecosystems may not be organisms or even "quasi-organisms." And the problem of definitively determining an ecosystem's boundaries may be so intractable as to require a postmodern resolution—an ecosystem is ecologically constructed by the questions ecologists ask and will be differently reified by different questions (Allen and Hoekstra 1992). Still, the ecosystem concept has not gone the way of phlogiston and the luminous ether in physics; it remains alive and well in ecology. Organisms may not be as tightly linked and functionally integrated in ecosystems as organs are in organisms, nor is every organism in an ecosystem strongly and equally connected to every other, but all are dependent on some others and none are wholly self-sufficient. Donald Worster (1994), who helped popularize what he calls the new "ecology of chaos," asserts "the principle of interdependency: . . . No organism or species of organism has any chance of surviving without the aid of others." And because actual predator-prey population dynamics may not be very accurately described by the Lotka-Volterra Logistic equation (Botkin 1990), and because in some cases such dynamics may exhibit mathematically chaotic behavior (May 1974), we should not leap to the conclusion that prey populations are not at all affected by predators and vice versa. The classical models may be too simplistic, but it's a question of developing more sophisticated descriptions, not of denying any relationship at all because the classically posited relationship is not confirmed by observation or experiment. Finally, while ecological succession may not follow a deterministic path toward a fixed climax and the temporal boundaries between seres may be blurry, ecological succession, however stochastic and variable, does occur in nature.

Robert McIntosh (1998), a leading neo-Gleasonian, offers the following caveat about too readily jumping to the conclusion that there is no ecological order in nature, however complex and disequilibrial it may now appear to be:

> The implication of anarchy, or lack of any order, is a common misrepresentation of Gleason's individualistic concept, which some have erroneously said is a random assemblage of species lacking any relations among the species. Neither Gleason nor any of his successors ever said that. Not *all* things are possible in an individualistic community, only some. The resulting pattern is more elusive than in a purported organismic community, but it is certainly not anarchy or random. . . . It is doubtful . . . that any ecologist envisioned a community as a merely chance aggregation of organisms and environment lacking discernible pattern. Gleason and his successors recognized patterns of gradual change of species composition in space and time, in contrast with the putative patterns of change of integrated groups of organisms.

Pickett and Ostfeld (1995), the leading architects of the flux-of-nature ecological paradigm offer a similar caveat:

> [T]he balance-of-nature metaphor can stand for some valid scientific ideas. The fundamental truth about the natural world that the idea may relate to is the fact that natural systems persist, and they do so by differential response to various components. The idea also points toward the ecological principle that there are limitations in natural systems. No component of a natural ecological system grows without limit. . . . Examples are density-dependent processes (i.e., the tendency of populations to grow when small and shrink when large) and the existence of successional trajectories.

These qualifications and caveats having been registered, what are the implications of the flux-of-nature paradigm in ecology for reconstructing conservation philosophy?

IMPLICATIONS OF THE FLUX-OF-NATURE PARADIGM FOR THE THREE TWENTIETH-CENTURY PARADIGMS IN THE PHILOSOPHY OF CONSERVATION

The new flux-of-nature paradigm in ecology creates different problems for the three historical paradigms in conservation philosophy.

PRESERVATIONISM

Most obviously, it forces rethinking preservationist policy. First, in an ever dynamic and non-teleological nature, what should we preserve? There are no "original" states of nature—no self-reproducing climax communities that will persist in perpetuity if only people do not disturb them—just multiple historic states of nature, temporarily persisting domains of ecological attraction. Nor are any ecologically recent historic states of nature free of anthropogenic influence; none are "pristine" any more than they are original. Preservationists therefore must consciously develop and defend criteria for determining which historic states of nature should be selected as worthy of preservation. And if they are to remain in the states that are selected, historic natural *and* anthropogenic disturbances will have to be simulated. More generally these inherently dynamic, ecologically open preserves will have to be actively and sometimes aggressively managed to hold back or redirect succession and to prevent invasion by weedy exotic species.

RESOURCISM

Resourcism may appear to be vindicated because nature is now thought to be less well integrated and organized than ecologists of Leopold's day believed. But, as just noted, the magnitude of the current paradigm shift in ecology can be exaggerated misleadingly. Single species of great utilitarian interest cannot be managed without regard to their relationships with other species and to the organization of the biotic communities and functionality of the ecosystems in which they exist.

HARMONIZATION

The virtual abandonment of the organismic model of ecosystems forces rethinking the harmonization conservation norm—land (or ecosystem) health—for only organisms can be said literally to be healthy. Frank recognition that land (or ecosystem) health is a metaphor combined with a clear articulation of the ecological conditions that the metaphor comprises might rescue this norm (Callicott 1992, 1995).

TOWARD AN INTEGRATED TWENTY-FIRST-CENTURY PHILOSOPHY OF CONSERVATION

The coincidence between the consolidation of the new flux-of-nature paradigm in ecology and the appearance of biodiversity as the norm for the new crisis transdisciplinary science of conservation biology in the mid-1980s is intriguing (Wilson 1988). Obviously, the felt need to conserve biodiversity is a response to the realization, growing more acute over the last quarter of the twentieth century, of a global species-extinction crisis. But is there more to it than that? Perhaps the neo-Gleasonian aspect of the new paradigm in ecology shifted concern away from conserving problematic ecological entities such as biotic communities and ecosystems and focused it instead on conserving individual species, regardless of either their utility as resources or ecological functionaries.

Clearly—in view of the fact that the Earth is fixing to endure only the sixth abrupt mass extinction event in its 3.5 billion year biography—the conservation of biodiversity should remain a central focus of conservation efforts and a cornerstone of any new, integrated conservation philosophy (Wilson 1992). Biodiversity conservation may provide a means of integrating preservationism into a new twenty-first-century philosophy of conservation through an answer to the central question forced on preservationists by the flux-of-nature paradigm in ecology—what should we preserve? Habitat for threatened species is certainly a leading candidate for Preservationist priority. Classic preservationism's beleaguered wilderness ideal might be replaced by the concept of a biodiversity reserve. Because of its long and tangled history and reversing value polarities, the very idea of wilderness preservation is confusing. Does wilderness exist for higher Transcendental uses by elect, materially vaporous human beings; or for noisy, virile outdoor recreation by sportsmen and -women; or as land laboratories for ecological study; or as habitat for species (such as interior obligate species of birds) that cannot adapt to urban, suburban, and exurban human disturbances or those (such as bears, wolves, and mountain lions) that do not coexist well with human habitations? Perhaps, in the "multiple use" spirit of Resourcism, wilderness might be preserved for all of these reasons. But calling preserved areas "biodiversity reserves" makes their highest use clear and unambiguous—habitat for threatened species. Other uses—Transcendental, recreational, and scientific—might be made of them as well, but only to the extent that such uses are compatible with and subordinate to their highest use.

If directionless successional change characterizes nature, as fluxy ecology would have it, to keep habitat fit for threatened species in biodiversity reserves requires constant management. This is a point of focus for integrating Resourcism into a new twenty-first-century philosophy of conservation. Resourcists traditionally focused on managing single species of great utilitarian value or interest. The lore of traditional forestry and wildlife and fishery management is being redirected to the conservation of threatened species irrespective of their utility as resources (Knight and Bates 1995). The only way to manage a species *in situ* is to manage its habitat, more particularly to try to preserve or restore the optimal environmental gradients in the habitat to which it is adapted. From the point of view of the new flux-of-nature paradigm, the principal way to do that is through the judicious manipulation or the fine-tuning of disturbance regimes (Pickett and White 1985). For example, to conserve native fishes of the Colorado River, such as the Colorado squawfish and humpback chub, whose populations had been declining since the Glen Canyon dam was constructed in 1963, fishery managers, who are now familiar with the concept of disturbance regimes, believe it is necessary to simulate the annual Colorado springtime flood by a pulsed high-volume release of water from the dam (Osmundson 2001). The first such simulated flood occurred in 1996 with successful results. The classic example is the conservation of the Kirtland's warbler which can nest only in immature stands of jack pine. But jack pine reproduction requires fire to open the tree's seeds. So only by simulating the historic fire regime in pine barrens can conservationists manage the vital habitat of the Kirtland's warbler and thus conserve the species (Kirtland's Warbler Recovery Team 1985).

The current emphasis in ecology on disturbance regimes and ecological processes in general has recently focused attention on ecosystem services—such as pollination, water purification, and regional climate modulation. Ecosystem services produce at least as much human utility as traditional natural resources—such as lumber and pulp, sport and meat—which we may reconceive as ecosystem goods (Costanza et al. 1997). Thus conserving them can be regarded as an up-dated Resourcist project. Biodiversity reserves are necessary for maintaining viable minimal metapopulations of threatened species that do not coexist well with human habitations and landscape fragmentations. And they may be sufficient for doing so if they are many, large, connected, judiciously located, well designed, and well managed (Noss and Cooperrider 1994). Such reserves are also necessary for maintaining ecosystem services, but are not sufficient for doing so. Pollination, for example, is as needed in farmed and gardened rural and urban landscapes as in national parks and forests; and few biodiversity reserves will be large enough to maintain favorable regional climatic conditions, or prevent soil erosion and flashy hydrodynamics. Conserving ecosystem services may thus be a point of focus for integrating classic Harmonization into a new twenty-first-century conservation paradigm. We might begin to explore ways in which human economic activities can be reformed so as not to be disruptive of vital ecosystem services, such as the retention, modulation, and purification of surface and ground waters. Indeed, conserving ecosystem services may so overlap or coincide with conserving land or ecosystem health that we can dispense with the latter, ecologically more problematic, term altogether, without any substantive (or pragmatic) change in the Harmonization philosophy of conservation.

Because Leopold believed that land health was positively correlated with native species diversity, his practical approach to harmonization did not match his conception. In fact his practice was to scale down the neo-Preservationist agenda. At best, the farmers with whom Leopold worked would, he supposed, continue to plow and plant, harvest and graze as they always had done. He mainly asked farmers to resist the ever-escalating industrialization of agriculture; the use of chemical fertilizers, pesticides, and herbicides; and such efficiency-inspired and production-driven modifications of the rural landscape as stream straightening and wetland draining. And he also asked them to farm as mindfully of soil and water conservation as conventional farming methods would permit. Finally, and most important, he urged the farmers with whom he worked to dedicate portions of their smallhold lands—fencerows, roadsides, woodlots, stream corridors, ponds, marshes, and bogs—to wildlife habitat (Leopold 1999). It was mostly in these "waste" lands (miniature biodiversity preserves) on the rural farmstead that native species would maintain land health or ecosystem services.

Leopold's vision of a traditional family farm oriented in these ways to conservation is still certainly laudable and beautiful (Leopold 1939). But the abandonment of the diversity-stability hypothesis in ecology frees us to think that ecosystem services may be maintained in ways supplementary to maintaining native species diversity. Wes Jackson (1987), for example, envisions a radical shift in agricultural emphasis from annual monocultures (of wheat, maize, soy beans, and the like) to "perennial polycultures." He and his associates are working to develop a mix of four groups of perennial plants—cool season grasses, warm season grasses, legumes, and sunflowers—bred to produce harvestable, edible, and processible seed. After the initial planting, no tillage would be necessary. In addition to contributing to the crop, the legumes would fix nitrogen from the air and make it available to the non-legumes as well; and the sunflowers would pull up minerals and moisture from deep in the soil, thus reducing or eliminating the need for artificial fertilizers and irrigation. Because the soil is not tilled annually, use of perennial polycultures may conserve many of the most valuable elements of ecosystem health and services—soil retention, nutrient recruitment and cycling, water-flow modulation and water purification, most obviously. While Jackson's model for perennial polycultures is the structure of native prairie communities, the perennial polycultures themselves would be highly artificial, consisting of select prairie plants (such as cutleaf sylphium, big bluestem, and lupine) bred to produce edible, harvestable, and processible seed or traditional seed-bearing crops (such as maize, wheat, and soy beans) bred to be perennials.

In conclusion, then, a viable reconstructed twenty-first-century philosophy of conservation would consist of an integration of central features of the three twentieth-century schools of conservation, informed and transformed by the contemporary flux-of-nature paradigm in ecology. Consistent with the neo-Gleasonian focus on individual species, a viable twenty-first-century philosophy of conservation would put a premium on the preservation of biodiversity, even if diversity is not vital to stability (which is problematic in any case). Existing systems of national parks and wilderness areas—the legacy of twentieth-century preservationism—might be enlarged, supplemented and connected by additions, and reconceived as biodiversity reserves. Complementing these preserves, we might work—in the spirit of twentieth-century harmonization—to reform human habitation and economic exploitation of lands outside biodiversity reserves so as to degrade their ecosystem health and services as little as possible. Both biodiversity reserves and the lands outside reserves that are productive of both goods (resources) and ecosystem services will require active management—the legacy of twentieth-century resourcism—largely by means of simulating or fine-tuning historic disturbance regimes.

REFERENCES

Allen, T. F. H. and T. Hoekstra. 1992. *Toward a unified ecology.* Columbia University Press, N.Y.

Ashworth, W. 1986. *The late, Great Lakes: An environmental history.* Knopf, New York.

Balogh, B. 2002. Scientific forestry and the roots of the modern American state: Gifford Pinchot's path to Progressive reform. *Environmental History* 7:198–225.

Botkin, D. B. 1990. *Discordant harmonies: A new ecology for the twenty-first century.* Oxford University Press, New York.

Callicott, J. B. 1992. Aldo Leopold's metaphor. Pages 42–56 in R. Constanza, B. G. Norton, and B. D. Haskell, editors. *Ecosystem health: New goals for environmental management.* Island Press, Washington, D.C.

Callicott, J. B. 1994. The value of ecosystem health. *Environmental Values* 5:345–361.

Callicott, J. B. and M. P. Nelson. 1998. *The great new wilderness debate.* University of Georgia Press, Athens.

Cittadino, E. 1980. *Ecology and the professionalization of botany in America, 1890–1905*. Studies in the History of Biology 4:171–198.

Clements, F. E. 1905. *Research methods in ecology.* University Publishing Company, Lincoln, Neb.

Clements, F. E. 1916. *Plant succession: An analysis of the development of vegetation.* Publication No. 242. Carnegie Institution, Washington, D.C.

Costanza, R. R. d'Arge, R. de Groot, S. Farber, M. Grasso, B. Hannon, K. Limberg, S. Naeem, R. V. O'Neill, J. Paruelo, R. G. Raskin, P. Sutton, and M. van den Belt. 1997. The value of the world's ecosystem services and natural capital. *Nature* 387:253–260.

Crosby, A. W. 1976. Virgin soil epidemics as a factor in the aboriginal depopulation of America. *William and Mary Quarterly* 33:289–299.

Curtis, J. T. and R. P. McIntosh. 1951. An upland forest continuum in the prairie-forest border region of Wisconsin. *Ecology* 32:476–496.

Davis, M. B. 1984. Climatic instability, time lags, and community disequilibrium. Pages 264–284 in J. Diamond and T. J. Case, editors. *Community ecology.* Harper and Row, New York.

Denevan, W. M. 1992. Native American populations in 1492: Recent research and a revised hemispheric estimate. Pages xvii–xxix in W. M. Denevan, editor. *Native populations of the Americas in 1492*, 2nd edition. University of Wisconsin Press, Madison.

Denevan, W. M. 1996. Pristine myth. Pages 1034–1036 in D. Levinson and M. Ember, editors. *Encyclopedia of cultural anthropology,* vol. 3. Henry Holt and Company, New York.

Edwards, J. 1978. *Basic writings.* New American Library, New York.

Elton, C. 1927. *Animal ecology.* Sidgewick and Jackson, London.

Emerson, R. W. 1836. *Nature.* James Monroe and Company, Boston.

Flader, S. L. 1974. *Thinking like a mountain: Aldo Leopold and the evolution of an ecological attitude toward deer, wolves, and forests.* University of Missouri Press, Columbia.

Forbes, W. and T. Haas. 2000. Leopold's legacy in the Rio Gavilan: Revisiting an altered Mexican wilderness. *Wild Earth* 10:61–67.

Fox, S. 1981. *John Muir and his legacy: The American conservation movement.* Little, Brown and Company, Boston.

Gleason, H. A. 1926. The individualistic concept of the plant association. *Bulletin of the Tory Botanical Club* 53:7–26.

Goodman, D. 1975. The theory of diversity-stability relationships in ecology. *Quarterly Review of Biology* 50:237–266.

Holling, C. S. 1973. Resilience and stability of ecology systems. *Annual Review of Ecology and Systematics* 4:1–23.

Hutchinson, G. E. 1940. Review of *Bio-ecology. Ecology* 21:267–268.

Hutchinson, G. E. 1959. Homage to Santa Rosalie: Or, why there are so many kinds of animals. *American Naturalist* 93:145–159.

Jackson, W. 1987. *Altars of unhewn stone: Science and the earth.* North Point Press, San Francisco.

Kay, C. 1994. Aboriginal overkill: The role of Native Americans in structuring Western ecosystems. *Human Nature* 5:359–398.

Kay, C. 1995. Aboriginal overkill and native burning: Implications for modern ecosystem management. *Western Journal of Applied Forestry* 10:121–126.

Kirtland's Warbler Recovery Team. 1985. *Kirtland's warbler recovery plan.* US Fish and Wildlife Service, Twin Cities, Minn.

Knight R. L. and S. F. Bates, editors. 1995. *A new century for natural resources management.* Island Press, Washington, D.C.

Leopold, A. 1918. Forestry and game conservation. *Journal of Forestry* 16: 404–411.

Leopold, A. 1924. Grass, brush, timber, and fire in southern Arizona. *Journal of Forestry* 22(6):1–10.

Leopold, A. 1925. Wilderness as a form of land use. *Journal of Land and Public Utility Economics* 1:348–350.

Leopold, A. 1936. Threatened species: A proposal to the Wildlife Conference for an inventory of the needs of near-extinct birds and animals. *American Forests* 42:116–119.

Leopold, A. 1939. The farmer as a conservationist. *American Forests* 45:294–299, 316, 323.

Leopold, A. 1941. Wilderness as a land laboratory. *Living Wilderness* 6:3.

Leopold, A. 1999. *For the health of the land: Previously unpublished essays and other writings.* J. B. Callicott and E. T. Freyfogle, editors. Island Press, Washington, D.C.

Lindeman, R. L. 1942. The trophic-dynamic aspect of ecology. *Ecology* 23: 399–418.

Lowenthal, D. 1958. *George Perkins Marsh: Versatile Vermonter.* Columbia University Press, New York.

Lowenthal, D. 2000. *George Perkins Marsh: Prophet of conservation.* University of Washington Press, Seattle.

McIntosh, R. P. 1983. Pioneer support for ecology. *BioScience* 33:107–112.

McIntosh, R. P. 1998. The myth of community as organism. *Perspectives in Biology and Medicine* 41:426–438.

May, R. M. 1974. Biological populations and nonoverlapping generations: Stable points, stable cycles, and chaos. *Science* 186:645–647.

Marsh, G. P. 1864. *Man and nature: Or, physical geography as modified by human action.* Charles Scribner, New York.

Meine, C. 1988. *Aldo Leopold: His life and work.* Madison, University of Wisconsin Press.

Miller, C. 2001. *Gifford Pinchot and the making of modern environmentalism.* Island Press, Washington, D.C.

Nash, R. *Wilderness and the American mind.* 1967. Yale University Press, New Haven, Conn.

Noss, R. F. and A. Y. Cooperrider. 1994. *Saving nature's legacy: Protecting and restoring biodiversity.* Island Press, Washington, D.C.

Odum, E. P. 1969. The strategy of ecosystem development. *Science* 164: 260–270.

Osmundson, D. B. 2001. *Flow regimes for restoration and maintenance of sufficient habitat to recover endangered razorback sucker and Colorado pikeminnow in the Upper Colorado River: Interim recommendations for the Palisade-to-Rifle Reach: Final report.* Colorado River Fishery Project, US Fish and Wildlife Service, Grand Junction, Colo.

Phillips, J. 1934. Succession, development, the climax and the complex organism, I. *Journal of Ecology* 22:554–571.

Phillips, J. 1935a. Succession, development, the climax and the complex organism, II. *Journal of Ecology* 23:120–246.

Phillips, J. 1935b. Succession, development, the climax and the complex organism, III. *Journal of Ecology* 23:554–571.

Pickett, S. T. A. and P. S. White, editors. 1995. *The ecology of natural disturbance and patch dynamics.* Academic Press, Orlando, Fla.

Pickett, S. T. A. and R. S. Ostfeld. 1995. The shifting paradigm in ecology. Pages 261–277 in R. L. Knight and S. F. Bates, editors. *A new century for natural resources management.* Island Press, Washington, D.C.

Pinchot, G. 1947. *Breaking new ground.* Harcourt, Brace, New York.

Plumwood, V. 1998. Wilderness skepticism and wilderness dualism. Pages 652–690 in J. B. Callicott and M. P. Nelson, editors. *The great new wilderness debate.* University of Georgia Press, Athens.

Pyne, S. J. 1982. *Fire in America: A cultural history of wildland and rural fire.* University of Washington Press, Seattle.

Simberloff, D. 1982. A succession of paradigms in ecology: Essentialism to materialism and probabilism. Pages 63–99 in Esa Saarinen, editor. *Conceptual issues in ecology.* Reidel, Boston.

Tansley, A. G. 1935. The use and abuse of vegetational concepts and terms. *Ecology* 16:284–307.

Thoreau, H. D. 1854. *Walden, Or life in the woods.* Ticknor and Fields, Boston.

Thoreau, H. D. 1970. *Huckleberries.* University of Iowa Press, Iowa City.

Turner, F. J. 1920. *The frontier in American history.* Henry Holt and Company, New York.

Walker, B. H. 1992. Biological diversity and ecological redundancy. *Conservation Biology* 6:18–23.

Whittaker, R. H. 1951. A criticism of the plant association and climatic climax concepts. *Northwest Science* 25:18–31.

Williams, G. C. 1966. *Adaptation and natural selection.* Princeton University Press, Princeton, N.J.

Wilson, E. O. 1988. *Biodiversity.* National Academy Press, Washington, D.C.

Wilson, E. O. 1992. *The diversity of life.* Harvard University Belknap Press, Cambridge, Mass.

Worster, D. 1988. Review of Michael P. Cohen, *The pathless way: John Muir and American wilderness. Environmental Ethics* 10:277–281.

Worster, D. 1990. The ecology of order and chaos. *Environmental History Review* 14:1–18.

Worster, D. 1994. *Nature's economy: A History of ecological ideas,* 2nd edition. Cambridge University Press, Cambridge, Mass.

Wendell Berry

Hell, No. Of Course Not. But . . . (2001)

I MADE A SORT OF VOW to myself some time ago that I wouldn't support any more efforts of wilderness preservation that were unrelated to efforts to preserve economic landscapes and their human economies. One of my reasons is that I don't think we can preserve either wildness or wilderness areas if we can't preserve the economic landscapes and the people who use them.

If the survival of the Arctic National Wildlife Refuge is now in crisis, should I make an exception? Well, maybe so. Do I want that refuge to be opened to oil exploration now that the Democrats, those redoubtable nature-protectors, are out of the way? Hell, no. Of course not. I would hate to see Alaska raped by the lords of timber and energy, as large sections of my own state have been. And so I add my vote to the votes of all the others who will be saying no.

But do I think that if we no-sayers "save" the refuge from the present threat we will have saved it? I will have to vote no again, and for the same reasons that I made my vow in the first place. You can't save wilderness preserves, refuges, and parks, if at the same time you let the economic landscapes and the land-using economies go to the devil. I can't look at the crisis of the Arctic Wildlife Refuge except as the result of a radical failure of the conservation movement over the last fifty or so years: its refusal to see that conservation as we have known it is not an adequate response to an economy that is inherently wasteful and destructive; its apparent belief that nature or wildness can be preserved merely by preserving wilderness; its inability to connect wilderness conservation with soil conservation or energy conservation or any form of frugality; its cherished contempt for ranchers, farmers, loggers, and other land users.

Suppose that fifty or sixty years ago conservationists had seen fit to cherish and protect that wildness that existed on the millions of small farms and ranches that we had then. If they had done so, we would have a lot more wildness than we have now, and a lot more farmers and ranchers, and the conservationists would have a lot more friends, even in the government. And think of the wildness that still might be preserved and nurtured, and the anguish that still might be prevented, if conservationists could recognize and support such a possibility even now. Think how much petroleum might be saved if more people were eating food produced by local farmers or ranchers. If the entire food economy is entirely dependent on long-distance transport, how can we avoid drilling for oil wherever we might find it?

The Arctic Wildlife Refuge is under threat now because policy may go wrong, because of greed and ignorance in high places, because corporations have no conscience. All that is true. But a lot more is true than that. The Refuge is also under threat because we have no energy policy, no agricultural policy, and no forestry policy that is not keyed to consumption rather than conservation. Why do we not have better policies? Because there is no organized public demand.

For this, I think, conservationists must bear a generous portion of the blame. They have cared too little for landscapes that were not describable as "wilderness" or "open space." They have too thoughtlessly "benefited" from cheap food and cheap fuel. When I think of the threat to the Arctic Wildlife Refuge, I think also of conservationists and wilderness lovers who fly or drive thousands of miles to walk a few hours or days in a certified wilderness. We have got to think of something better. If we don't, the government won't.

Scott Russell Sanders

Wilderness as a Sabbath for the Land (2001)

IF YOU HONOR THE SABBATH in any way, or if you respect the beliefs of those who do, or if you merely suspect there may be some wisdom bound up in this ancient practice, then you should protect wilderness. For wilderness represents in space what the Sabbath represents in time—a limit to our dominion, a refuge from the quest for power and wealth, an acknowledgment that Earth does not belong to us.

In scriptures that have inspired Christians, Muslims, and Jews, we are told to remember the Sabbath and keep it holy by making it a day of rest for ourselves, our servants, our animals, and the land. This is a day free from the tyranny of getting and spending, a day given over to the cultivation of spirit rather than the domination of matter. During the remainder of the week, busy imposing our will on things, we may mistake ourselves for gods. But on the Sabbath we recall that we are not the owners or rulers of this magnificent planet. Each of us receives life as a gift, and each of us depends for sustenance on the whole universe, the soil and water and sky and everything that breathes. The Sabbath is yet another gift to us, a respite from toil, and also a gift to the Earth, which needs relief from our appetites and ambitions.

Honoring the Sabbath means to leave a portion of time unexploited, to relinquish for a spell our moneymaking, our striving, our designs. Honoring wilderness means to leave a portion of space unexploited, to leave the minerals untapped, the soils unplowed, the trees uncut, and to leave unharmed the creatures that live there. Both wilderness and Sabbath teach us humility and restraint. They call us back from our ingenious machines and our thousand schemes to dwell with full awareness in the glory of the given world. By putting us in touch with the source of things, they give us a taste of paradise.

The instruction to honor the Sabbath appears as the fourth of the commandments announced by Moses after his descent from Mount Sinai, as reported in the Book of Exodus in the Hebrew Bible:

> Remember the Sabbath day, and keep it holy. Six days you shall labor and do all your work. But the seventh day is a Sabbath to the Lord your God; you shall not do any work—you, your son or your daughter, your male or female slave, your livestock, or the alien resident in your towns. For in six days the Lord made heaven and earth, the sea, and all that is in them, but rested the seventh day; therefore the Lord blessed the Sabbath day and consecrated it (Ex. 20: 8–11; all Biblical quotations from the New Revised Standard Version).

If the Lord quit shaping the Earth after six days, looked at what had been made, and saw that it was very good—as chronicled in the Book of Genesis—then who are we to keep on reshaping the Earth all seven days? On the Sabbath we are to lay down our tools, cease our labors, set aside our plans, so that we may enjoy the sweetness of *being* without *doing*. On this holy day, instead of struggling to subdue the world, we are to savor it, praise it, wonder over it, and commune with the Creator who brought the entire world into existence.

The Book of Deuteronomy provides another reason for resting on the Sabbath: "Remember that you were a slave in the land of Egypt, and the Lord your God brought you out from there with a mighty hand and an outstretched arm; therefore the Lord your God commanded you to keep the Sabbath day" (Deut. 5: 12–15). By reminding the Hebrew people of their own liberation from bondage, the Sabbath calls on them to re-enact that liberation every seventh day for the benefit of everyone and everything under their control.

Observing the Sabbath would not always have been easy for a farming people, as one can sense from another version of the commandment: "Six days you shall work, but on the seventh day you shall rest; even in plowing time and in harvest time you shall rest" (Ex. 34: 21). A delay in plowing or harvesting might mean the difference between a good crop and a poor one, so this was a severe discipline indeed. When I was a boy in rural Ohio some fifty years ago, I knew farmers who would not start a machine or harness a horse on the Sabbath, no matter the weather or the state of their crops. Nor would they take up saws or scythes to work by hand. The most they would do was walk the fields, scooping up handfuls of soil, inspecting corn or hay, listening for birds, all as a way of gauging the health of their place.

The link between honoring the Sabbath and honoring the Earth is spelled out elsewhere in Exodus:

> For six years you shall sow your land and gather in its yield; but the seventh year you shall let it rest and lie fallow, so that the poor of your people may eat; and what they leave the wild animals may eat. You shall do the same with your vineyard, and with your olive orchard. Six days you shall do your work, but on the seventh day you shall rest, so that your ox and your donkey may have relief, and your homeborn slave and the resident alien may be refreshed (Ex. 23: 10–12).

The great gift of the Sabbath is refreshment, renewal, a return to the state of wholeness. It is medicine for soil and spirit, a healing balm.

After every seventh cycle of seven years, according to the Book of Leviticus, the people of Israel were to celebrate the fiftieth year as a jubilee, when the land must be left fallow, all debts must be forgiven, all slaves and indentured servants must be freed, and all property must be returned to its original owners. "The land shall not be sold in perpetuity," God proclaims, "for the land is mine; with me you are but aliens and tenants" (Lev. 25: 23). This insistence that Earth belongs to God, not to humankind, echoes through the Bible, as in Psalm 24, which begins, "The earth is the Lord's and all that is in it, the world, and those who live in it; for he has founded it on the seas, and established it on the rivers" (Ps. 24: 1–2); or in Psalm 50, where God says, "I will not accept a bull from your house, or goats from your folds. For every wild animal of the forest is mine, the cattle on a thousand hills. I know all the birds of the air, and all that moves in the field is mine. If I were hungry, I would not tell you, for the world and all that is in it is mine" (Ps. 50: 9–12).

Whether celebrated every fiftieth year, every seventh year, or every seventh day, the Sabbath links an obligation to care for the poor—the great theme of Jesus and the Hebrew prophets—with an obligation to care for the land and all the creatures that depend on the land for shelter and food.

According to a pair of stories in the Gospel of Luke, Jesus embraced the liberating power of the Sabbath. Once, Jesus was teaching in a synagogue on the Sabbath when "there appeared a woman with a spirit that had crippled her for eighteen years. She was bent over and was quite unable to stand up straight." Jesus spoke to her and laid his hands on her, whereupon "she stood up straight and began praising God." When the Pharisees took him to task for healing on the day of rest, Jesus replied, "Does not each of you on the Sabbath untie his ox or his donkey from the manger, and lead it away to give it water? And ought not this woman, a daughter of Abraham whom Satan bound for eighteen long years, be set free from this bondage on the Sabbath day?" (Luke 13: 10–16). On another occasion, after curing a man of dropsy on the Sabbath, Jesus defended his action by asking the Pharisees, "If one of you has a child or an ox that has fallen into a well, will you not immediately pull it out on a Sabbath day?" (Luke 14: 5).

In both stories, Jesus interpreted the Sabbath as a day for the breaking of fetters. Instead of dwelling on what was forbidden, he dwelt on what was required—the relief of suffering, the restoring of health. The Gospel of Mark tells of another Sabbath when the Pharisees challenged Jesus for allowing his disciples to pluck heads of grain to relieve their hunger:

> And he said to them, "Have you never read what David did when he and his companions were hungry and in need of food? He entered the house of God, when Abiathar was high priest, and ate the bread of the Pres-

> ence, which it is not lawful for any but the priests to eat, and he gave some to his companions." Then he said to them, "The Sabbath was made for humankind, and not humankind for the Sabbath. . . ." (Mark 2: 25–27).

In that rousing last line, Jesus may seem to be turning the commandment on its head, yet he is actually recalling the spirit of freedom and jubilee implicit in the gift of the Sabbath. In his reading, the Sabbath becomes a foretaste of the kingdom of God, which is founded on compassion. Just as the universe, the Earth, and all living things arise out of the great unfolding of God, so the Sabbath is a reminder of this marvelous generosity. It is a day for deliverance not merely from toil but from whatever entraps us.

Our traps may be physical, as in the case of disease, but they may also be social or psychological. We may be trapped by poverty or by the relentless pursuit of wealth. We may be trapped by hatred or fear, by duties or lust. We may be trapped by the delusion that the world exists to satisfy our cravings. We may be trapped by addiction to chemicals or gadgets or noise. From all of these snares, and more, the Sabbath can help to release us.

And yet for most Americans, even those who attend church or synagogue or mosque, in recent decades the Sabbath has lost much of its serenity and nearly all of its meaning. Instead of being a day set aside for reflection and renewal, it has become a time for shopping, for catching up on chores, for watching television or movies, for mowing lawns or waxing cars, for burning up gas on the highways, for eating out or sleeping in. More and more jobs keep people on duty through the weekend. More and more stores, like the Internet, never close. Commerce and its minion, advertising, have spread around the clock, leaving scarcely any stretch of time unclaimed. In the same way, our machines and pollution have spread nearly everywhere on land and sea, leaving scarcely any stretch of Earth unclaimed.

This onslaught is squeezing out the wildness from our hearts and minds, as well as from the planet. The first word of the Sabbath commandment is *remember*—remember to rest, to limit your schemes, to relieve from toil all who depend on you. Remember that you were a slave and have been set free; remember that life itself is a gift from God; remember that Earth and its abundance belong to the Lord. Instead of remembering, we are quickly forgetting who we are, where we are, and how we ought to live.

In America today, the only lands with any chance of remaining wild are those we have deliberately chosen to protect. We need such lands, as we need respite from labor, as we need meditation and prayer, to call us back to ourselves, to remind us of who we are and of where we dwell.

On my journeys into the Boundary Waters Wilderness of northern Minnesota, my companions and I leave behind the rush of the highway, leave behind the clutter of stores, and launch our canoes into the glossy waters of Fall Lake near the town of Ely. As we paddle across to our first portage, we rock in the wake from motorboats, for gas-powered craft have recently been permitted to cruise the outermost lakes of the wilderness. The manufacturers of outboard motors, jet-skis, snowmobiles, and other loud machines are constantly pushing to open every last refuge to invasion by their products, and they are supported by people eager to use those machines, people too lazy or too addicted to power and speed to travel by means of their own muscles.

Crossing Fall Lake, we often hear the boom of radios above the snarl of engines. My companions and I talk loudly to make ourselves heard above the roar. After two portages, however, we drift into Pipestone Bay, which is free of motors, and here for the first time we're likely to see bald eagles, river otters, beavers, and other elusive creatures. In the stillness, we can hear the cries of loons, the splash of leaping fish. We can hear the lap of waves against the bows of the canoes. Here for the first time the buzz of the highway fades, our voices drop, the rhythm of our paddling slows, and we begin to see where we are.

The water is bounded on all sides by rocky shoreline fringed green with pines and hemlocks, white with birches, yellow with poplars. When we land, we find every crack in the granite brilliant with flowers and grass, every square foot of soil carpeted in lichens, liverworts, saplings, and moss. Bears have left black tufts of hair on the bark of trees, and raccoons have left their tracks like hieroglyphics in the damp sand. Not so long ago, this land was barren. It had been clear-cut, trapped out, mined. Then over the decades since being protected as wilderness, the land began to heal—the forest rising again, the animals returning, the streams running clear.

As we travel from lake to lake toward the heart of the Boundary Waters, I can feel my own mind running clear. By the second or third day, the frets and plans I carried from home have fallen away, and I sink into the peacefulness of this place, as into the depths of meditation or prayer. What I sense is not bland comfort, for the wind often blows in our faces as we paddle, cold rain often chills us, and mosquitoes lustily bite. Any one of us could break a leg on a portage, could go crashing over a waterfall, could spill into the water and drown. The peace of this watery wilderness is not the security and ease of a living room, a shopping mall, or any other space controlled by human beings. Wilderness restores our souls precisely because it is *not* controlled by us, because it obeys laws we did not write, because it reminds us of the vast, encompassing order that brought us into being and that moment by moment sustains us.

Even in the Boundary Waters, where every day feels like a Sabbath, my companions and I keep track of the calendar, for eventually we must go back home. As we draw near to our launching point, once again we encounter the raucous machines that are gnawing at the edges of the refuge. On the long drive to Indiana, the speed of our car over the pavement seems dizzying. The roadsides seem frantic with billboards and franchises. News from the radio speaks of a crazed and broken world I hardly recognize. Even as I slide back into my ordinary life, which is crowded with too many tasks and too many things, the peace of the wilderness lingers in me like a balm. Even if I never visit the Boundary Waters again, I am nourished by knowing it is there, following its ancient ways, unfettered, free. Every remnant of wilderness, like the Sabbath, is a reminder of our origins and our true home.

The Sabbath is one-seventh part of our days. Far less than one-seventh part of our land remains in wilderness. If we understand the lessons of restraint and liberation conveyed by the Sabbath, then we should leave alone every acre that has not already been stamped by our designs, and we should restore millions of acres that have been abused. We should build no more roads in our national forests. We should cut no more old-growth trees. We should drain no more wetlands. We should neither drill nor prospect in wildlife refuges, allowing those fragile places to be refuges in fact and not only in name. To set land free from serving us is to recognize that Earth is neither our slave nor our property.

Some people object that our economy will falter unless we open up these last scraps of wild land to moneymaking. They warn against the danger of "locking up" resources vital to our prosperity. But couldn't the same be said of the Sabbath? Why "lock up" a whole day of the week? Why spend time worshiping, why meditate or pray, when we could be using that time to produce more goods and services? If it is really true that our economy will fail unless we devote every minute and every acre to the pursuit of profit, then our economy is already doomed. For where shall we turn after the calendar and the continent have been exhausted?

Many of the politicians and industry lobbyists who call for the exploitation of our last remaining wild places also claim to be deeply religious. What sort of religion do they follow, if it places no limits on human dominion? What sort of religion do they follow, if it makes the pursuit of profit the central goal of life? If they believe in keeping the Sabbath holy, how can they reconcile this commandment with the drive to reduce every acre and every hour to human control? And if they do not believe in keeping the Sabbath, how do they pick and choose among the commandments?

To cherish wilderness does not mean that one must despise human works, any more than loving the Sabbath means that one must despise the rest of the week. Even if you do not accept the religious premise on which the Sabbath is based, as many people do not, then consider the wisdom embodied in the practice of restraint. Through honoring both Sabbath and wilderness, we renew our contact with the mystery that precedes and surrounds and upholds our lives. The Sabbath and the wilderness remind us of what is true everywhere and at all times, but which in our arrogance we keep forgetting—that we did not make the Earth, that we are guests here, that we are answerable to a reality deeper and older and more sacred than our own will.

John A. Vucetich and Michael P. Nelson

Distinguishing Experiential and Physical Conceptions of Wilderness (2008)

TWO OF THE MOST serious challenges for wilderness as a philosophical concept are its apparent fundamental dependence on culturally relative perspectives and the perpetuation of a dualism between humans and nature. This essay explains how both challenges might be accommodated by working through the consequences of recognizing that "wilderness" actually represents at least two distinct concepts. First, *physical* wilderness is conceived of as a large landscape where ecological processes are thought to operate largely in the absence of direct human influence. Physical wilderness may or may not be officially or legally labeled or designated. Second, *experiential* wilderness is a constellation of psychological phenomena that may be usefully reduced to a physical stimulus, the perception of that stimulus, and the reaction to the perception. Perceptual elements of a wilderness experience may be negative, positive, or some combination of both.

The consequences of distinguishing experiential wilderness from physical wilderness include recognizing three things:

1. *Instances* of wilderness experience are culturally and individually relativistic, but the concept (i.e., the concept of a wilderness experience) itself is not. An important goal for the analysis of wilderness experience is an understanding of *how* cultural and individual perspectives affect the nature of individual instances of wilderness experience.
2. Neither physical wilderness nor experiential wilderness depends fundamentally on the human-nature dualism being an objective reality; rather, wilderness experience treats the *perception* of the human-nature dualism. Some wilderness experiences reinforce that perception, while others cause it to dissipate.
3. Experiential wilderness focuses on the *subject* (i.e., the experienc*er*), and physical wilderness focuses on the *object* (i.e., that which is experienced). Experiential wilderness can be formally related to other generic psychological phenomena, specifically, empathy, deprivation, suffering, and coping. Such connections make wilderness importantly and interestingly connected to some readings of Buddhist metaphysics and expose wilderness to analysis by methods used to study religion, epistemology, traditional psychology, and neuropsychology.

1.0. THE TWO CONCEPTS OF WILDERNESS

Wilderness entails two distinct concepts, each representing a quite different domain of inquiry and knowledge, and each important for our relationship with nature in different ways. One conception of wilderness refers to a physical place, more specifically, a large tract of land (or even sea) where humans have at most only modest influence on ecological-evolutionary states (e.g., abundance and diversity of species) and processes (e.g., nutrient flows, natural selection, and organismal rates of birth, death, and dispersal). Although terms such as "large" and "modest" are importantly relative and subjective, the states and processes of the landscape in question are importantly analyzable by scientific methods. As such, physical wilderness is importantly but not exclusively assessed by environmental and ecological sciences. Some have suggested that the concept of physical wilderness be reassociated with ecoreserve, biodiversity reserve, or some such label.[1] No conceptual confusion would seem to arise from such relabeling. Regardless of the label, the concept associated with "physical wilderness" is fundamentally important for understanding the *physical* relationship between human *society* and the environment.[2] A second conception of wilderness is primarily experiential and not uniquely dependent on an objective state of affairs.

This experiential conception of wilderness is also critically important and is not easily or usefully reassociated with any other convenient label. Wilderness concepts have been widely criticized because they carry cultural bias and baggage that distract from effective and just use of the concept. However, confronting and understanding this baggage is essential for gaining a more mature *psychological* relationship with nature. Banishing the word "wilderness" to resolve problematic aspects of our relationship with nature would be like banishing the word "nigger" or "racism" in hopes that doing so would solve the problem of racism.[3] That a wilderness experience may be negative (from any perspective, cultural or individual) does not discredit the concept (of a wilderness experience), nor does it represent an occasion to banish the concept from discourse. On the contrary, more discourse and research are needed to (1) better understand the demarcation of a negative wilderness experience, (2) more clearly distinguish the causes from the symptoms of negative wilderness experiences, and (3) create methods for transforming negative wilderness experiences into positive ones. Analyzing, for example, the wilderness experiences characteristic of North Americans from the perspective of another culture would almost certainly help one to better understand the relationship between nature and North Americans. None of this would be achieved by simply banishing the concept from discourse.[4]

Experiential wilderness is important for affecting (positively or negatively) *psychological* relationships between *individuals* and nature. By contrast, the concept associated with "physical wilderness" is fundamentally important for understanding the *physical* relationship between human *society* and the environment.

2.0. AN EXPERIENTIAL CONCEPTION OF WILDERNESS

Let us suppose that the concept of wilderness is meaningful only if humans are in some way distinct from nature. Moreover, suppose that culture (not culture per se but the unique degree to which culture is developed and expressed in humans) is the feature that distinguishes humans from the rest of the natural world.[5] Taking this for granted, a simple and provisional concept of a wilderness experience would be *an experience deprived in some significant and general way of human culture.*

Given this conception, a wilderness experience may be had by any human, regardless of the historical or cultural context to which that human belongs (i.e., a wilderness experience would be a truly cross-cultural phenomenon). The physical conditions that stimulate a wilderness experience as well as the perception of and the psychological response to that experience may vary substantially among cultures and among individuals within a culture. Without disentangling physical wilderness from experiential wilderness this phenomenon is confused and even perhaps inaccessible. Confronting and understanding these variations is valuable because such understanding would likely lead to an understanding of and explanations for various cultures' relationships with nature (or the nonhuman world).

To illustrate this idea, compare and contrast these three wilderness experiences:

(1) a backpacker burdened with a backload of high-tech equipment, hiking in a second-growth forest parsed into pieces by active logging roads and experiencing emotions of peace and oneness;

(2) a seventeenth-century European colonist in North America with low-tech equipment, traveling through what by today's standards would constitute a first-rate physical wilderness area and experiencing emotions of angst and isolation;

(3) a young person from an indigenous tribe who, upon reaching an age of maturity, is sent by tribal elders from the community (into the "wilderness") with the expectation of learning

> something (perhaps via a bit of suffering) and returning as a hero, a rite of passage commonly referred to as a "vision quest."[6]

Each of these experiences differs greatly in terms of physical circumstances and psychological effects. Nevertheless, the salient characteristic of each experience is the deprivation of human culture, and each experience is thus usefully considered a wilderness experience.

Example 3 may represent a category of phenomena that is quite general. It is illustrated by narratives and traditions from many cultures, including, for example, North American Indian, South Asian, and Eastern European.[7] A brief exploration of one example might prove illustrative. Consider the famous Danish philosopher Søren Kierkegaard's *Eulogy on Abraham,* in which Kierkegaard interprets Abraham's near sacrifice of his own son Isaac.[8] En route to following God's commandment to kill his son, Abraham takes Isaac for three days through the desert wilderness to Mount Moriah. Kierkegaard asserts that the story represents Abraham's passing from the Sphere of the Ethical to the Sphere of the Religious, which requires abandoning human culture. Kierkegaard also claims that only by abandoning human culture could Abraham justify doing something as insane as planning to kill his son and then not doing it.[9] More generally, Kierkegaard says that passing from the Sphere of the Ethical to the Sphere of the Religious represents finding one's truly free self, which requires leaving society behind. The salient point is that Kierkegaard provides an important articulation of the value of experiential wilderness (i.e., deprivation of human culture) for human psychological development.

Goethe's *Faust,* written about fifty years before Kierkegaard's commentary while Enlightenment thought was being eclipsed by Romantic thought, provides a similar but distinct account of wilderness's effect on the human psyche. After becoming romantically attached to Gretchen, Faust retreats to a forest cavern. This short scene begins with Faust making extensive reference to nature's beauty as a means of expressing his lust-inspired joy. After Mephistopheles mocks Faust's expressiveness, Faust complains: "Can you not understand how my life's strength increases as I walk here in these wild places?" Mephistopheles concludes another round of berating with the words "How does the lofty intuition end?" which foreshadows the tragedy that Faust's infatuation for Gretchen becomes.[10] Goethe's Forest Cavern scene conspicuously portrays wilderness as a place where the rational (Faust's professorial nature) may be sacrificed, with tragic effect, to the irrational (Faust's bargain with the devil).

3.0. THE ANATOMY OF A WILDERNESS EXPERIENCE

The experiential notion of wilderness can be usefully decomposed into three elements: (1) a physical stimulus significantly characterized by some nonhuman element; (2) a perception of the stimulus that may be negative, positive, or some mixture of negative and positive; and (3) a psychological reaction to the perception that may also be positive, negative, or some combination of positive and negative.[11] The following sections explore each aspect of a wilderness experience.

3.1. PHYSICAL STIMULUS

Wilderness experiences seem to be triggered by a wide variation of kinds and intensities of physical stimuli. This variation seems to be associated with one's individual maturity, one's past personal experience, and the norms of one's culture and subcultures.

3.1.1. Cultural variation. During the seventeenth century the North American landscape generally stimulated wilderness experiences for European colonists but not North American Indians. Why? Not because North American Indians were Noble Savages but because they would have been surrounded by all the human culture they had ever known.[12] Perhaps one cannot be deprived of the customs of another culture if one has never conceived of them. North American Indians certainly could have had a wilderness experience. It was just more difficult for European colonists to avoid a wilderness experience once they left the narrow confines of their settlements.[13] This comparison seems applicable to the wilderness experiences of colonists and aboriginal peoples throughout the world and throughout history.

3.1.2. Subcultural variation. Consider a pleasant picnic on a remote lake arrived at by floatplane. This physical stimulus may not stimulate a wilderness experience for a life-long bush pilot, but it may do so for his or her urban client. Again, the primary difference may be the degree to which each is typically immersed in human culture. The urbanite is generally more sensitive to the deprivation of art museums, eight-lane highways, and Starbucks, whereas the bush pilot would tend not to be.[14] Such considerations may partly reveal and explain the nature of wilderness experience for urbanites, suburbanites, and persons whose primary experience is rural and agricultural (crops or trees).

3.1.3. Individual variation and past experience. Consider two people, each of whom believes that the intensity of a wilderness experience can be judged by the level of suffering caused by being deprived of human culture. Consider also both of these people experiencing the same cold, rainy day without shelter. Both would seem to be having the same wilderness experience, that is, being deprived of shelter from the rain. Nevertheless, the intensity of the experience could be much greater for one than the other if one is accustomed to such conditions and knows of behavioral (or attitudinal) responses that reduce the suffering. Perception and reaction (both physical and psychological) seem to be fundamental elements of wilderness experience.

3.1.4. Individual variation and sensitivity. Compare the experiences associated with the sight and sound of a herring gull in a physical wilderness (i.e., an ecoreserve) and of a gull at a public beach. Many people realize empathy more easily for the ecoreserve gull than for the beach gull. What accounts for the difference? Certainly, the difference is not in the experience of the gulls. Inasmuch as we are able to perceive the experience of any gull (i.e., empathize with the gull), perhaps our perception of a gull's experience ought to be importantly independent of a gull's environment, urban or wilderness. Both gulls are acquiring food, struggling to survive the elements, and expending great energy (against great odds) to reproduce and rear offspring. Most simply, both exist and continue to exist. The difference between viewing a gull in a physical wilderness area and viewing a gull in an urban area is the psychological condition of the person observing the gulls. In the ecoreserve, where human culture is sparse, some people find it easier to focus on nonhuman elements within their environment; it seems more difficult to do so at the beach.

3.2. PERCEPTION OF PHYSICAL STIMULUS

We consider an "experience" to include not only some physical stimulus but also the perception of that stimulus. In section 2.0 we provisionally defined wilderness experience as an experience of deprivation in some significant and general way of human culture. This definition represents wilderness experience too narrowly, that is, as a negative concept. A more general sense of "wilderness experience" would be *an experience for which the physical stimulus primarily entails the perception of nonhuman elements in one's environment.*[15]

From this more general definition two categorically distinct perceptions arise: deprivation of human elements (negative) and enriched awareness of nonhuman elements (positive). Wilderness experiences may be closely related to other so-called deprivation experiences. For example, if backpacking is virtuous for the wilderness experience it provides, perhaps vows of silence and poverty are virtuous in similar ways and for similar reasons. The experiences had by backpackers, silent retreatants, and street people may be more similar than is generally recognized. What are the similarities and differences between experiences entailing so-called deprivation in environments largely devoid of human infrastructure (i.e., a wilderness experience) and so-called deprivation in environments saturated with human infrastructure (i.e., nonwilderness deprivation experiences)?[16]

3.3. PSYCHOLOGICAL REACTION TO THE PERCEPTION

Although psychological reactions to wilderness perceptions are diverse and nuanced, some categorization might be useful. For example, suffering is one important reaction to the negative perception of *deprivation of human culture.* If "suffering" is the "inability to cope," then "coping" would be another important reaction to perceived deprivation. Coping may not be a positive reaction inasmuch as coping may represent the denial of suffering or the repression of psychological reactions that ought to arise from suffering. In any case, specifying a formal relationship between wilderness and suffering is useful because it highlights the strong association that some people make between wilderness and suffering (see section 4.2.1). However, our analysis indicates that suffering is not a fundamental aspect of wilderness. It merely demarcates one of many legitimate *reactions* to physical wilderness.

An important positive reaction to the positive perception of wilderness (i.e., an experience focused on something unrelated to human culture) is empathy for that which is focused upon, namely, the nonhuman elements. Focusing on an object is a prerequisite for empathizing with an object. Empathy does not arise spontaneously even in people predisposed to empathy. Empathy requires active engagement and focused experiences with objects. The relationship between wilderness and empathy is important because empathy is entwined with compassion, respect, and love.

However, empathy is not a necessary reaction to positive perceptions of wilderness. More selfish reactions are possible. Consider someone for whom human culture is perceived to be a primary source of stress and dissatisfaction in life. Such a person would tend to perceive a wilderness experience as a focused experience on nonhuman elements in his or her environment rather than as a deprivation of human culture. However, this reaction may not entail empathy. The reaction may be merely recreative. Although recreation is important, its value is limited.[17] At best, recreation generates respect for nonhuman elements, but only for their utilitarian or use value to the recreated self.

The relationship between empathy and wilderness may provide some explanation for the modern American wilderness experience. Since many average Americans are arguably not well endowed with (or even encouraged to develop) a rich sense of empathy, their wilderness experiences will tend to be negative.

4.0. IMPLICATIONS AND ELABORATIONS

4.1. HUMAN-NATURE DUALISM IN PHYSICAL AND EXPERIENTIAL WILDERNESS

As a preliminary, it seems useful to recognize a systematic relationship among various types of dualisms. First, consider the dualism between *us* and *them* as an instance of a general dualism. Then consider various types of dualisms to be distinguished by what is included in the "us" category. In the mind-matter dualism "us" refers to one's mind. In the subject-object dualism "us" refers to one's self, the mind and body, or the subject. In the human-nature dualism "us" refers to a collection of human selves. The relevant point is that the human-nature dualism is a type of subject-object dualism manifest at the cultural level.

Previous conceptions of wilderness have been criticized for appearing to depend fundamentally on a human-nature dualism. Because such a dualism is, at least, difficult to defend and more likely an illusion, conceptions such as wilderness that depend on a human-nature dualism inherit the criticisms laid against the dualism itself. However, this is no more than a superficial criticism. With respect to physical wilderness, which concerns the physical relationship between society and the environment, the human-nature dualism is not fundamental. Without considering it a metaphysical reality, dualism can be a pragmatically useful and sensible way of relating things. For example, thinking that humans can *cause* environmental alteration (destruction) is useful but requires a dualistic framework.

With respect to experiential wilderness, the human-nature dualism is useful, provisional, and perceived but not fundamental. Recall that the experiential concept of wilderness is about the psychological relationship between individuals and nature. Inasmuch as the concept "relationship" is useful, so is the dualism. In some sense a relationship may not even be perceptible without perceiving a dualism between the relating entities.[18] At its root the experiential concept of wilderness is about how one perceives the human-nature dualism. In this way wilderness does not depend upon the reality of a human-nature dualism; rather, it treats only the perception of a human-nature dualism. Wilderness experiences can lead to either the reinforcement or the dissolution of perceived human-nature dualisms.

Consider that a negative wilderness perception entails being deprived of human culture and that the root of the word "deprivation" is "private," which means "of, or pertaining to, the individual." By focusing on the self, the negative wilderness experience reinforces the human-nature dualism.[19] Positive wilderness perceptions also treat perceived human-nature dualism distinctively. Wilderness empathy, in particular, seems powerfully associated with the dissolution of perceived human-nature dualisms. Even dictionary definitions of empathy suggest a relationship between empathy and subject-object dualism: the "attribution of one's own feelings to an object" and "the imaginative projection of a subjective state into an object so that the object appears to be infused with it."[20]

From some perspectives, the ability to avoid illusions of duality between one's self and one's surroundings is a mark of psychological maturity. In this regard, perceiving wilderness in terms of deprivation may be less mature, and reacting to wilderness with empathy may be more mature.

4.2. A SURVEY OF WILDERNESS EXPERIENCES

In the sections below we highlight a few archetypal wilderness experiences. They further illustrate the impact of experiential wilderness on one's psyche and the importance of distinguishing between physical and experiential wilderness.

4.2.1. Adventure seekers. Some modern wilderness adventure seekers strongly associate wilderness experiences with suffering.[21] For reference, pick up any copy of popular magazines such as *Backpacker* and *Outside.* Adventure seekers judge their personal development by increasing their ability to suffer and cope. Consequently, each new wilderness experience must be more difficult and extreme than the previous, or personal development is stifled. The process is self-focused and nonempathetic, and it even includes characteristics of an addiction. The modern adventurer is like the Calvinist in strongly associating wilderness and suffering.[22] However, whereas the Calvinist may think he or she deserves to suffer (because of Original Sin), the adventurer seems to enjoy it or at least consider it worthwhile because it leads to personal fulfillment or social recognition. Despite these criticisms, this type of experience is not inconsistent with being an effective advocate for physical wilderness.[23]

Perhaps the wilderness adventurer seems relatively uninteresting because he or she is uncommonly exhibited in the simplified extreme degree portrayed above. However, if the extreme wilderness adventurer is at gross fault for strongly associating wilderness with suffering/coping, then might it follow that the adventurer would bear fault to the extent that he or she associates wilderness with suffering/coping even if that association is weak? If so, it would be illuminating to have an empirical understanding of the extent to which individuals within different cultures tend to associate wilderness and suffering. We suspect the association is significant for many North Americans.

4.2.2. Wilderness nihilism. Suppose a wilderness experience is judged positive to the degree one achieves feelings of unity and oneness with nature. As one has more wilderness experiences, one may become increasingly aware of the pervasiveness of human influence over virtually every landscape and seascape on the planet. For some, this awareness results in sadness and irritation with things like any sign, no matter how innocuous, that another human is or was nearby; jet contrails in the sky; testimonial-based knowledge of extirpated native species and exotic species in an ecosystem; and disruption caused to birds nesting too close to a hiking trail. These irritations can become serious obstacles to a positive wilderness experience (i.e., feelings of unity and oneness with nonhuman elements).

Attempts to empathize can result in a sense of despair. Wilderness experiences seem senseless or useless and hence nihilistic. Such experience may be driven by any of several processes: one may desire sentiment more than genuine empathy, or one may be unable to deal with the challenges that mature empathy sometimes poses. A peaceful wilderness experience can be difficult if not impossible to realize when one takes seriously the suffering of the object with which one empathizes. Although Buddhist principles maintain that empathy relieves rather than amplifies suffering, nihilism is a risk for naive practitioners of Buddhism.[24] Finally, one may have a greater capacity to empathize with nonhuman elements than with humans. Here one may be confusing empathy with pity and confusing empathy for humans with tacit support for unjust actions by humans (i.e., treating nature poorly).

4.2.3. Supersensitivity. The ability to focus on nonhuman elements in one's environment requires motivation and skill, which are achieved through practice. With limited skill, focusing on nonhuman elements may require being surrounded almost entirely by nonhuman elements. However, with increased skill, one can focus on nonhuman elements in human-dominated environments.[25] Empathy is similarly skill dependent.[26] At first, perhaps only extreme conditions stimulate empathy, such as being in a remote physical wilderness and witnessing some spectacular life drama in another species. However, with practice a deep sense of empathy might be triggered by far more subtle stimuli, such as merely seeing a single ant cross the sidewalk. Differences in motivation *and* skill account for much of the variation in one's ability to have a mature wilderness experience.[27]

Wilderness maturity might entail the ability to *experience* wilderness virtually anywhere. A person with mature wilderness experience skills would in fact find it difficult to avoid wilderness experiences. Under such circumstances there would be little connection between a wilderness experience and physical wilderness. Distinguishing these ideas is important. Although both are important, it is detrimental to confuse a wilderness experience (e.g., empathy for a squirrel in a suburban front yard) with a physical wilderness (e.g., the Brooks Range).

4.2.4. Buddhist thought. Although wilderness is not a formal concept in Buddhist thought, a formal conception might be developed from other aspects of Buddhist thought. For example, Buddhist thought includes the doctrine of karma, which indicates that suffering is deserved. However, in contrast to Calvinist and Puritan traditions, some Buddhist traditions teach that during this lifetime it is possible for one to be (at least partially) liberated from suffering, so suffering is not to be endured (as it is for Calvinists) but overcome. Moreover, an essential element for overcoming suffering is shifting one's focus away from one's self and onto others. Thus, an essential element in the Buddhist solution to suffering coincides with what has been portrayed here as a mature wilderness experience. According to Buddhist tradition, compassion and right relationships necessarily arise from genuine empathy. In this sense suffering, empathy, and wilderness are formally related.[28]

This connection does present some obvious challenges; for example, why should we be concerned with wilderness or environmental ethics when we routinely fail to treat people in an ethical manner? One Buddhist's comment points to an answer. The Dalai Lama has said: "One way you can develop empathy [for people] is to start with small sentient beings like ants and insects. Really attend to them and recognize that they too wish to find happiness, experience pleasure, and be free of pain. Start there with insects and really empathize with them, and then go on to reptiles and so forth. Other human beings and yourself will all follow."[29] The essential point here is that empathy for other humans does not begin with humans (and is not preempted by empathy for nonhumans) but rather that it begins by fostering the virtue of empathy.

Hence, what a Westerner would think of as a wilderness experience might really be a form of Buddhist practice. Though such practice might be prompted both in and by physical wilderness, it is certainly not necessarily dependent upon the existence of physical wilderness.

4.2.5. Judeo-Christian thought. The Judeo-Christian treatment of human-nature relationships has been discussed extensively.[30] Here we merely indicate how wilderness experience, as portrayed in this essay, represents a general framework from which such discussion may be considered.

Judeo-Christian thought, more than other worldviews, reinforces self-other dualism (individual salvation) and human-nature dualism (e.g., the Christian creation myth emphasizes the distinction between humans and other creations). Judeo-Christian thought also includes perceiving deprivation as virtuous because it is atonement for Original Sin and personal sin. Predispositions for dualism and deprivation seem to promote negative wilderness experiences.

Lynn White, Jr., famously suggested that Saint Francis represents a Christian solution to such problems. He described Francis's "view of nature and of man [as] rest[ing] on a unique sort of pan-psychism of all things animate and inanimate, designed for the glorification of their transcendent Creator."[31] White portrays Francis as fostering empathy by highlighting the "pan-psychism of all things" and transforming deprivation experiences into experiences that focus on something beside one's self (i.e., glorifying a transcendent Creator). This is consistent with the positive wilderness experiences portrayed here, especially if the transcendent Creator is in some significant way a self-expression of its nature. Viewing this solution from the perspective of wilderness experience described in this essay raises two significant questions: Does a God-humanity or God-creation dualism promote a negative wilderness experience? To what extent does a positive wilderness experience depend on any forms of pantheism?

4.3. WILDERNESS, IGNORANCE, AND KNOWLEDGE

Some wilderness advocates believe that ignorance is an essential component of a wilderness experience. According to these advocates, wilderness experiences are enhanced by lakes and mountains without names and a total lack of scientific or cultural interpretation.[32] Said experiences are, likewise, diminished in areas where lakes and mountains are named, studied, and interpreted. Justification for this position seems to be that knowledge invariably leads to control.[33]

This position seems absurd. While control might not be possible without knowledge, knowledge does not invariably lead to control. On the contrary, ignorance is not compatible with empathy, unity, compassion, or mature love. For example, although ignorance might arguably be the basis for romantic love, which is inherently ephemeral, mature love, like that found in healthy marriages, depends upon knowledge. Hence, promoting ignorance of nature might well run contrary to promoting empathy, unity, compassion, or a mature type of love for nature.

Some wilderness proponents actually advocate ignorance on the basis that knowledge spoils mystery, which is essential for wonderment and respect. Although Richard Feynman's aphorism is appropriate, it does not go far enough.[34] If knowledge were to spoil mystery, wonderment, and respect, then we would want to remain as ignorant as possible about the people we love. In fact, one might well claim that the kind of love found in marriage and rooted in knowledge is extremely mysterious.

We ought to be equally cautious about how we relate wilderness and wilderness preservation with knowledge. Although scientific research in a (physical) wilderness area may foster care and wonderment, it is not easy to judge when such research is too invasive or manipulative. Some knowledge is simply not worth the cost. No one could justify dissecting their spouse simply to better know him or her.

From a different perspective, increased knowledge may reduce one's valuation of an object. For example, one may value an ecosystem less upon learning that it is less pristine than originally thought. Although knowledge may cause reduction in value that is instrumental or conditional, knowledge is only likely to increase value when an object is intrinsically valued.

4.4. THE SUBJECTIVITY OF PHYSICAL WILDERNESS

The distinction between physical wilderness and experiential wilderness does not perfectly coincide with a distinction between subjective and objective elements of wilderness. Experiential wilderness, though importantly subjective, entails an important objective element—the physical stimulus (section 3.1). Although physical wilderness is importantly objective (i.e., it is about an object that exists independently of you and me), it is also importantly subjective. One must prescribe the conditions that represent physical wilderness. Is an ecosystem large enough or sufficiently uninfluenced by humans to qualify as physical wilderness? Which aspects of the ecosystem (e.g., species composition and nutrient cycling) are most important when judging human impact?

The depth to which objective and subjective elements seem entwined within both physical and experiential wilderness may be symptomatic of a more general false dichotomy between facts (objectivity) and values (subjectivity).[35] This complexity should not impair constructive use of either wilderness concept. It merely necessitates awareness and appropriate treatment.

CONCLUSION

Although experiential wilderness and physical wilderness are distinct, it is wrong to treat them as completely independent. For example, one can ask, To what extent does a mature sense of wilderness experience in a human community lead to increased physical wilderness? How does the existence of physical wilderness promote experiential wilderness? Is physical wilderness the only (or best) way to promote experiential wilderness? Having reasonable answers to these questions would be of great practical importance.

Experiential wilderness and physical wilderness do not always come together, and they might not always serve one another. If we can have wilderness experiences outside of physical wilderness areas, then do we really need physical wilderness? And if our rationale for the preservation of physical wilderness rests upon the provision of wilderness experiences, then we might have a weak foundation upon which to rest our need for physical wilderness.

The failure to recognize the distinction between experiential wilderness and physical wilderness can lead to practical problems in the management of landscapes designated as wilderness. Keeping these distinctions in mind might go some way toward helping us sort out a few of the conundrums that currently plague our thinking about the concept of wilderness. According to the Buddha,

> Mind is the forerunner of all actions.
> All deeds are led by mind, created by mind.
> If one speaks or acts with a corrupt mind, suffering follows, . . .
> If one speaks or acts with a serene mind, happiness follows.
>
> The Dhammapada

NOTES

1. This solution bears at least superficial similarity to an idea proposed by J. B. Callicott in his essay "Should Wilderness Areas Become Biodiversity Reserves?" in *The Great New Wilderness Debate,* ed. J. B. Callicott and Michael P. Nelson (Athens: University of Georgia Press, 1998); see also J. B. Callicott's "The Implication of the 'Shifting Paradigm' in Ecology for Paradigm Shifts in the Philosophy of Conservation" in this volume. Throughout this essay we use variously the terms *ecoreserve* and *physical wilderness* to refer to the same general concept. In some cases one label seems more appropriate than the other.

2. More generally, physical wilderness may also be considered a quantitative characteristic of an ecosystem with different dimensions and degrees. A physical wilderness may be particularly large but not very pristine, or it may be very pristine but not very large. The dimensions of physical wilderness may conflict with each other. For example, restoring a native species that had been extirpated by humans would increase the physical wilderness quality of an ecosystem by making its *state* less influenced by humans (compared to its original

state). However, restoration itself represents a significant human influence on an ecosystem's *process*. Also, compare two ecosystems, one where timber extraction has ceased (increasing the wilderness nature of ecosystem processes) but the forest's state is significantly altered and will be for many generations (the ecosystem state is not very wilderness-like) and another where forest management is intensified but for the purpose of returning the ecosystem to its former state. Which place is more wilderness-like? Although judging a place to be a physical wilderness or not entails a very important normative dimension, that which is being judged is an importantly objective circumstance.

3. This analogy to "racism" is appropriate if "wilderness" is a neutral label for the concept it represents—perhaps what we have referred to as physical wilderness. However, if "wilderness" is inherently and unalterably loaded with subjective or biased perspective, then simply banishing "wilderness" from discourse would be like banishing the words "babe" and "nigger" in the hope of resolving issues of sexism and racism. Banishing such terms may be a necessary condition for some sort of remediation, but it is not clear that it is a sufficient condition for remediation.

4. In fact, an important opportunity for this sort of conceptual therapy might be missed if the concept or word is simply banished. For example, realizing that the concept of wilderness in the Euro-American mind (e.g., as uninhabited) has some disastrous results when it is imported to other parts of the world might force us to reconceptualize our idea of wilderness and allow for human habitation in some form. This, then, can prompt important discussions about what forms of human habitation would or should be compatible with wilderness preservation or with the preservation of wilderness experiences. See also William Cronon's "The Riddle of the Apostle Islands: How Do You Manage a Wilderness Full of Human Stories?" in this volume.

5. J. B. Callicott and M. P. Nelson, *American Indian Environmental Ethics: An Ojibwa Case Study* (Upper Saddle River, N.J.: Pearson Prentice Hall, 2004).

6. Arguably, this is roughly what Jesus's wilderness experience was about. Jesus's wilderness experience was not about traveling over a landscape "untrammeled by man, where man himself is a visitor who does not remain" (Cronon, "The Riddle of the Apostle Islands"). To think so would miss the point of that parable and the point of this conception of wilderness.

7. Holy men who take retreat at remote locations in caves and under trees typify South Asian examples. Eastern Orthodox Christian Poustiniks who live as mystical hermits typify Eastern European examples.

8. Søren Kierkegaard, *Frygt og Bæven: Dialectisk Lyrik* (1843), translated in E. H. Hong and H. V. Hong, *Fear and Trembling/Repetition: Kierkegaard's Writings,* vol. 6 (Princeton, N.J.: Princeton University Press, 1983).

9. See also Marilynne Robinson, "Wilderness," in this volume.

10. Goethe, *Faust, Part I,* trans. David Luke (Oxford: Oxford University Press, 1998), ll. 3278, 3291.

11. On other conceptions of what an experience (and a wilderness experience) is see Karen M. Fox, "Navigating Confluences: Revisiting the Meaning of 'Wilderness Experience,'" in *Wilderness Science in a Time of Change Conference,* vol. 2: *Wilderness within the Context of Larger Systems,* ed. S. F. McCool, D. N. Cole, W. T. Borrie, and J. O'Loughlin, May 23–27, 1999, Missoula, Mont., Proceedings RMRS-P-15 (Ogden, Utah: U.S. Department of Agriculture, Forest Service, Rocky Mountain Research Station, 2000).

12. See, for example, Chief Luther Standing Bear, "Indian Wisdom," in Callicott and Nelson, *The Great New Wilderness Debate,* 201–6.

13. Is it sensible to consider a wilderness experience as entailing deprivation of human culture or deprivation of familiar culture? Can a country boy have a wilderness experience in the city? It may be useful to consider a set of distinct experiences, all sharing some element of deprivation. Deprivation of human culture is a wilderness experience. Deprivation of familiar culture is a foreign experience. We are not sure what kind of experience to call Valentine Smith's in Robert Heinlein's *Stranger in a Strange Land* (he was deprived of human culture but immersed in Martian culture). Regardless, the point is that wilderness experience may usefully be considered a deprivation of human culture and as such is not equivalent to other types of deprivation. Moreover, Valentine Smith's experience begs us to consider the question, If we could "experience" Martian culture, could we sensibly "experience" the culture of other nonhuman entities? Wolves, bees, beavers, ants, elephants all have relatively sophisticated cultures compared to, say, yeast cells. The existence of a wilderness experience may require at least some sense of such an experience (see section 3.3).

14. For a third (ancillary) illustration, consider a person exposed for his entire life to nothing but an urban environment. Such a person is *deprived of wilderness* (i.e., he is deprived of the deprivation of culture).

15. We intend "wilderness" to be defined by the nature of the physical stimulus, not by the perception or reaction to it. So while it would be sensible to say that what qualifies as a wilderness experience for you does not qualify as one for me, it would not be sensible to say that what is wilderness for you is not wilderness for me. For a very different conceptualization of the notion of wilderness experience see Fox, "Navigating Confluences."

16. The ways in which physical stimuli cause perceptions in wilderness experiences, as in other experiences, are complex. Explicitly recognizing the relevance of the relationship between stimulus and perception in wilderness experiences exposes the concept of wilderness to analysis by the tools of psychology, neurobiology, metaphysics, and epistemology.

17. We refer to the noblest sense of "recreation" (i.e., the re-creation of one's

spirit, not the brief escape from reality that might be provided by playing a video game or watching Monday night football). However, even the noblest sense of *recreation* (sensu John Muir, Aldo Leopold, and Kierkegaard; see section 2.0) entails a selfish (ego- or anthropocentric) interest. Regardless, this does not imply that only physical wilderness may have a nonanthropocentric value. Experiential wilderness that generates empathy for nonhuman elements has nonanthropocentric value.

18. Dualism may be problematic, primarily because it may imply a directionally causal relationship. This position was taken by W. D. Hart, *The Engines of the Soul* (New York: Cambridge University Press, 1988).

19. The sensibility of this idea depends on the extent to which one's self (identity) is defined by one's culture.

20. See www.empathy.com and www.m-w.com, respectively.

21. Wilderness adventurers are also thrill seekers. The thrill is in testing these adventurers' ability to suffer and cope.

22. The connections (somewhat profound) between wilderness and evangelical protestant religious traditions are well documented. See, for example, M. Stoll, *Protestantism, Capitalism, and Nature in America* (Albuquerque: University of New Mexico Press, 1997) and J. B. Callicott and P. S. Ybarra, "The Puritan Origins of the American Wilderness Movement," Teacherserve: An Interactive Curriculum Enrichment Service for Teachers, National Humanities Center, http://www.nhc.rtp.nc.us:8080/tserve/nattrans/ntwilderness/essays/puritan.html (2001).

23. In fact, wilderness adventurers are most often (but not always) just such advocates. This might also go some way in explaining why they are not as effective as others might be and why wilderness preservation often smacks of "my values or interests as opposed to yours" ("I like wilderness, but you like roads and lodges"), but both are seen as merely subjective preferences aimed at self-satisfaction.

24. Empathy could also cause increased suffering, but it would be an immature expression of empathy.

25. See William Cronon, "The Trouble with Wilderness, or, Getting Back to the Wrong Nature," in Callicott and Nelson, *The Great New Wilderness Debate,* 471–99.

26. See Daniel Goleman and the Dalai Lama, *Destructive Emotions* (New York: Bantam, 2003) and Goleman's *Emotional Intelligence* (New York: Bantam, 1997).

27. Recall from section 3.1.4 the comparison between the wilderness gull and the urban gull.

28. D. E. Cooper and S. P. James, *Buddhism, Virtue, and the Environment* (Hampshire, U.K.: Ashgate, 2005).

29. Goleman and the Dalai Lama, *Destructive Emotions,* 291. This idea also has precedence in contemporary Western ethics. See P. Singer, *Animal Liberation: A New Ethics for Our Treatment of Animals* (New York: Avon/New York Review, 1975, 1990).

30. A few examples include S. McFague, *The Body of God: An Ecological Theology* (Minneapolis, Minn.: Fortress Press, 1993); B. R. Hill, *Christian Faith and the Environment: Making Vital Connections* (Maryknoll, N.Y.: Orbis Books, 1998); M. D. Yaffe, *Judaism and Environmental Ethics: A Reader* (Lanham, Md.: Lexington Books, 2001); L. H. Sideris, *Environmental Ethics, Ecological Theology, and Natural Selection* (New York: Columbia University Press, 2003); J. B. Callicott, "Genesis and John Muir," in *Covenant for a New Creation,* ed. C. Robb and C. Casebolt (Maryknoll, N.Y.: Orbis Books, 1991); J. B. Callicott, *Earth's Insights: A Multicultural Survey of Ecological Ethics from the Mediterranean Basin to the Australian Outback* (Berkeley: University of California Press, 1994), 14–30.

31. Lynn White, Jr., "The Historical Roots of Our Ecological Crisis," *Science* 155 (1967): 1203–7.

32. See Z. Papanikolas, "The Unpaintable West," in *The World and the Wild: Expanding Wilderness Conservation beyond Its American Roots,* ed. D. Rothenberg and M. Ulvaeus (Tucson: University of Arizona Press, 2001), 24–25, 28 for a good example of this view.

33. The other rationale for this concern is that names have the effect of peopling the landscape, which, it is assumed, runs contrary to the idea of a wilderness. See, for example, P. Burnham, *Indian Country, God's Country: Native Americans and the National Parks* (Washington, D.C.: Island Press, 2000) and M. D. Spence, *Dispossessing the Wilderness: Indian Removal and the Making of the National Parks* (New York: Oxford University Press, 1999).

34. "It does not do harm to the mystery to know a little about it." Quoted in Robert S. and Michele M. Root-Bernstein, *Sparks of Genius: The Thirteen Thinking Tools of the World's Most Creative People* (Boston: Mariner Books, 2001), 132.

35. H. Putnam, *The Collapse of the Fact/Value Dichotomy and Other Essays* (Cambridge, Mass.: Harvard University Press, 2002).

William Cronon

The Riddle of the Apostle Islands (2003)

How Do You Manage a Wilderness Full of Human Stories?

THE APOSTLE ISLANDS are not on the way to anywhere. I managed to grow up in southern Wisconsin, and even to fall in love with the wild beauty of Lake Superior, without ever journeying to the northernmost tip of the state. There, the Bayfield Peninsula juts out into the cold waters of the lake and an archipelago of twenty-two small wooded islands lies just offshore. Not until a few years ago did I find myself, almost by accident, gazing out at those islands and realizing I had found one of the places on this good Earth where I feel most at home. I have been haunting them in all seasons ever since.

There is nothing especially dramatic about the Apostles. In some places, they meet the lake with narrow, pebble-covered beaches rising steeply to meet the forest behind. Elsewhere, they present low sandstone cliffs, brown-red in hue, that have been so sculpted by the action of wave and ice that one never tires of studying their beauty. In a few places where the geology is just right, the lake has widened crevices to form deep caves where kayakers can make their way into darkness and listen to the rise and fall of water on stone. Northern hardwood forest, swamp, marsh, and shore are the primary habitats, with nesting bird colonies in the cliffs and a peripatetic population of black bears that is surprisingly unfazed by the need to swim from island to island despite the notoriously cold temperatures of the lake.

For nearly thirty-five years, these lands and waters have been protected by the federal government as Apostle Islands National Lakeshore—a legacy of Wisconsin Senator Gaylord Nelson, father of Earth Day in 1970. Sometime later this year, the National Park Service will issue recommendations for future management of the park. Although the NPS study recommending wilderness designation for the Apostles (spearheaded by another Wisconsin senator, Russ Feingold) has not thus far attracted much attention, its implications reach far beyond the Apostle Islands. Anyone committed to rethinking human relationships with nature should pay attention to its findings.[1]

In the 1970 act that created it, the Lakeshore was dedicated to the "protection of scenic, scientific, historic, geological, and archaeological features contributing to public education, inspiration, and enjoyment." Since then, millions of Americans have come to appreciate the subtle, ever-changing beauty of the islands. Designating the Apostles as wilderness will be a milestone in the ongoing effort to protect them for future generations, and will constitute an important addition to our National Wilderness Preservation System in a region where far too little land has received such protection. Look at a map of legal wilderness in the United States, and for the most part you will see a vast blank expanse between the Appalachians and the Rockies. At a minimum, the Apostles can serve as a reminder that the Middle West also is a place of wildness, despite the common prejudice that nothing here deserves that label.

On the surface, there seems little reason to doubt that many of the Apostles meet the legal criteria specified by the 1964 Wilderness Act. Most visitors who wander these islands, whether by water or land, experience them, in the words of that Act, "as an area where the earth and its community of life are untrammeled by man, where man himself is a visitor who does not remain." Permanent improvements and human habitations are few, and those that do exist are often so subtle that many visitors fail to notice them. Whether one sails, kayaks, boats, hikes, or camps, opportunities for solitude are easy to find. Wild nature is everywhere.

And yet: the Apostle Islands also have a deep human history that has profoundly altered the "untouched" nature that visitors find here. The archipelago has been inhabited by Ojibwe peoples for centuries, and remains the spiritual homeland of the Red Cliff and Bad River Ojibwe bands whose reservations lie just across the water. Ojibwe people continue to gather wild foods here as they have done for centuries. The largest of the islands, Madeline, was the chief trading post on Lake Superior for French and native traders from the seventeenth century forward. Commercial fisheries have operated in these waters since the mid-nineteenth century, with small fishing stations scattered among the islands for processing the catch in all seasons. The islands saw a succession of economic activities ranging from logging to quarrying to farming. Most have been completely cut over at least once. The Apostles possess the largest surviving collection of nineteenth-century lighthouses anywhere in the United States. Finally, tourists have sought out the islands since the late nineteenth century, and they too have left marks ranging from lodges to cottages to docks to trails as evidence of the wilderness experience they came to find.

All of this would seem to call into question the common perception among visitors that the Apostles are "untouched," and might even raise doubts about whether the National Lakeshore should be legally designated as wilderness. But although most parts of these islands have been substantially altered by past human activities, they have also gradually been undergoing a process that James Feldman, an environmental historian at the University of Wisconsin–Madison who is writing a book about the islands, has evocatively called "rewilding." The Apostles are thus a superb example of a wilderness in which natural and human histories are intimately intermingled. To acknowledge past human impacts upon these islands is not to call into question their wildness; it is rather to celebrate, along with the human past, the robust ability of wild nature to sustain itself when people give it the freedom it needs to flourish in their midst.

Should the Apostle Islands National Lakeshore become part of the National Wilderness Preservation System? Emphatically yes.

But to answer the question so simply is to evade some of the most challenging riddles that the Apostle Islands pose for our conventional ideas of wilderness. In a much altered but rewilding landscape, where natural and cultural resources are equally important to any full understanding of place, how should we manage and interpret these islands so that visitors will appreciate the stories and lessons they hold? If visitors come here and believe they are experiencing pristine nature, they will completely misunderstand not just the complex human history that has created the Apostle Islands of today; they will also fail to understand how much the natural ecosystems they encounter here have been shaped by that human history. In a very deep sense, what they will experience is not the natural and human reality of these islands, but a cultural myth that obscures much of what they most need to understand about a wilderness that has long been a place of human dwelling.

If this is true, then the riddle we need to answer is how to manage the Apostle Islands as a *historical* wilderness, in which we commit ourselves not to erasing human marks on the land, but rather to interpreting them so that visitors can understand just how intricate and profound this process of rewilding truly is.

Among my favorite places for thinking about rewilding is Sand Island, at the extreme western end of the archipelago. Most visitors today disembark at a wooden pier on the eastern side of the island, and then hike more than a mile to reach the lovely brownstone lighthouse at the island's northern tip, constructed way back in 1881. Built of sandstone from another island, it is an artifact of an earlier phase of Apostles history that has now vanished except for the overgrown quarries one still finds in the woods. Gazing out at the lake from atop the tower, it is easy to imagine that this is a lone oasis of civilization in the midst of deep wilderness.

But the path you walk to reach this lighthouse is in fact a former county road. If you look in the right place you can still find an ancient automobile rusting amid the weeds. Frank Shaw homesteaded the southeastern corner of Sand Island in the 1880s, and by 1910 more than seventy people—most of them Norwegian immigrants—were living here year round. Sand Island had its own post office and general store. Island children had their own one-room school. There was even telephone service to the mainland, though it soon failed and was abandoned.

How did Sand Islanders support themselves in this remote rural settlement? Fishing was of course a mainstay. Logging went on occasionally, and from the 1880s forward the summer months saw a regular stream of tourists. But for several decades islanders also farmed. Few who visit this "pristine wilderness" today will recognize that the lands through which they hike are old farm fields, but such in fact they are. Indeed, look closely at the encroaching forest that was once Burt and Anna Mae Hill's homestead and you will quickly realize that the trees are not much more than half a century old. Indeed, some of the oldest are apple trees, offering mute evidence—like the lilacs and rose bushes that grow amid ruins of old foundations elsewhere on the island—of past human efforts to yield bounty and beauty from this soil.

The old orchards are in fact a perfect example of rewilding, since Burt Hill's farm still shapes the local ecology. As James Feldman describes the process, "In some areas of the clearing, willow, hawthorn, mountain ash, and serviceberry have moved into the sedge meadow in straight, regular lines, following the drainage ditches dug by Burt Hill when he expanded his farming operations in the 1930s." Nature alone cannot explain this landscape. You need history too.

The dilemma for the Park Service, then, is deciding how much of the Apostle Islands to designate as wilderness, and how to manage lands so labeled. More bluntly: should Burt Hill's orchard count as wilderness? And if it does, should park managers strive to erase all evidence of the Hills' home so visitors can imagine this land to be "pristine"?

What makes these questions so difficult is that the 1964 Wilderness Act and current National Park Service management policies draw quite a stark—and artificial—boundary between nature and culture. The implication of this boundary is that the two should be kept quite separate, and that wilderness in particular should be devoid of anything suggesting an ongoing human presence. Under the 1964 Act, wilderness is defined as a place that "generally appears to have been affected primarily by the forces of nature, with the imprint of man's work substantially unnoticeable." Strictly interpreted, this definition suggests that the more human history we can see in a landscape, the less wild it is. A curious feature of this definition is that it privileges visitors' perceptions of "untrammeledness" over the land's true history. It almost implies that wilderness designation should depend on whether we can remove, erase, or otherwise hide historical evidence that people have altered a landscape and made it their home.

Because this strict definition can exclude from the National Wilderness Preservation System too much land that might otherwise deserve protection, the less-well-known 1975 Eastern Wilderness Act offers an important counterpoint that is especially relevant to the Apostle Islands. It declares that wilderness areas can be designated east of the Hundredth Meridian even where land has been grazed, plowed, mined, or clear cut—land, in other words, that the 1964 Act would emphatically regard as "trammeled." Unfortunately, the implications of the 1975 Act have still not been fully appreciated, so that federal managers continue to remove historic structures and artifacts in a misguided effort to fool visitors into believing they are experiencing a "pristine" landscape.

For instance, current NPS management policies adopt a strict definition of wilderness comparable to the 1964 Act in declaring that "the National Park Service will seek to remove from potential wilderness the temporary, non-conforming conditions that preclude wilderness designation." The bland phrase "non-conforming conditions" generally refers to any human imprints that diminish the impression that a wilderness is "untouched"—imprints, in other words, that constitute the chief evidence of human history. As Laura Watt has suggested in her valuable study of Park Service management at Point Reyes in California, "The Trouble with Preservation, or, Getting Back to the Wrong Term for Wilderness Protection," NPS efforts to create the appearance of pristine wilderness—even in a heavily grazed and logged area like Point Reyes—have included the following:

- intentionally demolishing historic structures;
- promoting natural resources at the expense of cultural ones;

- implying that dramatically altered landscapes are much more pristine than they truly are;
- privileging certain historic eras over others; and
- refusing to interpret for park visitors the human history of places designated as wilderness.

At both Point Reyes and Apostle Islands National Lakeshore, Park Service managers have ironically become the principal vandals of historic structures—tearing down ranches at Point Reyes, removing farms, fishing camps, and cottages at Apostle Islands—in an effort to persuade visitors that land remains untrammeled. Park visitors deceived by this carefully contrived illusion not only fail to see the human history of the places they visit; they also fail to see the many features of present ecosystems that are inexplicable without reference to past human influence. As Laura Watt points out, although the Park Service has long opposed the reconstruction of historic buildings and sites as inherently false and misleading, it shows much less compunction about false and misleading reconstructions of "natural" landscapes.

NPS management policies do call for the protection of "significant" cultural resources even on lands designated as wilderness, but such resources must meet very high standards of significance—generally, listing on the National Register—to merit protection. As a result, NPS generally forces managers to choose between two mutually exclusive alternatives, wild and nonwild. One either designates an area as wilderness and tries to remove "non-conforming conditions" so as to manage it almost exclusively for wilderness values; or one designates an area as a cultural resource and manages it for values other than wilderness. The heretical notion that one might actually wish to protect and interpret a cultural resource in the very heart of wilderness so as to help visitors better understand the history of that wilderness is pretty much unthinkable under current regulations.

All of this may seem abstract and academic, but it has very practical implications for how Apostle Islands National Lakeshore and other parks are managed when designated as wilderness. Under NPS policies, "improvements" are to be held to a bare minimum in designated wilderness. This means that even if historic human structures and artifacts are permitted to remain (most would typically be removed or destroyed), the best one could hope for them would be stabilization, not active protection, restoration, or interpretation. Trails would be kept to a minimum, and their routes would emphasize nature over culture to encourage visitors' perception of untrammeled wilderness—even when, as at Sand Island, the trail is in fact an old road. Perhaps most importantly from the point of view of human history, interpretive signs would be removed altogether, so that historic features in the landscape that most visitors might otherwise miss could not be marked. Although one might hope that brochures, guidebooks, and displays in visitor centers would encourage visitors to look for evidence of these historic features, wilderness designation under current NPS policies would prevent them from being interpreted on the ground.

Why does this bother me so much? Because I can't help seeing the straight lines along which willows and serviceberries are invading Burt Hill's orchard. I can't help caring about all the dreams and hard work with which he planted these apple trees so long ago. For me, Burt and Anna Mae's story makes this wilderness all the more poignant, and I cannot understand why we think we need to annihilate the record of their lives so we can pretend to ourselves—pioneer-like—that no one before us has ever stood here.

What alternatives do we have? How might we combine designated wilderness with an equal and ongoing commitment to interpreting the shared past of humanity and nature? If we can answer this question for the Apostle Islands, I believe we can also answer it for many other landscapes whose histories also combine wildness with human dwelling. Among the suggestions I'd make would be the following:

Most importantly, we should commit ourselves to the notion that Apostle Islands National Lakeshore is and always will be a historical wilderness: for centuries in the past, and presumably for centuries still to come, human beings have played and will play crucial roles in these islands. Visitors should come away from the park with a deepened appreciation not just for the wild nature they find here, but for the human history as well.

The interpretive framework that can best integrate the natural and cultural resources of this park is James Feldman's concept of rewilding. It should be at the heart of what the park offers to visitors. Here is a natural landscape that has been utilized for centuries by different human groups for different human purposes: first by native peoples for subsistence, then for fur trading, then in turn for fishing, shipping, logging, quarrying, farming, touring, and other activities. Natural resources here have long been exploited as commodities, and island ecosystems have changed drastically as a result. The shifting composition of the forest, the changing populations of wildlife on the land and in the lake, the introduction of exotic species, the subtle alterations of geomorphology: all of these "natural" features also reflect human history. Visitors should come away with a more sophisticated understanding of them all.

Furthermore, these changes have not all been in one direction, which is why Feldman's narrative of rewilding can be a source of hope for all who support efforts at ecological restoration. Although parts of the Apostle Islands have been drastically altered by activities like clear cutting, wilderness is returning to such a degree that hikers can walk old logging roads and completely fail to realize that the woods through which they are traveling were stumps just half a century ago. I think they would learn more about restoration and rewilding if they could see those stumps in their mind's eye. We should be able to encounter an abandoned plow blade in the woods, or a rusting stretch of barbed-wire fence, or a neatly squared block of brownstone, without feeling that such things somehow violate our virginal experience of wilderness. We would do better to recognize in this historical wilderness a more complicated tale than the one we like to tell ourselves about returning to the original garden.

One of the most attractive features of Feldman's concept of rewilding is that it avoids the negative implication that past human history consists solely of exploiting, damaging, and destroying nature. As Feldman puts it, "rewilding landscapes should be interpreted as evidence neither of past human abuse nor of triumphant wild nature, but rather as evidence of the tightly intertwined processes of natural and cultural history." When we use words like "healing" to describe the return of wilderness to a place like the Apostles, we imply that past human history here should be understood mainly as "wounding" and "scarring."

Such words do no more justice to the complexity of human lives in the past than they do to our own lives in the present. They implicitly dishonor the memories of those like Burt and Anna Mae Hill who once made their lives here and who presumably loved these islands as much as we do.

In keeping with the principle that the Park Service should not be in the business of promoting illusions about a pristine wilderness with no human history, the default management assumption should be that existing human structures and artifacts will not be removed even from designated wilderness. No erasures should be the rule except where absolutely necessary. Even in instances where there are safety concerns about a collapsing structure, other solutions for protecting visitors should always be sought before resorting to destruction and removal. In a rewilding landscape, old buildings, tools, fencerows, and other such structures supply vital evidence of past human uses, without which visitors cannot hope to understand how natural ecosystems have responded to those uses. Moreover, such artifacts today stand as romantic ruins, haunting and beautiful in their own right. Far from diminishing the wilderness experience of visitors, they enhance and deepen it by adding complexity to the story of rewilding.

Moreover, not all structures and artifacts should be permitted to go to ruin. The Park Service has already worked hard (with far too little funding) to preserve the beautiful historic lighthouses that are among the most popular destinations on the islands. But a grave weakness of current Park Service interpretation is its extreme emphasis on lighthouses and fishing as if these constituted the sum total of past human activities in the islands. Equally important phases of island history remain almost invisible. Ojibwe and other native histories are only beginning to receive the attention they deserve, and the histories of later island residents often go entirely unmentioned.

An NPS commitment to interpreting all phases of Apostle Islands history would mean more than just tolerating the presence of romantic ruins in an otherwise wild landscape. Certain structures and artifacts are so important to visitor understanding of island history that at least a few need to be stabilized or restored, and actively interpreted. Nowhere can visitors now explore a former brownstone quarry with the benefit of informed interpretation to help them appreciate how important this industry was to the built environment of the United States during the closing decades of the nineteenth century. Visitors would look with entirely different eyes at the brownstone buildings in nearby towns if they were encouraged to see where that stone originally came from. The same goes for logging sites and especially for old farms. Visitors almost surely leave Apostle Islands National Lakeshore with no appreciation for farm families like Burt and Anna Mae Hill who once raised crops and children on these islands, even though the remnants of their farms are still visible on the ground and are still reflected in the ecology of the forests that now grow on abandoned fields.

The bias of historical interpretation in the Apostle Islands, like many other historic sites in the United States, is generally toward earlier, "pioneer" periods. One crucial human activity that goes almost entirely uninterpreted for tourists in the Apostle Islands is tourism itself. Many mid-twentieth-century tourist cottages have already been torn down as "non-conforming." So far, there has been no effort to preserve any of these structures as cultural resources in their own right, to help visitors understand how tourism has emerged over the past two centuries as one of the most potent cultural forces reshaping landscapes all over the world. (The designation of wilderness in Apostle Islands National Lakeshore is inexplicable without reference to this cultural force.) Interpreting the history of tourism should be just as important as interpreting the history of lighthouses and fishing, and at least a few early tourist structures need to be preserved if this goal is to be accomplished.

If I had my druthers, I would also permit limited signage and interpretation as tools for educating visitors and managers alike that the presence of cultural resources such as fishing camps and cottages in the midst of wilderness does not automatically degrade wilderness values or the wilderness experience. Does Aldo Leopold's shack or Sigurd Olson's cabin diminish the wild lands surrounding it? I honestly believe such cultural resources can enhance visitor appreciation of the complex history of rewilding landscapes. If we're to tell stories about ecological restoration, as surely we need to do if we're to envision a sustainable human future, we need to leave evidence on the ground that will bear witness to such stories.

I'm nonetheless willing to acknowledge that standardized bureaucratic rules and regulations may not easily accommodate the kind of interpretive ambiguities that I prefer. So the wiser, easier strategy is probably to think of wilderness in the Apostle Islands as existing along a continuum, from areas that will be treated as "pure" wilderness (even though they are full of historical artifacts that should not be removed) to highly developed sites like the lighthouses that are managed almost entirely for nonwilderness values.

I would argue for a few locations outside of the designated wilderness which, although still managed to protect wilderness values, could be modestly restored and actively interpreted so as to help visitors understand the historic landscapes of logging, quarrying, farming, and early tourism. One might consider designating them as "historical wilderness areas" to signal that they should be managed with an eye toward balancing natural and cultural resources more evenly than would typically be true in "designated wilderness."

Sand and Basswood islands are the obvious candidates to be designated as historical wilderness, because their histories are so rich and varied—encompassing fishing, logging, quarrying, farming, and tourism in addition to Ojibwe subsistence activities—and so can serve as microcosms for the whole archipelago. These islands could be regarded almost as classrooms for historical wilderness, where visitors can learn about the long-term cultural processes that have in fact shaped all of the Apostles. Then, when they visit the designated wilderness where much less interpretation is permitted, their eyes will be trained to see the rewilding process they will witness there.

What are the chances that this new approach to protecting wilderness might actually succeed in the Apostle Islands? Surprisingly good. The Park's superintendent, Bob Krumenaker, has been both visionary and eloquent in refusing to choose wilderness over history—or history over wilderness. "I don't think, if we do it right," he says, "that wilderness has to entail either balancing nature and culture—which suggests one gains while the other loses—or sacrificing one at the expense of the other. We can preserve both nature and culture at the Apostle Islands and should embrace the chance to do so."

Like Krumenaker, I favor educating visitors so they will recognize that wilderness can have a human history and still offer a flourishing home for wild nature. If we adopt such a strategy for managing wilderness in Apostle Islands National Lakeshore, the park can offer a truly invaluable laboratory, with implications far beyond its own boundaries, for rethinking what we want visitors to experience and understand when they visit a wilderness that is filled equally with human and natural histories.

Indeed, among the most precious experiences that Apostle Islands National Lakeshore can offer its visitors are precisely these stories. Management policy in the National Lakeshore should seek to protect wilderness values and historic structures, certainly, but it should equally protect stories—stories of wild nature, stories of human history. It is a storied wilderness. And it is in fact these stories that visitors will most remember and retell, even as they contribute their own experiences to the ongoing history of people and wild nature in the Apostle Islands.

NOTE

1. Editors' note: On December 8, 2004, 80 percent of the Apostle Islands National Lakeshore was designated as the Gaylord Nelson Wilderness Area. Gaylord Nelson died on July 3, 2005.

Rolf O. Peterson

Letting Nature Run Wild in the National Parks (1995)

> *"Natural" is a magician's word—and like all such entities, it should be used sparingly lest there arise from it, as now, some unglimpsed, unintended world, some monstrous caricature called into being by the indiscreet articulation of worn syllables.*
>
> Loren Eiseley, *The Firmament of Time,* 1960

THE SUN SHONE AUSPICIOUSLY on a bright April morning in 1988. I felt inappropriately bundled up in the warmest clothes I owned as our 26-foot National Park Service patrol boat nudged its way through a vast field of half-inch-thick ice in the middle of Lake Superior. I heard a sound like the tinkling of fragmented glass as broken shards of ice skittered across the frozen surface on both sides of the boat. But the "night ice" would be gone by midday, shattered by the slightest of breezes across the world's largest expanse of fresh water.

The fine weather was a good omen, yet our mission and our boatload of wolf traps engendered a sense of foreboding. For the first time, the wolves of Isle Royale were to become targets, and I, their longtime observer and admirer, would become "the hunter." In my heart, I didn't much care for the idea of capturing them—even for purposes of temporary study. But I felt compelled to expose them to possible risks, in order to ascertain the causes of their decline and possible extinction.

For years many people maintained that an aura surrounded the Isle Royale wolves, simply because they had never been handled by humans. We had considered leaving the wolves untouched; allowing them to live or die without an attempt on our part to understand why, but scientists and managers both inside and outside the National Park Service rejected this option. After thirty-odd years of observation, it was important to write the next, and perhaps the last, chapter with the best possible knowledge. Yet for the wolves and for me, nothing would ever again be quite the same.

I wondered if any of the 12 wolves remaining on Isle Royale were watching the approach of our boat, with its synchronous drone of twin engines and its smell of gasoline. The old animals among them surely knew that people and machines routinely appear in spring, and that wolves would once again have to yield the island's network of hiking trails. But surely no wolf could anticipate my plans to capture and examine them, and this thought bothered me. For almost 20 years, I had accorded them every privilege of complete freedom. My behavior, on behalf of science, had always been benign and predictable, and perhaps they had learned to trust me.

Time for reflection ended when the patrol boat dumped park staffer Bob Krumenaker and me on the rocky beach at our summer research cabin, along with a veritable mountain of gear. It was too late to change our minds.

Bob was the natural resource management specialist for Isle Royale National Park, and he and I together had borne much of the burden of response to the "wolf crisis" of the late 1980s. Now, for a few precious days, we were the only people on the big island—all 210 square miles of it. We allowed ourselves one day to walk the trails and choose a few short miles along which to set up our trapline. We enthused over each bit of fresh wolf sign, glad to find that there were still some wolves left and thankful that they were traveling the very trails we planned to trap.

Along the barren stretches and during lunch, Bob and I had the opportunity to chat. Over the previous six months, we had been consumed by all the details of our immediate task. Now we quickly dispensed with our mental checklists of procedures and wolf-trapping paraphernalia, which we had gone over countless times before, and went on to tackle larger matters. On a bright day in spring, it seems that one can solve all the problems of the world.

Bob and I agreed that it was wise to take the risk involved in handling the wolves, but we found ourselves on opposite sides when we addressed the question, "What should we do if all the wolves on Isle Royale die out?" For me, it was a disquieting discussion, for I then realized that NPS managers might allow wolves on Isle Royale to disappear, and might not welcome wolves back unless they returned of their own accord. The wolf-prey system that had worked so well for so long might simply end—period.

There is no land management agency in the United States that more earnestly seeks to preserve nature than the National Park Service. In recent decades, the Service has moved strongly toward nonintervention as a primary strategy, especially in large parks free of crushing outside influences. Bob Krumenaker was often more willing to consider alternatives than the agency he worked for, but in this case Bob was an able spokesman for the view that the wildness of Isle Royale would be diminished if, for any reason, humans tried to reverse wolf extinction on the island. We both agreed that these wolves played a vital role in a wild community. For me, the operative word was "vital," while Bob dwelt on "wild." According to Park Service management philosophy in the late 20th century, the *wild* portion of *wilderness* depends on minimizing overt human manipulation.

One person cannot hope to present objectively all the legitimate views on what the Park Service should do in the event of imminent wolf extinction on Isle Royale. Bob and I took a stab at this in a paper published in 1989, but the questions we raised were too hypothetical (or distressing) to be taken very seriously then.[1] At that time, we introduced most of the issues I again address here.

With the benefit of more knowledge and further exploration of Park Service policies, this chapter represents my own perspective on future options for wolves on Isle Royale. To the extent that I can, I've tried to enunciate all possible viewpoints, in the hope that the National Park

Service—and the public—will pay attention. This exploration of management possibilities was my own journey, and others may come to a conclusion different from mine. It is the journey itself, the honest evaluation of objectives for Isle Royale and the weighing of values, that I hope each reader of this book will undertake.

It is appropriate to state that my scientific career is not dependent on the wolves' existence. I could address problems on the island without wolves. As a scientist, however, it has been my professional preoccupation to explore the role of wolves in nature, and Isle Royale happens to be one of the best places on the planet to do this. Perhaps I am handicapped by a fascination with the wolves; I care deeply about their future, for I have learned that these animals help sustain life as we know it on Isle Royale.

In the spring of 1988, our immediate concerns took precedence over long-term considerations. Within five days Bob and I were handling wolves—gambling with the crown jewels, as it were. Over the next several years, almost every wolf alive on the island in 1988 was weighed, measured, examined, blood-sampled for disease and genetic studies, radio-collared and then promptly sent on its way. Much new knowledge was revealed by the more intensive study, and no wolves were sacrificed in the process. During this period, I concentrated my efforts on gathering data, expecting that the new information would produce a better vista for making management recommendations.

We learned that a suspected pathogen, canine parvovirus (CPV), had indeed invaded the island, but then died out. Its occurrence corresponded exactly to the wolf crash of 1980 to 1982, and to the chronic high mortality of 1982 to 1988. As an invisible agent of death for dogs both domestic and wild, CPV was carried inadvertently to Isle Royale, and to every other corner of the world, by people. All evidence—circumstantial though it was—pointed to CPV as the cause of the dramatic wolf decline and the high mortality of the 1980s.

Antibodies to CPV in the wolves' blood disappeared after 1988, but during the rest of their lives the surviving wolves usually failed to reproduce at normal rates. CPV was gone, so something else must explain the poor reproduction in Isle Royale's wolves. Food shortage did not seem to be an important factor, as the number of old moose present in

1990 was identical to the level of the 1970s, when twice as many wolves were present. Genetic losses remained a possible cause of their troubles, perhaps producing inbreeding depression, or poor early survival. However, we actually know little about the significance of lost genetic variability in isolated populations living in the wild.

Molecular studies confirmed the worst-case genetic scenario for Isle Royale wolves—they were heavily inbred. The high death toll of the 1980s linked to an introduced disease, had produced such a low point in wolf numbers that random events alone could easily snuff them out. Rapid genetic decay brought on by the passage of many generations might accentuate the risks. Scientific prediction and raw probability seemed to predict the demise of the wolves, and each year of poor reproduction was another nail in their collective coffin. Years went by, and the wolves failed to recover.

By 1993 the wolf population was top-heavy with old animals that seemed unable to replace themselves. Only three females, all quite old, were known to exist among the 13 wolves left, and only one of these had ever successfully reared young. But in the same year two of these females, in adjacent packs, reared four pups each—a normal litter size. This surprising change dramatically improved the odds of survival for the wolf population. A new generation had finally materialized, ready to take over when the parents succumbed to old age. Indeed, less than a year after we discovered the eight new pups, three of the four parents died.

Meanwhile, incriminating evidence from other areas linked CPV to the population dynamics of wild wolves. Historical analysis by Dave Mech showed that the ebb and flow of wolf numbers in northern Minnesota was correlated with exposure to CPV. In 1994, a wild wolf collared by Mech died from this disease.

In the same year, as evidence mounted that disease, not food shortage, had caused the wolf crash on Isle Royale in the previous decade, the NPS offered vaccination against parvovirus as an allowable intervention. After soliciting expert advice, I declined to vaccinate the wolves. Our goal, remember, had less to do with saving the current population than with learning as much as we could about the real-world dangers faced by a small population of wolves. We had completely missed the initial parvovirus outbreak, but now, remarkably, Isle Royale was parvo-free. Antibodies to CPV in the wolves themselves were our best indicator of disease exposure. Vaccination, which uses a modified version of the live virus to stimulate protective antibodies, would also make it impossible to determine whether CPV itself was present. The opportunity to learn more about disease risks outweighed other options, and with the concurrence of the NPS, I chose to leave the wolves untreated and unprotected.

The decision not to vaccinate was made for scientific reasons, but I sensed later that I had inadvertently assisted the NPS down a slippery slope of nonintervention. Passive observation can be an easy policy that doesn't require much expense or ecological understanding; perhaps that explains some of its appeal. But our national parks deserve better than rote adherence to tradition.

In the mid-1990s, with new data compiled, it is time to examine rigorously the policy options available to the NPS regarding the future of wolves on Isle Royale. As Winston Churchill once quipped, "At times it becomes necessary to do what is required." Managers could continue a hands-off tradition, allowing the wolves to die out and, should that occur, waiting for wolves to return on their own. If this is the inclination of Park Service managers, however, it should be stated openly and supported by scientific and aesthetic arguments, and the public should have a chance to voice its opinion.

As we consider the future of Isle Royale's wolves, we should think of our national parks as both laboratories and cathedrals.[2] There is genuine creative tension between science and soul, reason and myth. It is appropriate to celebrate this tension in our national parks, not to bury it in administrative rules or stale traditions. The wolves of Isle Royale have served both science and the human spirit for many years—ask any visitor. Science simply illuminates in a modest way that which invigorates the human soul. Let no park manager (or scientist) forget the importance of the latter!

Shelved in the headquarters of every national park are thick three-ring binders containing the "Management Policies of the National Park Service." After reading these tomes it would seem, on the surface at least, that the NPS is well-prepared to meet any challenge of the twenty-first century. However, the subtle yet serious problems our parks now face cannot be solved without thoughtful attention and creative action.

In recent decades, NPS management has been influenced by a certain worldview; an expedient myth summarizes a twentieth-century view of nature first developed in North America by descendants of European immigrants. Simply stated, it says "nature knows best," assuming that "nature" includes all life and processes apart from humans. Whenever humans run roughshod over the rest of nature, problems occur; one can learn more of this from Aldo Leopold, Rachel Carson, and dozens of later writers. However, leaving humanity out of nature is simply naive. Absolute wilderness (where the effects of humans are absent) is a myth; human influence pervades every corner of the earth. Natural events at Isle Royale in the last half of the twentieth century have helped reinforce a noninterventionist policy of management within the NPS, but the limitations of this approach may become most clear at Isle Royale. Perhaps here, the human animal can find its proper place. In the words of Paul Tillich, "Mankind becomes really human only at the time of decision."

For over four decades, Isle Royale embodied the notion of the forest primeval, a world in ecological equilibrium—that beautiful and elusive "balance of nature." This balance was effected most readily in "absolute wilderness," in the complete absence of human direction. For decades, people said, "After all, look at Isle Royale."

The hands-off philosophy that directed the prevailing NPS management notion fit hand in glove with a powerful idea in Western civilization—that the human species is distinct from other forms of "lower" animal life, that we have somehow risen above nature, that nature will operate properly only if we are kept out of the picture. It is an old idea that can be traced back to treatises by Greek and Roman scholars. But it is a perception that ecologist Daniel Botkin believes to be "one of the main impediments to progress on environmental issues."[3] The view that modern humans have no legitimate role as players in the natural world of wilderness is a pervasive one. I understand it, and I have sympathized with it in some instances. But in fact, we are natural creatures rooted in the earth, and it is by our unique mental capacities that "nature" and "natural" are defined. Ralph Waldo Emerson considered nature to be the "shadow of man," and Loren Eiseley felt that "no word bears a heavier or more ancient, or more diverse array of meanings."

In his book *Discordant Harmonies,* Botkin states that we must recognize that humans already play a role in every ecosystem on earth, and that we have the capacity to intervene softly on behalf of all life. He argues that the smaller the size of a "natural" area, the greater the need for human involvement to maintain important ecological processes. We must mitigate for the constrictions we have already imposed on nature.

With the laboratory and cathedral metaphor in mind, let us look for guidance from the management policies of the NPS. In our national parks, for example, there is a clear dictate to favor the conservation of native species, as opposed to those that arrived with the assistance of humans. Native species belong to the primeval communities present before European civilization arrived to mess things up. However, one must deal with the surprising fact that moose were evidently not present on Isle Royale prior to the twentieth century. Both the absence of moose on the island in the nineteenth century and the increased numbers of moose in northern Ontario, which precipitated their arrival on Isle Royale, had everything to do with the spread of European civilization. Europeans first provided firearms to native humans, then virtually eliminated the natives and logged and burned vast regions, creating favorable conditions for moose.

Careful archaeological work by the NPS has revealed much evidence over the past 4,000 years of Native Americans, caribou, and beaver on Isle Royale, but no indication that moose or wolves inhabited Isle Royale before 1900. Thus the NPS policy of maintaining "native" species cannot clearly guide us in our current quandary. In an ironic blend of tradition and history, one might argue that neither the wolf nor the moose are purely "native" species at Isle Royale.

Another NPS management policy states that natural processes will be relied upon as much as possible to regulate wildlife populations. At Isle Royale, it is abundantly clear that wolf predation has helped control the moose population, thereby influencing the entire forest community. One can actually find indirect evidence of the influence of wolves in growth rings of the island's trees.[4] Wolf predation certainly qualifies as an important natural process that could be maintained, according to policy. However, is not extinction also a natural process? Animal populations on islands are naturally prone to wild swings and high rates of extinction. There is a large element of chance in the makeup of an island's fauna because of the limited number of species. Should we favor predation, or species extinction? NPS policy does not tell us which natural process should take precedence.

If wolves were extirpated naturally and no human causation was detected, could they also return naturally, or has human activity altered the recolonization possibilities for wolves? This is a relatively easy question to answer, as there are well over 100,000 people in the city of Thunder Bay on the mainland shore. The city comes complete with its network of highways and rail lines, which reduce the likelihood that wolves would inhabit the shoreline and make the run to Isle Royale.

That run to the island, of course, *must* be done in winter, across the ice of Lake Superior. Given the long-term prospect of global warming, can we really assume that ice will form on Lake Superior with a "natural" frequency? Throughout the warm decade of the 1980s, which followed a decades-long warming trend, there were almost no ice bridges from Isle Royale to the mainland. There is a broad scientific agreement that the warming of the global environment can be attributed to the build-up of CO_2 and other greenhouse gases, resulting from the combustion of fossil fuels.[5] Must we be able to assess the local impacts of historic warming on the entire northern hemisphere before we decide whether wolves have a future on Isle Royale? The more one is drawn into this question, the more convoluted the answers become.

In one area, at least, NPS policy would seem to provide a clear path. NPS management recognizes the need to eliminate exotic species, and to mitigate past human disturbance. This is an enlightened view, for exotic organisms transported by humans around the earth have been responsible for most species extinctions in recorded history.

In the case of invisible disease organisms, although there is no practical way to wall off the parks, it will sometimes be possible to mitigate after the fact. In view of the strong circumstantial evidence linking the collapse of the wolf population from 1980 to 1982 (and the decline thereafter) to an introduced disease, it seems legitimate to restore the wolves after the mayhem passes. Yet the National Park Service has not faced this particular combination of circumstances elsewhere, and it is understandably reluctant to step in.

When the advice of Fraser Darling, a noted wildlife scientist, was sought concerning elephant "overpopulation" in Kenya's Tsavo National Park, he responded that, "The surest road to the right answer usually lies along a simple path uncovered by common sense." Yet E. O. Wilson, also a noted scientist, considered common sense to be "that overrated capacity composed of the set of prejudices we acquire by the age of 18."

Common sense may imply an appeal to mass opinion, which could be ill-advised when applied to the little-understood complexity of nature within our national parks. I prefer to use the term "ecological realism" as a guideline, under which options are laid out based on our best understanding of ecological relationships. In that vein, we can assume that moose are likely to be part of the fauna that inhabit Isle Royale National Park for the foreseeable future—for at least another century. We have already seen that wolves have reduced moose density for long periods of time, enabling plant communities to develop that would not even appear if moose population growth was unchecked by wolves. There can be no natural process more closely aligned with the behavior and evolution of moose than wolf predation, that inscrutable agent of natural selection.

Moose are what they are due to eons of close shepherding by wolves. In the words of the poet Robinson Jeffers,

> What but the wolf's tooth whittled so fine
> The fleet limbs of the antelope.

It seems only prudent to maintain wolf predation on Isle Royale as long as moose continue to inhabit the place. No one can say how long that might be; over the span of centuries, few things in nature are permanent. According to the Roman poet Lucretius, writing in the first century B.C., "not one thing is like itself forever."

Adherence to policy directives or appeals to common sense do not address all the values at stake at Isle Royale. The unique characteristics of this park demand specific consideration, and both science and soul should be part of the formula.

In the 1990s, at the urging of the National Academy of Sciences, the NPS began to prepare a new vision—one that included an explicit role for science in the management of our parks. The motto "Science for parks, and parks for science" was bandied about, and there were calls for institutional change. These ideas were reminiscent of those expressed 30 years before in the 1963 Leopold Report, which reviewed the role of science in national park management: "We recommend that the NPS recognize the enormous complexity of ecologic communities and the diversity of management procedures required to preserve them."[6] A similar recommendation was made by an earlier National Academy of Sciences report, which was also released in 1963. According to more than a dozen additional official reviews issued since that time, an absence of adequate science and monitoring to ensure long-term ecological integrity will reduce national parks to scenic pleasuring grounds with an uncertain future.

Allowing nature free rein in national parks is not as easy as one might suppose, even when the effort is backed by good science and determination. One scientific success within our national parks involved demonstrating the essential role that fire plays in maintaining diverse types of forest communities. Because of this new understanding, the rigid fire suppression programs that had been part of park management since the NPS was established were slowly relaxed in the 1970s.

But nature dishes out extremes on a grand scale, as the world discovered during the Yellowstone fires of 1988. Such large natural processes are not easily accommodated in and around most national parks, regardless of policy. Before the smoke had cleared, the U.S. Congress demanded a thorough review of fire management in national parks. For the next five years, until Isle Royale National Park produced a revised fire management plan that had been signed off at all levels of the bureaucracy, all lightning-started fires were suppressed on the island.

As fate would have it, a single dry thunderstorm in August of 1991 started four fires on Isle Royale in one evening, something I had not observed in the previous two decades. If they had not been extinguished by fire crews, these fires might have profoundly changed the character of the island. We will never know. A former Interior Department official sent a brief note acknowledging this news in my annual report, volunteering that it was "unconscionable" for the NPS to put out natural fires on Isle Royale. After 1993, with an approved fire management plan on line, lightning-started fires were once again allowed to burn, under prescribed conditions. The catch? It may be more expensive to allow a fire to burn naturally than to extinguish it, because of the demanding monitoring schedule stimulated by Congressional demands. So it goes.

As a tool, science is moot on the question of whether wolves belong aboard the Isle Royale ark. Yet science has given us a glimpse of what we might expect in their absence. The wolf reduction of the 1980s illustrated how wolf predation had previously kept moose in check and allowed the forest to grow. In the complete absence of wolves, moose might so destroy their own resource base that they would face extinction themselves. This almost happened to reindeer on St. Matthew, a tundra island in the Bering Sea.

At Isle Royale, where plant life is much more productive, it is likely that moose would, through a series of increases and crashes, simply dig in for the long haul. The successful moose would be those that were adapted to extreme resource scarcity. This adaptation is accomplished by growing to a smaller size, with a correspondingly smaller demand for resources. Such developments are evident in island fauna scattered across the earth—caribou on the Slate Islands in Lake Superior, black-tailed deer on Alaska's Coronation Island, red deer on the island of Rhum off the coast of Great Britain and, in the southern hemisphere, reindeer on the island of St. Georgia. An extreme example, a truly miniaturized elephant only a few feet tall, existed on several Mediterranean islands during the Pleistocene. Striking fossil evidence from red deer on the British island of Jersey, off the coast of present-day France, indicates that such miniaturization might be accomplished in as little as 6,000 years—a mere 2,000 generations.[7]

The scientific value of allowing Isle Royale to remain wolfless would involve the study of a runaway population of large herbivores and its effects on plants. How high would the moose population go with only starvation to stop them? What particular combination of physical deterioration and severe winter weather would precede moose die offs? How much would annual production of calves be reduced as moose density increases? What new direction might the plant community and all other life take in the absence of a top carnivore? Better answers would be possible now than could be obtained in the 1930s, when all this happened before, because our tools have improved. However, general answers to these questions are already available from other areas.

The frequency of wolf recolonization and extinction in the absence of human interference is possibly of scientific interest. The mechanics of extinction for small populations are certainly important to understand, and this quest explains the consensus that allowed the wolf population to flounder in the late 1980s. In the event of extinction, the schedule for wolf reappearance would depend on human development on the mainland, ice cover on Lake Superior, and the dispersal idiosyncrasies of wolves inhabiting the mainland shore—all circumstances unique to Isle Royale. Larger questions of broad scientific interest would be unapproachable.

I believe we stand to gain more, scientifically, by furthering the existence of wolves on Isle Royale, and by propping them up when necessary. A minimal maintenance program would probably suffice, as a recent mathematical simulation suggests that the mean time before extinction for small, isolated wolf populations is at least several decades. One important question to explore is the robustness or repeatability of the wolf-moose relationship at Isle Royale. When wolves entered the scene in the late 1940s, moose were evidently at a very low level. Could wolves also successfully regulate an overpopulation of moose? Could an infusion of new genes from mainland wolves perk up a wolf population suffering from genetic isolation and inbreeding? What about the long-term dynamics of wolf and moose populations? Population fluctuations must occur on a time scale of decades for such long-lived animals, and there is no other locale known where the nature of their fluctuations is likely to be discovered.

Scientific value aside, however, the most influential arguments regarding the future of any national park will be spiritual, or inspirational. The Hubble telescope project had support from the public because people were inspired by the prospect of visible images from the edge of the universe, not by millions of bits of data stored in the computer banks of scientists. While the scientific perspective can be distilled to black and white choices, more subjective values have legitimate appeal.

Spirit is a powerful influence in the management of national parks and wilderness. For many wilderness advocates, an island left to the devices of nature, not overrun by technological humans, conjures up powerful images. For some people, even well-intentioned human intervention may degrade a wilderness. Isolated in a tempestuous lake, little visited by humans (by NPS standards), closed in winter, and replete with large carnivores as well as their prey, Isle Royale has clearly been left to nature, not man. Who would care to change that?

The wolf is more than simply a member of the Isle Royale fauna. Ecologically, it plays a disproportionately important role as a top carnivore. More significantly, the wolf is the enabler of a successful marriage between cathedral and laboratory. It is an important image—one that allowed Isle Royale management to escape the angst and attention associated with other national parks, where natural areas have been more significantly compromised by the ills of modern civilization. Since the wolves arrived, the NPS has been in an enviable position at Isle Royale, because the island's fauna and flora got along rather nicely without human help. A very capable park superintendent once confided to me that his biggest management challenge at Isle Royale was directing his own staff.

In avoiding the bumpy road that other parks have been forced to follow, Isle Royale has stimulated major advances in public understanding of natural areas. In 1931, when Isle Royale was designated a national park, the NPS itself was less than 20 years old, the 1916 legislation that established the agency set forth a vision for our national parks that has undergone little change in the decades that followed:

> ". . . to conserve the scenery and the natural and historic objects and the wildlife therein, and to provide for the enjoyment of the same in such manner and by such means as will leave them unimpaired for the enjoyment of future generations."

Early parks were opened up as attractive recreation areas and managed as facades of scenery, with little or no understanding of their ecological underpinnings. Because of the utilitarian attitudes of early twentieth-century Americans, wolves and mountain lions were eliminated to protect game animals, bears were fed in public sideshows, laundry operations were set up in natural hot springs, and resorts were placed in the middle of key natural features. Society is still paying dearly for some of these early mistakes.

In 1929, as momentum grew to create a national park at Isle Royale, the Michigan state legislature funded a study of the island's natural resources and archaeology. The $15,000 dedicated to this task may be, in constant dollars, the greatest single financial commitment ever made to study the resources of Isle Royale.

Enter Adolph Murie, a wildlife biologist with a new PhD degree, who came to Isle Royale in the summers of 1929 and 1930 in order to study its fauna. Murie would eventually conduct the first scientific study of wolves in Alaska's Mt. McKinley National Park; at Isle Royale, he launched what became one of the first studies of moose conducted anywhere.

After two to three decades of unrestricted growth, the moose population at Isle Royale had reached an unheard of density in the late 1920s, and Murie concluded that both the ecosystem and the moose were suffering. Murie was a staunch preservationist whose wilderness vision for Isle Royale was extreme, but he nonetheless felt that the runaway moose population had to be limited through human action. He recommended a moose reduction through public hunting, state-authorized shooting, live trapping and removal, or the introduction of large carnivores such as wolves. The "land should not be teeming only with moose," Murie wrote, "but teeming with all of nature."[8] Such holistic thinking was atypical within the national parks of the 1930s, but it was a hallmark of philosophy for Adolph Murie and his older brother Olaus, a wildlife biologist who gained prominence for his studies of park wildlife and his advocacy of wilderness.

When Murie's study and recommendations were finally published, after a several-year delay caused by lack of funding, Isle Royale was mostly public land in limbo. Its national park status had been approved by Congress in 1931, but there was no money appropriated for land acquisition. Through most of the 1930s, the island was owned primarily by the state of Michigan, a custodian whose interest waned as full park status slowly became a reality. Murie's recommendations sat on the shelf, and his prediction of an inevitable die off of moose was realized in 1934. Later, the Michigan Department of Conservation arranged to move 71 moose by boat from Isle Royale to the mainland, hoping to relieve population pressures and reestablish moose on the mainland. It was a heroic effort, but of no consequence to the island's huge moose population.

In 1936 Adolph Murie took a job with the NPS, directing Civilian Conservation Corps projects in western parks. He was asked to summarize his vision for managing Isle Royale as a national park. His comments were direct and uncompromising when it came to the wilderness potential of the island. Wilderness was, he felt, the greatest value of the park. In Murie's view, the park should not be developed as a popular summer resort; he even argued against the establishment of a system of hiking trails. He felt the NPS should "secure a personnel which has a feeling for wilderness and an understanding of wilderness values."[9]

Murie would probably have considered the term "wilderness management" an oxymoron. He suggested advising NPS administrators that their success depended not on "projects accomplished, but by projects sidetracked." Such candor in official reports was rare, even in those days. The previous year Adolph's brother Olaus, together with the likes of Robert Marshall and Aldo Leopold, formed the Wilderness Society, an organization dedicated to the preservation of wilderness.

The NPS took a more traditional tack than the one recommended by Murie, allowing some development and suppressing "unsightly" natural fires. Wildfires that began in logging slash had raged through the island in 1936. Nevertheless, the wilderness character of Isle Royale was an influential ideal, unique among national parks of the day.

After the "catastrophic" moose decline during the spring die off of 1934 and perhaps additional declines after the fire of 1936, moose numbers rebounded in the 1940s. Another biologist, the U.S. Fish and Wildlife Service's Laurits Krefting, sounded an alarm by calling for moose controls—either via the gun or through introduction of a large carnivore such as the wolf. The "moose question" at Isle Royale was again a major issue. The NPS response was another study, conducted this time by James Cole, a Park Service biologist. Cole reported that winter browse was abundant, and believed the moose population needed no public assistance. The Park Service encouraged private efforts to release wolves on Isle Royale in 1952, as a partial solution to the concerns over moose numbers, but this failed when wolves released from the Detroit Zoo proved to be too tame. In any case, wild wolves got there first, and the rest is history.

By the 1960s, when the NPS began to undertake the challenge of managing living landscapes, Isle Royale was widely recognized as an ecosystem with all of its parts intact. Historian Al Runte claimed that among national parks, Isle Royale "had come closest to the ideal ecological preserve."[10] No one can take much credit for this; it emerged from the geography of the park. The presence of wolves completed a classic food chain, inspiring a scientific quest and helping to rehabilitate the wolf's reputation in the mind of the public. The "bloodthirsty demon" became a saint in the wilderness.

I sense a rocky road for the NPS should it select, unilaterally or with rudimentary review, a hands-off option for management of wolves on Isle Royale. The interested public—the real constituency of our national parks—is well-educated and supportive of complete ecosystems. For the past three decades, through its interpretive programs and materials for public education, the Park Service has passed along research findings that demonstrate the vital role of wolf predation in the ecology of Isle Royale. Most visitors to the park have heard of the boom and bust pattern of the moose population that prevailed before the arrival of wolves. It was a true story and a good one, and the public accepted it.

If wolves are to be excluded on the grounds that it would be unnatural to bring them back, then it would seem that the interpretive message has been wrong all these years. Perhaps the significant aspect of the wolves on Isle Royale was not their creative importance as an evolutionary force, not their health and welfare program for moose, not their indirect role in maintaining plant life that would otherwise be diminished, not their symbolic importance to the wilderness aura of Isle Royale. Instead, the salient feature of Isle Royale wolves—what really mattered—was simply their method of arrival. Perhaps NPS interpreters could get that message across to a perplexed public, but I could not.

In an earlier era that we both knew, Durward Allen wrote, "The moose and wolf need no one to lead them . . . only a place to be left alone."[11] Isle Royale has been that place for 40 years, serving as both cathedral and laboratory. But of what use is a cathedral without sacred imagery? Of what value is a laboratory without subjects?

If wolves are to be assured a future on Isle Royale, their greatest sanctuary, Park Service managers must courageously face their toughest resource management decision—one that will set a precedent for other parks and wilderness areas. It is a question to be embraced, not avoided. If the wolves of Isle Royale are threatened by insularity, can the grizzlies of Yellowstone be far behind? Enlightened by 35 years of scientific research and sensitive to an informed public, humans have a magnificent opportunity to use intellect in sustaining nature. The risk is not great, and there will always be opportunities—perhaps once or twice a century—to reverse the decision.

Someday, when I am long gone, animal and plant life on Isle Royale may be so changed that wisdom will call for a different approach. But this time around, at the dawn of a new millennium, I must vote for the wolves.

NOTES

1. Peterson, R. O., and R. J. Krumenaker. 1989. "Wolf Decline on Isle Royale: A Biological and Policy Conundrum." *The George Wright Forum* 6:10–15.

2. For these metaphors I am indebted to Tom McNamee. 1987. *Nature First: Keeping Our Wild Places and Wild Creatures Wild.* Robert Rinehart, Inc. Boulder, Colo. 54 pp.

3. Pages 8, 91, 124 in Botkin, D. B. 1990. *Discordant Harmonies.* Oxford University Press. 241 pp.

4. McLaren and Peterson. 1994. *Science* 266:1555–1558.

5. See, for example, *Climate Change: The IPCC Scientific Assessment* ("IPCC Report"), published in the summer of 1990—as well as subsequent IPCC reports.

6. Leopold, A. S., S. A. Cain, C. M. Cottam, I. N. Gabrielson, and T. L.

Kimball. 1963. "Wildlife Management in the National Parks." *Transactions of the North American Wildlife and Natural Resources Conference* 28:28–45.

7. Lister, A. 1994. *Natural History* 6:60–61.

8. Little, John J. 1980. "Adolph Murie and the Wilderness Ideal for Isle Royale National Park." Pages 97–114 in Lora, Ronald (ed.). *The American West: Essays in Honor of W. Eugene Hollan.* University of Toledo, Toledo, Ohio.

9. Murie, A. 1935. Preservation of Wilderness on Isle Royale. Excerpted in pages 14–15 of the Summer 1991 issue of *Horizons, the Newsletter of the Sigurd Olson Environmental Institute,* Ashland, Wis.

10. Page 147 in Runte, A. 1987. *National Parks: The American Experience.* University of Nebraska Press, Lincoln. 335 pp.

11. Allen, D. L. 1979. *Wolves of Minong: Isle Royale's Wild Community.* Houghton Mifflin Co., Boston. 499 pp.

Kurt Jax and Ricardo Rozzi

Ecological Theory and Values in the Determination of Conservation Goals (2004)

Examples from Temperate Regions of Germany, United States of America, and Chile

INTRODUCTION

A SIGNIFICANT NUMBER AND diversity of people and institutions today agree about the necessity of conservation (Primack et al. 2001). However, in spite of this general agreement, defining conservation goals is a complex issue and there is much disagreement on the question of what to conserve and, moreover, how this should be done. At the same time, current globalization and large-scale ecological, economic and social problems make it necessary to set precise goals for conservation actions (Figueroa & Simonetti 2003).

Setting aside some areas and leaving them alone, protecting them by drawing lines or even fences around them is not enough (Pickett et al. 1997, Armesto et al. 1998, Bruner et al. 2001, Liu et al. 2001). First, environmental problems are no longer purely local or regional, but they have now an important global dimension (Chapin & Sala 2001). For example, atmospheric changes induced by humans, such as the Antarctic ozone hole or global warming due to greenhouse gases, are problems that affect and concern the planet and society as a whole (Vitousek 1994). Even more, the causes and consequences of those changes are often spatially uncoupled, they do not stop at national boundaries, and they may lead to sequels of hitherto unknown dimensions (Rozzi & Feinsinger 2001).

A second reason to argue about the direction of conservation efforts is that it is becoming increasingly evident that conservation cannot be done against the will of the people and/or by completely excluding them from protected areas, but only by including them (Alcorn 1991, Shaxson 1991, Toulmin 1991, Rozzi et al. 2000). The lack of participation by local communities has been a major cause of failure in many conservation projects (Abu Sin 1991), and at the same time, the rights of indigenous people and the value of traditional ecological knowledge has gained increasing recognition (Mark 2001), especially after the 1992 Earth Summit (Jardin & Kares 2000).

We argue that conservation questions cannot be delegated to science alone because they are also questions of values for at least three reasons: (1) humans are affected by conservation actions (Alcorn 1991, Armesto et al. 2001), (2) the role of humans within conservation must be discussed in the face of conflicting social interests (Jardin & Kares 2000, Rozzi et al. 2000), and (3) conservation essentially concerns our moral attitudes toward human and non-human nature (Callicott & Nelson 1998, Callicott 1999).

This paper analyzes the role of ecological science and social values in the definition of conservation goals and discusses the difficulties of this definition. In particular, we discuss why nature alone cannot provide unequivocal guidelines and how ecological theory can contribute to defining conservation units and criteria. Going beyond the traditional role of ecology as a provider of empirical data and predictions, we emphasize the hitherto neglected heuristic role of ecological theory in clarifying conservation goals and connecting facts and values.

We provide a historical introduction on the origins of protected areas in two Northern Hemisphere temperate countries, Germany and the United States, as two contrasting models. Following this, we examine conservation criteria and policies involved in the protected areas of the southernmost forests of the world, the Magellan archipelago of Chile. We compare the Chilean case with the Northern Hemisphere cases, as well as with more recent conservation approaches involving zoning and regulation of human activities within protected areas. These examples display a wide range of possible conservation approaches and the values implied within them. Building on these experiences, we finally discuss a novel approach for defining conservation goals, derived from ecological theory and ecosystem management concepts, which may help clarify the goals and the interface between societal decisions and scientific knowledge.

EXPERIENCES FROM THE NORTH: HISTORICAL AND RECENT CONSERVATION GOALS IN GERMANY AND THE UNITED STATES

Conservation efforts and the establishment of the first protected areas started in both Germany and the United States in the nineteenth century. However, the main emphasis of conservation and the kinds of areas that were protected differed strongly on the two sides of the north Atlantic (Table 1).

The first protected area in Germany, established during the 1830s, was the Drachenfels, a hill with an old castle ruin towering above the banks of the Rhine south of Bonn. The reason to protect it as a natural monument (Naturdenkmal) was the danger of a complete destruction of the castle and the mountain side pointing toward the Rhine by a quarry, which had already caused part of the old ruin to collapse. Later the area was greatly extended to include the surrounding hills in the nature protection area (Naturschutzgebiet) in Siebengebirge. Both the hills of the Siebengebirge and the Drachenfels ruin, however, had a high symbolic value in the context of romanticism and the search for national identity in Germany, which at that time was divided into many small more or less independent states.

The broader conservation movement in Germany was articulated and driven toward practical and political relevance most effectively by

Table 1 The beginnings of nature conservation in Germany and the USA.

	Germany	USA
First protected area	1830s: first natural monument (Drachenfels) Later extended to first nature conservation area (Naturschutzgebiet Siebengebirge)	1872: first national park (Yellowstone)
Main emphasis of early conservation	Cultural landscapes, Protection of resources	Wild landscapes, Protection of resources
Role of humans	equilibrium including humans	equilibrium excluding ("modern") humans

the musician Ernst Rudorff. Inspired by the traditions of romantic art and skilled in writing, Rudorff became the major spokesman of the new idea of conservation (Knaut 1990). This conservation idea started not as a movement to protect "wild" landscapes, but as "Heimatschutz" (Dominick 1992, Knaut 1993), which meant the protection of the home country or home landscape (the "Heimat"). This was essentially the protection of cultural landscapes, that is of landscapes molded by centuries of extensive use practices.

"Heimatschutz" was an explicit reaction against the rise of industrialization and urbanization in Germany. It expressed the desire to secure what was conceived of as the historical identity of the German nation, which during Rudorff's time had already existed as a unified state since 1871. Thus in its first decades, conservation was mainly Heimatschutz and the conservation of natural monuments, a word coined explicitly as a parallel to cultural monuments, meaning extraordinary singular features of nature like particular old trees or remarkable rock assemblages.

An additional emphasis of early conservation in Germany was the protection of natural resources, e.g., birds (but only "useful" birds; see Berlepsch 1899) or game (Rozzi et al. 2001). Human beings were not excluded from conservation but, as major agents of the development of the rural landscapes, they were included in the idea of "Heimatschutz," however only as far as they dwelled in traditional, non-industrial lifestyles.

In contrast to the German model of "Heimatschutz," conservation efforts in the United States emphasized the protection of "wild," "untouched" landscapes, pursuing the "wilderness" ideal of Henry David Thoreau and John Muir (Nash 1982, Oelschlaeger 1991). The first park in the United States (state park at that time) was the Yosemite Valley in the Sierra Nevada of California, established in 1864 by the state of California. Later, in 1890, Yosemite was declared a national park (Runte 1997).

The first national park in the United States was established in 1872, namely Yellowstone National Park, which also constituted the first national park of the world. Moreover, Yellowstone can be considered the prototype of all national parks and has shaped this notion (Runte 1997, Sellars 1997). The area is situated in the northern Rocky Mountains of the United States, mostly in the state of Wyoming, and covers an area of almost 9,000 km^2. It protected the wild landscape, which was perceived as not used and altered by humans. The main features which led to the establishment of this park were its magnificent landscapes, including many geothermal features—geysers and hot springs—and abundant wildlife, including attractive large mammals, such as grizzly bears (*Ursus arctos*) and elk (*Cervus elaphus*).

Following the idea of wild landscapes, humans were explicitly excluded or at least considered irrelevant for the current appearance of the protected landscapes. This does not, however, imply that national parks were meant to exclude human visitors. The founding law of Yellowstone stated explicitly that the Park was created "for the benefit and enjoyment of the people." Still today the criteria of the International Union for Conservation of Nature and Natural Resources (IUCN 1994) for the establishment of national parks explicitly require restricted public access. Besides protecting wild nature, another emphasis of early American conservation was—as in Germany—the protection of natural resources, particularly forests, a current connected with the name of the forester Gifford Pinchot (see Norton 1991).

The American idea of preserving wild nature has become very popular and has been the inspiration of conservation systems in southern

South America (see below). However, what is often forgotten is the fact that national parks were never meant to completely exclude people.

SIMILARITIES BETWEEN GERMAN AND NORTH AMERICAN MODELS OF CONSERVATION

Evident differences place early conservation strategies in Germany and the United States at opposite ends of a gradient: culturally molded nature versus wild nature. At the same time, however, there are also important similarities between conservation approaches in both Northern Hemisphere regions, and those similarities have even become more apparent as knowledge about the conditions that prevailed in nineteenth century North America increase.

As in Germany, the establishment of protected areas in the United States was a reaction to the growing impact of humans on the landscape. In North America, human impact was not as much industrialization and urbanization, but the extensive land use that reached the "untouched" western areas of the continent. In addition, the natural heritage of the wild and magnificent landscapes, protected in parks, was considered part of the identity of the American nation, as a substitute for the longer cultural heritage of the European nations (Nash 1982, Runte 1997).

During the twentieth century the early conservation aims were criticized in many respects (e.g., as being too narrow or too conservative), and were changed in that course. The German tradition of conserving cultural landscapes was soon extended to particular (rare) species of plants and animals, which often depended on these habitats, and later to the protection of wild landscapes. In 1970, almost 100 years after the establishment of Yellowstone National Park, the establishment of national parks began also in Germany with the creation of the Bavarian Woods National Park, the first German national park. Today, there are 13 national parks in Germany, with a few more in the stage of planning or negotiation. It is important to note, however, that to date most German parks do not fulfill the strict IUCN-criteria, which demand that at least 75% of the area should be completely free of human use.

Especially in the last decade, the traditional approach of German conservation has been criticized as being too conservative and antiquated, turning nature into a museum with species. Further, many of those species would not occur in those protected areas without human influence, and they will not survive without the perpetuation of these old practices or their substitution by other forms of active management. This debate is still prevailing.

On the other hand, in the United States the notion of "untouched" nature has been seriously challenged (see Callicott & Nelson 1998). Recent ecological, anthropological, and geological research has demonstrated that the landscapes of North America were not in a "pristine" state when Europeans arrived (Russell 1980, Callicott 1999). First, the North American indigenous population was on the order of millions (Diamond 1999). Secondly, the notion of American Indians as "noble savages" or "homo oecologicus" which had no significant impact on the natural setting has turned out to be an idealization. That simplified notion is as false as that of an almost "empty" country, waiting to be taken over by the white intruders (Mann 2002). The pendulum has swung back so far that some scholars see almost every landscape as influenced by land use practices of American Indians (e.g., Kay 1994; see Vale 1998 for a criticism). Similar doubts about the factual basis of the western wilderness idea have also been expressed for other parts of the Americas (e.g., Gómez-Pompa & Kaus 1992).

Under this perspective, Yellowstone or other American national parks would also be "cultural landscapes," if they are to be protected as "vignettes of primitive America," i.e., in the state which the first Europeans found them (as proposed in an influential paper by Leopold et al. 1963). In this case, humans, with their traditional land use practices would be included in American national parks as much as in traditional German "Naturschutzgebieten." However, both the American and the German ideal would exclude modern man as a valid actor, avoiding industrial and urban development in Europe, and non-indigenous European settlers in the United States.

In contemporary conservation strategies the seeming (and sometimes real) contradiction between the German and United States contrasting conservation philosophies becomes even less relevant. Several concepts have been developed aimed at reconciling conservation and human needs, which propose the design of protected areas including different zones subject to different intensity and type of human use. Hence, different conservation concepts—such as those of the contrasting German and United States traditions—would apply to different zones of a protected area.

The zoning criterion is an essential component of the Biosphere Reserve concept launched by UNESCO through its Man and Biosphere (MAB) program in the 1970s. Each biosphere reserve includes three distinct zones: (1) core zone, strictly dedicated to protect "wilderness," which involves complete exclusion of human activities (except regulated scientific research); (2) surrounding buffer areas, which are defined to permit or even foster traditional forms of land use which, in turn, may be essential to conserve the culturally-founded diversity of habitats and species associated with those traditional practices; (3) transition areas, where productive and other economic activities and infrastructure are permitted (Jardin & Kares 2000).

In southern South America, zoning criteria have been implemented as a means to reduce user conflicts by the Argentinean administration of national parks in Patagonia (Martín & Chehébar 2001, Salguero 2001). Each Argentinean national park includes five zones: (1) strict conservation areas, where human activity (except for scientific research) is forbidden; (2) extensive public use zones, where extensive uses such as scientific, educational, tourist and recreational are permitted; (3) intensive public use zones, which are relatively small areas where intensive tourism and recreation is allowed, including associated service infrastructure such as hotels, lodges, restaurants, camping facilities; (4) natural resource use zones, where sustainable productive activities and indigenous people's residences are allowed; (5) special use areas, which are small areas for administration, services or human settlement not related to public use.

Strategies based on zoning criteria can provide a valuable bridge between opposite notions associated with the wilderness–United States or the cultural-landscape-German conservation traditions. The zoning approach seems to us particularly suited for regions, such as southern South America, which maintain heterogeneous mosaics of landscapes regarding the degree of human influence. The extreme south of Chile, for example, includes a broad diversity of ecosystems that range from pristine (i.e., wild) to completely man-modified (i.e., cultural) landscapes (Rozzi 2002).

CONSERVATION AND PROTECTED AREAS IN SOUTHERN CHILE

Only four years after the creation of Yellowstone National Park in United States, the first Latin American protected area was established in Mexico. The creation of the Mexican Reserva Forestal Desierto de los Leones, was followed by the Reserva Perito Moreno in Argentina (1903), and the Reserva Forestal Malleco in Chile (1907) (Ormazabal 1988). Since then the number of national parks, state and private reserves has significantly increased in Chile (Armesto et al. 2001) and throughout Latin America (Primack et al. 2001). Today, the Chilean state maintains 92 protected areas, which includes 32 national parks, 47 reserves, and 13 national monuments (Table 2). The area protected by these 92 units represents 19% of the Chilean land surface, which almost triples the mean of 6.4% for South American countries (Armesto & Smith-Ramírez 2001).

Among Chilean administrative regions, Magallanes exhibits an outstanding 7,079,285 ha of protected land, which represents roughly 50% of the region. National parks cover 4,732,785 ha, which represent 53% of the total area devoted to public national parks in Chile. Magellanic reserves comprise 2,346,189 ha, i.e., 42.6% of the area of reserves in the entire country. Therefore, Magallanes has the highest rank of protection in Chile, concentrating nearly 50% of the country's protected land. At the same time, such a large amount of protected land emphasizes the importance of the Magellanic region as a reservoir of non-fragmented temperate ecosystems for Chile and the world.

In spite of the large proportion of protected land, current figures and conservation approaches in Magallanes present several problems. First, the country's distribution of protected areas is very biased toward the extreme south (Armesto et al. 1998). Administrative regions Eleventh (Aysén) and Twelfth (Magallanes), which extend between 44° and 56° S,

Table 2 Protected Areas in the southernmost Administrative Region of Chile, Magallanes.

Category	Name	Province	Area (ha)	Percentage relative to total Protected Area in Chile
National Park	Bernardo O'Higgins	Ultima Esperanza	*2,962,420	33.2%
(Total in Chile:	Torres del Paine	Ultima Esperanza	242,242	2.7%
N=32; 8,912,724 ha)	Pali Aike	Magallanes	5,030	0.1%
	Cabo de Hornos	Antarctica	63,093	0.7%
	Alberto d Agostini	Antarctica	1,460,000	16.4%
	Sub-total		4,732,785	53.1%
Reserve	Alacalufes	Ultima Esperanza	2,313,875	42.0%
(Total in Chile:	Laguna Parrillar	Magallanes	18,814	0.3%
N=47; 5,503,499 ha)	Magallanes	Magallanes	13,500	0.2%
	Sub-total		2,346,189	42.6%
National Monument	Cueva del Milodon	Ultima Esperanza	189	1.1%
(Total in Chile:	Los Pinguinos	Magallanes	97	0.5%
N=13; 17,669 ha)	Laguna de los Cisnes	Tierra del Fuego	25	0.1%
	Sub-total		311	1.8%
	Total Magallanes		7,079,285	49.0%
	Total Chile		14,433,892	100.0%

For each category the total numbers (N) and total area in Chile are given in parentheses. The extreme right column calculates the percentage that each Magellanic protected area represents relative to the entire country.

*This figure corresponds to the area of the National Park Bernardo O'Higgins included in the Region of Magallanes. The total area of this national park is 3,525,901 ha, but 563,481 ha are included in the Region of Aysen, north of Magallanes (Data from Muñoz et al. 1996).

include more than 80% of the Chilean protected land. Hence, large protected areas in Magallanes should not hide the lack of protection in other critical regions of Chile.

A second problem arises from the scarcity of park personnel: less than 20 park rangers work permanently in Magallanes. This yields a mean of one park ranger per 3,540 km^2. This is a common problem in Latin America, where a dramatic situation also occurs in the Brazilian Amazon, which has only 23 permanent park rangers for the whole basin, i.e., an average of one park ranger per 6,053 km^2 of protected land (Primack et al. 2001). This situation contrasts with the United States, which has 4,002 permanent park rangers, that is an average of one park ranger per 82 km^2. The majority of protected areas in Magallanes also lack proper infrastructure, such as means of transportation, which are indispensable in this archipelago region. This lack of transport and personnel determines that not a single park ranger works in the diverse habitats included in the 1,460,000 ha of the National Park de Agostini—the second largest of Chile. Therefore, most protected land in Magallanes, as is the case in other regions of Latin America, would fall within the label of "paper parks" (Rozzi & Silander unpublished results). In fact, Magellanic national parks do not fulfill the requirements and the criteria of IUCN (1994) for this category of protected areas.

A third problem in the Magellanic region arises from the almost complete disregard for local people living close to protected areas, and in some cases indigenous residents have been displaced from their land (Rozzi et al. 2000, Rozzi 2002). The United States preservationist paradigm, sketched above, has had a strong influence on the conservation approach in the extreme south of Chile. The debate about the influence that pre-Columbian cultures had on their local ecosystems and regional landscapes, and the integration of indigenous people into conservation areas is as intense and controversial in South America as in North America. This discussion involves two extreme positions: (1) one that idealizes aboriginal people as living in harmony with nature; and (2) another that considers native people as threats that should be removed from "pristine" or "natural" landscapes. Both are misleading oversimplifications (Alcorn 1991). Regarding the first position, it would be interesting to evaluate the work done in southern Chile by the German missioner and anthropologist Martin Gusinde, who was deeply concerned about the future of the Fuegian Indians. In his monumental ethnographic work, Gusinde describes in detail several concepts and practices of traditional ecological knowledge of Kaweskar, Yahgan, and Selknam, indigenous people at the austral extreme of South America (see Gusinde 1946, 1961). Regarding the second position, it follows a preservationist approach identified with John Muir (see Norton 1991), which has been strongly influential for conservation designs in Latin America during the last 130 years (Rozzi et al. 2001). In southern Chile, indigenous populations have been excluded from national parks. For example, the national parks of Chiloé, Bernardo O'Higgins, and Cape Horn have respectively excluded Huilliche, Kaweskar, and Yahgan communities. Interestingly, today the general trend of abandonment and human exclusion in protected areas of southern Chile is changing due not only to conceptual changes about the role of humans as ecosystem components (McDonnell & Pickett 1993, Rozzi et al. 1994), but also to a growing interest in ecotourism.

Ecotourism is promoting a shift, which instead of emphasizing a preservationist approach, underlines the statement "parks are created for the benefit and enjoyment of the people," asserted in the founding law of Yellowstone National Park. In the extreme south of Chile, this statement (which is closer to the United States conservation tradition identified with Gifford Pinchot, see Norton 1991), is acquiring a prevalent role today. This shift toward ecotourist activities requires, however, careful examination in order to achieve a sustainable compatibility between conservation and human needs or benefits (di Castri & Balaji 2002, Figueroa et al. 2003).

Between Yellowstone National Park and the Magellan national parks, in particular Torres del Paine National Park, some remarkable similarities exist. In the Magellan Region, Torres del Paine National Park constitutes an area that, like Yellowstone, possesses marvelous landscapes (including glaciers and mountain peaks), and attractive megafauna (including species like rheas, *Pterocnemia pennata,* and guanacos, *Lama guanicoe*). Also, like Yellowstone, Torres del Paine is visited by a large number of tourists. Although the number of visitors to Torres del Paine (43,624 in 1995) ranks two orders of magnitude below Yellowstone (more than 3 million in 1995), for Chile it holds the largest number of foreign visitors and it has a substantial impact on the development of the nearby city of Puerto Natales (Villarroel 1996). Of the visitors to Torres del Paine, 62% are from overseas, coming from Europe (37%), North America (15%), and Oceania (10%) (Ferrer 2001).

Torres del Paine National Park was created in 1959, and was designated as a Biosphere Reserve in 1978. Like Yellowstone National Park, the Torres del Paine landscape shows signs of human influence. The austral landscape exhibits the marks left mainly by European colonists that arrived at Magallanes at the beginning of the twentieth century (Dollenz 1991). Before the Chilean government acquired the park, it belonged to German ranchers who burned large expanses of forests to increase pasture area, which was later overgrazed (see Martinic 1984). Therefore, in spite of the goal to protect pristine or "wild" areas, the imprints of both indigenous and European settlers, are present even in the remote austral regions of the American continent.

Within this context ecotourism poses complex puzzles to conservation biology. On the one hand, it seems to favor a larger integration between society and protected areas. On the other hand, with current deficiencies in the planning and regulation of ecotourism within parks, such as Torres del Paine, undesirable environmental impacts may follow (Villarroel 1996, Massardo et al. 2001). Hence, a close collaboration among government offices, tourism agencies, and academic institutions is required for the planning of protected areas, and defining their conservation goals.

HUMAN VALUES, SCIENCE, AND THE DETERMINATION OF CONSERVATION GOALS

The short overview of conservation strategies in Germany (protection of cultural landscapes), the United States (wilderness ideal), and southern Chile (preservation paradigm, and the more recent interest in ecotourism as a potentially sustainable economic activity) illustrates the broad spectrum of conservation goals and the different role of humans within conservation. Consequently, it is not always clear what exactly should be protected within reserves or national parks. However, with increasing human pressure on nature, especially in a period of a rapidly growing global economy, and an increasing probability of human-induced global changes, the necessity for a conscious decision about conservation aims and measures becomes greater. Confronted with this scenario, and a broad range of conservation goals: what should we protect? What roles should humans play in this context? Where can we find guidelines? What is the role of science?

Answering these questions requires us to systematically integrate multiple aspects that influence any conservation strategy, aspects that hitherto have in part been developed independently from each other (Jentsch et al. 2003). Such integration has not been achieved, and it challenges the prevailing trend of specialization that dominates science and other disciplines since the second half of the twentieth century (Rozzi et al. 1998). Hence, to interconnect diverse aspects of conservation, such as empirical data, ecological theory, human values and worldviews, represents an urgent and important task. At the same time, this task demands novel theoretical and practical approaches.

Values enter the determination of conservation goals in many different ways: in our images of nature (Ahl & Allen 1996, Rozzi 1999, Rozzi 2003), in our economic values (Daly & Townsend 1994, Daily 1997), in our political preferences (Norton 1991), in our moral attitudes toward human and non-human nature (Rolston 1990), and even in our decisions about what is important in science. However, values are often not explicit and remain hidden behind seemingly objective scientific facts or economic necessities.

The provision of empirical data is one of the basic tasks of ecological research within conservation. It is necessary to describe the current conditions of an area or—by means of, e.g., paleoecological analyses—to restore its "original" or "natural" conditions, e.g., in terms of plant cover or animal life. However, criteria for selection of particular areas and their subsequent management are not purely based on scientific knowledge.

What kinds of data are collected and what kinds of questions are asked is already a matter involving value decisions. Although many people argue that, for conservation purposes, ecology should simply identify the "natural" condition, this task is far from a purely "objective" scientific enterprise. For example, the concept of what is natural plays a major role when deciding which role humans should play within protected "natural" areas. Both the German and American early conservationists wished to protect "natural" landscapes although their images of what is "natural" differed considerably. In addition, these concepts have changed during the following decades and they still have different meanings for different groups of people. Particularly difficult questions related directly to value decisions arise today with respect to alien plants and animals (invasive species) entering an area and spreading there. Should they be considered as "natural"?

Ecological theory is a third important and often neglected ingredient in the determination of conservation goals, which can serve two main purposes. First, ecological theory allows us to go beyond a purely static description of an area, by providing insights into the interactions between the elements of ecological systems, their dynamics, and the ways they might respond to external changes. Ideally, ecological theory should provide the means for predicting the development of ecological systems.

A second, much less considered role of ecological theory is its heuristic use in the formulation of research questions and conservation goals (Jax 2003). Ecological theory can help identify gaps in our knowledge and expose uncertainties. Even more important within conservation, ecology can help clarify our questions, forcing us to be more precise about the concepts we use. Although this remains a difficult task, ecological theory can also help distinguish between values and facts and promote their integration in the definition of conservation goals. We illustrate this point using one of the currently most discussed approaches to conservation, the strategy of ecosystem management.

PRESERVING ECOSYSTEMS: THE SOLUTION TO CURRENT CONSERVATION DILEMMAS?

Ecosystem management represents an increasingly popular strategy, which is compatible with a dynamic view of nature (Christensen et al.

1996). It recognizes ecosystems as permanently changing and, at the same time, promotes a multiple use perspective.

The management of whole ecosystems—in contrast to that of single "commodities"—seems to be an elegant solution to many conservation problems. By protecting the whole ecosystem, we avoid protecting only certain parts of an area at the cost of others. This approach, to our knowledge, was first applied systematically in Yellowstone National Park, starting in the late 1960s (Jax 2001, 2002b). During the 1990s the notion of ecosystem management experienced a rapid rise in North American environmental policy (Grumbine 1994, Christensen et al. 1996, Boyce & Haney 1997, Jax 2002b).

In contrast to its beginnings, in which ecosystem management was mainly a particular way of dealing with complex natural settings, the notion has now been extended to an ambitious societal program (Jax 2002b). Although the ecosystem approach in the United States means very different things to different people (Yaffee 1999), some common ground is emerging. The ecosystem is used here as a cipher for the treatment of "the whole," a whole that also includes humans, their societies and resource use practices. Moreover, it emphasizes interagency management and a focus on natural boundaries in contrast to administrative ones (Grumbine 1994, Carpenter 1995, Szaro et al. 1998). In this context, the ecosystem and ecosystem management concepts are becoming strongly value-laden, departing from the perspective of "value-neutral" science.

It is this ecosystem approach which is applied by the Convention on Biological Diversity (CBD). The Fifth Conference of the Parties of this convention, which took place in Nairobi in 2000, passed a resolution that recommended the "ecosystem approach" as a cross-cutting issue for the CBD and obliged all parties to implement this approach within their conservation policies (resolution COP V/6). Based on the so-called Malawi-Principles, the approach emphasizes the social dimensions of management and that societal choices have to be made. It also acknowledges the changing nature of ecological systems (Botkin 1990, Pickett & Ostfeld 1995, Plachter 1996).

The ecosystem approach is considered a major tool for implementing the three basic goals of the CBD, namely biological conservation, sustainable use of natural resources, and equitable sharing of benefits (Smith & Maltby 2001). However, the implementation of such an approach is far from simple. First of all, it is an illusion that we would really be able to grasp the whole. This is an epistemological problem (Pickett et al. 1994, Rozzi et al. 1998). To investigate anything in nature, we have to select and isolate a particular characteristic of interest, from which we mentally form the system which we then describe and analyze. This has direct consequences for the scientific perception of the ecosystem as the very object of the ecosystem approach. In spite of some "naive-realistic" attitudes, an ecosystem is not a natural entity that can be identified in nature without reference to particular interests and selection criteria (Jax et al. 1998). It is defined in a task-specific manner. Definitions of ecosystems are manifold (Jax 2002a), and those that are commonly accepted as embracing the many and contrasting meanings are, in consequence, very general, too general to provide clear criteria for defining the goals of ecosystem management.

To implement an effective approach to ecosystem management it is necessary to: (1) set a baseline, (2) define what an ecosystem is, and (3) have criteria to decide when it is "destroyed" or deviates significantly from a baseline condition.

For example, the case of southern Chile might involve questions such as: are the subantarctic evergreen rainforest ecosystems characterized by a particular species composition or just by a particular physiognomy of plant and animal types? Has the invasion of the North American beaver (*Castor canadensis*)—which started in the late 1940s on Tierra del Fuego, Navarino island and other areas of the Cape Horn archipelago (Lizarralde & Venegas 2001)—created new and more diverse ecosystems? Today, is *Castor canadensis* part of the "old ecosystem" or is it the destroyer of the "original ecosystems"? Will we say that an ecosystem has become "another" ecosystem if some native undergrowth species are lost (or replaced by alien species) or will the ecosystem only be "another" if its physiognomy is also changed?

The ways in which ecosystems are defined must be communicated in a clear manner. However, this is still frequently not done, generating difficulties at different levels. To serve this communication purpose, Jax and collaborators (Jax et al. 1998, Jax 2002a) have recently devel-

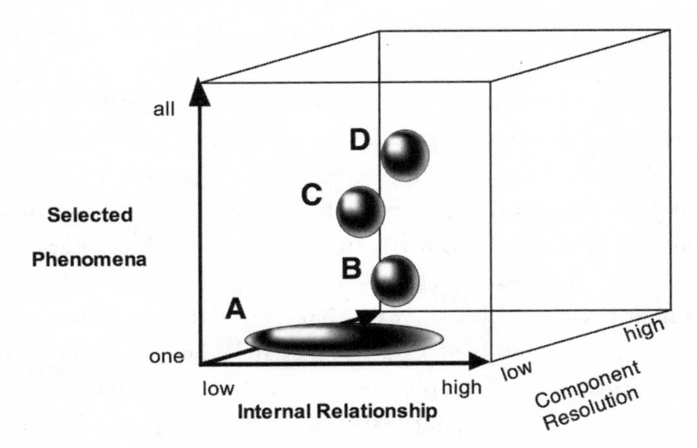

Figure 1. Representation of different definitions of "ecosystem" that are applied in ecosystem management strategies (see text).

oped an approach to clarify and provide an unambiguous definition and specification of any ecological unit. To do this, statements are needed about: (1) whether the unit is bounded topographically or functionally, (2) which kind of relationships among the components are minimally required, (3) which phenomena (i.e., components and internal relations) are selected for the definition of the unit, and (4) what is the degree of resolution of the unit's components.

The first criterion represents an essential distinction. It describes an element (e.g., organism) either seen as a part of a unit by virtue of being in a particular spatial location or by virtue of being functionally (i.e., by interactions) related to other elements. For example, within the bounds of a *Nothofagus* forest on an austral island, are all species components of one ecosystem or are there several separate ecosystems characterized by specific functional connections within these topographical bounds?

The remaining criteria apply to both spatially and functionally bounded units. They can be seen as gradients, which can be assembled as three axes into a graphical scheme that allows visualizing the different definitions (Fig. 1).

The axis of selected phenomena displays which and how many phenomena (kinds of objects and/or processes) are included in the definition of the ecosystem. The axis of internal relationship indicates the degree of intensity and specificity that internal relationships are required to have in order to call a unit an ecosystem, or even an intact ecosystem. In some definitions, for example, the requirements to call a system an ecosystem are such that interactions have to be very specific and lead to an equilibrium state, or that feedback-loops are present which lead to the self-regulation of the system (high internal relationships). In other definitions any interactions between the organisms (low internal relationships) are sufficient to call the system an ecosystem. The axis of component resolution describes to what degree the components of the ecosystem must be resolved, e.g., whether the system parts are considered at the species or just trophic levels. Based on the initials of the three axes (selected phenomena, internal relationship, component resolution) this scheme was named the "SIC-scheme" (Jax 2002a). Fig. 1 displays this general scheme to illustrate some of the most common meanings of ecosystem in the context of ecosystem management.

These meanings vary according to the conservation aims. The most general definition is depicted by the ellipse "A." Here the ecosystem is preserved simply as a system of interacting natural objects. Indeed, the interactions themselves may be the focus of management (which can also mean to refrain from active management). The kinds (i.e., species) of organisms are not of special importance here (low component resolution), and the degree of required internal relationships might also vary. For example, particular feedbacks may be demanded within the system to call it an ecosystem, such as the criterion that most primary production must occur within the system itself. This kind of ecosystem may be useful for the management of wilderness areas, even in regions that have been strongly impacted by humans but where now "nature can take its course." This is an especially interesting concept for ecosystem management in central European countries, where completely "pristine" areas no longer exist.

Sphere "B" (Fig. 1) depicts another frequently applied definition of ecosystem, which focuses on particular interactions and processes. Here, the ecosystem is described by particular functional compartments, interacting in a manner that particular services—such as primary production, clean air or waters—are provided by the system. Component resolution is thus slightly higher than in type A, but still particular species are not of interest, only functional types. The degree of interaction is higher than in many other definitions because interactions, and particular feedbacks, between specific functional elements are essential for the definition. This kind of definition is sufficient when the aim of ecosystem management is to provide benefits for humans in the form of "ecosystem services" (Costanza et al. 1997).

Sphere "C" (Fig. 1) depicts a third type of ecosystem definition that demands a higher resolution in the three axes. For example, a *Nothofagus* forest ecosystem or a *Sphagnum* bog and the essential interactions that perpetuate such systems are to be protected. The aim is to protect a large ecosystem which is "typical" for the area, without the necessity that all constituent species have to be preserved in the long run, except for some conspicuous and dominant taxa such as *Nothofagus* trees and *Sphagnum* mosses. Particular types of taxa (indicator species, keystone species or "umbrella species"; see Simberloff 1998) are thus already part of the definition. This—physiognomic—view of ecosystems is perhaps the most common one in the practice of conservation and resource management.

Finally, sphere "D" (Fig. 1) illustrates a concept of ecosystem defined by all species occurring in a setting. Interactions themselves are protected mostly for the sake of conserving the interacting components. These may be those species which are present in a protected area at a date t (e.g., the date at which the measures start) or—much more difficult to determine—all species which are considered as "typical" for a particular site. The aim here is to perpetuate all species, without fixing particular growth rates or dwelling places, abundances, or specific ratios between species. Everything, besides the species composition, is in a condition of waxing and waning, including local disturbances and recolonizations (within the system). This aim is formulated, for example, in some national parks and corresponds to the current strategy of ecosystem management in Yellowstone national park (Jax 2001).

CONCLUDING REMARKS

Based on the concise examination of conservation approaches that have taken place in temperate regions of Germany, the United States, and Chile, followed by the analysis of conservation units based on the SIC-scheme, what can we learn for conservation in the austral Magellanic region?

First, conservation traditions encompass interests for the preservation of both natural and cultural heritages. Even more, these two dimensions are mutually dependent, as shown by the "natural areas" of Yellowstone and Torres del Paine, which have been molded in part by humans. Therefore, nature and humans are brought together in the object of conservation, as well as in the processes occurring in the protected units. Consequently, the dichotomies between nature and culture, and between protected areas and human presence, become irrelevant.

Second, in the context of current global change it is impossible to completely isolate protected areas from human influences (Primack et al. 2001). Human impacts can arise as much from local populations (for example, firewood extraction) as from remote populations inhabiting a different hemisphere, as in the case of the austral ozone hole caused by the emissions of chemicals in Northern Hemisphere industrialized countries. Moreover, in the three temperate regions considered, humans as components of ecosystems may be a "keystone species." In addition, a dynamic view of nature—the "flux of nature"—points out that biotas and ecosystems will change over time, even within "protected areas." Hence, to preserve species or habitats it is not enough to "isolate" protected areas, but often it requires active management and conservation.

The two former conclusions invite us to revise the conservation approach undertaken in the extreme south of Chile, where local people have been excluded from protected areas, and where the National Forestry Service (CONAF, the organization responsible for these areas) has serious logistic and financial limitations to carry out conservation and/or management programs.

Third, our analyses demonstrate that conservation goals involve not only scientific criteria, but also philosophical, political and broader cultural, social and economic dimensions. Hence, interdisciplinary and interagency cooperation is urgently needed. None of these actors can see or understand the "whole" by themselves. Therefore, operational definitions of the units and goals of conservation need to be jointly defined.

This process requires explicitly presenting the goals, methods and values involved in conservation or management of species and ecosystems.

Finally, we consider that the ecosystem approach to conservation, as currently conceptualized within the guidelines of the CBD, represents an extremely valuable tool. It allows integrating solid empirical research, sound ecological theory and human value dimensions. However, it is important to avoid the pitfalls that these approaches can have, when unproductive and improper mingling of facts and values involve a fuzziness of basic and practically relevant theoretical concepts, such as the ecosystem concept. These problems could undermine the usefulness of the ecosystem approach, concealing the issues really at stake. In this context, ecological theory, embedded in interdisciplinary work and social participatory processes, represents an indispensable key element for determining conservation goals.

ACKNOWLEDGEMENTS

The authors thank Juan J. Armesto and an anonymous reviewer for valuable comments on the manuscript. The work that initially led to this paper has been kindly supported by travel grants of the Deutscher Akademischer Austauschdienst (DAAD) to Kurt Jax. Ricardo Rozzi acknowledges the support of the Department of Ecology & Evolutionary Biology, University of Connecticut, and the Millennium Center for Advanced Studies in Ecology and Biodiversity (CMEB). This article is part of the BIOKONCHIL project (FKZ 01 LM 0208, German Ministry of Education and Research, BMBF), and the ongoing research and conservation activities conducted by the Omora Foundation and the Universidad de Magallanes, at the Omora Ethnobotanical Park, Puerto Williams, Chile.

LITERATURE CITED

Abu Sin ME (1991) Community-based sustainable development in central Butana, Sudan. In: Baxter PTW (ed) When the grass is gone: 152–161. The Scandinavian Institute of African Studies, Uppsala, Sweden.

Ahl V & TH Allen (1996) Hierarchy theory: a vision, vocabulary, and epistemology. Columbia University Press. New York, New York, USA. 206 pp.

Alcorn JB (1991) Ethics, economics and conservation. In: Oldfield ML & JB Alcorn (eds) Biodiversity: culture, conservation and ecodevelopment: 317–349. Westview Press, Boulder, Colorado, USA.

Armesto JJ & C Smith-Ramírez (2001) Importancia de la distribución de las áreas protegidas: el caso del bosque chileno. In: Primack R, R Rozzi, P Feinsinger, R Dirzo & F Massardo (eds) Elementos de conservación biológica: perspectivas latinoamericanas: 454–456. Fondo de Cultura Económica, Ciudad de México, México.

Armesto JJ, C Smith-Ramírez & R Rozzi (2001) A strategy for managing Chilean ecosystems for conservation and indigenous people. Journal of the Royal Society of New Zealand 31: 365–877.

Armesto JJ, R Rozzi, C Smith-Ramírez & MTK Arroyo (1998) Effective conservation targets in South American temperate forests. Science 282: 1271–1272.

Berlepsch HV (1899) Der gesamte Vogelschutz, seine Begründung und Ausführung. Verlag Eugen Köhler, Gera, Germany. 89 pp.

Botkin DB (1990) Discordant harmonies. A new ecology for the 21st century. Oxford University Press, Oxford, United Kingdom. 241 pp.

Boyce MS & A Haney (eds) (1997) Ecosystem management. Applications for sustainable forest and wildlife resources. Yale University Press, New Haven, Connecticut, USA. 361 pp.

Bruner AG, RE Gullison, RE Rice, G Da Fonseca (2001) Effectiveness of parks in protecting tropical biodiversity. Science 291: 125–128.

Callicott B (1999) Beyond the land ethic: more essays in environmental philosophy. State University of New York Press, Albany, New York, USA. 427 pp.

Callicott JB & MP Nelson (1998) The great new wilderness debate. University of Georgia Press, Athens, Georgia, USA. 696 pp.

Carpenter RA (1995) A consensus among ecologists for ecosystem management. Bulletin of the Ecological Society of America 76: 161–162.

Chapin T & O Sala (eds) (2001) Future scenarios for biological diversity. Springer Verlag, Berlin, Germany. 392 pp.

Christensen NL, AM Bartuska, SR Carpenter, C D'Antonio, R Francis, JF Franklin, JA Machamon, RF Noss, DJ Parsons, CH Peterson, MG Turner & RG Woodmansee (1996) The report of the Ecological Society of America committee on the scientific basis for ecosystem management. Ecological Applications 6: 665–691.

Costanza R, R D'Arge, R De Groot, S Farber, M Grasso, B Hannon, K Limburg, S Naeem, RV O'Neill, J Paruelo, RG Raskin, P Sutton & M Van Den

Belt (1997) The value of the world's ecosystem services and natural capital. Nature 387: 253–260.

Daily G (ed) (1997) Nature's services: societal dependence on natural ecosystems. Island Press, Washington, District of Columbia, USA. 392 pp.

Daly HE & KN Townsend (1994) Valuing the earth: economics, ecology, ethics. MIT Press, Cambridge, Massachusetts, USA. 384 pp.

Di Castri & V Balaji (2002) Tourism, biodiversity and information. Backhuys Publishers, Leiden, The Netherlands. 501 pp.

Diamond J (1999) Guns, germs, and steel: the fates of human societies. Norton & Company, New York, New York, USA. 480 pp.

Dollenz O (1991) Recolonización de un coironal incendiado en el Parque Nacional Torres del Paine, Magallanes, Chile. Master Thesis, Facultad de Ciencias, Universidad de Chile, Santiago, Chile. 92 pp.

Dominick RHI (1992) The environmental movement in Germany: prophets and pioneers 1871–1971. Indiana University Press, Bloomington, Indiana, USA. 290 pp.

Ferrer M (2001) Turismo en Torres del Paine. Tesis de Maestría, Departamento de Geografía de la Universidad Autónoma de Madrid (UAM), Madrid, España. 154 pp.

Figueroa E & J Simonetti (eds) (2003) Biodiversidad y globalización. Editorial Universitaria, Santiago, Chile. 327 pp.

Figueroa E, C Bravo & R Álvarez (2003) Biodiversidad y turismo: oportunidades para el de sarroilo económico y la conservación en Chile. In: Figueroa E & J Simonetti (eds) Biodiversidad y globalización: 285–323. Editorial Universitaria, Santiago, Chile.

Gómez-Pompa A & A Kaus (1992) Taming the wilderness myth: environmental policy and education are currently based on western beliefs about nature rather than on reality. BioScience 42: 271–279.

Grumbine RE (1994) What is ecosystem management? Conservation Biology 8: 27–38.

Gusinde M (1946) Urmenschen im Feuerland. Paul Zsolnay Verlag, Berlin, Germany. 389 pp.

Gusinde M (1961) The Yamana: the life and thought of the water nomads of Cape Horn. Volumes I–V, translated by F. Schutze. New Haven Press, New Haven, Connecticut, USA. 415 pp.

International Union for Conservation of Nature (IUCN) (1994) Guidelines for protected area management categories. IUCN, Gland, Switzerland. 261 pp.

Jardin M & C Kares (2000) Solving the puzzle: the ecosystem approach and biosphere reserves. United Nations Educational, Scientific and Cultural Organization (UNESCO), Paris, France. 32 pp.

Jax K (2001) Naturbild, Ökologietheorie und Naturschutz: zur Geschichte des

Ökosystemmanagements im Yellowstone-Nationalpark. Verhandlungen der Gesellschaft für Geschichte und Theorie der Biologie 7: 115–134.

Jax K (2002a) Die Einheiten der Ökologie. Analyse, Methodenentwicklung und anwendung in Ökologie und Naturschutz. Peter Lang, Frankfurt, Germany. 249 pp.

Jax K (2002b) Zur Transformation ökologischer Fachbegriffe beim Eingang in Verwaltungsnormen und Rechtstexte: das Beispiel des Ökosystem-begriffs. In: Bobbert M, M Düwell & K Jax (eds) Umwelt, Ethik & Recht: 69–97. Francke-Verlag. Tübingen, Germany.

Jax K (2003) Wofür braucht der Naturschutz die wissenschaftliche Ökologie? Die Kontroversen urn den Hudson River als Testfall. Natur und Landschaft 78: 93–99.

Jax K, CG Jones & STA Pickett (1998) The self identity of ecological units. Oikos 82: 253–264.

Jentsch A, H Wittmer, K Jax, I Ring & K Henle (2003) Biodiversity. Emerging issues for linking natural and social sciences. Gaia 12: 121–128.

Kay CE (1994) Aboriginal overkill. The role of native Americans in structuring western ecosystems. Human Nature 5: 359–398.

Knaut A (1990) Der Landschafts- und Naturschutzgedanke bei Ernst Rudorff. Natur und Landschaft 65: 114–118.

Knaut A (1993) Zurück zur Natur! Die Wurzeln der Ökologiebewegung. Jahrbuch für Naturschutz und Landschaftspflege (Supplement) 1: 1–480.

Leopold AS, SA Cain, CM Cottham, IM Gabrielson & TL Kimball (1963) Wildlife management in the national parks. Transactions of the North American Wildlife and Natural Resources Conference 28: 28–45.

Liu J, M Linderman, Z OuyangG, L An, J Yang & H Zhang (2001) Ecological degradation in protected areas: the case of Wolong nature reserve for giant pandas. Science 292: 98–101.

Lizarralde M & C Venegas (2001) El castor: un ingeniero exótico en las tierras más australes del planeta. In: Primack R, R Rozzi, P Feinsinger, R Dirzo & F Massardo (eds) Elementos de conservación biológica: perspectivas latinoamericanas: 233–235. Fondo de Cultura Económica, Ciudad de México, México.

Mann C (2002) 1491. The Atlantic 289: 195–211.

Mark A (2001) Symposium: managing protected natural areas for conservation, ecotourism, and indigenous people. Journal of the Royal Society of New Zealand 31: 811–812.

Martín C & C Chehébar (2001) The national parks of Argentinian Patagonia—management policies for conservation, public use, rural settlements, and indigenous communities. Journal of the Royal Society of New Zealand 31: 845–864.

Martinic M (1984) Última esperanza en el tiempo. Ediciones Universidad de Magallanes, Punta Arenas, Chile. 289 pp.

Massardo F, O Dollenz & R Rozzi (2001) Ecoturismo en ci Cono Austral de America. In: Primack R, R Rozzi, P Feinsinger, R Dirzo & F Massardo (eds) Elementos de conservación biológica: perspectivas latinoamericanas: 303–305. Fondo de Cultura Económica. Ciudad de México, México.

McDonnell MJ & STA Pickett (eds) (1993) Humans as components of ecosystems. Springer Verlag, New York, New York, USA. 364 pp.

McKirahan RD (1994) Philosophy before Socrates. Hackett Publishers, Indianapolis, Indiana, USA. 436 pp.

Muñoz M, H Núñez & J Yáñez (eds) (1996) Libro rojo de los sitios prioritarios para la conservación de la biodiversidad biológica en Chile. Corporación Nacional Forestal, Santiago, Chile. 203 pp.

Nash R (1982) Wilderness and the American mind. Yale University Press, New Haven, Connecticut, USA. 425 pp.

Norton BG (1991) Toward unity among environmentalists. Oxford University Press, New York, New York, USA. 287 pp.

Oelschlaeger M (1991) The idea of wilderness. Yale University Press, New Haven, Connecticut, USA. 477 pp.

Ormazabal C (1988) Sistemas nacionales de áreas protegidas en América Latina. Oficina Regional de la FAO (Food and Agriculture Organization of the United Nations) para América Latina y el Caribe, Santiago, Chile. 205 pp.

Pickett STA, J Kolosa & CG Jones (1994) Ecological understanding: the nature of theory and the theory of nature. Academic Press, Orlando, Florida, USA. xiii + 206 pp.

Pickett STA & RS Ostfeld (1995) The shifting paradigm in ecology. In: Knight RL & SF Bates (eds) A new century for natural resources management: 261–278. Island Press, Washington, District of Columbia, USA.

Pickett STA, M Shachak, RS Ostfeld & GE Likens (1997) Toward a comprehensive conservation theory. In: Pickett STA, RS Ostfeld, M Shachak & GE Likens (eds) The ecological basis of conservation: heterogeneity, ecosystems, and biodiversity: 384–399. Chapman & Hall, New York, New York, USA.

Plachter H (1996) Bedeutung und Schutz ökologischer Prozesse. Verhandlungen der Gesellschaft für Ökologie 26: 287–303.

Primack R, R Rozzi, P Feinsinger, R Dirzo & F Massardo (2001) Elementos de conservación biológica: perspectivas latinoamericanas. Fondo de Cultura Económica, Ciudad de México, México. 797 pp.

Rolston H (1990) Biology and philosophy in Yellowstone. Biology and Philosophy 5: 241–258.

Rozzi R (1999) The reciprocal links between evolutionary-ecological sciences and environmental ethics. BioScience 49: 911–921.

Rozzi R (2002) Biological and cultural conservation in the archipelago forest ecosystems of southern Chile. Ph.D. Thesis, Department of Ecology and Evolutionary Biology, University of Connecticut, Storrs, Connecticut, USA. 359 pp.

Rozzi R (2003) Biodiversity and social wellbeing in South America. Encyclopedia of Life Support Systems (EOLSS). UNESCO-EOLSS. http://www.eolss.net

Rozzi R & P Feinsinger (2001) Desafíos para la conservación biológica en Latinoamérica. In: Primack R, R Rozzi, P Feinsinger, R Dirzo & F Massardo (eds) Elementos de conservación biológica: perspectivas latinoamericanas: 661–688. Fondo de Cultura Económica, Ciudad de México, México.

Rozzi R, JJ Armesto & J Figueroa (1994) Biodiversidad y conservación de los bosques nativos de Chile: una aproximación jerárquica. Bosque 15: 55–64.

Rozzi R, E Hargrove, JJ Armesto, STA Pickett & I Silander (1998) "Natural drift" as a post-modern metaphor. Revista Chilena de Historia Natural 71: 9–21.

Rozzi R, I Silander, JJ Armesto, P Feinsinger & F Massardo (2000) Three levels of integrating ecology with the conservation of South American temperate forests: the initiative of the Institute of Ecological Research Chiloé, Chile. Biodiversity and Conservation 9: 1199–1217.

Rozzi R, R Primack, P Feinsinger, R Dirzo & F Massardo (2001) ¿Qué es la conservación? In: Primack R, R Rozzi, P Feinsinger, R Dirzo & F Massardo (eds) Elementos de conservación biológia: perspectivas latinoamericanas: 35–58. Fondo de Cultura Económica, Ciudad de México, México.

Runte A (1997) National parks: the American experience. University of Nebraska Press, Lincoln, Nebraska, USA. 335 pp.

Russell HS (1980) Indian New England before the Mayflower. University Press of New England, Hanover, New Hampshire, USA. 284 pp.

Salguero J (2001) Integración social en los parques nacionales andino-patagónicos. In: Primack R, R Rozzi, P Feinsinger, R Dirzo & F Massardo (eds) Elementos de conservación biológica: perspectivas latinoamericanas: 499–501. Fondo de Cultura Económica, Ciudad de México, México.

Sellars RW (1997) Preserving nature in the national parks: a history. Yale University Press, New Haven, Connecticut, USA. 380 pp.

Shaxson TF (1991) National development policy and soil conservation programs. "Conservation for Sustainable Hillslope Farming," International Workshop Proceedings. FAQ, Masera, Lesotho, South Africa.

Simberloff D (1998) Flagships, umbrellas, and keystones: is single-species management passé in the landscape era? Biological Conservation 83: 247–257.

Smith RD & E Maltby (2001) Using the ecosystem approach to implement the CBD. A global synthesis report drawing lessons from three regional pathfinder workshops. UNESCO/MAB (http//www.unesco.org/mab/docs/Report.pdf), Paris, France. 69 pp.

Szaro RC, WT Sexton & CR Malone (1998) The emergence of ecosystem management as a tool for meeting people's needs and sustaining ecosystems. Landscape and Urban Planning 40: 1–7.

Toulmin C (1991) Bridging the gap between top-down and bottom-up in natural resource management. In: Baxter PTW (ed) When the grass is gone: 152–161. The Scandinavian Institute of African Studies, Uppsala, Sweden.

Vale TR (1998) The myth of the humanized landscape: an example from Yosemite National Park. Natural Areas Journal 18: 231–236.

Villarroel P (1996) El caso de Puerto Natales—Torres del Paine, XII Región: efecto del turismo en el desarrollo local. Ambiente y Desarrollo XII 4: 58–64.

Vitousek PM (1994) Beyond global warming: ecology and global change. Ecology 75: 1861–1876.

Yaffee SL (1999) Three faces of ecosystem management. Conservation Biology 13: 713–725.

Kathleen Dean Moore

Wilderness as Witness (Cape Perpetua) (2002)

IT TAKES A STRONG STOMACH to drive over the Coast Range from my house to the Pacific Ocean. The road goes the way of the rivers, following tight curves between the hills. Logging trucks crowd the turns, going the other way. They downshift to hold heavy loads against the grade. Over the crest of the range, in the green tumble of hills that form the headwaters for the coastal salmon streams, each curve uncovers another square of bare-ass mountainside, clear-cut to the mud. There's hardly a green leaf left in the cut—only grey dirt, shattered tree trunks lying every which way, and rootwads bulldozed into muddy piles. Even the rivers are grey, muddied by rain that erodes the raw draglines.

I drive as fast as I can through this part, keeping my eyes on the single row of alders that the loggers left along the road to hide the carnage. I know that on the coast, just south of Cape Perpetua, I'll come finally to remnant patches of ancient rain forest, somehow saved from the crosscut saws—six-hundred-year-old Sitka spruce and western redcedar that grow, dark and mossy, down the slope to the edge of the sea. I push through the scarred hills, trying to concentrate on how the ancient forest will smell—all damp earth and cedar—and how surf sounds, far away through deep ferns.

South of the Cape, I walk a trail under Sitka spruce to the edge of the cliff, where the forest cracks off into the sea. On the headland, the air is suddenly salt-thick and cold, the wind ferocious. In wild surf scuds of sea-foam spring up like startled birds, and logs shoot ten feet in the air. A few children run shouting along the cliff edge, holding their hats against the gale, ducking under sheets of spray, changing course simultaneously, like sanderlings. I pull my windbreaker tight around me and sit on a bench overlooking the sea.

The place I sit is a memorial bench. Someone who deeply loves the coast must have chosen the site, just above the wild collision of coastal stream and cobbles. I read the inscription on the brass plaque: *Mother, when you hear a song or see a bird, please do not let the thought of me be sad, for I am loving you just as I always have. It was heaven here with you.*

The note confuses me. At first, I think it's the mother who has died. And then the unthinkable works its way into my mind. A living, grieving mother must have written this note, as if her child were not dead, but was speaking to her through the sea of her pain. And the heaven they shared? I'm thinking it must be here, in this exact spot, where the sea surges into the river at high tide, and gulls stand hip-deep shouldering fresh water across their backs, as they must have done for centuries.

I imagine a mother pulling rainpants on a child already dancing to go. A last pat on his wool hat, and he runs across the grass in too-big boots. She pulls on her own raincoat and follows him down the trail. At the cliff edge, she stands beside him in the wind, looking out to sea.

How can she live with the sorrow?

We're told by psychologists that there is a pattern to grief: everyone must make the same terrible five-stage journey, putting one foot in front of the other, step by step in air suddenly gone cold and thick. My friend Katherine, who knows many kinds of sorrow, thinks that people experience the same five stages of grief, no matter whether it's a person who is mourned, or a part of the world—a forest, a salmon run, a species, a stream. The quality of the pain may be different, and its intensity, but all the stages of grief are there, in people who loved the devastated land.

The first stage of grief is denial. Maybe the forest isn't really dead. All those seeds hiding in the bulldozed ground—they might grow into a forest someday. And if it's too late to save this forest, isn't there still time to save the forests on the other side of the mountains? And maybe the salmon runs aren't extinct; the salmon might be waiting in the ocean until the rivers clear and silt washes off the spawning beds. "Look around," my neighbor says, trying to lift my spirits. "It's still a beautiful world. The environmental crisis is just a protest-industry fundraising scam."

Step two. Anger. What kind of person can cut an ancient forest to bloody stumps, bulldoze the meadows to mud, spray dioxin over the mess that's left, and then set smudge fires in the slash? And when the wounded mountainside slumps into the river, floods tear apart the waterfalls and scour the spawning beds, and no salmon return, what kind of person can blame it on an act of God—and then wave the bulldozers through the stream and into the next forest, and the next? I hope there's a cave in hell for timber industry executives like this, where an insane little demon hops around shouting, "jobs or trees, jobs or trees," and buries an axe-blade in their knees every time they struggle to their feet.

Step three. Bargaining. Look, we're rational people. Let's work this out. Destroy this forest if you have to, but plant new seedlings in the slash. Drain this wetland and build your stupid Kmart, but dig a new swamp next to the highway. Let cattle trample this riverbank and crap in this headwater, but fence them from this spawning bed. Kill the smolts in your turbines, but buy new fish for another stream. Then let's try to create some community. Let's study the issue again in five years.

Step four. Depression. Hopelessness deep and dark enough to drown in.

And gradually, disastrously, grief's final step: acceptance. On the Oregon coast, the children know mostly fish-poor flood-stripped streams. Here, estuaries are fouled, and no river water is safe to drink. That's the way it is. Why should they think it could be any different? Children who have never seen an ancient forest climb the huge, crumbling, blood-red stumps, as they might climb onto the lap of a vacant-faced grandfather. They look out over the ferns and hemlock seedlings, unable to imagine what used to be. They don't remember waking up to birdsong. How can they miss a murrelet if they've never seen one? It's not just their landscape that has been clear-cut, but their imaginations, the wide expanse of their hope.

And when their grandparents' memories of unbroken forests fade, and the old stories get tedious—the streams of red salmon pushing upriver—and the photograph albums hold only dry images of some other place, some other time, then another opening in the universe shuts, a set of possibilities disappears forever.

This is what we must resist: gradually coming to accept that a stripped down, hacked up, reamed out, dammed up, paved over, poisoned, bulldozed, impoverished landscape is the norm—the way it's supposed to be, the way it's always been, the way it must always be. This is the result we should fear the most.

I turn away from the ocean and hike up the creek into a forest that's never been logged. It's dark here, and noisy with wind and distant surf. Shadows sink into the whorls of maidenhair ferns and shaggy trunks of cedars centuries old. The decaying earth is a black granite wall bearing the names of all that has been lost and forgotten on the far side of the mountain: the footprints of cougar and elk, yellow-bellied salamanders pacing across dark duff, swordferns unfurling, the sweet flute of the varied thrush, the smell of cedar and soil; the wild coastal river, its headwaters buried in mossy logs, its waters leaping with salmon, its beaches dangerous with surf and swaying bears. Kneeling, I trace a heron's tracks engraved in black soil at the edge of the stream.

Into the shadows, light falls like soft rain. It shines on every hemlock needle and huckleberry, each lifted leaf of sorrel. There's a winter wren singing somewhere in the salal, and a raven calling from far away. I lean against an ancient Douglas-fir that soars to great height and disappears into the overcast.

The wild forest is a witness, standing tall and terrible in the storm at the edge of the sea. It reminds us of what we have lost. And it gives us a vision of what—in some way—might live again.

INDEX